NATO ASI Series

Advanced Science Institutes Series

A series presenting the results of activities sponsored by the NATO Science Committee, which aims at the dissemination of advanced scientific and technological knowledge, with a view to strengthening links between scientific communities.

The Series is published by an international board of publishers in conjunction with the NATO Scientific Affairs Division

A Life Sciences **B Physics**	Plenum Publishing Corporation London and New York
C Mathematical and Physical Sciences **D Behavioural and Social Sciences** **E Applied Sciences**	Kluwer Academic Publishers Dordrecht, Boston and London
F Computer and Systems Sciences **G Ecological Sciences** **H Cell Biology**	Springer-Verlag Berlin Heidelberg New York London Paris Tokyo

The ASI Series Books Published as a Result of Activities of the Special Programme on CELL TO CELL SIGNALS IN PLANTS AND ANIMALS

This book contains the proceedings of a NATO Advanced Research Workshop held within the activities of the NATO Special Programme on Cell to Cell Signals in Plants and Animals, running from 1984 to 1989 under the auspices of the NATO Science Committee.

The books published as a result of the activities of the Special Programme are:

Vol. 1: **Biology and Molecular Biology of Plant-Pathogen Interactions.** Edited by J. A. Bailey. 1986.

Vol. 2: **Glial-Neuronal Communication in Development and Regeneration.** Edited by H. H. Althaus and W. Seifert. 1987.

Vol. 3: **Nicotinic Acetylcholine Receptor: Structure and Function.** Edited by A. Maelicke. 1986.

Vol. 4: **Recognition in Microbe-Plant Symbiotic and Pathogenic Interactions.** Edited by B. Lugtenberg. 1986.

Vol. 5: **Mesenchymal-Epithelial Interactions in Neural Development.** Edited by J. R. Wolff, J. Sievers, and M. Berry. 1987.

Vol. 6: **Molecular Mechanisms of Desensitization to Signal Molecules.** Edited by T. M. Konijn, P. J. M. Van Haastert, H. Van der Starre, H. Van der Wel, and M. D. Houslay. 1987.

Vol. 7: **Gangliosides and Modulation of Neuronal Functions.** Edited by H. Rahmann. 1987.

Vol. 9: **Modification of Cell to Cell Signals During Normal and Pathological Aging.** Edited by S. Govoni and F. Battaini. 1987.

Vol. 10: **Plant Hormone Receptors.** Edited by D. Klämbt. 1987.

Vol. 11: **Host-Parasite Cellular and Molecular Interactions in Protozoal Infections.** Edited by K.-P. Chang and D. Snary. 1987.

Vol. 12: **The Cell Surface in Signal Transduction.** Edited by E. Wagner, H. Greppin, and B. Millet. 1987.

Vol. 19: **Modulation of Synaptic Transmission and Plasticity in Nervous Systems.** Edited by G. Hertting and H.-C. Spatz. 1988.

Vol. 20: **Amino Acid Availability and Brain Function in Health and Disease.** Edited by G. Huether. 1988.

Series H: Cell Biology Vol. 20

Amino Acid Availability and Brain Function in Health and Disease

Edited by

Gerald Huether

Max-Planck-Institut für experimentelle Medizin
Hermann-Rein-Str. 3, 3400 Göttingen, FRG

Springer-Verlag
Berlin Heidelberg New York London Paris Tokyo
Published in cooperation with NATO Scientific Affairs Division

Proceedings of the NATO Advanced Research Workshop on Amino Acid Availability and Brain Function in Health and Disease held in Göttingen, FRG, September 14–18, 1987

ISBN-13: 978-3-642-73177-8 e-ISBN-13: 978-3-642-73175-4
DOI: 10.1007/978-3-642-73175-4

Library of Congress Cataloging-in-Publication Data. NATO Advanced Research Workshop on Amino Acid Availability and Brain Function in Health and Diseases (1987 : Göttingen, Germany) Amino acid availability and brain function in health and disease / edited by Gerald Huether. p. cm.—(NATO ASI series. Series H., Cell biology ; vol. 20) "Proceedings of the NATO Advanced Research Workshop on Amino Acid Availability and Brain Function in Health and Disease, held in Göttingen, FRG, 14–18 September 1987"—T.p. verso. Includes index.
ISBN-13: 978-3-642-73177-8
1. Neurochemistry—Congresses. 2. Brain—Metabolism—Congresses. 3. Amino acids—Metabolism—Congresses. 4. Brain—physiology—congresses. I. Huether, Gerald, 1951-. II. Title. III. Series. [DNLM: 1. Amino Acid—physiology—congresses. 2. Amino Acid—therapeutic use—congresses. 3. Behavior—drug effects—congresses. 4. Brain—physiology—congresses. 5. Brain Chemistry—congresses. 6. Brain Diseases—drug therapy—congresses. 7. Neuroregulators—metabolism—congresses. QU 60 N279a 1987]
QP356.3.N36 1987 612'.8042—dc 19 DNLM/DLC for Library of Congress 88-12368

Softcover reprint of the hardcover 1st edition 1988

2131/3140-543210

FOREWORD

The picture on the following page is being reproduced here, at the request of the participants in the Advanced Research Workshop "Amino Acid Availability and Brain Function in Health and Disease". I displayed this limewood carving, entitled "Neurochemistry", during my closing remarks to this extraordinarily stimulating and productive workshop so ably organized by my collaborator Dr. Gerald Huether.

We scientists need two sturdy legs to carry us through all the twists and turns of our academic careers. We should also have, as it were, a reserve leg handy, to help us stay upright when this career ends. My "third leg" is wood carving. The idea for "Neurochemistry" came to me in the plane carrying me to the congress of the International Neurochemical Society in Jerusalem. We need the hands for our meticulous experimental work, and at least one ear to listen to the messages our neurons send us.

A few years ago it would have been premature to hold a workshop on this subject. Now, however, the time was just right to allow an overview of the status of current research, and to point out the promising new openings it has created. There is no doubt that the book to be published as a result of this workshop will be, for the next years at least, the standard text on the subject. Hearty thanks to all speakers for their brilliant contributions and to all participants for the lively, uninhibited and stimulating discussion.

It is surely a philosophical question whether the human brain can understand the functioning of the human brain. Descartes's Ncogito ergo sum Amay at first glance seem a purely neurochemical statement, but surely it is much more. I am certain the coming years will bring us more and yet more discoveries and explanations concerning the functions of our brain. And yet I somehow nourish the hope and desire that there will remain something beyond the explanation and scientific discovery. For it will be seen that this small residue makes us what we are: human souls with our individual thoughts, feelings and behavior.

Volker Neuhoff

Göttingen, December 1987

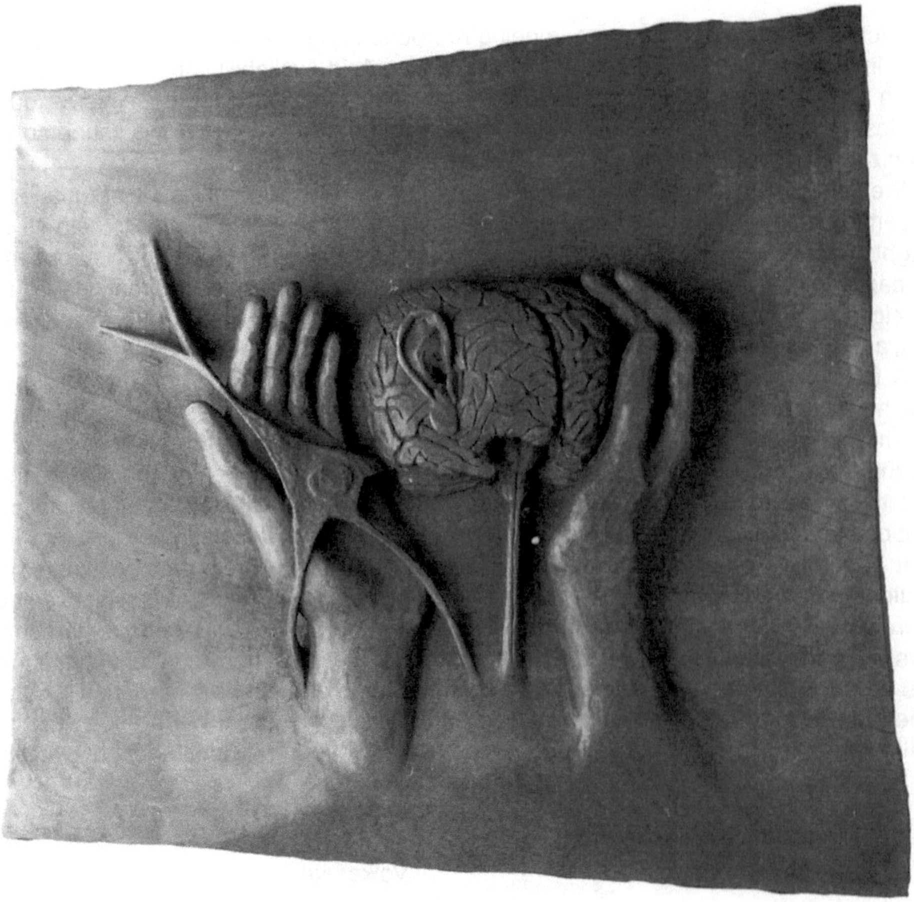

Volker Neuhoff: "Neurochemistry"
Lime wood, original size 45×46 cm

PREFACE

It is now about 20 years since the first reports on the effects of diet-related alterations of transmitter metabolism were published. Neurochemists were the first to realize that the formation of amino acid-derived transmitters in the brain can be affected by an altered supply of their respective precursor amino acids. The potential impact of this finding for the modulation of neural brain function and behaviour was soon recognized by neurobiologists and physicians, and workers from a growing number of different research disciplines became involved in various aspects of the relationship between amino acid availability and brain function. Thus, an interdisciplinary research programme was born, this area attracting, among others, neurobiologists, physiologists, nutritionists, endocrinologists, neurologists, psychiatrists, pharmacologists, and, most recently even embryologists. Various brain functions have been reported to be modulated, either through changes in the brain's amino acid supply or through impairment of peripheral amino acid metabolism. Successful treatment of various mental dysfunctions by precursor alteration has been described. New research strategies have been initiated, and the results (both expected and unexpected) of this research have contributed to a better understanding of the ways in which our brains function and of the bio-organisation upon which these functions rely. Scientists from related areas, as well as practitioners from various fields, and even the public to some extent, have become increasingly interested in the outcome of this new research area.

The groups involved in studies of the interrelationship between amino acid availability and brain function have expanded in their number as well as in the diversity of their approach at the risk of losing contact and becoming isolated from each other. Their reports are scattered in a wide array of journals and a general and comprehensive view is difficult to obtain. It was therefore of substantial interest and urgent necessity for the scientists in these various fields to be given an opportunity to get together, to exchange their ideas, to critically assess the existing knowledge and to develop future research strategies. This opportunity was given at the Advanced Research Workshop "Amino Acid Availability and Brain Function in Health and Disease", which took place at the Max-Planck-Institute for Experimental Medicine in Göttingen, Germany. Leading scientists were invited to discuss recent progress in this field, to address open questions, and to outline research strategies for the future.

It was generally felt by the participants that the research on the influence of altered amino acid availability on brain function is currently entering a turning point with regard to three important aspects: Firstly, there existed certain long-standing controversies between individual groups, e. g., whether bound or free tryptophan would be more important for the regulation of serotonin synthesis, or whether carbohydrate or protein appetite is regulated via serotonergic mechanisms. However, it has since become apparent that these questions must be answered on an "as well as" rather than an "either/or" basis. Hence, the Workshop did represent the end of certain scientific disagreements. Secondly, it was obvious that owing to the refinement of the experimental tools at our disposal (in vivo dialysis and voltammetry) we are currently entering an exciting phase of expansion and substantiation of our knowledge on the relationship between precursor availability and brain function. Hence, a state is being reached in which it is possible to measure now in the living animal what previously could only be assumed on a theoretical basis. And, as a third aspect, it has become clear that the early enthusiasm about the perspectives of precursor treatment of various behavioural, mental, or neurologic dysfunctions is now being replaced by a more realistic judgment of the limits and problems that arise when rather mechanistic approaches are applied to a homeostatic system that is maintained by multiple, feedback loops arranged in networks.

This Proceedings Volume contains the individual contributions presented at the seven sessions of the Workshop. A summary and a comprehensive discussion of the most important problems and questions of general interest are included at the end of each chapter.

It is hoped that the book will provide thus a survey of the present state of the art and of the most recent developments in the field of amino acid availability and brain function. Hence, it is intended to serve as a qualified work of reference and consultation for researchers, clinicans, nutritionists, and other groups of biomedical and biological research working in this area.

The book has been edited in close cooperation with the chairmen of the individual sessions of the Workshop who prepared the Summary and the Discussion-Report at the end of each chapter. Much more than the individual contributions, these parts will not only reflect some of the stimulating atmosphere and interchange of ideas we enjoyed at the Workshop but they will outline also the problems and difficulties that arose with regard to the implications of precursor availability in brain function.

In order to ensure a most rapid publication of this Proceedings Volume, proofreading of the retyped manuscripts was made by me instead of the individual authors. Typing errors which may (and certainly will) have escaped my inspections are a regrettable consequence of this procedure and I apologize for these mistakes.

I would like to express my sincere thanks to all colleagues who invested part of their time and energy in the Workshop and the book. In the name of all participants I gratefully acknowledge the financial support from the NATO, Scientific Affairs Division, making this workshop independent from interests other than those of the participants: frankly to discuss their findings and to exchange their ideas.

Gerald Huether

Göttingen, December 1987

AMINO ACID AVAILABILITY AND BRAIN FUNCTION IN HEALTH AND DISEASE

September, 14—18, 1987
Göttingen, FRG

Acworth, Ian N.
Dpt. of Brain and Cogn. Sci.
E 25-637
Mass. Inst. Technol.
Cambridge, Mass. 02139
USA

Anderson, G. Harvey
Dpt. of Nutr. Sci.
University of Toronto
Toronto, Ontario M55 1A8
Canada

Balazs, Robert
MRC Dev. Neurobiol. Unit
1 Wakefield Street
London WC1N 1 PJ
U.K.

Bernasconi, Raymond
Ciba Geigy AG
K-125.08.01
Klybeckstr.
4002 Basel
Switzerland

Bessman, Samuel P.
Dpt. of Pharmacology
University of Southern California
School of Medicine
Los Angeles, CA 90033
USA

Blundell, John E.
Dpt. of Psychology
University of Leeds
Leeds, LS2 9JT
U.K.

Bradberry, Charles W.
Dpt. of Pharmacol.
Yale University
Medicine School
333 Cedar Street
New Haven, CT 06510
USA

Brown, Sandra A.
10 Larkhall St.
St. John's
Newfoundland, A1B2C3
Canada

Bruinvels, Jaques
Inst. of Pharmacology
Erasmus-University
Postbus 1738
3000 DR Rotterdam
The Netherlands

Ceyhan, Mehmet
Hacettepe University
Etimesgut Bölge Hastenesi
Etimesgut-Ankara
Turkey

Commissiong, John W.
Physiology/Mc Intyre 1236
McGill University
3655 Drummond Street
Montreal, PQ H3G 1Y6
Canada

Cowen, Philip J.
MRC Clinical Pharmacol. Unit
Res. Unit
Littlemore Hosp.
Oxford OX 4 4XN
U.K.

Curzon, Gerald
Dpt. of Neurochemistry
Inst. of Neurology
1 Wakefield Street
London WC1N 1PJ
U.K.

Demling, Joachim
Psychiatrische Klinik
Universität Erlangen
Schwabachanlage 6
8520 Erlangen
Germany

De Feudis, Francis V.
3 Spanish River Road
Grafton, Mass. 01519
USA

Diederich, Jochem
Pfrimmer & Co.
Pharmazeutische Werke Erlangen
Hofmannstr. 26
8520 Erlangen
Germany

Dilmen, Ugur
Turkish Health and Therapy Fdn.
Dpt. of Clinical Sci.
Senyuya Misket Sok. No. 30
06510 Emek-Ankara
Turkey

Dudley, Robert
The NutraSweet Company
Preclinical Res. Department (C-8)
4711 Golf Road
Skokie, Ill. 60076
USA

Enwonwu, Cyril O.
Meharry Medical Coll.
Center for Nutrition
Box A 73
Nashville, Tn. 37208
USA

Fekkes, Durk
Dpt. Pharmacol.
Fac. of Med.
Erasmus University
Postbox 1738
3000 DR Rotterdam
The Netherlands

Fernstrom, John D.
Dpts. of Psychiatry
& Pharmacology
University of Pittsburgh
Western Psychiatric Institute
and Clinic
3811 O'Hara Street
Pittsburgh, PA 15213-2593
USA

Franklin, Keith B. J.
Psychology Dpt.
1205 Dr. Penfield Av.
McGill University
Montreal,
PQ H3A 1B1
Canada

Gallo, Vittorio
Instituto Superiore Di Sanita
Viale Regina Elena 299
00161 Roma
Italy

Gaull, Gerald E.
Nutrition and Medical Affairs
The Nutra Sweet Company
Box 1111
4711 Golf Road
Skokie, Ill. 60076
USA

Gibson, E. Leigh
Dpt. of Psychology
University of Birmingham
P. O. Box 363
Birmingham B15 2TT
U.K.

Gietzen, Dorothy W.
Dpt. of Physiol. Sci.
School of Vet. Med.
University of California
Davis, Calif. 95616
USA

Growdon, John H.
Dpt. of Neurology
Massachusetts General Hospital
Suite 730
Boston, MA 02114
USA

Hamberger, Anders
Inst. of Neurobiology
University of Göteborg
P. O. Box 33031
S-400 33 Göteborg
Sweden

Harper, Alfred, E.
Dpt. of Biochemistry
420 Henry Mall
University of Wisconsin-Madison
Madison, WI 53706
USA

Hjelle, Jerry
The NutraSweet Company
Preclinical Res. Department (C-8)
4711 Golf Road
Skokie, Il. 60076
USA

Holder, Mark D.
Dpt. of Psychology
Memorial University of Neufoundland
St. Johns
Newfoundland A1B 3X9
Canada

Huether, Gerald
Max-Planck-Institut für
experim. Medizin
Hermann-Rein-Str. 3
3400 Göttingen
Germany

Hyland, Keith
Inst. of Child Health
University of London
30 Guilford Str.
London WC1 W 1 EH
U.K.

Jobe, Phillip C.
University of Illinois
Coll. of Med.
Dpt. of Basic Sci.
Box 1649
Peoria, Ill. 61656
USA

Johansen, Liv
Neurochem. Lab.
University Oslo
P. O. Box 1115 Blindern
Oslo 3
Norway

Kvamme, Elling
Neurochem. Lab.
University of Oslo
P. O. Box 1115 Blindern
Oslo 3
Norway

Lajtha, Abel
State Center for Neurochem.
Ward's Island
New York, NY 10035
USA

Langer, Klaus
Forschungsinst.
f. exp. Ernährung
Langemarckplatz 5a
8520 Erlangen
Germany

Leathwood, Peter D.
Research Dpt.
Avenue Nestle 55
CH-1800 Vevey
Switzerland

Lefauconnier, Jeanne-Marie
INSERM Reserche Toxicologie Exp.
Hopital Fernand Widal
200 Rue du Faubourg Saint-Denis
75475 Paris CEDEX 10
France

Lehnert, Hendrik
Abt. f. Endokrinologie
Med. Universitäts-Klinik
Langenbeckstr. 1
6500 Mainz
Germany

Lookingland, Keith J.
Dpt. of Pharmacol. and Toxicol.
Michigan State University
East Lansing
48824-1317 Michigan
USA

Martin, Friedrich
Pfrimmer & Co.
Pharmazeutische Werke Erlangen
Hofmannstr. 26
8520 Erlangen
Germany

Maher, Timothy J.
Dpt. of Pharmacol.
Massachusetts Coll.
of Pharmacy
179 Longwood Avenue
Boston, MA 02115
USA

Meldrum, Brian S.
Inst. of Psychiatry
Dpt. of Neurology
Denmark Hill
London SE 5 8AF
U.K.

Møller, Svend E.
Sct. Hans Mental Hospital
Research Lab.
4000 Roskilde
Denmark

Neuhoff, Volker
Max-Planck-Institut
für experim. Medizin
Hermann-Rein-Str. 3
3400 Göttingen
Germany

Pepplinkhuizen, Lolke
Univ. Hospital Dyksigt
Dr. Molewaterplein 40
3015 GD Rotterdam
The Netherlands

Pratt, Oliver E.
Dpt. of Neuropathology
Inst. of Psychiatry
University of London
Denmark Hill
London SE5 8AF
U.K.

Raleigh, Michael J.
University of California
Neuropsych. Inst.
Center for Health Sci.
760 Westwood Plaza
Los Angeles, CA 90024
USA

Rao, Marie-Luise
Universitäts-Nervenklinik und
Poliklinik
Sigmund-Freud-Str. 25
5300 Bonn 1
Germany

Riederer, Peter
Psychiatrische Klinik u. Poliklinik
Universitäts-Nervenklinik
Füchsleinstr. 15
8700 Würzburg
Germany

Rogers, Quinton R.
Dpt. of Physiol. Sci.
School of Vet. Med.
University of California
Davis, CA 95616
USA

Salter, Mark
Dpt. of Biochemistry
Wellcome Res. Lab
Langley Court
Beckenham, Kent BR3 BS
U.K

Sawatzki, Günther
Milupa AG
Bahnstr. 14-30
6382 Friedrichsdorf/Ts
Germany

Schousboe, Arne
Dpt. of Biochemistry A
The Panum-Inst.
University of Copenhagen
Blegdamsvej 3 C
2200 Copenhagen N
Denmark

Smith, Isabel
Inst. of Child Health
University of London
30 Guilford Str.
London WC1 N 1EH
U.K.

Smith, Quentin R.
Dpt. of Health and Human Services
Building 10-6C103
Nat. Inst. of Health
Bethesda, Maryland 20205
USA

Van Gelder, Nico M.
Centre de Recherches en Sciences Neurologiques
University de Montreal
110, Ave. des Pins ouest
Montréal H2W 1R7
Canada

Vernadakis, Antonia
University of Colorado
Health Sci.Center
4200 East Ninth Ave.
Denver, Col. 80262
USA

Wallace, James A.
School of Medicine
University of New Mexico
Albuquerque, New Mexico 87131
USA

Wolff, Joachim R.
Zentrum Anatomie
Universität Göttingen
Kreuzbergring 36
3400 Göttingen
Germany

Wurtman, Richard J.
Dpt. of Brain and Cognitive Sci.
Massachusetts Inst. of Technology
Room E25-604
Cambridge, Mass. 02139
USA

Young, Simon N.
Dpt. of Psychiatry
McGill University
1033 Pine Ave. West
Montreal, PQ H3A 1A1
Canada

Youdim, Moussa B. H.
Faculty of Medicine
Technion Israel Inst. of Technol.
Efron Str., P.O.B. 9697
Haifa 31096
Israel

Zanchin, Giorgio
Instituto di Clinica delle Malatti
Nervose e Mentali
University of Padova
Via Gustiniani 5
35128 Padova
Italy

TABLE OF CONTENTS

I. PERIPHERAL CONTROL OF BRAIN AMINO ACID SUPPLY

II. UPTAKE, COMPARTMENTATION AND UTILIZATION OF BRAIN AMINO ACIDS

III. INFLUENCE OF ALTERED PRECURSOR AMINO ACID SUPPLY ON TRANSMITTER METABOLISM AND BRAIN FUNCTION

IV. INFLUENCE OF ALTERED PRECURSOR AMINO ACID SUPPLY ON FUNCTIONAL NEUROTRANSMISSION

V. INFLUENCE OF ALTERED PRECURSOR AMINO ACID SUPPLY ON PHYSIOLOGY AND BEHAVIOUR

VI. AMINO ACID AVAILABILITY AND BRAIN DYSFUNCTION

Chapter I.

PERIPHERAL CONTROL
OF
BRAIN AMINO ACID SUPPLY

NUTRITIONAL & METABOLIC CONTROL OF BRAIN AMINO ACID CONCENTRATIONS

A. E. Harper & J. K. Tews

Department of Biochemistry
University of Wisconsin Madison
420 Henry Mall
Madison, WI 53706 U.S.A.

INTRODUCTION

Stability of brain total free pools of indispensable amino acids (IAA) of rats with widely differing intakes of good quality protein is impressive. For example, protein intakes of young rats that were fed for 4 hr diets containing from 5 to 55 % of casein, ranged from 0.4 to 3.9 g/100 g body weight, a 7.5-fold difference. Plasma total IAA concentrations reflected these differences in protein intake and ranged from 1.3 to 5.5 μmol/ml, a 4.5-fold difference. At the same time, brain IAA concentrations differed by only 32 %, ranging from 1.5 to 2.1 μmol/g brain (1). Coefficients of variation of fasting brain IAA concentrations are about ± 20 %. This degree of stability in concentrations of blood constituents, e. g. glucose or sodium, is usually maintained only through actions of specific and highly responsive regulatory systems.

Stability of concentrations of body fluid constitutents despite fluctuations in the external environment was perceived over 100 years ago by Claude Bernard as being essential for the survival and efficient functioning of free-living organisms. Through the investigations of Bernard, and subsequently of Henderson, Cannon, and Barcroft, this phenomenon, homeostasis, was recognized as a basic biological phenomenon (2). Henderson emphasized that concentrations of blood and body fluid constituents were not constant, but fluctuated considerably in response to changes in the external environment. He concluded that restoration of the stable state depended upon mechanisms which would respond when deviations occurred, by an action that tended to correct the deviation. Cannon, who proposed the term homeostasis, concluded that internal disturbances in such systems are normally kept within narrow limits by adjustments brought about by feedback mechanisms. Barcroft emphasized that acquisition of increasingly effective systems for maintenance of homeostasis was a major mechanism of evolution. He noted that for many controlled variables the initial effect of deviations beyond the limits of tolerance was loss of mental coordination. He concluded that higher mental and intellectual development, which permit the greatest freedom in a fluctuating external environment, depend upon accuracy of control of the internal environment achievable only through complex, integrated control mechanisms.

As many amino acids (AA) are, or are precursors of, neurotransmitters, stability of brain AA concentrations is critical for effective mental performance. Abnormal accumula-

NATO ASI Series, Vol. H20
Amino Acid Availability and Brain Function in Health and Disease. Edited by G. Huether

tions of AA or AA metabolites or severe depletion of AA from brain pools will cause behavioral effects and mental impairment. Mechanisms that limit both depletion and overloading of brain AA pools are required to prevent such effects. Plasma AA concentrations may differ by 3- to 5-fold after meals differing widely in protein content, but within a few hours are restored to the stable, fasting state; brain AA pools are affected much less. The control systems that provide this stability have limits; if the limits are exceeded, or the systems are defective, impairment of function will occur. Although brain AA concentrations of rats remain quite stable despite wide differences in their intake of proteins with relatively well-balanced IAA patterns, consumption of diets containing disproportionate amounts of IAA will lead to depletion or overloading of specific AA pools (3).

Overview of Systems that Control Brain AA Concentrations

Brain AA concentrations are influenced, at least indirectly, by many factors, but ultimately would appear to be determined by plasma AA concentrations and proportions and rates of transport of AA between blood and brain. A critical question, then, is: are brain AA concentrations regulated in any way through feedback systems in the brain itself which are responsive to changes in brain AA pools, or are they controlled only indirectly through mechanisms external to the brain? Systems contributing to control of brain AA concentrations are summarized below (see Fig. 1).

Plasma AA concentrations depend on protein intake, rates of stomach emptying, intestinal absorption and metabolism, and uptake and metabolism in liver and other organs and tissues. Uptake of AA from blood into brain depends upon the characteristics of the blood-brain barrier (BBB) and the degree of competition among AA for uptake; this depends in turn on plasma AA concentrations and proportions (4). Brain metabolism, cell membrane transport, and rates of efflux of AA from brain will also influence brain AA concentrations.

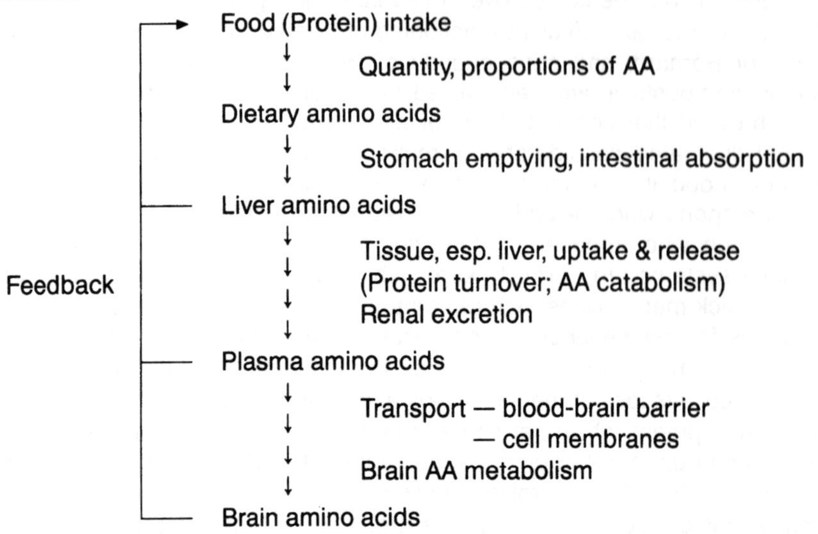

FIGURE 1

Depression of food intake and preference for a diet containing a moderate amount of protein with a balanced AA pattern occur rapidly if a single IAA is depleted from the brain pool, as occurs after consumption of a diet that is deficient in, or imbalanced with respect to, one IAA (5). Similar effects on feeding behavior occur when the diet contains a large excess of a single IAA or when protein intake is extremely low or high (6). These responses are associated with brain mechanisms that respond to deviations in blood or brain AA pools (7).

Based on this analysis, it would appear that control of brain AA concentrations is exerted indirectly, mainly by mechanisms that are external to the brain, and that there are three levels of control: 1) regulation of plasma AA concentrations; 2) limitations of brain transport systems and competition among AA for uptake into the brain; 3) control of food or protein intake. These processes are all influenced by diet.

Regulation of Plasma AA Concentrations

Plasma AA concentrations tend to reflect protein or AA intake during the period after a meal, but are quite stable in the post absorptive (fasting) state. They are regulated by systems that respond appropriately to correct deviations from the stable, fasting concentrations.

Gastric emptying, which is under complex hormonal control, is the initial process regulating rate of entry of AA into blood. Emptying of low protein diets from the stomach is rapid, but as protein intake increases, emptying is increasingly delayed (8). AA, especially TRP and PHE, trigger release of the intestinal hormone, cholecystokinin (CCK), which acts to delay stomach emptying (9). CCK is only one of several factors which control the rate of flow of AA to the intestine and, thereby rate of entry of AA into the blood stream when protein intake is increased.

Protein synthesis is the primary process for which AA are needed. Influx of AA into liver stimulates synthesis and suppresses degradation of tissue proteins (10, 11). Preferential use of AA for protein synthesis is assured by the relative K_m's of AA activating and AA degrading enzymes; the former, being low, are saturated with usual tissue AA concentrations whereas the latter, with K_m's about 10 times fasting plasma values, are rarely saturated even when tissue AA concentrations are high (2, 12). Ingestion of a meal also stimulates release of insulin which, in turn, stimulates uptake of AA and synthesis of proteins by muscle and heart (13), thus removing AA from blood. When protein intake is low, most of the ingested amino acids are used for protein synthesis, and plasma AA concentrations do not rise above, or even to, fasting values until these needs have been met. These effects are more striking in the growing organism which is accumulating protein than in the adult in which protein synthesis and degradation are in balance.

Amino acid catabolism, which occurs mainly in liver, represents the major mechanism for control of plasma AA concentrations when protein intake exceeds the amount needed for protein synthesis (2). Because K_m's of AA degrading enzymes are high, rates of AA oxidation increase with increasing plasma and tissue AA concentrations and, therefore, with increasing protein or AA intake over a wide range. This results solely from the increase in supply of substrate, without any increase in enzyme content. Responses of this type, due to the physicochemical properties of these enzymes, restore plasma AA concentrations to the stable, post-absorptive state within a few hours after a meal that does not contain a great excess of protein. This reduces the likelihood of overloading of tissues, including brain, with large amounts of AA.

A high intake of protein stimulates hepatic cyclic AMP synthesis which increases transport of AA to sites of oxidation in liver (14). Also, amino acid-degrading enzymes in liver are adaptive. Activities are low if protein intake is chronically low and can rise several-fold if protein intake is chronically high (2). Increased activities of these enzymes increase the capacity of organisms for catabolism of AA. After adaptation is complete, animals fed 60 % protein diets eat three to four times as much protein as those consuming an adequate (15 % protein) diet but their plasma concentrations of IAA, except for branched-chain AA (BCAA: LEU, ILE and VAL), will be lower than those of animals eating a low protein diet (15).

The BCAA, which are competitors for uptake into the brain of neurotransmitter precursors such as TRP, PHE and TYR (4), are exceptions. Their concentrations rise with increasing protein intake and do not decline during the adaptive period. Catabolism of BCAA differs in several respects from that of most other AA. BCAA, unlike other IAA, are not degraded mainly in the liver but in most organs and tissues, including brain. Also, the initial step in their degradation is a readily reversible transamination which provides substrate for the next and limiting step, an irreversible oxidative decarboxylation. Branched-chain α-ketoacid (BCKA) dehydrogenase which catalyzes this reaction is not adaptive, but is activated in both muscle and liver when protein intake is high. Despite increased activity of this limiting enzyme, rate of degradation of BCAA will increase little until the concentration of its substrate, BCKA, increases (16). As the initial transamination is reversible, a portion of the BCKA formed is reaminated; therefore, when protein, and hence BCAA, intake is high, steady-state levels of BCAA will rise until the rate of oxidative decarboxylation of BCKA increases. The end result can be seen by comparing relationships between plasma total IAA and BCAA concentrations as protein intake increases in studies (1) done before adaptation has occurred (e. g., four hr) and in others (17) done following adaptation (seven to 14 days). In short-term studies all IAA concentrations, including that of TRP, rise more or less in parallel with increasing protein intake. As a result, plasma TRP to large neutral AA (TRP/LNAA) ratios, which determine brain TRP and serotonin levels (18), do not differ much in rats with widely different protein intakes. After adaptation, concentrations of IAA other than BCAA, including TRP, rise very little with increasing dietary protein level above 20 %, but BCAA concentrations tend to rise steadily as protein intake increases. Thus, plasma TRP/LNAA ratios of rats that have adapted to different dietary levels of protein tend to decline as protein intake increases.

Transport of AA from Blood to Brain

AA entry into brain depends on the characteristics of BBB transport systems for the various classes of AA (neutral, acidic, basic). Concentrations of AA for half-maximal rates of transport (K_m) across the BBB approximate "usual" plasma AA concentrations (4), or may be below these values (19). Hence, rates of uptake of blood AA into brain should be highly responsive to changes in plasma AA concentrations; however, competition among AA within a single transport group will greatly influence uptake of any other AA in that group. Extent of competition depends upon relative concentrations of AA in blood and relative affinities of individual AA for the transport system. Therefore, in using kinetic analyses to evaluate effects of dietary modifications on rate of transport of AA into brain in intact animals, K_m's(app), in which effects of competing AA on the K_m of the transport carrier for each AA are taken into account, must be substituted for absolute K_m's.

Reliable values for K_m's(app) depend upon accurate determinations of absolute K_m; Smith (19) has reported that K_m's determined by Oldendorf's single injection method (4, 20) may be up to 5-fold too high, and that K_m's(app) are also somewhat higher than presently proposed values. Regardless of the true value, K_m's(app) will be considerably higher than usual plasma AA concentrations and will vary widely depending on plasma AA concentrations. For example, in rats adapted to low (6 %), moderate (18 %) or high (50 %) casein diets then given a single, 4—7 hr feeding of one of them or a protein free diet, K_m's(app) for THR and VAL differed by about 3-fold (21, 22).

In order to investigate effects of these treatments on rates of THR or VAL entry into brain, rats from the study of K_m's(app) were briefly infused i. v. with a tracer amount of ^{14}C-labeled THR or VAL (constant infusion technique (23)). Fluxes of these AA into brain differed greatly, primarily because, in rats adapted to a high protein intake, tissue enzymes for catabolism of VAL and THR respond differently. Threonine dehydratase activity increases greatly, leading to rapid degradation of THR and striking decreases in plasma THR concentration to values below those for rats fed normal or low amounts of protein (17). In contrast, enzymes for catabolism of VAL and other BCAA respond differently, as described above, when protein intake increases, and plasma concentrations of these AA increase several-fold in rats fed a high protein diet (16).

As a result of the lowered plasma THR concentrations (21), influx of THR into brain of rats adapted to the high protein diets was depressed. When rats adapted to the high protein diet were fed a non protein meal, rate of THR entry into brain was about 25 % of that observed when such a meal was fed to rats adapted to the 6 % casein diet. After ingestion of the high protein meal, a significant trend toward increased THR influx occurred only in rats adapted to the moderate (18 %) or high (50 %) dietary levels of casein; in these, influx increased by about 50 and 100 %, respectively. Brain THR concentrations were clearly affected more by the adaptation diet than by the final meal; they were generally several-fold higher in rats adapted to the low protein diet than in those accustomed to high protein.

Rates of VAL entry into brain and plasma and brain VAL concentrations invariably increased as protein content of the final meal was raised; the adaptation diet had only minor effects on these variables (22).

When rates of entry of THR and VAL into brain are considered in relation to calculated K_m's(app) for these amino acids, changes in influx evidently reflect the balance between changes in K_m's(app) and in plasma AA concentrations. Thus, the low plasma THR levels occurring in rats adapted to the high protein diet cannot maintain a high rate of entry of THR into brain in the presence of high levels of several other neutral AA. Conversely, although K_m(app) for VAL increased several-fold in rats fed the high protein diet, VAL influx increased because plasma VAL concentrations were elevated enough to counteract, at least in part, effects of elevated concentrations of competing AA in plasma.

These results illustrate a major difficulty in predicting accurately rates of AA influx into brain. Such predictions are based on the AA composition of plasma which changes over time after ingestion of a meal. Plasma AA concentrations of rats fed a chow diet will differ greatly from those of rats fed low or high protein diets or diets having unusual proportions of amino acids. Inclusion or omission some of the potential competitor amino acids in the calculation will also lead to different predictions. For example, K_m(app) for VAL increased as much as 50 % if THR concentrations were included with the LNAA in our calculations described above. Including SER and ALA which compete with THR for transport (24) raised the K_m(app) for THR by about 15 %. Our calculations of predicted fluxes based on the changing concentrations of THR or VAL and values for other AA commonly used to calculate K_m's(app) (4), indicate maximum influxes expected if concentrations of other AA remain unchanged. The size of the discrepancies between predicted and observed fluxes, especially under dietary conditions that led to high plasma THR and VAL concentrations, emphasize the importance of changes in concentrations of AA other than THR or VAL on rates of entry of these AA into brain (21, 22).

Concentrations of AA in brain and plasma are often similar so the distribution ratio is not above 1; thus, as for VAL in our studies, LNAA uptake into neurons and glia is unlikely to occur against a concentration gradient. However, the ratio is often greater than 1 for THR, implying that its transport into brain cells has involved an active process. AA transport across the BBB is believed to occur by facilitated diffusion, a non concentrative process not requiring energy; in contrast, entry into cells is often energy requiring and concentrative. For example, in the presence of glucose, brain slices concentrate THR in the intracellular space by as much as 15 fold after a 15 min incubation (25). When rates of THR or VAL passage across the BBB (21, 22), are compared with their absolute concentrations in brain, other factors besides influx into either extracellular or intracellular spaces would seem to be involved in control of brain AA concentrations; otherwise they would soon surpass those actually observed. It seems evident that efflux from brain must be an important control mechanism. Other factors affecting brain AA levels include protein synthesis and degradation, synthesis of neurotransmitters, and other metabolic reactions such as degradation of BCAA.

Rates of entry of the dispensable AA into brain are usually very low (24). Many of these AA are synthesized readily in brain from glucose (26) and, in general, their concentrations are far above those of the IAA.

Responses of Brain AA Concentrations to Dietary Changes

Regulation of plasma AA concentrations and limited transport of AA through the BBB keep fluctuations of brain AA concentrations in animals consuming diets within the usual range of protein content and AA composition within tolerable limits. If diet modifications are extreme, however, the limits of these systems to maintain stability of brain AA concentrations can be exceeded. This can be illustrated by changes observed in brain AA concentrations in response to consumption of diets that differ greatly in protein content or AA composition.

Brain total IAA concentration of rats, measured 20 to 330 min after they started consuming 35 to 55 % casein diets, increased only 17 % above the value of 1.78 μmol/g for rats consuming a diet that was just adequate in protein (15 % casein) (1). Responses of individual AA differed considerably, however, with THR, VAL, and TYR increasing by 30 to 50 %, whereas LYS and ARG remained almost constant. Brain total IAA concentration declined by only about 10 % in rats fed a 5 % casein diet. Again, responses of individual brain AA differed considerably; LYS, ARG, THR and PHE values changed little, whereas values for TRP tended to increase, and those for BCAA, MET and TYR declined by 19 to 38 %. Increases in plasma AA concentrations, when casein content of the diet was increased from 15 to 55 %, ranged from 65 to 195 %, average 113 %. When casein content was reduced to 5 %, individual plasma AA concentrations declined by 26 to 62 %, average 47 %. Brain AA pools thus fluctuate much less than plasma pools under these conditions; presumably brain uptake is limited by mutual competition among plasma AA when plasma AA pattern is not greatly distorted.

In studies of rats adapted to diets containing from 5 to 75 % of casein, the situation was different (17). After 11 days, brain total concentration of IAA other than BCAA of rats fed the 45, 50 or 55 % casein diets fell 20 % below that for rats fed only 15 % of casein, and with higher protein diets the decline was somewhat greater. At the same time, brain total BCAA concentration increased 75 %. These changes presumably reflect effects of metabolic adaptations on plasma AA concentrations which, in turn, affect competition for uptake at the BBB. Plasma concentrations of IAA other than BCAA declined about 6 %, whereas those of BCAA increased by 75 %; hence, competition from other neutral AA for uptake of BCAA declined while competition from BCAA for uptake of other neutral AA increased.

In contrast to the relative stability of brain AA concentrations when rats consume widely different amounts of high quality protein, substantial deviations in AA pattern of diets cause striking changes in brain AA pools (5). Ingestion of a diet low in protein (6 % casein) to which a mixture of IAA lacking HIS (6 %) has been added, depletes brain HIS pool by about 80 % within 3 to 6 hr. The IAA added to create this imbalance compete with HIS for uptake into brain, but also apparently compete with each other for uptake, so that brain IAA concentrations are not elevated, despite the large increases in plasma concentrations.

Ingestion of a similar low protein diet to which has been added a large surplus of PHE results in unusually high elevations of PHE and TYR in plasma (7- and 9-fold, respectively) and in brain (4 and 4.5-fold, respectively). Brain VAL, ILE, LEU, MET and HIS pools are depressed by 35—40 % (6). PHE and TYR presumably compete strongly with other large neutral AA for uptake into brain, resulting in depletion of brain LNAA pools; at the same time other plasma LNAA, which are in low concentrations, do not compete effectively with PHE and TYR so brain concentrations of PHE and TYR rise abnormally. This resembles somewhat the situation in phenylketonuria, in which lack of the ability to degrade PHE results in elevation of plasma PHE by 15- to 30-fold.

The limits of tolerance of changes in brain AA concentrations before impairment of function occurs have not been established, but it is important to remember that coefficients of variation of fasting brain AA concentrations are about ± 20 %.

Control of Protein (Food) Intake

When protein intake is greatly increased or decreased, or when the dietary AA pattern is distorted, plasma concentrations of some AA may approach the limits of tolerance of the rat, and potential for altering brain AA concentrations is created. In these conditions, food intake is depressed (27). The depression is transitory in rats adapting to a high protein diet. Rapid and severe depressions are observed, however, when rats are fed diets in which the AA pattern is distorted by moderate additions of all but one IAA or a large surplus of one IAA, both of which cause plasma IAA patterns to resemble those of the diet. In these situations, if given the choice, the rat will select a diet containing a moderate amount of protein with a balanced AA pattern in preference to the one causing the distortion of plasma AA pattern.

Distorted plasma AA patterns are not restored to the standard, stable state by depressions of food intake, as they are by selection of a balanced diet, but more severe distortion of body fluid AA patterns is prevented. Rats will adjust gradually, through metabolic adaptations, to diets in which the AA distortion is not too extreme and will survive and grow. The responses in feeding behavior appear to represent reactions of homeostatic mechanisms that come into play, and contribute to stability of brain AA concentrations, when metabolic systems are overloaded or impaired. The nature of the signals initiating these responses has been elusive, but ultimately control of feeding behavior is a function of the brain (7, 27).

SUMMARY AND CONCLUSIONS

Brain AA concentrations are not regulated directly by feedback mechanisms in brain, but are controlled indirectly through a complex of processes, predominant among which are regulation of blood AA concentrations and characteristics of blood-brain AA transport systems. Blood AA concentrations are regulated, mainly through AA-degrading systems. Rates of AA degradation change rapidly in response to changes in tissue AA concentrations; capacities of degradative systems change in response to changes in protein intake. BBB transport systems, with low K_m's, have limited capacity, and rates of uptake of AA into brain are influenced by changes in competition among AA for carriers when plasma AA concentations change. If systems for regulation of plasma AA are overloaded, or if plasma AA patterns are distorted so that brain AA pools are depleted or expanded, food intake is depressed and AA intake is, thereby, reduced.

Immediately after rats begin to consume a diet containing a large amount of high quality protein, plasma AA concentrations rise, mutual competition among AA for uptake into brain prevents extreme changes in brain AA pools, rates of AA catabolism increase without any change in enzyme activity, and brain AA concentrations remain relatively stable. If protein intake remains high food intake is depressed, AA degrading-systems undergo adaptation, rates of AA degradation increase further, AA other than BCAA are cleared rapidly from plasma, food intake increases, and brain AA concentrations again remain relatively stable, but at new steady-state levels with somewhat elevated BCAA concentrations and somewhat depressed concentrations of AA other than BCAA.

After rats consume a low protein diet having imbalanced proportions of AA, the brain pool of any IAA in low concentration in the diet is depressed; if the diet contains a large excess of one IAA, the pool of that AA rises greatly. The limits of systems for control of brain AA are exceeded by these conditions. With protein intake being low, the capacity of AA degrading systems is low, so surpluses of AA are cleared from blood slowly. AA provided in the diet in excess compete with those in low concentration for uptake into brain. With AA imbalances in which all IAA but one are elevated in plasma, the brain pool of the one is depressed; those in surplus in blood compete with each other for uptake into brain, and their brain concentrations do not rise. With one AA in excess, the situation is similar to that with genetic defects of AA catabolism in which the ability to degrade a particular AA is impaired. The one AA in excess competes with AA in low concentration for uptake into brain and pools of the latter are depleted. AA in low concentration do not compete effectively with the one in surplus for uptake into brain, so its concentration in brain rises.

This complex of control systems ensures stability of brain AA concentrations despite changes over a wide range in protein intake and plasma AA concentrations if the dietary protein is of high quality. If the diet contains disproportionate amounts of some AA, alterations in brain AA pools can be severe. The body does not appear to have a regulatory system that will correct such deviations. Under these conditions, however, food intake is depressed and preference for a diet with a balanced pattern of amino acids becomes evident. Further overloading or depletion of brain and body pools with amino acids is thus curtailed.

REFERENCES

1) PETERS, J. C., HARPER, A. E. (1987). Acute effects of dietary protein on food intake, tissue amino acids, and brain serotonin. *Am. J. Physiol.* **252:** R902—R914.

2) HARPER, A. E. (1974). Control mechanisms in amino acid metabolism. In: *The Control of Metabolism* (Sink, J. D., ed.), Pennsylvania State University Press, University Park Pennsylvania, pp. 49—74.

3) HARPER, A. E., BENEVENGA, N. J., WOHLHUETER, R. M. (1970). Effects of ingestion of disproportionate amounts of amino acids. *Physiol. Rev.* **50:** 428—558.

4) PARDRIDGE, W. M. (1983). Brain metabolism: A perspective from the blood brain barrier. *Physiol. Rev.* **63:** 1481—1535.

5) PENG, Y., TEWS, J. K., HARPER, A. E. (1972). Amino acid imbalance, protein intake, changes in rat brain and plasma amino acids. *Am. J. Physiol.* **222:** 314—321.

6) PENG, Y. GUBIN, J., HARPER, A. E., VAVICH, M. G., KEMMERER, A. R. (1973). Food intake regulation: amino acid toxicity and changes in rat brain and plasma amino acids. *J. Nutr.* **103:** 608—617.

7) ROGERS, Q. R., LEUNG, P. M. B. (1977). The control of food intake: when and how are amino acids involved? In: *The Chemical Senses and Nutrition*, Academic Press New York, pp. 213—249.

8) ROGERS, Q. R., HARPER, A. E. (1964). Transfer rates along the gastrointestinal tract. In: *The Role of the Gastrointestinal Tract in Protein Metabolism* (Munro, H. M., ed.), Blackwell Scientific Publications Oxford, pp. 3—24.

9) DAVENPORT, H. W. (1977). Physiology of the digestive tract, fourth ed. *Year Book Medical Publishers Chicago,*pp. 178—180.

10) JEFFERSON, L. S., KORNER, A. (1969). Influence of amino acid supply on ribosomes and protein synthesis of perfused rat liver. *Biochem. J.* **111:** 703—712.

11) MORTIMORE, G. E., SURMACZ, C. A. (1984). Liver perfusion: an in vitro technique for the study of intracellular protein turnover and its regulation in vivo. *Proc. Nutr. Soc.* **43:** 161—177.

12) ROGERS, Q. R. (1976).The nutritional and metabolic effects of amino acid imbalances. In: *Protein Metabolism and Nutrition* (Cole, D. J. A., Boorman, K. N., Buttery, P. J., Lewis, D., Neale, R. J., Swan, H., eds.),Butterworths London, pp. 279—301.

13) JEFFERSON, L. S., RANNELS, D. E., MUNGER, B. L., MORGAN, H. E. (1974). Insulin in the regulation of protein turnover in heart and skeletal muscle. *Fed. Proc.* **33:** 1098—1104.

14) TEWS, J. K., WOODCOCK, N. A., HARPER, A. E. (1972).Effect of protein intake on amino acid transport and adenosine 3',5'-monophosphate content in rat liver. *J. Nutr.* **102:** 409—417.

15) ANDERSON, H. L, BENEVENGA, N. J., HARPER, A. E. (1968). Associations among food and protein intake, serine dehydratase and plasma amino acids. *Am. J. Physiol.* **214:** 1008—1013.

16) HARPER, A. E., MILLER, R. H., BLOCK, K. P. (1984). Branched-chain amino acid metabolism. *Ann. Rev. Nutr.* **4:** 409—454.

17) PETERS, J. C., HARPER, A. E. (1985). Adaptation of rats to diets containing different levels of protein: effects on food intake, plasma and brain amino acid concentrations and brain neurotransmitter metabolism. *J. Nutr.* **115:** 382—398.

18) FERNSTROM, J. D. (1983). Role of precursor availability in control of monoamine biosynthesis in brain. *Physiol. Rev.* **63:** 484—546.

19) SMITH, Q. R. (1987). Kinetic analysis of neutral amino acid transport across blood-brain barrier. In: *Amino Acids in Health and Disease: New Perspectives* (Kaufman, S., ed.), Alan R. Liss, New York, pp. 65—85.

20) PARDRIDGE, W. M., OLDENDORF, W. H. (1975). Kinetic analysis of blood-brain barrier transport of amino acids. *Biochim. Biophys. Acta* **401:** 128—136.

21) TEWS, J. K., GREENWOOD, J., PRATT, O. E., HARPER, A. E. (1987). Threonine entry into brain after diet-induced changes in plasma amino acids. *J. Neurochem.* **48:** 1879—1886.

22) TEWS, J. K., GREENWOOD, J., PRATT, O. E., HARPER, A. E. (1987). Valine entry into rat brain after diet-induced changes in plasma amino acids. *Am. J. Physiol.* **252:** R78—R84.

23) PRATT, O. E. (1985).Continuous-injection methods for the measurement of flux across the blood brain barrier. The steady-state, initial-rate method. *Res. Methods Neurochem.* **6:** 117—150.

24) OLDENDORF, W. H. (1971). Brain uptake of radiolabeled amino acids, amines, and hexoses after arterial injection. *Am. J. Physiol.* **221:** 1629—1639.

25) TEWS, J. K., GOOD, S. S., HARPER, A. E. (1978). Transport of threonine and tryptophan by rat brain slices: relation to other amino acids at concentrations found in plasma. *J. Neurochem.* **31:** 581—589.

26) STONE, W. E., TEWS, J. K., WHISLER, K. E., BROWN, D. J. (1972). Incorporation of carbon from glucose into cerebral amino acids, proteins and lipids, and alterations during recovery from hypoglycaemia. *J. Neurochem.* **19:** 321—332.

27) HARPER, A. E., PETERS, J. C. (1983). Amino acid signals and food intake and preference: relation to body protein metabolism. In: *Nutritional Adequacy, Nutrient Availability and Needs* (Mauron, J., ed.), Birkhauser Verlag Basel, pp. 107—134.

ESSENTIAL NATURE OF THE "DISPENSABLE" AMINO ACIDS AND THEIR POSSIBLE INFLUENCE ON THE DEVELOPMENT AND FUNCTION OF THE BRAIN, WITH A NOTE ON THE QUESTION OF WHETHER THE FETUS REALLY CONCENTRATES AMINO ACIDS FROM THE MOTHER'S BLOOD

Samuel P. Bessman, Phyllis, Acosta, Rita Harper, Molly Towell

Dr. Samuel P. Bessman
Department of Pharmacology & Nutrition
University of Southern California
2025 Zonal Avenue
Los Angeles, CA 90022
United States

Dr. Phyllis Acosta
Ross Laboratories
1287 Jimstone Square West
Westerville, Ohio 43081
United States

Dr. Rita Harper
Division of Perinatal Medicine
Cornell University Medical College
300 Community Drive
Manhasset, NY 10130
United States

Dr. Molly Towell
Department of Obstetrics & Gynecology
1200 Main Street West
McMaster Medical Centre
Hamilton, Ontario L8S 479
Canada

INTRODUCTION

Since the discovery that only certain amino acids must be ingested, usually eight by adults, and possibly nine or even ten by growing children, there has been little interest in those amino acids which the normal individual can make, except for some pioneering work in general imbalances of amino acids by Professor Harper (1).

The thesis which concerns us at present is that the conventional belief that the non-essential amino acids can be made by the individual and are, therefore, of little or no dietary consequence should be replaced by a more stringent rule — the individual *must* make a certain minimum amount of each non-essential amino acid or there will be more or less severe consequences to his protein metabolism.

This conclusion is based on two general considerations — the first is a purely theoretical genetic one, and the second is experience with hereditary amino acid diseases involving cerebral dysfunction. These include the deficiency of phenylalanine hydroxylase, commonly called phenylketonuria, and the several disorders involving deficiency of one or another of the urea cycle enzymes. By analysis of these problems and study of the possible effects of heterozygosity for the synthesis of any non-essential amino acid, we may be able to gain insight into those milder syndromes or complications which could

NATO ASI Series, Vol. H20
Amino Acid Availability and Brain Function in Health and Disease. Edited by G. Huether

occur as a result of the increased demand for protein synthesis which is a characteristic of many more or less severe illnesses. Deficiencies might result from genetic heterozygosity for one or more enzymes which are required for the total synthesis of all of the non-essential amino acids. It is also of major importance that the brain synthesizes all of the "dispensable" amino acids, including those with neurotransmitter function, glutamic acid, glycine, serine, glutamine and, in the fetus, arginine and histidine.

A. Theoretical Basis

There are at least twenty-seven discrete enzymes formally required for the synthesis of all 12 non-essential amino acids. This counts only once those enzymes like the transaminases which are common to most amino acid syntheses.

Organ evolution is, *inter alia,* a process of loss of some function of groups of cells and resulting hypertrophy of those functions which remain. The earliest discrete cells must have been autotrophs, capable of synthesizing all of their requirements from elementary materials such as CO_2, ammonia and water. As multicellular organisms evolved, this specialization resulted in separate organs with appropriate functions, and loss of the multipotentiality of all cells.

The vitamins are an interesting case in point. The lower organisms synthesize their own vitamins, but the higher organisms do not. How did this incapability survive and flourish in the animal phyla? Perhaps their counterparts are the parasitic and saprophytic plants. What advantage did loss of capability to synthesize vitamins confer? Most likely, the animals which lost the enzymes necessary for vitamin synthesis survived because they could obtain these vitamins from their diet and could use the energy saved from vitamin synthesis to competitive advantage. Although no animals can survive without vitamins, all of the species clearly do survive on their natural diet.

Let us consider the essential and the non-essential amino acids in this light. The normal individual requires only about seven grams per day of a "good" assortment of essential amino acids and requires about ten grams of any mixture of non-essential amino acids, or only a couple of grams of ammonia and some carbon source such as glucose or citrate, to achieve adequate protein nutrition. Nevertheless, there is no doubt that the normal individual breaks down and synthesizes more than two hundred grams of protein per day. Is this discrepancy due to a remarkably efficient recycling system for amino acids, or is there a large synthesis of non-essential amino acids going on?

There is little information on the rate of synthesis of most non-essential amino acids. It is known that glycine synthesis occurs at the rate of at least about 50 grams per day, from tracer dilution studies, but there has not been much interest in pursuing the rate of synthesis of the non-essential amino acids. The only amino acid on which considerable work has been done from this standpoint is tyrosine, which occupies a unique place between essential and non-essential amino acids.

B. Amino Acid Disorders

We have made some calculations on the synthesis of arginine, for we can obtain unequivocal minimum information about its degradation rate by observing the irreversible formation of urea. The normal adult excretes about thirty grams of urea per day. This requires the breakdown of eighty-seven grams of arginine. If we include a few more grams of arginine for creatine synthesis and guanidination of neurotransmitters, the normal individual must synthesize ninety or more grams of arginine every day. Even the few grams of arginine necessary for creatine and guanidination could not be obtained from the normal diet, which contains about one-half to one gram per day of arginine. A fetus homozygous for any defect in the arginine cycle would be hard pressed to find more than a few micromoles of arginine in his mother's blood to form a properly functioning brain, particularly since the mother would be an obligate heterozygote with a diminished rate of synthesis. This matter was investigated by Jack Schaefer in our laboratory. He found that the brain of the rat fetus synthesizes arginine before birth. This falls essentially to zero in the first weeks of life. Experience with arginine cycle illness in children shows clearly that the brain deficits must be present before birth. The ammonia poisoning of some of the urea cycle syndromes is not the cause of the mental deficit, for there is no evidence of hyperammonemia before birth.

In the case of phenylketonuria, the evidence for a tyrosine deficiency prior to birth has accumulated.

1) The only chemical difference between phenylketonuric children and normal children at birth is a deficiency of melanin which is a polymerized product of tyrosine. It is manifest as blue eyes and lighter than normal hair. The hair color deficit can be alleviated simply by the oral administration of tyrosine, which was first shown by Penrose.
2) In phenylketonuria, the newborn has a normal blood level of phenylalanine. The elevation in serum phenylalanine occurs characteristically in the first week after birth. This shows that prenatal phenylalanine levels are not elevated.
3) It was first proposed by Kutter (2) that heterozygosity for phenylketonuria might have associate brain damage and he presented evidence to support this proposition. Berman's group (3) found extensive evidence to support this thesis. These findings were verified by our group (4) which was able to show that the heterozygous siblings of the phenylketonuric child are fifteen IQ points lower than the homozygous normal children born of the same mother. There is further evidence that the maternal obligate heterozygote is not as much affected in IQ as her heterozygous offspring. These two findings were predicted in the first publication of the "Justification Theory" (5).
4) High coincidence of heterozygosity for PKU was found in child-mother pairs of children with non-specific mental retardation (6).
5) Further information on the effects of tyrosine deficiency on the fetal brain was reported by Tomaszewski, from our laboratory, who showed that the structural protein, tubulin, was not formed normally in rat fetuses of dams on a tyrosine deficient diet.
6) An unexpected, but most important, observation was made by Dr. Phyllis Acosta who showed that a normal pregnant mother, at the end of the second and third trimesters of pregnancy had a moderate rise in blood phenylalanine and tyrosine one hour after a test meal of hamburger. A pregnant hyperphenylalaninemic mother, given the same meal at the end of her first, second, and third trimesters had an expectedly larger rise in blood phenylalanine, but a significant fall in tyrosine of 13 %, 39 % and 31 % respectively in all of these tests. This was in spite of the fact that this mother was receiving 1, 2, and 3 grams of supplementary tyrosine per day in each trimester, respectively.

If these observations are found to be generally true for all hyperphenylalaninemic patients, it will contribute much to the evidence for prenatal tyrosine deficiency. It has been reported by several researchers (7, 8) that a fall in a particular amino acid after a meal indicates that that amino acid is available only in limited amounts which are inadequate for the protein synthesis stimulated by the meal itself.

7) On the basis of the findings on blood tyrosine in heterozygotes after a meal, some preliminary experiments were carried out with Dr. Rita Harper on newborn infants at the North Shore University Hospital. The working hypothesis was that failure to thrive might be caused by a developmental deficiency in the enzymes for synthesizing one or more non-essential amino acids. Each child was given a normal feeding of milk at time zero and blood samples for plasma amino acids were taken at zero, 30', 60', and 90', and two hours. Figure 1 shows the data obtained from three apparently normal children and two who were diagnosed as failure to thrive, based on a poor weight gain. Cases 1, 2, 3 were clinically normal. The data are reported qualitatively. The in-

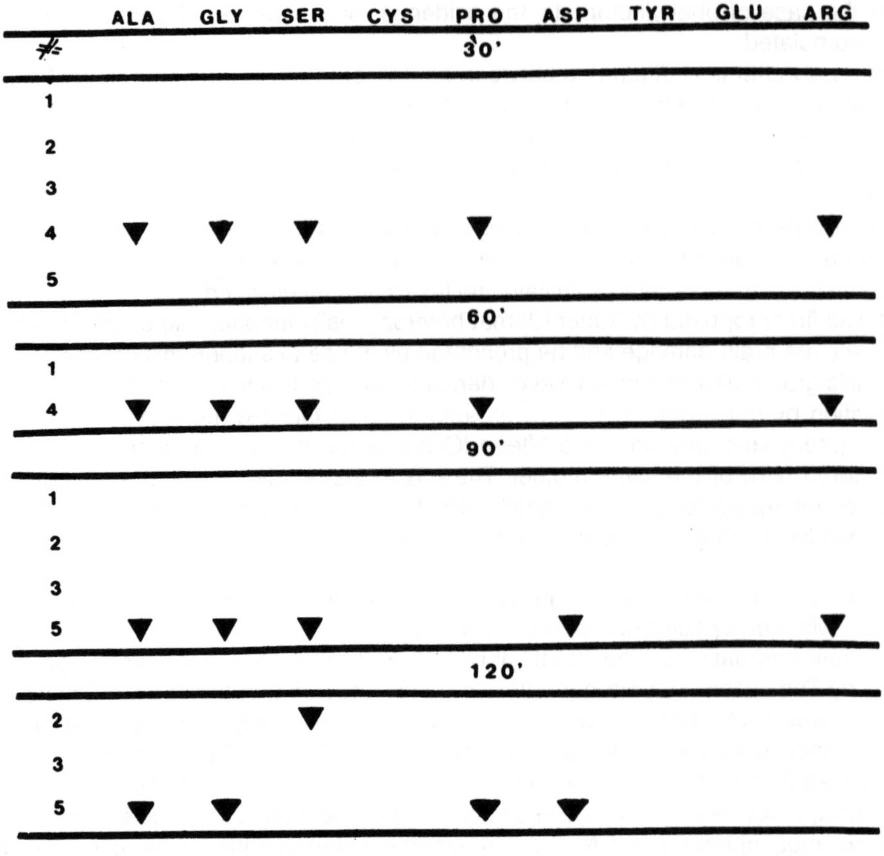

FIGURE 1

Drop of plasma amino acids in newborns after feeding. Each child was given a normal feeding of milk at time zero and blood samples for amino acid analyses were taken at zero, 30, 60, 90 and 120 min. Inverted triangles signify a drop of more than 20 % in the plasma level of the particular amino acids (see text).

verted triangle signifies a drop of more than 20 % in the plasma level of the particular non-essential amino acid compared to zero time. The data are incomplete because blood specimens were not obtained for all times for any patients. It is noteworthy that only once did any one of the three normal patients show a significant drop in amino acid level after a meal and that was #2, who showed a fall in serine at 120 minutes. Numbers 4 and 5 were the two cases of failure to thrive on which we obtained at least two specimens in addition to the zero values. This suggests that heterozygosity in any amino acid synthesis may be fairly common. This kind of experiment should be repeated with animal brain formation of amino acids.

8) The offspring of maternal PKU pregnancies have been almost invariably retarded even though they are usually heterozygotes. This was first explained as damage to the fetus from the mother's high blood level of phenylalanine. It was subsequently recognized that the children were severely damaged even though the mother's phenylalanine levels were kept low by a restricted phenylalanine diet. This was then explained by a misinterpretation of cord blood levels of phenylalanine. It was stated, on the basis of cord blood values, that the placenta concentrates phenylalanine from the maternal blood so that the fetus is subjected to blood levels of phenylalanine much higher than maternal levels. It seemed to us that the data on cord blood did not take into consideration two important facts. The first is that the fetus disposes of amino acids slowly because its major route is protein synthesis, whereas the mother has many pathways in addition to protein synthesis to dispose of an amino acid load. The second is that there is a lag in movement of amino acids, as indeed of most solutes, into or out of a closed space, in this case, the fetus.

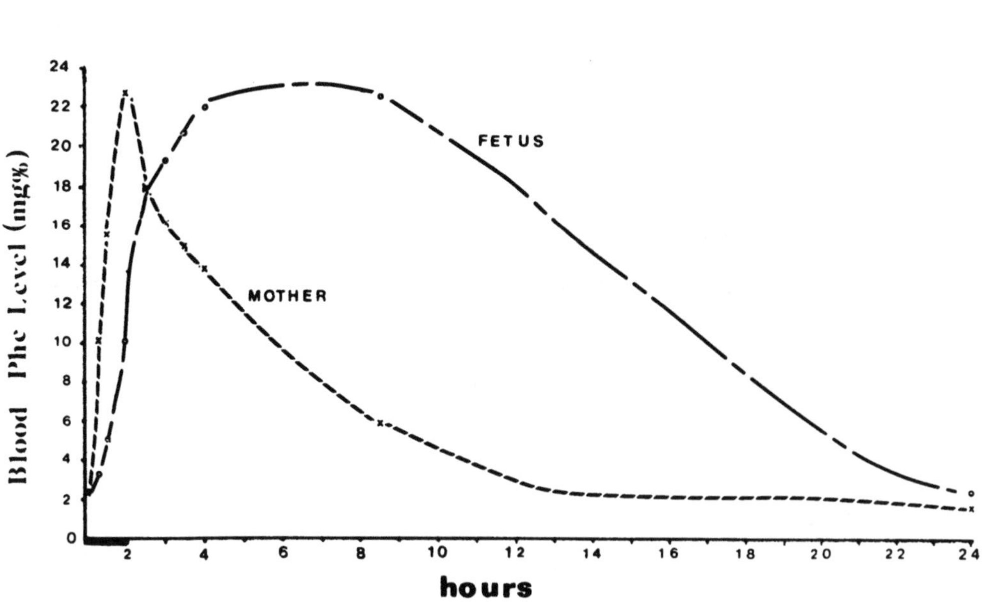

FIGURE 2

Relation between maternal and fetal Phe level in the plasma of goats during and after an one hour infusion (i. v.) of Phe to the mother.

We studied the uptake and loss of phenylalanine by the blood of a near term goat fetus during and following the intravenous administration of 250 mg/kg of phenylalanine to the mother. Indwelling catheters had been placed in maternal and fetal blood vessels several weeks before the experiment. Figure 2 shows the relation between the maternal and fetal blood levels during the one hour intravenous infusion to the mother and for the next 24 hours. The first thing we note is that the fasting maternal level is higher than the fetal level. The second is that the fetus certainly does not concentrate phenylalanine from the mother during the infusion. The mother's levels until the cessation of the infusion are 50 to 75 % higher than the fetal blood levels. The fetal blood level rises as the maternal level falls, and in about two hours after the maternal level begins to fall, the fetal blood level rises to just below the highest level attained by the mother. The fetal level never exceeds the highest level attained by the mother. This was also seen in experiments with pregnant rats which were performed in our laboratory many years ago. Never did the fetal blood level exceed the highest level attained in the mother's blood.

The lag between the maternal and fetal maxima we attribute to the buffer effect of the placenta which "loads up" and slowly discharges into the fetus. After about eight hours, the fetal level begins to fall and approaches the maternal level. By zero hours the fetal level has not yet fallen below the maternal level.

The same lag, on both rise and fall, can be seen in a glucose tolerance curve when spinal fluid and blood levels are followed. As the blood level of glucose rises after a test dose, there is a slow rise of spinal fluid glucose. A single comparison at this time would lead to the "conclusion" that the spinal fluid is relatively impermeable to the blood glucose. When the blood glucose falls and meets the spinal fluid level the interpretation of single equal values would be that the spinal fluid is freely permeable to glucose. If a sample is taken as the spinal fluid glucose falls slowly and the blood glucose falls rapidly, the interpretation could be that the spinal fluid concentrates glucose from the blood. This is all due to a failure to consider the inertia of changes between blood and tissues.

10) Professor Robert Fisch (9) has reported data which show that the tyrosine deficiency in the PKU fetus must be operative as early as the first month of gestation. The higher than normal incidence of congenital abnormalities includes equal contributions from types of abnormality which occur in all months.

I have presented these ideas from our research on metabolic issues in brain metabolism that we might discuss some of the work to be presented in the next sessions in the light of genetic issues and the general question of the complexities of the metabolic relation between the brain and the rest of the organism.

REFERENCES

1) HARPER, A. E. (1958). Balance and imbalance of amino acids. *Ann. N. Y. Acad. Sci.* **69:** 1025—1041.

2) KUTTER, D., HUMBEL, R., METZ, H. (1970). Fetale Hirnschäden bei mütterlicher heterozygoeter Phenylketonurie. *Medizinische Klinik* **65:** 653—656.

3) FORD, R. C., BERMAN, J. L. (1977). Phenylalanine metabolism and intellectual functioning among carriers of phenylketonuria and hyperphenylalanemia. *Lancet* **1:** 767—771.

4) BESSMAN, S. P., MOHAN, C., ZAIDISE, I. (1986). Intracellular site of insulin action: Mitochondrial Krebs cycle. *Proc. Natl. Acad. Sci. USA* **83:** 5067—5070.

5) BESSMAN, S. P. (1972). Genetic failure of fetal amino acid "justification": A common basis for many forms of metabolic, nutrition, and "nonspecific" mental retardation. *J. Pediatrics* **81:** 834—842.

6) FUJIMOTO, A., CRAWFORD, R., BESSMAN, S. P. (1979). Relative ability of mother and child to convert phenylalanine to tyrosine — a possible cause of nonspecific mental retardation. *Biochem. Med.* **21:** 271—276.

7) OZALP, I., YOUNG, V. R., NAGCHAUDHURI, J., TONTISIRIN, K., SCRIMSHAW, N. S. (1972). Plasma amino acid response in young men given diets devoid of single essential amino acids. *J. Nutr.* **102:** 1147—1158.

8) GRAHM, G. G., PLAKO, R. P. (1973). Postprandial plasma free methionine as an indicator of dietary methionine adequacy in human infant. *J. Nutr.* **103:** 1347—1351.

9) FISCH, R. O., DOEDEN, D., LANSKY, L. L., ANDERSON, J. A. (1969). Maternal phenylketonuria: deterimental effects on embryogenesis and fetal development. *Am. J. Dis. Child.* **118:** 847—858.

THE IMPORTANCE OF THE LIVER AS A REGULATOR OF AMINO ACID SUPPLY TO THE BRAIN

M. Salter and C. I. Pogson

Biochemistry Department
Wellcome Research Laboratories
Langley Court
Beckenham Kent BR3 3BS
U.K.

INTRODUCTION

In mammals, tryptophan and phenylalanine are essential components of the diet, dietary tyrosine being supplemented by hydroxylation of phenylalanine in the liver. The concentrations of the amino acids in the blood are regulated by the rates of dietary intake, protein turnover and peripheral metabolism (occurring mostly in the liver; see Figure 1). The aromatic amino acids are of particular interest because of their role as precursors of several neurotransmitters, i. e., dopamine, noradrenaline and serotonin. The aromatic amino acids share, with many other neutral amino acids, a common transport system into the brain and it is clear that the ratio of their concentrations, as well as their absolute concentrations, is important for neurotransmitter synthesis. The synthesis of serotonin has been shown to be regulated by the supply of its precursor, tryptophan. Tryptophan hydroxylase is considered to be the rate-limiting enzyme for serotonin synthesis and is unsaturated

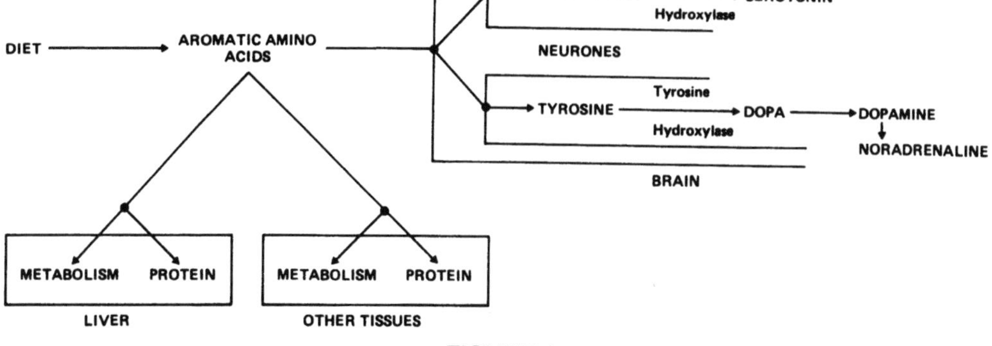

FIGURE 1
Metabolism of the aromatic amino acids

NATO ASI Series, Vol. H20
Amino Acid Availability and Brain Function in Health and Disease. Edited by G. Huether

with respect to its substrate, tryptophan (1). Increases or decreases in brain tryptophan will therefore increase or decrease the synthesis of serotonin; the normal concentration of brain tryptophan approximates to the K_m of tryptophan hydroxylase for tryptophan (1). The control of brain serotonin is of great importance because of the role of this neurotransmitter in the cerebral functions associated with mood, sleep, sensitivity to pain and appetite.

The evidence for tyrosine-induced increases in catecholamine synthesis is not quite so clear cut. Until recently it was believed that tyrosine hydroxylase, which is considered to be rate-limiting for catecholamine synthesis, was normally saturated with tyrosine. However, it has now been shown that increases in the synthesis of dopamine and noradrenaline can occur after tyrosine administration, the increases becoming more apparent with the frequencing of neuronal firing. These findings are of interest because of the possible therapeutic role of tyrosine in depression.

Quantitatively, the liver has been shown to be the most important site of aromatic amino acid metabolism in the body; changes in the hepatic metabolism of the aromatic amino acids are therefore likely to be reflected in the supply of the amino acids to the brain for subsequent neurotransmitter synthesis. Regulation of this hepatic metabolism is thus clearly important for the synthesis in the brain of aromatic amino acid-derived neurotransmitters.

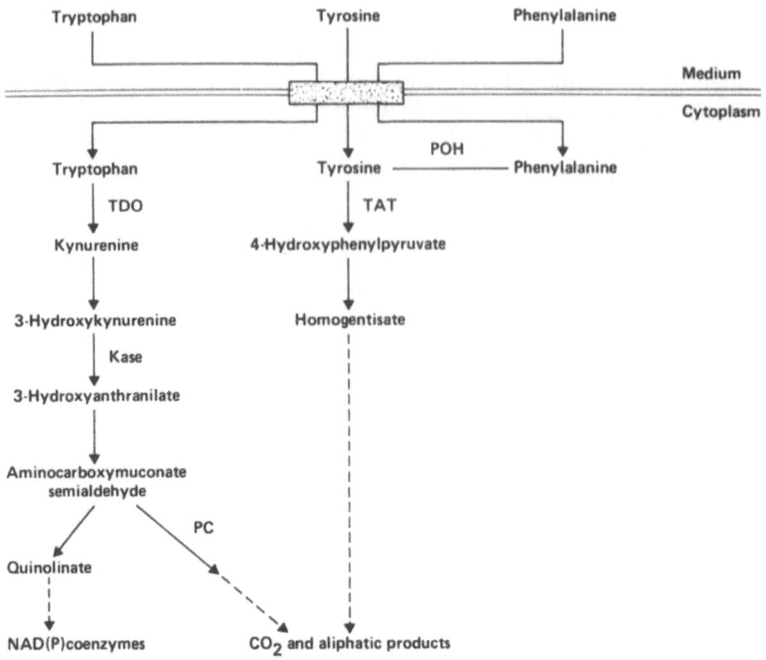

FIGURE 2
Pathways of hepatic aromatic amino acid metabolism

Hepatic Transport of the Aromatic Amino Acids

The aromatic amino acids are transported passively across the liver plasma membrane by two transport systems, L and T. The ubiquitous L-system has been shown to be subdivided into L1 and L2 in rat liver cells, both inhibited by the nonmetabolizable amino acid analogue, 2-aminobicyclo-(2.2.1)-heptane-2-carboxylic acid (BCH), and will transport most of the neutral amino acids (2). System L2 is quantitatively the most important transport system in freshly-prepared rat liver cells; in primary culture of rat liver cells, however, transport mediated by system L1 increases, especially at low substrate concentrations (2). Until recently, the aromatic D- and L-amino acid specific system T had only been observed in human (3) and rat (4) red blood cells; recent work, however, has demonstrated the presence of the T-system in freshly-prepared rat liver cells (5). With systems L1 and L2, the T-system is the only Na^+-independent transport system for neutral amino acids seen in eukaryotic cells. Although the T-system is a comparatively low K_m, low V_{max} system in comparison with the L2-system, at physiological concentrations of the aromatic amino acids the two systems have an approximately equal quantitative role in the transport of their substrates (5). The T-system is relatively resistant to regulation and is unaffected by dexamethasone, glucagon, starvation of rats for 48 hr or by 24 hr culture of rat liver cells (5).

Hepatic Phenylalanine Metabolism

The initial step of intracellular phenylalanine metabolism in the liver is catalysed by phenylalanine hydroxylase (see Figure 2). Haley and Harper (6) have estimated that 75 % of a phenylalanine load is metabolised by phenylalanine hydroxylase, which is found predominantly in the liver, and its physiological importance is amply demonstrated by the result of the virtual absence of its activity in phenylketonuria.

Many detailed studies have been carried out upon the regulation of phenylalanine hydroxylase *in vivo, in vitro,* and in liver cells. The enzyme is subject to short- and long-term regulation. Glucocorticoids increase the synthesis of mRNA specific for phenylalanine hydroxylase. The rat liver enzyme is phosphorylated *in vitro* by the cyclic AMP-dependent and calmodulin-dependent protein kinases (7). This phosphorylation is also seen in liver cells exposed to glucagon and α- or β-adrenergic agonists (7, 8). Phenylalanine hydroxylase is activated by low concentrations of its substrate after phosphorylation. Increases in the phosphorylation state of the enzyme correlate well with increases in flux through the enzyme at physiological concentrations of phenylalanine in isolated rat liver cells (7). The phosphorylation-dependent increase in the activity of phenylalanine hydroxylase can be reversed with insulin or spermine (7).

Hepatic Tyrosine Metabolism

The product of the hydroxylation of phenylalanine, tyrosine, is transaminated by tyrosine aminotransferase (see Figure 2). This pyridoxal-phosphate-dependent enzyme is present predominantly in the liver and is responsible for most of tyrosine metabolism in the body. Tyrosine aminotransferase has been claimed to be the rate-limiting enzyme for degradation of intracellular tyrosine (see below) and under most conditions, therefore, the steps 'beyond' the transaminase exert no influence on flux through the pathway (9). The enzyme is subject to short-term regulation and is induced by glucagon and glucocorticoids (10). Although insulin increases enzyme activity in Hepatoma cells, it inhibits the glucocorticoid-mediated induction of tyrosine aminotransferase in cultured liver cells. Maintenance of rats on a pyridoxine-deficient diet for 6 weeks causes decreases in flux through tyrosine aminotransferase in isolated liver cells as well as decreases in the cofactor saturation of the isolated enzyme (11). Tyrosine aminotransferase activity is increased by tryptophan *in vivo* (12) and tryptophan rather than glucocorticoid has been suggested as maintaining the diurnal rhythm of the enzyme in the rat.

Hepatic Tryptophan Metabolism

In mammals more than 90 % of total body tryptophan metabolism has been believed to occur through the kynurenine pathway of the liver (13). The first enzyme of the pathway is tryptophan 2,3-dioxygenase (see Figure 2) and this has been considered, until recently, to catalyse the rate-limiting step of tryptophan metabolism (see below). The fate of tryptophan-derived carbon is metabolism either to CO_2 and C_2 units through the citric acid cycle or to the nicotinamide coenzymes. The distribution of tryptophan carbon between these two pathways depends on the activity of picolinate carboxylase, which lies at the branch of the two pathways. Although changes in the activity of this enzyme can change the distribution of flux within the kynurenine pathway under normal conditions, only changes in the activity of tryptophan 2,3-dioxygenase can change the absolute flux through the pathway.

Tryptophan 2,3-dioxygenase is unsaturated with respect to its haem cofactor in rat liver homogenates (14, 15) and manipulation of haem synthesis *in vivo* will vary the subsequent degree of saturation (15). However, recent work in this laboratory with isolated rat liver cells and perfused liver (14), has shown that, in the presence of physiological concentrations of tryptophan and haem, the enzyme is likely to be already fully saturated with its cofactor. Regulation of tryptophan 2,3-dioxygenase by changes in haem concentration is, therefore, unlikely to occur *in vivo*.

Glucocorticoids have been shown to elevate tryptophan 2,3-dioxygenase mRNA and thus enzyme protein *in vivo* (16) and the diurnal rhythm of enzyme activity in the mammal is thought to be due to changes in the concentration of circulating glucocorticoid. Adrenalectomy prevents the diurnal rhythm but enzyme activity does not fall to levels lower than those seen at the minimum of the diurnal rhythm (16). As well as stabilising tyrosine aminotransferase, tryptophan also decreases the rate of degradation of tryptophan 2,3-dioxygenase, thus increasing its activity (12). This effect is reasonably rapid because of the relatively short half-life of the enzyme, 2—2.5 hours. Recent work has demonstrated that glucagon will increase the translation of tryptophan 2,3-dioxygenase-specific mRNA and increases the activity of the enzyme in cultured rat liver cells (17); insulin reverses the glucagon-dependent increase (17). Tryptophan 2,3-dioxygenase activity is elevated in acutely and chronically diabetic rats; this effect is likely to be mediated by glucagon and reversed by insulin administration (16). Although, under normal conditions, the enzymes beyond tryptophan 2,3-dioxygenase are unable to affect the absolute rate of flux through the kynurenine pathway, in pyridoxine deficiency (where kynureninase activity is lowered considerably), significant decreases in flux through the pathway are observed (11).

Quantification of the Control of Metabolism

As previously mentioned, the observed changes and diverse mechanisms of regulation of phenylalanine hydroxylase, tyrosine aminotransferase and tryptophan 2,3-dioxygenase have been interpreted as evidence for rate-limiting roles of these enzymes. However, these observations are insufficient for such conclusions and more exacting and direct measurements are needed to quantify the importance of each step in the pathways. This has been achieved by determination of the control coefficients of the enzymes using the metabolic control theory of Kacser and Burns (18) and Heinrich and Rapaport (19). These control coefficients express the change in flux through a pathway produced by a unit change in the activity of any enzyme in that pathway, and it has been shown that the sum of all control coefficients in a pathway must be 1. Salter *et al.* (20) have shown that a plot of kynurenine pathway flux *vs* tryptophan 2,3-dioxygenase activity is hyperbolic rather than linear. The extent to which flux is increased or decreased through the pathway by a unit change in the activity of tryptophan 2,3-dioxygenase depends, therefore, upon the absolute activity of the dioxygenase. Control coefficients for tryptophan 2,3-dioxygenase, calculated from this relationship, range from 0.75, under conditions of non-induced dioxygenase activity, to 0.25, under conditions of maximally-induced dioxygenase activity. Description of tryptophan 2,3-dioxygenase as the rate-limiting enzyme of tryptophan metabolism is, therefore, clearly inappropriate. Further calculations, shown in Salter *et al.* (20), demonstrate, that the remainder of the control coefficient resides with the transport systems L and T. Further work (20) has demonstrated that a similar distribution of control coefficients occur between phenylalanine hydroxylase, tyrosine aminotransferase and the transport systems L and T for liver phenylalanine and tyrosine metabolism respectively. The importance of the control residing with the transport proteins is emphasised by the observation that a physiological mix of amino acids inhibits aromatic amino acid metabolism by 25 % (2). Under normal conditions, enzymes catalysing steps 'beyond' tryptophan 2,3-dioxygenase, tyrosine aminotransferase and phenylalanine hydroxylase, for tryptophan, tyrosine and phenylalanine metabolism respectively, do not possess significant control coefficients and therefore contribute little to the regulation of these pathways. However, under conditions of pyridoxine deficiency (where large decreases in kynureninase activity are observed), kynureninase has been shown to possess a significant control coefficient for flux through the kynurenine pathway (11).

A Further Pathway of Liver Tryptophan Metabolism

Although it has been thought that over 90 % of total body tryptophan metabolism occurs through the kynurenine pathway, recent evidence from this laboratory (Salter and Pogson, unpublished observations) suggest that this may not be the case and that tryptophan 2,3-dioxygenase may in fact play, quantitatively, a lesser role than previously believed in total body tryptophan metabolism. The evidence for this is as follows.

(1) Intraperitoneal injection of tryptophan (including [ring-2-^{14}C]-tryptophan) at a dose of 750 mg/kg body wt. results in the appearance of over 40 % of the ^{14}C in the urine and 30 % in the rat body as an aromatic metabolite of tryptophan. Cleavage of ^{14}C from the 2-position of the pyrrole-ring of tryptophan by tryptophan 2,3-dioxygenase and formamidase makes it impossible for metabolites of the kynurenine pathway to contain the label. Less than 10 % of the ^{14}C administered was recovered as $^{14}CO_2$ (an indication of trypto-

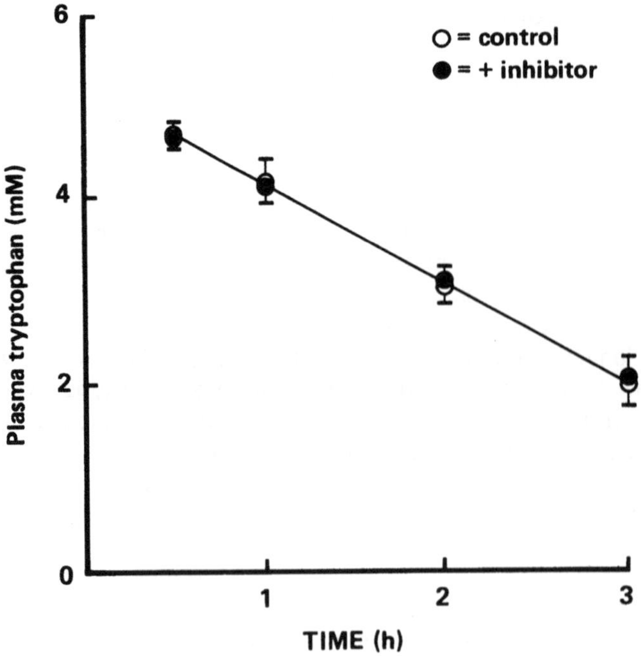

FIGURE 3

Effect of tryptophan 2,3-dioxygenase inhibitor on tryptophan clearance *in vivo*

phan 2,3-dioxygenase activity). Similar results have been obtained with a tracer dose of [ring-2-^{14}C] tryptophan.

(2) Specific and potent inhibitors of tryptophan 2,3-dioxygenase *in vitro* and in isolated rat liver cells had no effect on the concentration of plasma tryptophan *in vivo* (see Figure 3) (in the presence or absence of a tryptophan load), even though the circulating concentration of inhibitor was high enough to produce over a 90 % inhibition of flux through the kynurenine pathway.

(3) The rate of clearance of tryptophan from rat plasma *in vivo* after a tryptophan load is approximately 6—7 times the V_{max} of tryptophan 2,3-dioxygenase (measured either spectrophotometrically or radiometrically *in vitro* or in liver cells. See Table 1). The rate of clearance *in vivo* is elevated by pre-treatment of the rat with a tryptophan load (750 mg/kg body wt.). Although tryptophan 2,3-dioxygenase is also elevated by tryptophan pretreatment, the increase in the rate or clearance *in vivo* is far in excess of the increased dioxygenase activity.

Measurement of whole-body indoleamine 2,3-dioxygenase has shown that this tryptophan-cleaving activity is much too low to explain the rate of clearance *in vivo* (Salter and Pogson, unpublished observations). However, if the rate of tryptophan disappearance (as opposed to appearance of products) is measured in incubations of isolated liver cells, the rate is found to be far greater than that predicted solely from the simultaneously measured rate of tryptophan 2,3-dioxygenase but, is close to the rate of tryptophan clearance for the whole liver *in vivo* (Salter and Pogson, unpublished observations; Table 1). Even at physiological concentrations of tryptophan, tryptophan 2,3-dioxygenase activity is responsible for only approximately 30 % of the total tryptophan metabolism in rat liver cells

TABLE 1
Rate of tryptophan metabolism and tryptophan 2,3-dioxygenase activity

	[Tryptophan] (mM)	Tryptophan removal	Tryptophan 2,3-dioxygenase
		(μmol/h per rat)	
Liver cells	0.05	7.20 ± 0.50	2.40 ± 0.10
	0.5	43.50 ± 2.50	10.50 ± 0.90
	5	195 ± 10	23.50 ± 1.50
In vivo	4	160 ± 15	

(Table 1). At higher concentrations found in the blood after a large tryptophan load, the dioxygenase assumes, quantitatively, an even smaller role (Table 1). Protein synthesis is not responsible for tryptophan removal in liver cells because the only amino acids available are those endogenously produced by proteolysis. Flux in rat liver cells with [carboxyl 1-^{14}C] tryptophan is quantitatively accounted for solely by the flux through kynureninase of the kynurenine pathway (12, 21) and therefore other pathways of tryptophan decarboxylation are unlikely to be of quantitative significance in the metabolism of tryptophan by liver cells.

Similar rates of tryptophan removal can also be demonstrated in rat liver homogenates (Salter and Pogson, unpublished observations). The activity appears to be located in the cytosol and is unaffected by 1 mM-aminooxyacetate, 100 mM-leucine or -lysine, making it unlikely, therefore, that tryptophan removal is catalysed by a transaminase or amino acid oxidase.

SUMMARY AND CONCLUSION

The liver has been shown to play a quantitatively major role in the part of peripheral aromatic amino acid metabolism and regulation of this metabolism, therefore, may well influence supply of aromatic amino acids to the brain. The liver metabolism of the aromatic amino acids has been extensively studied and their pathways of metabolism are subject to a number of diverse methods of regulation. For tyrosine and phenylalanine metabolism, the control of metabolism is shared between tyrosine aminotransferase and phenylalanine hydroxylase respectively and the transport systems L and T. Although a similar scenario for tryptophan 2,3-dioxygenase and transport is also true, it appears that tryptophan 2,3-dioxygenase plays only a quantitatively minor role in liver tryptophan metabolism. The nature and properties of the remaining tryptophan metabolism have yet to be fully elucidated. However, it is clear that the regulation of this process may be of great significance for the supply of tryptophan to the brain.

REFERENCES

1) KNOWLES, R. G., POGSON, C. I. (1984). Tryptophan uptake and hydroxylation in rat forebrain synaptosomes. *J. Neurochem.* **42:** 677—684.

2) WEISSBACH, L., HANDLOGTEN, M. E., CHRISTENSEN, H. N., KILBERG, M. S. (1982). Evidence for two Na^+-independent neutral amino acid transport systems in primary cultures of rat hepytocytes. *J. Biol. Chem.* **257:** 12006—12011.

3) ROSENBERG, R., YOUNG, J. D., ELLORY, J. C. (1980). L-Tryptophan transport in human red blood cells. *Biochim. Biophys. Acta.* **598:** 375—384.

4) JAYDUTT, V. V., CHRISTENSEN, H. N. (1985). Discrimination of Na^{+}-independent transport systems, L, T and asc in erythrocytes. *J. Biol. Chem.* **260:** 2912—2921.

5) SALTER, M., KNOWLES, R. G., POGSON, C. I. (1986). Transport of the aromatic amino acids into isolated rat liver cells. *Biochem. J.* **223:** 499—506.

6) HALEY, C. J., HARPER, A. E. (1978). The importance of transamination and decarboxylation in phenylalanine metabolism *in vivo* in the rat. *Arch. Biochem. Biophys.* **189:** 524—530.

7) POGSON, C. I., DICKSON, A. J., KNOWLES, R. G., SALTER, M., SANTANA, M. A., STANLEY, J. C., FISHER, M. J (1986). Control of phenylalanine and tyrosine metabolism by phosphorylation mechanisms. *Adv. Enz. Regul.* **25:** 309—328.

8) FISHER, M. J., POGSON, C. I. (1984). Phenylalanine hydroxylase in liver cells. *Biochem. J.* **219:** 79—85.

9) DICKSON, A. J., MARSTON, F. A. O., POGSON, C. I. (1981). Tyrosine aminotransferase as the rate-limiting step for tyrosine catabolism in isolated rat liver cells. *FEBS Lett.* **127:** 28—32.

10) STANLEY, J. C., FISHER, M. J., POGSON, C. I. (1985). The metabolism of L-phenylalanine and L-tyrosine by liver cells isolated from adrenalectomized rats and from streptozotocin-diabetic rats. *Biochem. J.* **228:** 249—255.

11) STANLEY, J. C., SALTER, M., FISHER, M. J., POGSON, C. I. (1985). The effect of pyridoxine deficiency on the metabolism of the aromatic amino acids by isolated rat liver cells. *Arch. Biochem. Biophys.* **240:** 792—800.

12) SALTER, M., STANLEY, J. C., FISHER, M J., POGSON, C. I. (1984). The influence of starvation and tryptophan administration on the metabolism of phenylalanine, tyrosine and tryptophan in isolated rat liver cells. *Biochem. J.* **221:** 431—438.

13) YOUNG, S. N., ST. ARNAUD-McKENZIE, D., SOURKES, T. L. (1978). Importance of tryptophan pyrrolase and aromatic amino acid decarboxylase in the catabolism of tryptophan. *Biochem. Pharmacol.* **27:** 763—767.

14) SALTER, M., POGSON, C. I. (1986). The role of haem in the regulation of rat liver tryptophan metabolism. *Biochem. J.* **240:** 259—263.

15) WELCH, A. N., BADAWY, A. A.-B. (1980). Tryptophan pyrrolase in haem regulation. *Biochem. J.* **192:** 403—410.

16) SALTER, M., POGSON, C. I. (1985). The role of tryptophan 2,3-dioxygenase in the hormonal control of tryptophan metabolism in isolated rat liver cells. *Biochem. J.* **229:** 499—504.

17) NAKAMURA, T., SHINNO, H., ICHIHARA, A. (1980). Insulin and glucagon as a new regulator system for tryptophan oxygenase activity demonstrated in primary cultured rat hepytocytes. *J. Biol. Chem.* **255:** 7533—7535.

18) KACSER, H., BURNS, J. A. (1973). The control of flux. *Symp. Soc. Exp. Biol.* **32:** 65—104.

19) HEINRICH, R., RAPAPORT, T. A. (1974). A linear steady-state treatment of enzymatic chains. *Eur. J. Biochem.* **42:** 89—95.

20) SALTER, M., KNOWLES, R. G., POGSON, C. I. (1986). Quantification of the importance of individual steps in the control of aromatic amino acid metabolism. *Biochem. J.* **234:** 635—647.

21) SMITH, S. A., CARR, F. P. A., POGSON, C. I. (1980). The metabolism of L-tryptophan by isolated rat liver cells. *Biochem. J.* **192:** 673—686.

BIOAVAILABILITY OF AMINO ACIDS AND AMINO ACID PRECURSORS FOR NEUROTRANSMITTER ACTION: THE ROLE OF HORMONES

Marie Luise Rao

Nervenklinik und Poliklinik, Psychiatrie
Rheinische Friedrich-Wilhelms-Universität
Sigmund-Freud-Straße 25
5300 Bonn 1
Federal Republic of Germany

INTRODUCTION

In the context of neurotransmitter action, amino acids serve as inhibitory (taurine and glycine) and excitatory neurotransmitters (glutamate and aspartate), precursors of neurotransmitters (tyrosine and tryptophan), and transducers of transmitter components (glycine and serine in transmethylation processes). These amino acids derive exogenously from nutrients, and endogenously from the intermediary metabolism, and from the degradation of protein stores. They circulate in blood in the free from and bound to blood proteins (tryptophan), and blood cells (taurine). Hormones influence the amino acid bioavailability for neural actions in that they regulate the amino acid release from muscle proteins, their metabolism, and their transport across the plasma membrane. The latter step comprises several loci of action, e. g., influx into cellular systems of the gastrointestinal tract, efflux from muscle and metabolizing organs, and transport across the blood-brain barrier.

Each step involves the action of one or several proteins and is thus liable for regulatory processes at the molecular level. On account of the half saturation constants of the enzymes involved, the steps in the intermediary metabolism are regulated by substrate concentration, i. e., by the amino acid concentrations; they are also influenced by hormones. The transmembrane translocating processes are controlled by concentration gradients of the amino acid transported or its competitors, by the presence of small co- and counter ions (Na^+, K^+, Cl^-), by the energy charge of the cell, and by hormones. The bioavailability of tryptophan and tyrosine in the periphery as precursors for neurotransmitters depends on their and on their competitor's plasma concentration; these factors regulate the concentration in the central nervous system.

In rats a carbohydrate meal brings about an increase in plasma tryptophan. Likewise in healthy subjects, i. v. glucose administration elicits an augmentation of the plasma tryptophan level. Studies in our laboratory have shown that in healthy subjects serum tryptophan and whole blood serotonin correlate, albeit weakly ($r = 0.295$, $P < 0.001$, $n = 136$; Ref. 1). In the feedback control of amino acid metabolism, neurotransmitter ac-

NATO ASI Series, Vol. H20
Amino Acid Availability and Brain Function in Health and Disease. Edited by G. Huether

tion and mood, the role of hormones is still a disputed subject. In the following, examples are given for a possible interrelationship between hormones and amino acid metabolism; these are steps, which may be regarded as key elements in this chain of reaction.

Insulin

This hormone is a key regulator of protein balance in muscle and thus indirectly of circulating amino acid levels. After a meal the elevated plasma insulin level promotes the transmembrane translocation of amino acids by muscle and their subsequent incorporation into proteins. Levels of branched-chain amino acids (valine, leucine, and isoleucine) are responsive to changes in circulating insulin levels (2). Upon fasting the drop in the insulin level elicits a net release of amino acids from muscle. Insulin stimulates protein synthesis and inhibits protein degradation (3—8). Chronic hyperinsulinism in humans is generally accompanied by a decrease in the plasma level of most amino acids. However, glycine, serine, alanine, and tyrosine are increased (9). When rats receive a subconvulsive dose of insulin their plasma tryptophan level rises while the concentration of other amino acids in plasma decreases (10). On the other hand the glucose-induced insulin release is strongly affected by tryptophan (11), thus closing the loop. Disturbances of this loop may manifest themselves in mild diabetes and during the later stages of pregnancy (12), when glucose intolerance is observed. In untreated diabetes mellitus branched chain amino acids are increased, whereas most other amino acids are lowered or in the normal range (13); since these amino acids, i. e., valine, leucine, and isoleucine are increased and compete with tyrosine and tryptophan for the L-amino acid uptake system across the blood brain barrier into the central nervous system (14), the concomitant decrease of tyrosine or tryptophan causes a drop in catecholamine and indoleamine synthesis (15). An imbalance of the net changes in this chain of reactions due to the influence of hormones may well be regarded as precipitating factors for mood swings observed in this endocrinological disease, exacerbation of depression or schizophrenia during pregnancy, and post partition psychosis.

Although the transmembrane transport of the L-system for amino acids is insulin-insensitive (14), the net transport of these amino acids into the brain depends largely on the peripheral amino acid levels. The latter in turn are partially regulated by the influence of insulin in that the hormone stimulates amino acid uptake into liver cells, thymocytes and muscle cells (16—20), thus pointing to an indirect action of insulin with respect to the blood brain barrier. The peripheral concentration of amino acids may also feed-back regulate the rate of liberation of amino acids from peripheral tissue; lowering of glycine and leucine facilitates the liberation of amino acid from muscle (21). In the context of amino acid feedback regulation, it is interesting to note that the peripheral branched chain amino acid level may exert a tonic influence on β-cell secretory activity (22).

Glucagon

The insulin-antagonist's level increase in plasma is accompanied by a drop of total plasma free amino acid concentrations (23, 24). The reaction is due to an enhancement of the hepatic utilization of amino acids without a urinary amino acid nitrogen increase. Glucagon infusion in fasted obese subjects, however, results in an increase in valine, leucine and isoleucine levels (22). Thus the liver regulates protein anabolism and catabolism and plays its role in the amino acid bioavailability for central nervous action. There is a remarkable parallelism between the action of glucagon and progesterone. Both hormones enhance protein catabolism and lower the plasma free amino acid concentration (23).

Growth Hormone

Growth hormone stimulates the accumulation of amino acids into cells (21) and the ensuing protein synthesis (25). Growth hormone exerts a direct anabolic action by stimulating ribosomal activity and synthesis of messenger and ribosomal RNA (26, 27). Lack of growth hormone is responsible for lowered protein synthesis (28). Hypophysectomy produces a reduction in the overall protein catabolism in muscle. On the other hand in patients with cerebral gigantism peripheral amino acid concentrations are generally lower than in control subjects. However, branched-chain amino acid levels are higher than in healthy subjects (29). The influence of growth hormone on the plasma amino acid level depends on the nature of the amino acid and on the acute versus chronic administration: one hour after growth hormone administration, a marked decrease in the plasma concentrations of threonine, alanine, valine, methionine, isoleucine, leucine, tyrosine and phenylalanine was observed (28); whereas 3—5 days of administration of growth hormone to hypopituitary dwarfs leads to an increase in plasma threonine, serine, glycine and methionine levels (30). Generally the peripheral concentration of essential amino acids in hypopituitary dwarfs is lower compared to that of normal children. One may speculate about the role of the increase in branched-chain amino acids seen during rapid growth on account of the competitive nature of these compounds with respect to their influence on the influx of tryptophan and tyrosine into the brain. It is known that social stress leads to retarded growth in children. It would be interesting to see the aromatic amino acid to competing amino acid ratio in those children compared to children with normal growth. Such studies, however, have not yet been conducted.

Glucocorticoids

As we have seen, insulin, growth hormone, and glucagon decrease plasma amino acid levels; it is interesting to note that administration of the catabolic glucocorticoids increases the urinary excretion of several amino acids (threonine, serine, asparagine, glutamine, alanine, histidine, glycine, lysine, cystathionine, and ornithine (31, 32)). Most notably the levels of valine, leucine, isoleucine and tyrosine are increased. The rise in amino acid excretion seems to be a direct effect of the glucocorticoid, dexamethasone, on renal amino acid transmembrane transport processes (31). One may speculate whether these steps contribute to the increased availability of the aromatic amino acids, tryptophan and tyrosine. However, the plasma amino acids decrease only to a slight extent. A severe defect in renal transmembrane transport as seen during uremia, also changes drastically the serum amino acid pattern. We observed in a patient with uremic encephalopathy and seizure activity a three-fold increase in the ratio of serum tryptophan and tyrosine to competing amino acid levels compared to healthy subjects (Rao, unpublished observation). The level of ammonia was in the reference range. It appears that impaired uptake at the renal level shifts serum amino acid profiles in that they are comparable to those seen during hepatic encephalopathy.

Hyperalaninemia observed after the administration of dexamethasone is characteristic of chronic glucocorticoid excess and may contribute to the steroid-induced hyperglucagonemia (32). Chronic glucocorticoid excess may be the cause for the increase in brain tyrosine aminotransferase in Cushing patients (33), which may cause reduction in the availability of tyrosine and concomitantly catecholamines.

Catecholamines

Glucocorticoids and catecholamines are termed stress hormones. Catecholamines inhibit the release of alanine and glutamine from muscle (34). Catecholamine-induced inhibition of amino acid release from muscle requires β-adrenergic receptor participation of muscle. This effect is mediated by a β-adrenergic receptor (adenylate cyclase) system. There is an inverse relationship between glycogenolysis and amino acid release which points to the fact that muscle glucose metabolism and alanine/glutamine formation are unrelated events (34). It is claimed that the implication may be a „fight or flight" response (34). It involves a temporary cessation of insulin secretion, release of glucagon, and elevated levels of lactate and glycerol.

Depending on the nature of the stress, there is a biphasic response in peripheral amino acid levels. Concomitant with an observed amino acid (tyrosine, phenylalanine, leucine and isoleucine) increase during physical stress (35), the entry of tryptophan into the brain may be reduced (36). Immobilization stress leads to increases in glutamic and aspartic levels in serum and brain (36). The mechanism of epinephrine action does not involve the inhibition of amino acid efflux. The physiological role of this adrenergic modulation of the amino acid release from proteins by the increased epinephrine secretion lies in the provision of precursor for gluconeogenesis. In this context is it important to note that during stress the interorgan flux of amino acids is changed in that a decrease in plasma amino acid levels makes itself evident (37). Infusion of adrenaline or a combination of adrenaline, glucagon or cortisol decreases the plasma concentration of most amino acids except that of alanine. Besides stimulating the release of glucagon, alanine acts as an important transporter of amino groups from the periphery to the liver and as glucose precursor.

In exercising man splanchnic uptake of amino acids increases (35). An increase in the net amino acid release by muscle during moderate exercise is only observed for alanine (35). However, during heavier work loads, increases between 8—35 % are noted for isoleucine, leucine, methionine, tyrosine and phenylalanine. This is attributed to the altered splanchnic exchange rather than augmented peripheral release. One might speculate that during physical stress the increase in plasma neutral amino acid level may cause a reduction in brain aromatic amino acid levels. In fact it was shown that administration of low doses of epinephrine causes a reduction in circulating amino acids in man (38) and high doses of epinephrine and norepinephrine to rats increase plasma tyrosine levels (39). Since this effect is completely inhibited by the α-receptor blocker phenoxybenzamine (39), the epinephrine and norepinephrine effects might be mediated by adrenergic α-receptor's change of total blood flow of the liver, since stimulation of α-adrenergic receptors causes vasoconstriction and reduced hepatic blood flow which may lead to reduced metabolism and increases in plasma amino acid levels. This in turn changes the blood flow and subsequently the liver metabolism, turning the key for changes in the peripheral amino acid levels. The ensuing changes in hormone and amino acid levels control the amino acid transmembrane translocation and eventually the amino acid precursor bioavailability for neurotransmitter synthesis in the brain. To sum it up stress and so called „stress hormones" lead to increases in peripheral taurine, aspartate, glutamate, alanine and tyrosine levels. Valine, tryptophan and arginine are decreased (36). The ratio of tryptophan to competing amino acids decreases and tyrosine to competing amino acid increases. Teleologically this makes sense. You want the catecholamine level to rise, and so it does. In fact during stressful events the increased tyrosine to competing amino acid serum level ratio corresponds to the increased brain norepinephrine turnover.

In this context it is interesting to note that during stress, plasma taurine levels exhibit the most pronounced increase of all amino acids (36); the heart might be the major source of the increased plasma taurine level. Norepinephrine causes a drop of taurine in cardiac tissue (40). Oral administration of taurine to rats prevents the stress-induced decrease in adrenal epinephrine levels. There may be other taurine-hormone short-circuits (40), and many of these actions may be linked to insulin (41): taurine ameliorates disturbances in carbohydrate metabolism (42).

Thyroid Hormones

Taurine's transmembrane transport system is also influenced by triiodothyronine and insulin (43). Thus on the one hand taurine potentiates the action of insulin and on the other, insulin and triiodothyronine enhance the transport of taurine into the cerebral cortex of the rat.

Lack of thyroid hormones is responsible for the decrease in protein turnover (44—47). Thyroid hormones parallel the effect of growth hormone in muscle (25). Hypothyroidism increases brain taurine content in neonatal rats and brain growth is delayed (48). Although hypothyroidism is prevalent in many areas, it is not known whether in humans the inhibitory neurotransmitter taurine is involved in precipitating depression and lack of energy in these patients.

Sex Hormones

In the context of hormones and amino acid metabolism some of the most interesting observations are naturally those in human females and their menstrual cycle. During the phase of the cycle when progesterone levels are highest, amino acid levels are lowest (49). This finding has been attributed to the catabolic effect of progesterone and in turn to increased amino acid utilization by the liver. The total plasma amino acid concentration is lower between days 19 and 22 of the menstrual cycle than between days 2 and 5, and total tryptophan increases (50); it is interesting to note that the ratio of tyrosine to competing amino acids is decreased (0.076 ± 0.009) in subjects on estrogen-progestogen therapy compared to that of controls (0.109 ± 0.014, $P < 0.001$), whereas that of tryptophan to competing amino acids is increased during said therapy (0.103 ± 0.017 to 0.117 ± 0.017, $P < 0.01$). It was reasoned that the reduction in tyrosine availability may contribute to the occurrence of depressive symptoms in susceptible individuals on oral contraceptive therapy (50—52). In a different context it was indeed observed that in susceptible women during those days of the menstrual cycle when the change-over-time of the hormone concentrations is most pronounced (i. e., the increase in serum estrogen and progesterone levels are largest), most extensive premenstrual symptoms (dysphoric mood, irritability, anxiety, abdominal discomfort, breast pain, less sexual interest, muscle and joint pain, low energy and increased appetite) are observed (53).

CONCLUSIONS

Insulin, glucagon, growth hormone and thyroid hormones decrease peripheral amino acid concentrations albeit by different mechanisms. Catecholamines decrease the tryptophan/competing amino acid ratio and increase that of tyrosine/competing amino acids. During the phase of the most pronounced increase in hormone concentration-over-time of the menstrual cycle, the tyrosine/competing amino acid ratio decreases which may be associated with premenstrual symptoms. Changes in hormone metabolism as seen in patients presenting with diabetes, Cushing's disease, hypothyroidism, during the menstrual cycle, and during pregnancy cause an imbalance in peripheral amino acid concentrations. The latter may be responsible for altered neurotransmitter synthesis in the central nervous system and for the depressive of manic episodes seen in some of these patients.

REFERENCES

1) RAO, M. L., FELS, K. (1987). Beeinflussen Tryptophan und Serotonin beim Menschen die Melatonin-Ausschüttung und damit die Funktion des „Regulators der Regulatoren“ (Zirbeldrüse)? In: *Fortschr. Pharmakother.*, **Vol. 3.** (Stille, G., Wagner, W., eds.). Karger, Basel, pp. 87—99.

2) CAHILL, G. F. Jr., AOKI, T. T., MARLISS, E. B. (1972). Endocrinology. In: *Handbook of Physiology* (Steiner, D., Freinkel, N., eds.), **vol. 1,** Williams & Wilkins, Baltimore, p. 563.

3) FULKS, R., LI, J. B., GOLDBERG, A. L. (1975). Effects of insulin, glucose, and amino acids on protein turnover in rat diaphragm. *J. Biol. Chem.* **250:** 290—298.

4) GOLDBERG, A. L. (1979). Influence of insulin and contractile activity on muscle size and protein balance. *Diabetes* **28:** 18—24.

5) LI, J. B., GOLDBERG, A. L. (1976). Effects of food deprivation on protein synthesis and degradation in rat skeletal muscle. *Am. J. Physiol.* **231:** 441—448.

6) MORTIMORE, G. E., MONDON, C. F. (1970). Inhibition by insulin of valine turnover in liver. *J. Biol. Chem.* **245:** 2375—2383.

7) RANNELS, D. E., KAO, R., MORGAN, H. E. (1975). Effect of insulin on protein turnover in heart muscle. *J. Biol. Chem.* **250:** 1694—1701.

8) RANNELS, D. E., McKEE, E. E., MORGAN, H. E. (1977). Regulation of protein synthesis and degradation in heart and skeletal muscle. In: *Biochemical actions of hormones* (Litwack, G., ed.), **vol. 4,** Academic Press, New York, pp. 135—195.

9) ADIBI, S. A., MORSE, E. L. AMIN, P. M. (1971). Interrelationships between level of amino acids in plasma and tissues during starvation. *Am. J. Physiol.* **221:** 829—838.

10) FERNSTROM, J. D., WURTMAN, R. J. (1972). Elevation of plasma tryptophan by insulin in rat. *Metabolism* **21:** 337.

11) LINDSTRÖM, P., SEHLIN, J. (1986). Aromatic amino acids and pancreatic islet function: a comparison of L-tryptophan and L-5-hydroxytryptophan. *Mol. Cell. Endocrinol.* **48:** 121—126.

12) PHELPS, R. L., METZGER, B. E., FREINKEL, N. (1981). Carbohydrate metabolism in pregnancy. XVII. Diurnal profiles of plasma glucose, insulin, free fatty acids, triglycerides, cholesterol, and individual amino acids in late normal pregnancy. *Am. J. Obstet. Gynecol.* **140:** 730—736.

13) GOLDBERG, A. L., CHANG, T. W. (1978). Regulation and significance of amino acid metabolism in skeletal muscle. *Fed. Proc.* **37:** 2301—2307.

14) PARDRIDGE, W. M. (1977). Kinetics of competitive inhibition of neutral amino acid transport across the blood-brain barrier. *J. Neurochem.* **28:** 103—108.

15) FERNSTROM, J. D., WURTMAN, J. R. (1974). Nutrition and the brain. *Sci. Am.* **230:** 84—92.

16) MANCHESTER, K. L. (19). Insulin in protein synthesis. In: *Biochemical actions of hormones* (Litwack, G., ed.), **Vol. 1,** Academic Press, New York, pp. 267.

17) SANDERS, R. B., RIGGS, T. R. (1967). Modification by insulin of the distribution of two model amino acids in the rat. *Endocrinology* **80:** 29—37.

18) GOLDFINE, I. D., GARDNER, J. D., NEVILLE, D. M. (1972). Insulin action in isolated rat thymocytes. I. Binding of 125I-Insulin and stimulation of amino butyric acid transport. *J. Biol. Chem.* **247:** 6919—6926.

19) ELSAS, L. J., ALBRECHT, O. S., KOEHNE, W., ROSENBERG, L. E. (1967). Effect of puromycin on insulin-stimulated amino acid transport in muscle. *Nature* **214:** 916—917.

20) ELSAS, L. F., ALBRECHT, I., ROSENBERG, L. E. (1968). Insulin stimulation of amino acid uptake in rat diaphragm. Relationship to protein synthesis. *J. Biol. Chem.* **243:** 1846—1853.

21) HJALMARSON, A., AHREN, K. (1967). Sensitivity of the rat diaphragm to growth hormone. 1. In vivo and in vitro effects of growth hormone on amino acid transport. *Acta Endocrinol.* **54:** 645—662.

22) AOKI, T. T., MÜLLER, W. A., BRENNAN, M. F., CAHILL, G. F. Jr. (1974). Effect of glucagon on amino acid and nitrogen metabolism in fasting man. *Metabolism* **23:** 805—814.

23) LANDAU, R., LUGIBIHL (1969). Effect of glucagon on concentration of several free amino acids in plasma. *Metabolism.* **18:** 265—276.

24) FLOYD, J. C. Jr., FAJANS, S. S., PEK, S. et al. (1971). Synergistic effect of essential amino acids and glucose upon insulin secretion in man. *Diabetes* **19:** 109—115.

25) GOLDBERG, A. L., TISCHLER, M., DeMARTINO, G., GRIFFIN, G. (1980). Hormonal regulation of protein degradation and synthesis in skeletal muscle. *Fed. Proc.* **39:** 31—36.

26) FLORINI, J. R., BREUER, C. B. (1966). Amino acid incorporation into protein by cell-free systems from rat skeletal muscle. V. Effects of pituitary growth hormone on activity of ribosomes and ribonucleic acid polymerase in hypophysectomized rats. *Biochemistry* **5:** 1870—1976.

27) TATA, J. (1974). The biochemical basis of STH-action. *Nova Acta Leopoldina* **40:** 123.

28) STAHNKE, N., PLETTNER, C., BLUNCK (1977). Effects of growth hormone and protein metabolism. Acute changes in plasma amino acids in growth retarded patients with and without growth hormone deficiency. *Acta Paediatr. Scand.* **66:** 153—159.

29) BEJAR, R. L., SMITH, G. F., PARK, S., SPELLACY, W. N., WOLFSON, S. L., NYHAN, W. L. (1970). Cerebral gigantism: concentrations of amino acids in plasma and muscle. *J. Pediatrics* **105:** 105—111.

30) ZACHMANN, M. (1969). Influence of human growth hormone (HGH) on plasma and urine amino acid concentrations in hypopituitary dwarfs. *Acta Endocrinol.* **62:** 513—520.

31) ZINNEMANN, H. H., JOHNSON, J. J., SEAL, U. S. (1963). Effect of short-term therapy with cortisol on the urinary excretion of free amino acids. *J. Clin. Endocr.* **23:** 996—1000.

32) ZISCHKA, R., ORTH, E., CASTELLS, S. (1970). Effects of short-term administration of dexamethasone on urinary and plasma free amino acids in children. *J. Clin. Endocr.* **31:** 95—97.

33) SNAPE, B. M., BADAWY, A. A.-B. (1981). Enhancement of rat brain tyrosine aminotransferase activity by cortisol. *Biochem. J.* **198:** 417—420.

34) GARBER, A. J., KARL, I. E., KIPNIS, D. M. (1976). Alanine and glutamine synthesis and release from skeletal muscle. VI. β-Adrenergic inhibition of amino acid release. *J. Biol. Chem.* **251:** 851—857.

35) FELIG, P., WAHREN, J. (1971). Amino acid metabolism in exercising man. *J. Clin. Invest.* **50:** 2703—2714.

36) MILAKOFSKY, L., HARE, T. A., MILLER, J. M., VOGEL, W. H. (1985). Rat plasma levels of amino acids and related compounds during stress. *Life Sci.* **36:** 753—761.

37) WERNERMAN, J., BRANDT, R., STRANDELL, T., ALLGÈN, L.-G., VINNARS, E. (1985). The effect of stress hormones on the interorgan flux of amino acids and on the concentration of free amino acids in skeletal muscle. *Clin. Nutr.* **4:** 207—216.

38) WESTCOTT, K. R., LAPORTE, D. C., STORM, D. R. (1979). Resolution of adenylate cyclase sensitive and insensitive to Ca and Ca-dependent regulatory proteins (CDR) by CDR-sepharose affinity chromatography. *Proc. Nat. Acad. Sci. USA* **76:** 204—208.

39) ERIKSSON, T., CARLSSON, A. (1982). Adrenergic influence on rat plasma concentration of tyrosine and typtophan. *Life Sci.* **30:** 1465—1472.

40) LAMPSON, W. G., KRAMER, J. H., SCHAFFER, S. W. (1983). Potentiation of the actions of insulin by taurine. *Can. J. Physiol. Pharmacol.* **61:** 457—463.

41) MACALLUM, A. B., SIVERTZ, C. (1942). The potentiation of insulin by sulfones. *Can. Chem. Process.* **26:** 569.

42) KURIYAMA, K., OHKUMA, S., MURAMATSU, M. (1981). Effect of taurine of calcium paradox and ischemic heart failure. *Am. J. Physiol.* **240:** H238—H246.

43) GONG, J.-H. (1983). Wirkung von Tri-Iodothyronin und Insulin auf den Taurintransport bei Neocortex-Kulturen des Rattenhirns. *Diss. University of Bonn.*

44) FLAIM, K. E., LI, J. B., JEFFERSON, L. S. (1978). Effects of thyroxine on protein turnover in rat skeletal muscle. *Am. J. Physiol.* **235:** E231—E236.

45) INGBAR, S. H. (1986). The thyroid gland. In: *Textbook of Endocrinology* (Williams, R. H., ed.), 3rd ed., W. B. Saunders Co., Phiadelphia.

46) GOLDBERG, A. L., GRIFFIN, G. E., DICE, J. F. (1977). Influence of food deprivation and adrenal steroids on DNA synthesis in various mammalian tissues. *Am. J. Physiol.* **228:** 310—317.

47) GRIFFIN, G. E., GOLDBERG, A. L. (1977). Hormonal control of protein synthesis and degradation in rat skeletal muscle. *J. Physiol. Comm.:* 54P—55P.

48) HEINONEN, K. (1975). Effects of hypothyroidism and thyroxine substitution on the metabolism of L-methionine, L-cystathionine and taurine in developing rat brain. *Acta Endocrinol.* **80:** 487—500.

49) CRAFT, I. L., WISE, I. J. (1969). Changes in amino acid metabolism during the menstrual cycle. *J. Obstet Gynaec. Brit. Cwlth.* **76:** 928—933.

50) CRAFT, I. L., PETERS, T. J. (1971). Quantitative changes in plasma amino acids induced by oral contraceptives. *Clin. Sci.* **41:** 301—307.

51) MØLLER, S. E. (1981). Effect of oral contraceptives on tryptophan and tyrosine availability: evidence for a possible contribution to mental depression. *J. Neuropsychobiology* **7:** 192—200.

52) OEPEN, H., OEPEN, I., FUCHS, G. (1969). Über den Einfluß von Ovulationshemmern auf die Serum-Aminosäuren im Vergleich mit geschlechts-, cyclus- und schwangerschaftstypischen Befunden. *Arch. Gynäk.* **208:** 33—43.

53) HALBREICH, U., ENDICOTT, J., GOLDSTEIN, S., NEE, J. (1986). Premenstrual changes and changes in gonadal hormones. *Acta Psychiatr. Scand.* **74:** 576—586.

FEEDING, STRESS, EXERCISE AND THE SUPPLY OF AMINO ACIDS TO THE BRAIN

G. Curzon

Department of Neurochemistry
Institute of Neurology
Queen Square
London WCIN 3BG
UK

INTRODUCTION

This chapter concerns some effects of food deprivation, feeding and stress on brain amino acids. In particular, it concerns amino acids which are precursors of transmitter amines. In the past, most attention has been paid to two of these, tyrosine and tryptophan, which through the action of tyrosine hydroxylase and tryptophan hydroxylase respectively are converted to 3,4-dihydroxyphenylalanine (dopa) and 5-hydroxytryptophan by reactions which are rate-limiting for catecholamine and 5-hydroxytryptamine (5-HT) synthesis. Various classical precursor loading experiments suggest that brain tyrosine hydroxylase is close to saturation with its substrate and is therefore relatively insensitive to altered substrate availability but that tryptophan hydroxylase is about 50 % saturated (1, 2) and therefore more sensitive. Less attention has been paid to the fact that histidine decarboxylase is normally far below saturation with its precursor amino acid (3) and thus the synthesis of histamine is far more responsive to changes of precursor availability than are the syntheses of 5-HT or the catecholamines.

Until the early 1970's it was largely assumed that the rather moderate ranges of tryptophan and tyrosine availabilities to the brain which were apparent under physiological conditions were unlikely to have much influence on the synthesis and hence the functional activity of transmitter amines. Indeed, much data suggests that despite the recent considerable interest in the possible effects of diet-induced changes of 5-HT synthesis on appetite, mechanisms controlling the effects of diet (on brain tryptophan at least) may normally serve mainly to ensure sufficient supplies of tryptophan for central 5-HT synthesis. However, other evidence points to large interindividual differences in the responsiveness of transmitter synthesis to precursor supply. It is therefore conceivable that there are subgroups in which 5-HT functional activity responds directly to feeding. Also neuronal activity may affect relationships between precursor availability and transmitter synthesis. The present chapter contains some comments on the above possibilities but primarily deals with how dietary supply, stress and exercise affects the availability of amino acids to the brain. Special attention is paid to amino acids which are transmitter precursors, in particular to tryptophan. It also contains some comments on possible behavioural implications.

NATO ASI Series, Vol. H20
Amino Acid Availability and Brain Function in Health and Disease. Edited by G. Huether

EFFECTS OF FEEDING

Fasting for 24—120 h has surprisingly little effect on plasma concentrations of most amino acids in the rat, apart from phenylalanine which increases somewhat and alanine, arginine, glutamate and threonine which show moderate decreases (4). Brain values, on the whole also do not alter significantly and some of the changes that do occur (5) are readily explicable in terms of the plasma findings i. e. the rises of brain phenylalanine and the falls of arginine and alanine. Plasma glutamate falls but this is likely to have little effect on brain values as the uptake site is normally saturated (6, 7). Two amino acids, however, show marked and highly significant discrepancies between the effects of fasting on their plasma and brain values i. e. threonine which despite a considerable fall in the plasma on 24 h starvation (4) remains unaltered in the brain and tryptophan which despite unaltered total plasma values (i. e. free tryptophan plus tryptophan loosely bound to albumin) increases in the brain (8). Another seemingly paradoxical finding is that brain tryptophan increases if rats are given a large tryptophan-free high carbohydrate meal (9, 10).

Thus, brain tryptophan increases in two circumstances in which decreases might have been expected. These results are largely explicable as follows. (a) Fasting increases plasma unesterified fatty acids. These displace tryptophan from albumin to make it more available to the brain (11, 12) and (b) a high carbohydrate meal causes insulin secretion with resultant increased uptake by muscle and decreased plasma levels of the large neutral amino acids (LNAAs). These compete with tryptophan for transport to the brain so that the ratio of plasma tryptophan to competers rises (13).

The above mechanisms imply that how rat brain tryptophan concentration is affected by a meal depends not only on its composition but also on how much brain tryptophan has already been elevated by previous food deprivation. The greater this elevation, the smaller will be any additional rise due to a carbohydrate meal as there will be an opposing fall of plasma free tryptophan due to the inhibitory effect of feeding on lipolysis. Conversely, a protein-containing meal which does not alter the ratio of plasma tryptophan to competers should decrease brain tryptophan according to the extent to which previous food deprivation has elevated plasma free tryptophan.

How good is the evidence for these mechanisms? Unfortunately, few of the numerous reports on the effects of feeding on brain tryptophan include data on all the relevant variables. Also, some earlier investigations suffered from methodological defects (14). As a result, the relative importance of mechanisms (a) and (b) above were often disputed. However, detailed analyses suggest that in both rats (10) and humans (15) both mechanisms can be involved in the effects of feeding on brain tryptophan.

How important are the mechanisms? It is relevant that an investigation of fasting psychiatric patients undergoing subcaudate tractotomy showed that the variation of cortical and ventricular tryptophan values within the group correlated well with plasma free tryptophan values (16) but that correlations with plasma total tryptophan were poorer and, in general, correlations became weaker if the plasma tryptophan values were (as in the rat experiments) (17) divided by the sum of the plasma concentrations of the LNAAs. The LNAA values varied over a 1.5 fold range which implies that moderate plasma LNAA changes may have no more than minor effects on tryptophan transport to the human brain. This agrees well with calculations from influx experiments using rats (18, 19).

When considering the role of dietary effects on LNAA levels in the control of brain tryptophan (and other amino acids), it is pertinent that total plasma LNAA concentrations in a group of freely feeding healthy adults following their own dietary preferences were essentially unaffected by breakfast and only rose about 15 per cent above fasting values after lunch (20). Similarly, other workers found no significant diurnal variation of plasma amino acids in human subjects under normal British dietary conditions (21). Even when meals of considerably different compositions were given to humans, plasma tyrosine and tryptophan concentrations varied approximately in proportion with the values for other LNAAs. Thus, plasma competer ratios of subjects given carbohydrate or 20 per cent protein meals (22, 23) altered moderately if at all. The reported changes would be expected to have little if any effect on the availability of 5-HT to receptors (24).

There is no doubt that the *presence* of the group of LNAAs in plasma alters the availability of tryptophan and other individual LNAAs to the brain and that certain extremes of meal composition can have substantial effects on concentrations of amino acids therein (25—27). However, more usual acute dietary variations of the ratio of plasma tryptophan to LNAAs probably have rather small effects. Indeed, in the above studies (22, 23) the ratio increased considerably only when a tryptophan supplement was added to the meal.

Even when the ratio changes appreciably, this must be interpreted with caution. In human studies, plasma and brain data is, for obvious reasons, rarely obtained on the same subject. Therefore, ratios of plasma tryptophan to LNAAs are not infrequently used as the sole indices of brain tryptophan level. In some papers, the ratio changes simply because plasma tryptophan concentration has changed. In others, the ratio is reported without data on its numerator and denominator. It is relevant here that, in a number of non-dietary studies, correlations between plasma and brain tryptophan values were not strengthened but weakened by taking competing LNAAs into account (16, 28, 29). Also, a recent study (30) indicated that physiologically relevant differences in chronic protein intake did not cause dose-related changes in the ratio of plasma tryptophan to LNAAs and that under these conditions, the ratios did not, on the whole, predict brain values.

EFFECT OF STRESS

Early work (31) showed that brain 5-HT metabolism increased on 2 h immobilization stress and this was subsequently shown to be associated with a rise of brain tryptophan (8). Initially, the latter effect was attributed to the associated increase of plasma free tryptophan (11, 32) which resulted from a stress-induced rise of plasma non esterified fatty acids. However, a recent study (33) described below indicated that it was more convincingly explained as part of a general alteration of the plasma-brain relationship for LNAAs. The investigation showed that immobilization decreased plasma amino acid values by 20—50 per cent. The only exception was free tryptophan, which as before, rose significantly (11) while total tryptophan fell significantly (in common with 12 of 15 of the other amino acids measured). Other workers (34) obtained similar but less marked changes on immobilization for 30 min. The decrease could be due to stress-provoked increases in plasma catecholamines, as isoprenaline infusion increases plasma free tryptophan but decreases total tryptophan and tyrosine (35) and adrenaline is reported to decrease most amino acids in human plasma (36).

The plasma changes on immobilization (33) were not associated with parallel changes in the brain. On the contrary, brain tryptophan and four other LNAAs (phenylalanine, valine, leucine and isoleucine) rose significantly and no amino acid fell significantly. The above increases were not explicable by brain influx rates, calculated using plasma amino acid values and published kinetic data (7, 37). This is strikingly shown in Figure 1 as (with the possible exception of histidine) brain LNAA values of immobilized rats are higher than predictable from calculated influxes. Thus, immobilization appears to alter relationships between plasma and brain LNAA concentrations in favour of the brain. A similar failure of predicted influx values to explain brain LNAA concentrations occurs following portocaval anastomosis (28, 38).

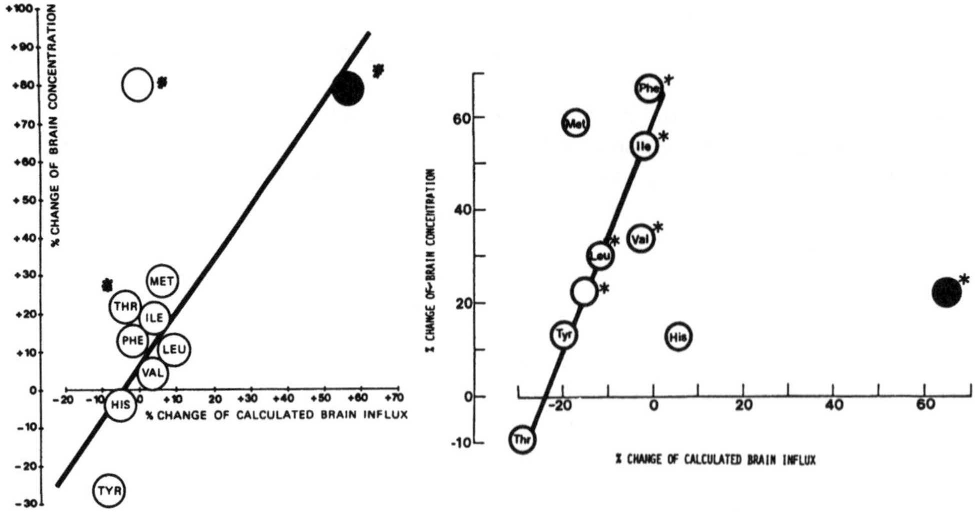

FIGURE 1

Relationships between percentage changes of influx of LNAAs into the rat brain (calculated from published K_m and V_{max} values as described by Pardridge (37) and percentage increases of brain LNAA concentrations. The effects of 2 h trained running are shown on the left. The effects of 2 h immobilization are shown on the right. Standard abbreviations are used for amino acids apart from tryptophan. ○ and ● refer to plasma total and free tryptophan respectively. Significant increases of brain amino acid concentrations on exercise or immobilization are indicated by asterisks. The regression line for the effect of exercise was calculated using influx values for all the LNAAs including plasma free (but not total) tryptophan ($r = 0.93$, $n = 9$, $p < 0.001$). The line for the effect of immobilization was calculated using influx values for 7 of the 9 LNAAs (including all 5 for which the increase in brain concentration was significant) and data for plasma total (but not free) tryptophan ($r = 0.94$, $n = 7$, $p < 0.01$). Results from refs. (33, 52).

The above effects of immobilization might reflect various mechanisms, e. g. breakdown of the blood-brain barrier as indicated by the penetration of albumin into the brain during immobilization (39). However, this was detected in very few brain regions while the amino acid changes were determined on whole brain (33). Other mechanisms that cannot be excluded could involve altered brain amino acid metabolism or efflux. Changes in cerebral blood flow are not likely to be responsible, as they have no more than moderate effects on amino acid influx (40) and changes of flow on immobilizing conscious normal rats are only slight (41). Another, more likely possibility is that the kinetics of transport across the barrier are altered, perhaps because of stress-provoked hormonal changes. For example, thyroid hormones, plasma levels of which are initially increased in stress (42) increase amino acid transport to the brain (43). The enhancement of brain uptake of LNAAs by isoprenaline (35, 44) suggests that increased catecholamine secretion could be involved. Plasma growth hormone increases in stress (45), but data on its effect on transport are equivocal (46, 47).

In our investigation (33), a 22 per cent rise in brain tryptophan was associated with an almost identical rise of plasma free tryptophan. However, Figure 1 shows that the relationship between the effects of immobilization on brain tryptophan concentration and influx is highly consistent with the relationships for other LNAAs if plasma total tryptophan is used to calculate tryptophan influx but not if plasma free tryptophan is used. This is explicable in terms of a postulated effect of immobilization on the extraction of plasma tryptophan by the brain. The transport site, under these conditions may compete effectively with plasma albumin for tryptophan, stripping the amino acid off from the protein (48), so that the brain „sees" all the plasma tryptophan instead of only the free component. Stripping off would be facilitated by the decreased K_m values for uptake of tryptophan and other LNAAs which are consistent with Figure 1.

A special situation occurs when food-deprived rats are stressed. Under these circumstances, sympathetic activation of lipolysis and hence of the liberation of plasma tryptophan from its binding to albumin is more readily provoked. Thus, relatively mild stresses such as the removal of cage-mates can be sufficient to increase plasma free tryptophan and hence brain tryptophan (49).

As well as the LNAAs, brain levels of the other amino acids were maintained on immobilization despite their decreased plasma concentrations (33). Results for alanine and serine may reflect the same mechanism as that responsible for the disposition of the LNAAs, as they are at least partly transported by the same site. Results for the basic amino acids, arginine and lysine, which are transported by a different site, are essentially explicable in terms of influx calculated as indicated in Figure 1 and agree qualitatively with previous data (50). Plasma glutamate was decreased by stress (33), but this is unlikely to affect brain values as the uptake is normally saturated (37).

Thus, various mechanisms ensure that the general decrease of plasma amino acid concentrations on immobilization for 2 h does not lead to decreases in the brain. Instead, brain values are maintained or increased. The increase of tryptophan must play a part in the increased brain 5-HT metabolism which occurs (8, 11) and the maintenance of brain tyrosine level despite a fall of plasma tyrosine is also of interest as immobilization alters the relationship between its level in the brain and dopamine synthesis (51).

EFFECT OF EXERCISE

Exercise (running) acutely and substantially increases brain and CSF tryptophan concentration in rats trained to run on a treadmill. These changes were associated with increased plasma free (but not total) tryptophan (52, 53) which was due to exercise-induced lipolysis as indicated by elevated concentrations of plasma unesterified fatty acids. The significant correlation between plasma free and brain tryptophan concentrations suggested a causal relationship (52) and despite the evidence against the immobilization stress induced increase of brain tryptophan being explained in this way (see above), a detailed investigation (54) confirmed that plasma free tryptophan was the sole determinant of the rise of brain tryptophan on exercise. Running for 2 h at 20 m/min almost doubled both plasma free and brain values. Although increases arterial concentrations of LNAAs are reported following exercise in man (55, 56) other rat plasma LNAAs were essentially unaffected.

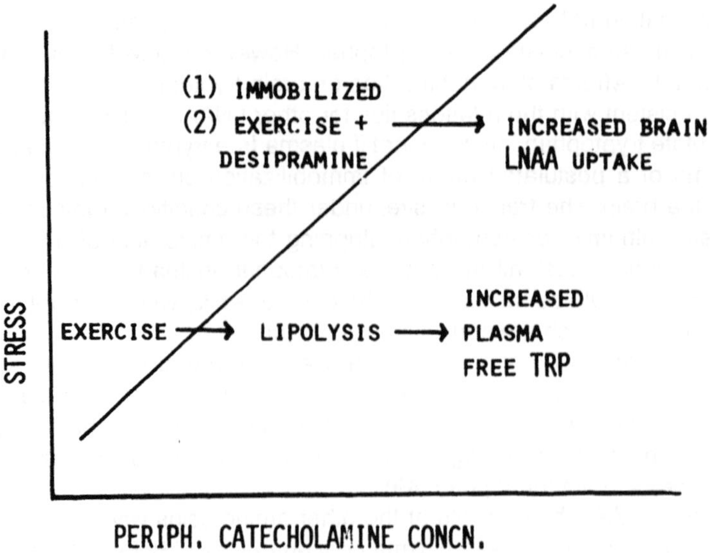

FIGURE 2
Proposed brain uptake mechanisms for tryptophan and other LNAAs activated by exercise and immobilization.

Apart from tryptophan, exercise only increased one other rat brain LNAA (threonine). This was not explicable in terms of the plasma values and is of some interest as a similar threonine-specific discrepancy between plasma and brain changes was noted on fasting (4). Exercise also increased brain lysine whereas it decreased glycine, alanine and γ-aminobutyric acid (GABA). Brain tryptophan could conceivably increase as a result of hyperammonemia as this is reported to occur on prolonged exercise (57) and could lead to increased brain tryptophan (58) as the result of a rise of brain glutamine (59). However, the transport of other LNAAs to the brain would also be affected (60). As a general increase in brain LNAAs did not occur in the exercise study (54) it seems unlikely that brain tryptophan increased because of hyperammonemia.

The different mechanisms by which exercise and immobilization influence brain amino acids may reflect greater stress and resultant greater sympatho-adrenal stimulation in the immobilized animals (Fig. 2) causing increased brain uptake of LNAAs (35, 44). In agreement with this suggestion, exercise did increase the ratio of brain tryptophan to plasma free tryptophan if the rats were pretreated with the catecholamine reuptake inhibitor desipramine (52). Immobilization is likely to be experienced as highly stressful as the animal cannot control the stress (61, 62) while trained running is likely to be less stressful as shock is avoided by running.

COMMENT

Tryptophan transport to the brain is under multiple influences but specific ones predominate in different circumstances. Thus, food deprivation (10, 11) and exercise (54) increase brain tryptophan by increasing plasma free tryptophan; immobilization can increase it by altering LNAA transport kinetics (33) and a large carbohydrate meal increase it by creasing plasma concentrations of competing LNAAs (9, 10). In other situations, e. g. ageing (29) or hepatic failure (28), combinations of these influences are involved.

Much of the work described in this chapter suggests that early ideas on the functional importance of physiologically relevant changes of brain transmitter precursor concentrations require some revision. For example, it now seems more likely that the effects of feeding or of moderate periods of fasting on the disposition of tryptophan ensure adequate brain stores of the amino acid when dietary supplies are temporarily withdrawn (8—10) but do not *normally* have much effect on 5-HT concentrations at receptors which mediate appetite. Other recent findings agree as they indicate that evidence is against normal dietary changes of the ratio of plasma tryptophan to LNAAs having much effect on food intake (63, 64). Furthermore, there is, as yet no data from in vivo dialysis studies suggesting that the administration of tryptophan, even at pharmacological dosage, increases the availability of 5-HT to receptors.

Nevertheless, as pharmacological evidence indicates that actions at 5-HT receptors can affect food intake (65—67) it is conceivable that, in certain subjects, brain tryptophan changes following dietary alterations, stress or exercise are sufficient to affect appetite or other behavioural components. This possibility seems quite likely as even laboratory rats of a single strain, kept under standard conditions show remarkable neurochemical individuality in the effects of tryptophan injection on their central tryptophan concentration and 5-HT metabolism (68). Furthermore, considerable individual variation has been noted in the effects of serotonergic drugs on feeding in humans (69).

These findings suggest that disorders of feeding could occur, in vulnerable subjects, through inappropriate effects of altered food intake on transmitter metabolism. For example, as administered tryptophan is reported to depress appetite in humans (70) increases of brain tryptophan on food deprivation (8, 11) might be sufficiently large in susceptible individuals to cause anorexia. Somewhat similarly, increases of brain tryptophan in stress (8, 11) could have a causal role in some stress-induced anorexias. Evidence that increased transport of tryptophan to the brain in stress can lead to altered brain function comes from work showing that the competing LNAA valine opposed stress-induced rises of corticosterone (71, 72) and that this effect of valine was largely prevented by tryptophan (71). These findings suggest that when brain tryptophan rises in stress there is a resultant increase of 5-HT at hypothalamic 5-HT receptors (73, 74) thought to mediate corticosterone secretion. There is also evidence that stress-induced increases of brain tryptophan augment morphine analgesia (75). This implies that exercise-induced increases of brain tryptophan (52—54) could enhance the postulated mood elevating effects of enkephalins released during exercise.

REFERENCES

1) CARLSSON, A., LINDQVIST, M. (1978). Dependence of 5-HT and catecholamine synthesis on concentrations of precursor amino acids in rat brain. *Naunyn Schmiedebergs Arch. Pharmacol.* **303:** 157—164.

2) SVED, A. F. (1983). Precursor control of the function of monoaminergic neurones. In: *Nutrition and the Brain* (Wurtman, R. J., Wurtman, J. J., eds.), **vol. 6,** pp. 224—263, Raven Press, New York.

3) PARDRIDGE, W. M. (1986). Potential effects of the dipeptide sweetener aspartame on the brain. In: *Nutrition and the Brain* (Wurtman, R. J., Wurtman, J. J., eds.), **vol. 7,** pp. 141—204, Raven Press, New York.

4) ENWONWU, C. O. (1987). Differential effect of total food withdrawal and dietary protein restriction on brain content of free histidine in the rat. *Neurochem. Res.* **12:** 483—487.

5) KNOTT, P. J., JOSEPH, M. H., CURZON, G. (1973). Effects of food deprivation and immobilization on tryptophan and other amino acids in rat brain. *J. Neurochem.* **20:** 249—251.

6) OLDENDORF, W. H., SZABO, J. (1976). Amino acid assignment to one of three blood-brain amino acid carriers. *Am. J. Physiol.* **230:** 94—98.

7) PARDRIDGE, W. M. (1979). Kinetics of competitive inhibition of neutral amino acid transport across the blood-brain barrier. *J. Neurochem.* **28:** 103—118.

8) CURZON, G., JOSEPH, M. H., KNOTT, P. J. (1972). Effects of immobilization and food deprivation on rat brain tryptophan metabolism. *J. Neurochem.* **19:** 1969—1974.

9) FERNSTROM, J. D., WURTMAN, R. J. (1971). Brain serotonin content: increase following ingestion of carbohydrate diet. *Science* **171:** 1023—1025.

10) SARNA, G. S., KANTAMANENI, B. D., CURZON, G. (1984). Variables influencing the effect of a meal on brain tryptophan. *J. Neurochem.* **44:** 1575—1580.

11) KNOTT, P. J., CURZON, G. (1972). Free tryptophan in plasma and brain tryptophan metabolism. *Nature* **239:** 452—453.

12) CURZON, G., FRIEDEL, J., KNOTT, P. J. (1973). The effects of fatty acids on the binding of tryptophan to plasma protein. *Nature* **242:** 198—200.

13) FERNSTROM, J. D., WURTMAN, R. J. (1973). Brain serotonin content: physiological regulation by plasma neutral amino acids. *Science* **178:** 414—416.

14) CURZON, G. (1979). Methodological problems in the determination of total and free plasma tryptophan. *J. Neurol. Trans. Suppl.* **15:** 221—226.

15) PEREZ-CRUET, J., CHASE, T. N., MURPHY, D. L. (1974). Dietary regulation of brain tryptophan metabolism by plasma ratio of free tryptophan and neutral amino acids in humans. *Nature* **248:** 693—695.

16) GILLMAN, P. K., BARTLETT, J R., BRIDGES, P. K., HUNT, A., PATEL, A. J., KANTAMANENI, B. D., CURZON, G. (1981). Indolic substances in plasma, cerebrospinal fluid and frontal cortex of human subjects infused with saline or tryptophan. *J. Neurochem.* **37:** 410—417.

17) FERNSTROM, J. D., LARIN, F., WURTMAN, R. J. (1973). Correlations between brain tryptophan and plasma neutral amino acid levels following food consumption in rats. *Life Sci.* **13:** 517—524.

18) CURZON, G. (1985). Effects of food intake on brain transmitter amine precursors and amine synthesis. In: *Psychopharmacology and Food* (Sandler, M., Silverstone, T.), pp. 59—70, Oxford.

19) MANS, A. M., BIEBUYCK, J. F., SAUNDERS, S. J., KIRSCH, R. E., HAWKINS, R. A. (1979). Tryptophan transport across the blood-brain barrier during acute hepatic failure. *J. Neurochem.* **33:** 409—418.

20) SCRIVER, C. R., GREGORY, D. M., SOVETTS, D., TISSENBAUM, G. (1985). Normal plasma free amino acid values in adults: the influence of some common physiological variables. *Metabolism* **34:** 868—873.

21) MILSON, J. P., MORGAN, M. Y., SHERLOCK, S. (1979). Factors affecting plasma amino acid concentrations in control subjects. *Metabolism* **28:** 313—319.

22) ASHLEY, D. V., BARCLAY, D. V., CHAUFFARD, F. A., MOENNOZ, D., LEATHWOOD, P. D. (1982). Plasma amino acid responses in humans to evening meals of differing nutritional composition. *Am. J. Clin. Nutr.* **36:** 143—153.

23) ASHLEY, D. V. M., LIARDON, R., LEATHWOOD, P. D. (1985). Breakfast meal composition influences plasma tryptophan to large neutral amino acid ratios of healthy lean young men. *J. Neural. Trans.* **63:** 271—283.

24) LEATHWOOD, P. D. This volume.

25) MOLLER, S. E. (1985). Effect of various oral protein doses on plasma neutral amino acid levels. *J. Neural. Trans.* **61:** 183—191.

26) YOKOHOSHI, H., WURTMAN, R. J. (1986). Meal composition and plasma amino acid ratios: effect of various proteins or carbohydrates and of various protein concentrations. *Metabolism* **35:** 837—842.

27) GLAESER, B. S., MAHER, T. J., WURTMAN, R. J. (1983). Changes in brain levels of acidic, basic and neutral amino acids after consumption of single meals containing various proportions of protein. *J. Neurochem.* **41:** 1016—1021.

28) BLOXAM, D. L., CURZON, G. (1978). A study of proposed determinants of brain tryptophan concentration in rats after portocaval anastomosis or sham operation. *J. Neurochem.* **31:** 1255—1263.

29) SARNA, G. S., TRICKLEBANK, M. D., KANTAMANENI, B. D., HUNT, A., PATEL, A. J., CURZON, G. (1982). Effect of age on variables influencing the supply of tryptophan to the brain. *J. Neurochem.* **39:** 1283—1290.

30) FERNSTROM, J. D., FERNSTROM, M. H., GRUBB, P. E. (1987). Twenty-four-hour variations in rat blood and brain levels of the aromatic and branched-chain amino acids: chronic effects of dietary protein content. *Metabolism* **36:** 643—650.

31) BLISS, E. L., AILION, J., ZWANZIGER, J. (1968). Metabolism of norepinephrine serotonin and dopamine in rat brain with stress. *J. Pharm. Exp. Ther.* **164:** 122—134.

32) CURZON, G., KNOTT, P. J. (1974). Fatty acids and the disposition of tryptophan. In: *Aromatic Amino Acids in the Brain* (Ciba Foundation Symposium 22), pp. 217—229, Elsevier.

33) KENNETT, G. A., CURZON, G., HUNT, A., PATEL, A. J. (1986). Immobilization decreases amino acid concentrations in plasma but maintains or increases them in brain. *J. Neurochem.* **46:** 208—212.

34) MILAKOFSKY, L., HARE, T. A., MILLER, J. M., VOGEL, W. H. (1985). Rat plasma levels of amino acids and related compounds during stress. *Life Sci.* **36:** 753—761.

35) HUTSON, P. H., KNOTT, P. J., CURZON, G. (1980). Effect of isoprenaline infusion on the distribution of tryptophan, tyrosine and isoleucine, between brain and other tissues. *Biochem. Pharmac.* **29:** 509—516.

36) SHAMOON, H., JACOB, R., SHERWIN, R. S. (1980). Epinephrine induced hypoaminoacidemia in normal and diabetic subjects: effects of blockade. *Diabetes* **11:** 875—881.

37) PARDRIDGE, W. M. (1977). Regulation of amino acid availability to the brain. In: *Nutrition and the Brain* (Wurtman, R. J., Wurtman, J. J., eds.), **vol. I,** pp. 141—204. Raven Press, New York.

38) MANS, A. M., BIEBUYCK, J. F., DAVIS, D. W., HAWKINS, R. A. (1984). Portocaval anastomosis: brain and plasma metabolite abnormalities and the effect of nutritional therapy. *J. Neurochem.* **43:** 697—705.

39) BELOVA, T. I., JONSSON, G. (1982). Blood-brain barrier permeability and immobilization stress. *Acta. Physiol. Scand.* **116:** 21—29.

40) HAWKINS, R. A., MANS, A. M., BIEBUYCK, J. E. (1982). Amino acid supply to individual cerebral structures in awake and anaesthetized rats. *Am. J. Physiol.* **242:** E1—E11.

41) OHATA, M., FREDERICKS, W. R., SUNDARAM, V., RAPOPORT, S. I. (1981). Effect of immobilization stress on regional cerebral blood flow in the conscious rat. *J. Cereb. Blood Flow Metab.* **1:** 187—194.

42) LANGER, P., FOLDES, O., KVETNANSKY, L., CALMAN, J., TORDA, T., EL DAHER, F. (1983). Pituitary-thyroid function during acute immobilization stress in rats. *Exp. Clin. Endocrinol.* **82:** 51—60.

43) DANIEL, P. M., LOVE, E. R., PRATT, O. E. (1975). Hypothyroidism and amino acid entry into brain and muscle. *Lancet* **2:** 872.

44) ERIKSSON, T., CARLSSON, A. (1982). Isoprenaline increases brain concentrations of administered L-DOPA and L-tryptophan in the rat. *Psychopharmacology (Berlin)* **77:** 98—100.

45) KANT, G. J., LENOX, R. H., BUNNELL, B. N., MOUGEY, E. H., PENNINGTON, L. L., MEYERHOFF, J. L. (1983). Comparison of stress responses in male and female rats: pituitary cyclic AMP and plasma prolactic growth hormone and corticosterone. *Psychoneuroendocrinology* **8:** 421—428.

46) COCCHI, D., DI GIULIO, A., GROPPETTI, A., MANTEGAZZA, P., MULLER, E. E., SPANO, P. F. (1975). Hormonal imputs and brain tryptophan metabolism: the effect of growth hormone. *Experientia* **31:** 384—385.

47) TANG, L. C., COTZIAS, G. C. (1976). Modification of the actions of some neuroactive drugs by growth hormone. *Arch. Neurol.* **33:** 131—134.

48) PARDRIDGE, W. M. (1979). Tryptophan transport through the blood-brain barrier: in vivo measurement of free and albumin-bound amino acid. *Life Sci.* **25:** 1519—1528.

49) KNOTT, P. J., HUTSON, P. H., CURZON, G. (1977). Fatty acid and tryptophan changes on disturbing groups of rats and caging them singly. *Pharmacol. Biochem. Behav.* **7:** 245—252.

50) BANOS, G., DANIEL, P. M., MOORHOUSE, S. R., PRATT, O. E. (1974). Inhibition of entry of some amino acids into the brain with observations on mental retardation in the aminoacidurias. *Psychol. Med.* **4:** 262—269.

51) MARCOU, M., KENNETT, G. A., CURZON, G. (1987). Enhancement of brain dopamine metabolism by tyrosine during immobilization: an in vivo study using repeated cerebrospinal fluid sampling in conscious rats. *J. Neurochem.* **48:** 1245—1251.

52) CHAOULOFF, F. ELGHOZI, J. L., GUEZENNEC, Y., LAUDE, D. (1985). Effects of conditioned running on plasma, liver and brain tryptophan and on brain 5-hydroxytryptamine metabolism of the rat. *Br. J. Pharmacol.* **86:** 33—41.

53) CHAOULOFF, F., LAUDE, D., GUEZENNEC, Y., ELGHOZI, J. L. (1986). Motor activity increases tryptophan, 5-hydroxyindoleacetic acid and homovanillic acid in ventricular cerebrospinal fluid of the conscious rat. *J. Neurochem.* **46:** 1313—1316.

54) CHAOULOFF, F., KENNETT, G. A., SERRURRIER, B., MERINO, D., CURZON, G. (1986). Amino acid analysis demonstrates that increased plasma free tryptophan causes the increase of brain tryptophan during exercise in the rat. *J. Neurochem.* **46:** 1647—1650.

55) AHLBORG, G., FELIG, P., HAGENFELDT, L., HENDLER, R., WAHREN, J. (1974). Substrate turnover during prolonged exercise in man. *J. Clin. Invest.* **53:** 1080—1090.

56) FELIG, P., WAHREN, J. (1971). Amino acid metabolism in exercising man. *J. Clin. Invest.* **50:** 2703—2714.

57) MUTCH, B. J. C., BANISTER, E. W. (1983). Ammonia metabolism in exercise and fatigue: a review. *Med. Sci. Sports. Exerc.* **15:** 41—50.

58) CHAOULOFF, F., LAUDE, D., MIGNOT, E., KAMOUN, P., ELGHOZI, J. L. (1985). Tryptophan and serotonin turnover rate in the brain of genetically hyperammonaemic mice. *Neurochem. Int.* **7:** 143—153.

59) HAWKINS, R. A., MILLER, A. L., NIELSEN, R. C., VEECH, R. M. (1973). The acute action of ammonia in rat brain metabolism in vivo. *Biochem. J.* **134:** 1001—1008.

60) RIGOTTI, P., JONUNG, T., PETERS, J. C., JAMES, J. H., FISCHER, R. E. (1985). Methionine sulfoximine prevents the accumulation of large neutral amino acids in brain of portocaval shunted rats. *J. Neurochem.* **44:** 929—933.

61) SWENSON, R. M., VOGEL, W. H. (1983). Plasma catecholamine and corticosterone as well as brain catecholamine changes during coping in rats exposed to stressful footshock. *Pharmacol. Biochem. Behav.* **18:** 689—693.

62) WEISS, J. M., GOODMAN, P. A., LOSITO, B. C., CORRIGAN, S., CHARRY, J. M., BAILEY, W. H. (1981). Behavioural depression produced by an uncontrollable stressor: relationship to norepinephrine, dopamine and serotonin levels in various regions of rat brain. *Brain Res. Rev.* **3:** 167—205.

63) PETERS, J. C., HARPER, A. E. (1987). Acute effects of dietary proteins on food intake, tissue amino acids and brain serotonin. *Am. J. Physiol.* **252:** R902—R914.

64) FERNSTROM, J. D. (1987). Food induced changes in brain serotonin synthesis: is there a relationship to appetite for specific macronutrients? *Appetite* **8:** 163—182.

65) SUGRUE, M. F. (1987). Neuropharmacology of drugs affecting food intake. *Pharmacol. Ther.* **32:** 145—182.

66) KENNETT, G. A., DOURISH, C. T., CURZON, G. (1987). 5-HT_{1B} agonists induce anorexia at a postsynaptic site. *Eur. J. Pharmacol.* **141:** 429—435.

67) HUTSON, P. H., DOURISH, C. T., CURZON, G. (1988). Evidence that the hyperphagic response to 8-OH-DPAT is mediated by 5-HT_{1A} receptors. *Eur. J. Pharmacol.* (in press).

68) HUTSON, P. H., SARNA, G. S., KANTAMANENI, B. D., CURZON, G. (1985). Monitoring the effect of a tryptophan load on brain indole metabolism in freely moving rats by simultaneous cerebrospinal fluid sampling and brain dialysis. *J. Neurochem.* **44:** 1266—1273.

69) SILVERSTONE, T., GOODALL, E. (1986). Serotonergic mechanisms in human feeding: the pharmacological evidence. *Appetite (Suppl.)* **7:** 85—97.

70) HRBOTICKY, N., LEITER, L. A., ANDERSON, G. H. (1985). Effects of L-tryptophan on short term food intake in lean men. *Nutrition Res.* **5:** 595—607.

71) JOSEPH, M. H., KENNETT, G. A. (1983). Corticosteroid response to stress depends upon increased tryptophan availability. *Psychopharmacology* **79:** 79—81.

72) YEHUDA, R., MEYER, J. S. (1984). A role for serotonin in the hypothalamic-pituitary adrenal response to insulin stress. *Neurochemistry* **38:** 25—32.

73) ALOI, J. A., INSEL, R. T., MUELLER, E. A., MURPHY, J. A. (1984). Neuroendocrine and behavioural effects of m-chlorophenylpiperazine administration in rhesus monkeys. *Life Sci.* **34:** 1325—1331.

74) KOENIG, J. I., GUDELSKY, G. A., MELTZER, H. Y. (1987). Stimulation of corticosterone and B-endorphin secretion in the rat by selective 5-HT receptor subtype activation. *Eur. J. Pharmacol.* **137:** 1—8.

75) KELLY, S. J., FRANKLIN, K. B. J. (1984). Evidence that stress augments morphine analgesia by increasing brain tryptophan. *Neurosci. Lett.* **44:** 305—310.

76) RANSFORD, C. P. (1982). A role for amines in the antidepressant effect of exercise: a review. *Med. Sci. Sports Exerc.* **14:** 1—10.

SUMMARY AND DISCUSSION-REPORT OF CHAPTER I

Alfred E. Harper

SUMMARY

The general emphasis throughout the contributions presented here was on the stability of brain amino pools. Brain amino acid concentrations were shown to be maintained within relatively narrow limits despite fluctuations in blood amino acid concentrations caused by nutritional deprivation, substantial changes in the protein content of the diet, and stress from immobilization or exercise. Severe disproportions among the amino acids in diets were, however, acknowledged as conditions that caused marked overloading or depletion of both blood and brain amino acid pools. Genetic impairment of enzyme systems for catabolism of amino acid caused by inborn errors of metabolism, such as phenylketonuria, are well known to result in overloading and depletion of brain pools. Prolonged protein malnutrition leads to depletion of brain amino acid pools.

Amino acid pools in brain fluctuate much less generally than those in plasma. The high degree of control of brain amino acid concentrations was attributed to: (1) the physico-chemical and kinetic properties of the blood-brain barrier transport systems which have limited capacity, and K_m-values equal to or below plasma amino acid concentrations, and are, therefore, highly responsive to competition among plasma amino acids for uptake into the brain; and (2) effective regulation of amino acid degrading systems, mainly in the liver, which have K_m-values well above plasma amino acids concentrations, and therefore rates of these catabolic reactions change rapidly in response to changes in blood and tissue amino acid concentrations; in addition, these systems are adaptive and increase or decrease in capacity in response to prolonged changes in amino acid intake of animals, and also to changes in endocrine state. It was recognized that endocrine and pathological abnormalities can alter the effectiveness of these control systems. Control of brain amino acid concentrations was considered to be effected mainly indirectly through the properties of the blood-brain-barrier and the amino acid-degrading systems with little evidence being presented that homeostatic mechanisms in the brain itself contribute to normal physiological control of brain amino acid pools. If brain amino acids pools deviate greatly from the stable state, however, responses in feeding behaviour are observed. These are mediated by brain systems which may result in curtailment of food consumption, and thereby prevent deviations in brain pools from becoming more severe, or in preference for another diet, if one is available, which will reduce or correct the deviation.

NATO ASI Series, Vol. H20
Amino Acid Availability and Brain Function in Health and Disease. Edited by G. Huether

DISCUSSION-REPORT

Do the recurring short periods of change in plasma amino acids concentrations after meals alter rates of uptake of amino acids into brain and brain amino acids concentrations significantly and do they produce rhythmic changes in brain amino acid concentrations of the type observed in blood?

Leathwood: Based on the finding that, in rats, brain tryptophan tends to rise during a period of starvation, Gerald Curzon suggested that the increase in brain tryptophan following a low-protein, high-carbohydrate meal should become less and less marked with duration of starvation.

This interesting idea does not, unfortunately, fit with observations. We have shown (Ashley, Leathwood and Moennoz, 1984) that with shorter fasts the increase in brain tryptophan is less and less until after 3 hours fasts a carbohydrate meal produces no change at all in brain tryptophan. This implies that the claim "that carbohydrate meals increase brain tryptophan and hence brain serotonin" cannot be extrapolated to the free feeding rat.

Wurtman: Perhaps Swiss rats differ from American rats — and people. We find perfectly-respectable (and statistically-significant) increases in the plasma tryptophan ration when rats or people consume a carbohydrate-rich, protein-poor "lunch" three hours after a nutrient-mixed breakfast. Of course, this effect depends in part on the relative sizes of "lunch" and "breakfast": If the "breakfast" — the meals that precedes the 3-hour fast — is sufficiently large, and especially if it is rich in fat, a lot of it will still be left in the stomach at the time that "lunch" is eaten, and the breakfast protein may sufficiently dilute the lunchtime carbohydrate to block the carbohydrate's effect on the plasma tryptophan ration. Point: The investigator has to use his head in designing experiments if he is not to be plagued by false negatives.

Curzon: The increase of brain tryptophan after a high carbohydrate meal must surely be opposed by any previous starvation-induced change of plasma free tryptophan concentration and effects of starvation must in general (by definition) have a dependency on the period of food withdrawal. The findings of Ashley et al. however suggest that the effect of a carbohydrate meal on brain tryptophan also has its own relationship with the length of the preceding period in which the rat has no access to food. Another point worth making is that there is an interaction between the effects of handling stress and of food deprivation on the disposition of tryptophan.

What is the contribution of insulin-related effects to the changes of tryptophan-availability seen after carbohydrate ingestion?

Pratt: With regard to the regulation of plasma amino acid levels, the effects of hormones such as insulin are complex. To help understand what happens we should consider the role of the muscles of the body (which often constitute around half of the body mass) as a storage organ. Even through generally agreed that their protein acts as a store, it is not simple passive storage of amino acids. Measurements of input and output of amino acids from muscle (Daniel et al. 1973, Pozetsky et al. 1969, Daniel, Pratt and Spargo, 1975) show not only that insulin causes a rapid uptake of blood amino acids but also that release of amino acids takes place in response to hormonal control and need for gluconeogenetic substrate. A large part of the amino acids released are converted to glutamine or alanine so that the overall effect is to reduce the relative proportion of indispensable large neutral amino acids and increase the proportion of the dispensable compounds. On a low protein diet this protein store is drawn upon over a long period. The increase in dispensable amino acids may reduce the metabolism of threonine to glycine and account for the high levels of threonine found during fasting.

Other tissues may well handle amino acids in a similar manner to the muscles, except for the liver which is mainly responsible for the breakdown of amino acids except for the branched chain ones, of which the muscles are a major catabolic site.

Harper: The elevation in brain tryptophan in rats consuming high-carbohydrate, protein-free diets is usually attributed to release of insulin causing a differential effect on muscle uptake of tryptophan and other amino acids so that the plasma tryptophan/large neutral amino acids ratio is increased. As this effect is eliminated by including only between 5 and 10 % of protein in the diet, is it not more likely that the brain tryptophan response is the result of differential effects of protein intake on the metabolism of different amino acids when protein intake is altered over the lower end of the intake range?

Wurtman: I think it has been shown clearly that insulin does produce a lowering of the large neutral amino acids in the blood and this effect is certainly amplified when low amounts of neutral amino acids are coming in from the diet.

Fernstrom: The release of insulin is key to the elevation in brain tryptophan level seen after carbohydrate ingestion by fasting rats. The mechanism of insulin's effects is an indirect one, following from the insulin-induced reductions in the serum levels of tryptophan's major competitors for transport across the blood brain barrier. The simplest way to illustrate these notions is with the observation that the ingestion of carbohydrates by fasting, diabetic rats does not lower the serum levels to tryptophan's transport competitors, nor does it produce a rise in brain tryptophan levels (Crandall and Fernstrom, 1980, Diabetes 29: 460—466).

Stress and metabolic strain affect the amino acid supply to the brain: Are such effects adequately considered in experiments designed to study the influence of precursor treatment on transmitter formation?

Huether: The effects of immobilization stress on the free amino acid pools in blood and brain reported by Gerald Curzon are rather impressive. I wonder if minor stress, e. g. transferring rats from the animal house to the lab, isolating them from group members etc. would also affect precursor availability endogeneously? And if so, how would this interfere with studies on the effects of precursor loading? We were surprised to find in the blood of chicks, after only 30 min of visual isolation from their group-mates, a massive decrease of many blood amino acids including all large neutral-ones. In the brain however, some free amino acids, e. g. tryptophan, were increased to a different extent in different brain regions (Huether and Neuhoff, Neurochem. Res., in press).

We must not forget, that precursor loading is not only accompanied by stress but often by additional severe metabolic strain of various kinds. Two examples are narcosis and injection of large volumes of saline. Ketamine narcosis slowly decreases the plasma concentrations of all glucogenic amino acids, up to 15 hrs in guinea pigs that were narcotized for only 1 hr. Saline injections — due to the low solubility of tryptophan or tyrosine often administered in large volumes — affect not only the availability of precursor amino acids but also the rate of synthesis and degradation of the respective monoamines in the rat brain (Huether et al. 1985, Neurochem. Int. 7 , 725—730).

What we normally do is to compare between, e. g., tryptophan treated and control animals. We overlook, that these controls underwent some sort of treatment (stress, strain etc.). What we measure under such circumstances is the effect of tryptophan on the ability of the animal to compensate for the metabolic shifts caused by the control-treatment. But we interpret our results as consequences of an altered precursor availability on normal metabolism and physiology rather than what they really are: effects on a more or less deranged and shifted metabolism and physiology.

Curzon: Gerald Huether's point that so-called "control" treatments may involve exposing animals to highly abnormal procedures is very important. This is something that has been neglected. More information on the effects of vehicle injection on naive, previously handled and previously injected rats would be valuable. Whether animals withdrawn for injection were single or group-housed also can potentially affect results. In the latter case, order of withdrawal can make a difference. These variable may be especially critical in studies on food-deprived rats as disturbing them appears to affect sypathetic activity more than in replete animals. Also, sympathetic activity has more lipolytic effect in starved animals (Brodie, Krishna and Hynie, 1969, Biochem. Pharmacol., 18: 1129—1134). Therefore removing group-housed fasted (but not fed) rats from cages can increase plasma unesterified fatty acid and hence alter plasma tryptophan binding. These effects become greater with successive removal (Curzon and Knott, 1975, Brit. J. Pharmacol. 54: 389—396), presumably because of the cumulative effect of stress associated with previous removal of their cage mates. Plasma corticosterone also rises. In some cases the change of plasma tryptophan binding was sufficient to increase brain tryptophan and 5-HIAA values (Knott, Hutson and Curzon, 1977, Pharmacol. Biochem. Behav., 7: 245—252).

Chapter II.

UPTAKE, COMPARTMENTATION AND UTILIZATION OF BRAIN AMINO ACIDS

REGULATION OF AMINO ACID TRANSPORT AT THE BLOOD-BRAIN BARRIER

Quentin R. Smith

Laboratory of Neurosciences
National Institute on Aging
National Institutes of Health
Bethesda, Maryland 20892
USA

INTRODUCTION

Amino acids that are not synthesized in brain but are required for cerebral metabolism must be obtained from plasma by transport across the blood-brain barrier. Morphological evidence has established that the barrier is located at the cerebral capillaries and is formed by a continuous layer of endothelial cells that are joined by tight junctions (Fig. 1).

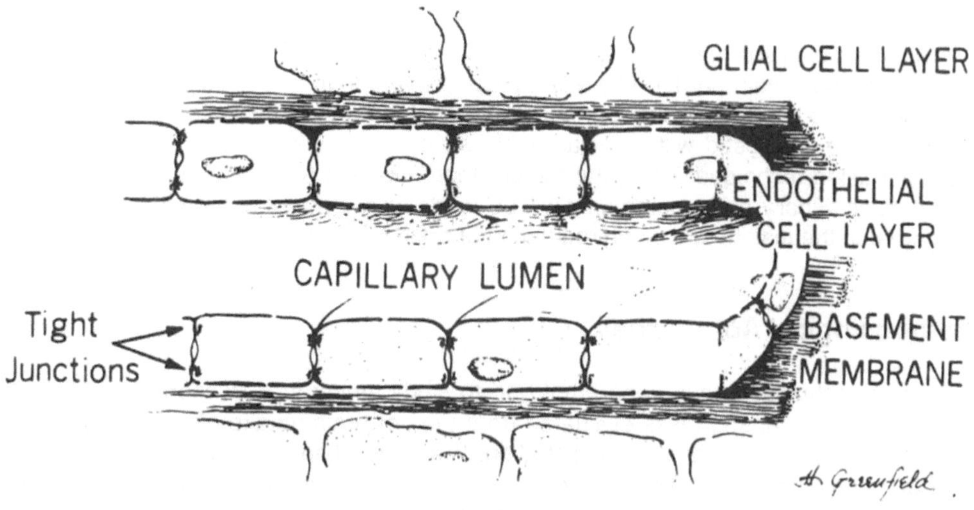

FIGURE 1

Diagram of the blood-brain barrier at the cerebral capillary. Tight junctions between endothelial cells limit intercellular diffusion. Facilitated transport systems are located in the endothelial cell membranes.

NATO ASI Series, Vol. H20
Amino Acid Availability and Brain Function in Health and Disease. Edited by G. Huether

These junctions essentially fuse the plasma membranes of adjacent endothelial cells and markedly restrict diffusion through the interendothelial cleft (1). As there are no alternative paracellular pathways, the brain capillary forms a considerable diffusion barrier to the free exchange of hydrophilic solutes between plasma and brain extracellular fluid (2).

Essential nutrients that are required for cerebral metabolism are transported across the blood-brain barrier by specific saturable carriers in the endothelial cell membranes. There are two carriers at the blood-brain barrier for the essential amino acids; one for those with neutral side chains and one for those with cationic (basic) side chains (3, 4). Together these two transport systems mediate the uptake of the ten amino acids that cannot be synthesized within the central nervous system. Evidence indicates that these systems are equilibrative and not concentrative or active. They appear to function primarily to facilitate amino acid exchange across brain capillaries. As transport across the blood-brain barrier is the "rate-limiting" step in amino acid movement from plasma to brain intracellular fluid (5), these transport systems have a central role in the determination of brain amino acid concentrations.

Interest in amino acid transport into brain has been stimulated in recent years by the discovery that several key metabolic pathways in brain are sensitive to alterations in brain amino acid supply. For example, the rates of synthesis and brain concentrations of the monoamine neurotransmitters, serotonin, dopamine and norepinephrine, depend in part on the brain concentrations of their precursor amino acids, tryptophan and tyrosine (6). In addition, the rates of synthesis of S-adenosyl methionine from methionine and histamine from histidine vary with the brain concentration of precursor amino acids (5). These metabolic pathways are substrate dependent because the rate-limiting enzymes are not saturated with amino acid at normal brain concentrations. As a result, changes in brain levels of precursors are directly reflected in the formation rates of the products. Since many of these pathways are involved in neurotransmission, the possibility arises that brain amino acid availability may influence neuronal activity and brain function.

In this article I shall review the current knowledge of neutral amino acid (NAA) transport at the blood-brain barrier. The NAAs include tryptophan and tyrosine, precursors for the monoamine neurotransmitters. In the first part of the review, the properties of the NAA carrier will be discussed with special emphasis on our recent results using the in situ brain perfusion technique. Then, the factors that influence NAA influx into brain will be summarized. Finally, a few comments will be made on the possible "homeostatic" role of the carrier in the maintenance and regulation of brain NAA levels.

Methodological Approaches

Numerous techniques have been employed to study amino acid transport into brain. In the first studies, transport was evaluated from the net brain uptake of an amino acid at various times after peripheral administration of a large dose of the amino acid, with or without a competitor (7, 8). These studies, which were published in the 1960's, established the two cardinal features of NAA transport into brain; that transport is mediated by a saturable carrier, not by passive diffusion, and that brain transport is uniquely sensitive to competitive inhibition. These studies, however, did not identify the exact location of the carrier (blood-brain barrier or brain cell membrane) and provided little quantitative information on the kinetics of amino acid transport.

In the 1970's, the properties of the NAA carrier were extensively characterized in detailed studies by Oldendorf and Pardridge in California and by Pratt and coworkers in England using two separate methods that measured selectively transport across the blood-brain barrier (for reviews, see 9, 10). With Oldendorf's technique, a solution containing a labelled amino acid and [3H]H_2O is injected into the carotid artery, and then the brain uptake of the amino acid is measured, relative to that of water, after a single pass of the solution through the brain capillaries. This method, which is referred to as the brain uptake index technique, is fast, simple to perform and allows determination of amino acid influx in the absence normal physiological concentrations of competing amino acids. However, there are problems in that the method has limited sensitivity, provides only estimates of influx from plasma and is subject to errors because of a "mixing" artifact (11, 12). In contrast, with Pratt's technique, the influx of an amino acid is determined from the initial unidirectional uptake of a labeled amino acid from plasma during intravenous infusion of the tracer at a rate that maintains a constant plasma concentration. The intravenous infusion method allows accurate determination of NAA influx from plasma and is ideally suited for studies of regional influx rates using quantitative autoradiography (13). However, the method allows only limited control of plasma amino acid concentrations and therefore cannot be used to measure the transport of a NAA in the absence of competitors.

Recently, we developed a new technique to measure blood-brain barrier transport that overcomes many of the limitations of previous methods (11, 14). With this technique, one cerebral hemisphere of an anesthetized rat is perfused briefly (5—300 s) by retrograde infusion of fluid into the external carotid artery. Influx is determined from the net brain uptake of labelled amino acid during perfusion. The method, unlike the intravenous infusion technique, allows absolute control over perfusate composition for studies of influx in the absence or presence of competing amino acids. Furthermore, the method provides more reliable measures of influx than the brain uptake index technique because perfusion time can be extended beyond that of a single pass (3—7 s). Over the past few years we have utilized the brain perfusion technique to further characterize the cerebrovascular NAA transporter and to obtain new, accurate estimates of the kinetic constants that describe NAA transport across the blood-brain barrier (11, 15, 16).

Cerebrovascular Neutral Amino Acid Transport

NAA transport across the blood-brain barrier in adult animals is mediated primarily by a single facilitated system that is saturable, stereospecific and energy independent (4, 5, 15). This transport system has measurable affinity for virtually all plasma NAAs, with the possible exception of glycine and proline, but has greatest affinity for those essential amino acids that cannot be synthesized within the central nervous system. This includes the aromatic amino acids (TRP, PHE, TYR, HIS), the branched-chain amino acids (LEU, ILE, VAL) and MET and THR.

Approximately seven separate transport systems for NAAs have been identified in mammalian tissues based on transport selectivity, sodium dependence and inhibition by specific substrates (17, 18). Of these transport systems, the cerebrovascular NAA transporter corresponds most closely to the L system of Christensen. Both transporters are sodium independent and prefer amino acids with large, hydrophobic side chains, such as leucine and phenylalanine (4, 15, 16). In addition, both transport the L system specific substrate, BCH (2-aminobicyclo-[2.2.1]heptane-2-carboxylic acid)(19). The L system is present in most all tissues of the body and appears to mediate equilibrative, bidirectional transport across cell membranes. It's transport activity is notably unresponsive to hormonal regulation or starvation (17). Evidence supporting the presence of the L system at the blood-brain barrier has been obtained not only from in vivo studies, but more recently from in vitro studies using isolated brain capillaries (20, 21, 22).

Two other NAA transport systems have been identified in isolated brain capillaries of mature animals. Betz and Goldstein demonstrated the presence of the A system, which transports small NAAs, is sodium dependent and is inhibitable by the A system specific analogue, MeAIB (methylaminoisobutyric acid)(20). In addition, Tayarani *et al.* presented evidence for the ASC system, which is sodium dependent and transports primarily alanine, serine and cysteine (22). These transport systems, however, contribute little to NAA uptake into brain in vivo. Transfer rates for small NAAs, such as alanine, serine, proline and glycine, are the lowest of all the NAAs at the blood-brain barrier. And, as first shown by Oldendorf, the small components of uptake that are present are sensitive in most cases to inhibition by large NAAs, suggesting transfer by the L system carrier (23). Results in our laboratory support this conclusion. With the brain perfusion technique, we find that brain influx rates for several NAAs can be accounted for entirely by L-system transfer (15, 24). Furthermore, we have been unable to demonstrate a saturable component of uptake for MeAIB, the A system specific substrate. Since the A and ASC systems are present at the capillaries but do not contribute to NAA transfer into brain, it is likely that they are located only on the abluminal membrane of the endothelial cell (20). Their function may be to mediate active NAA efflux from brain or to accumulate NAAs in endothelial cells for capillary metabolism and protein synthesis.

Most studies of NAA transport into brain report a small "nonsaturable" component that contributes significantly only at very high amino acid concentrations (> 1 μmol/ml) (9, 10, 16). The "nonsaturable" component may reflect passive diffusion of amino acids across the endothelial cell membranes. Alternatively, the "nonsaturable" component may reflect the presence of a low-affinity transport system or may arise as an artifact of the measurement technique (24, 25). Regardless of the cause or mechanism, the "nonsaturable " component contributes minimally (< 10 %) to influx for most amino acids at normal plasma concentrations.

The concentration dependence of influx for an amino acid in the absence of competitors can be described by the equation

$$Jin = VmaxC/(Km + C) + KdC \qquad (1)$$

where Jin is the unidirectional influx rate, Vmax is the maximal transport rate of the saturable component, Km is the half-saturation concentration and Kd is the constant of nonsaturable diffusion. Table 1 lists best fit values of Vmax, Km and Kd for 10 NAAs that were studied with the in situ brain perfusion technique.

Among the 10 amino acids, Vmax values differ by ~ 5 fold whereas Km values differ by ~ 80 fold. Variability in Vmax was noted by Christensen as one characteristic of L system transport (17). Amino acids with large hydrophobic side chains, such as phenylalanine and tryptophan, have greatest affinity (lowest Km) for the cerebrovascular NAA transport system. Small NAAs, such as cysteine, asparagine, alanine and serine, have low but measurable affinity with Km values, estimated from inhibition analysis, that equal 0.48 ± 0.03, 3.8 ± 0.5, 6.6 ± 0.9 and 7.9 ± 0.9 μmol/ml, respectively (16).

TABLE 1

Vmax, Km and Kd values for NAA transport across the blood-brain barrier of the pento-barbital-anesthetized rat

Amino Acid	V_{max} (μmol/s/g × 10^4)	Km (μmol/ml)	Kd (ml/s/g × 10^4)
Phenylalanine	6.9	0.011	1.27
Tryptophan	9.1	0.015	0.74
Leucine	9.9	0.029	0.62
Methionine	4.2	0.040	0.40
Isoleucine	10.0	0.056	—
Tyrosine	16.1	0.064	—
Histidine	10.2	0.100	0.48
Valine	8.2	0.21	0.77
Threonine	2.8	0.22	0.64
Glutamine	7.2	0.88	0.48

From Smith et al. (16).

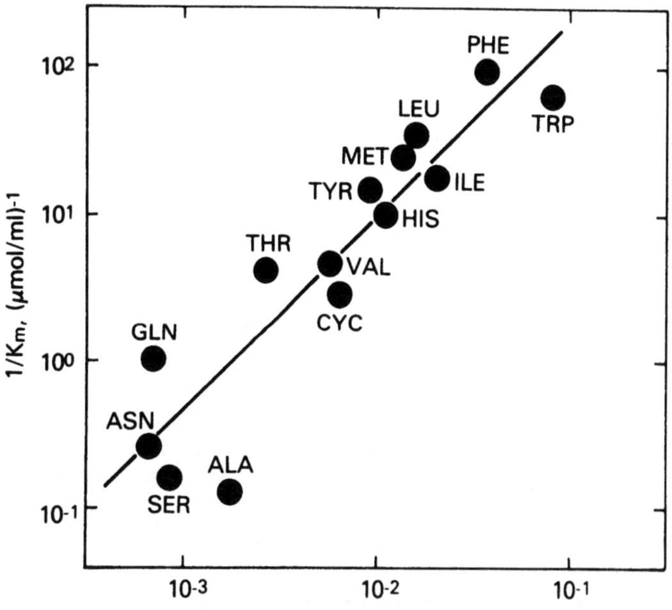

FIGURE 2

Relation of apparent affinity (1/km) to octanol/water partition coefficient for 14 NAA's. CYC, cycloleucine. The octanol/water partition coefficient for an amino acid is one measure of side chain hydrophobicity. The correlation coefficient equaled 0.91.
From Smith *et al.* (16).

The variation in Km among the amino acids can be attributed primarily to differences in side chain hydrophobicity. Figure 2 shows that the apparent affinity (1/Km) of a L-α-NAA is linearly related over ~ 3 orders of magnitude to the octanol water partition coefficient, a measure of side chain hydrophobicity (16). Other factors influence affinity as well. The transporter is stereoselective and prefers L, as compared to D, amino acids (5). In addition, for affinity substrates must have a free carboxyl group with an unsubstituted amino group on the α-carbon.

Although the order of affinity among the NAAs for the cerebrovascular NAA carrier is similar to that of the L system, the two transporters differ in one important area. Km values for cerebrovascular transport of most NAAs are only 1—10 % of values for L system transport in other tissues. For instance, the Km for phenylalanine at the blood-brain barrier is 0.01 μmol/ml, whereas Km values for phenylalanine in intestine, kidney, heart, liver and red cell range from 1—5 μmol/ml (26, 27). This difference could arise from allosteric modification of a single transport system. Alternatively, it could occur if there are two separate isozymes or variants of the L system. Evidence in favor of the latter hypothesis was presented by Weissbach *et al.* (25) who found two separate "L system" transporters, one high affinity and one low affinity, that were differentially expressed in rat hepatocytes during the first 24 hours of primary culture. The high affinity form, tentatively designated L1, showed minimal activity in freshly isolated cells but increased in activity with time in culture. The increase in activity could be blocked by inhibitors of mRNA and protein synthesis (25). If there are two separate L system variants, the cerebrovascular NAA transporter may correspond to the high affinity L1 form. The high affinity of the cerebrovascular transporter has important functional implications because it makes brain NAA transport sensitive to competition effects.

Factors that Influence NAA Transport into Brain

Brain concentrations of the essential NAAs are determined by plasma concentrations and transport across the blood-brain barrier. Influx of a NAA from plasma is given by the equation

$$Jin = VmaxC/[Km(1 + \sum Ci/Ki) + C] + KdC \qquad (2)$$

where C is the plasma concentration of the amino acid of interest, Vmax, Km and Kd are the transport constants of that amino acid, and Ci and Ki are the plasma concentration and corresponding half saturation constant of each competing amino acid (9, 10, 16). Eq. 2 differs from Eq. 1 only in that the Km is multiplied by $(1 + \sum Ci/Ki)$, which corrects the true Km for the presence of competing NAAs. The term, $Km(1 + \sum Ci/Ki)$, is referred to as the apparent Km (Km (app)) and is defined as the plasma concentration of the amino acid that gives an influx rate equal to 50 % of Vmax (Fig. 2).

Table 2 lists mean values of plasma concentration, Km(app) and influx from plasma for the 10 NAAs with the highest affinity for the carrier. Influx values from plasma are markedly less than those predicted by Eq. 1 because of transport inhibition by competing NAAs. Cerebrovascular NAA influx is sensitive to competition effects because > 10 NAAs share the same transport system and because the Km for each NAA is low, comparable to the respective plasma concentration. As a result, $\sum Ci/Ki > 10$, and thus the system is nearly saturated with NAAs as a group at normal plasma concentrations. The percent saturation can be estimated as 96 % using the equation $[C/Km + \sum Ci/Ki]/[1 + C/Km + \sum Ci/Ki]$ (16). With transport saturation, individual NAAs must compete for available transport sites. Competition reduces the influx rates by > 75 % and increases the Km(app) by 10—30 fold. In contrast, competition is not significant in most other tissues because transport is mediated by high Km processes (Km ≥ 1 μmol/ml) and thus $\sum Ci/Ki < 1$ (26, 27).

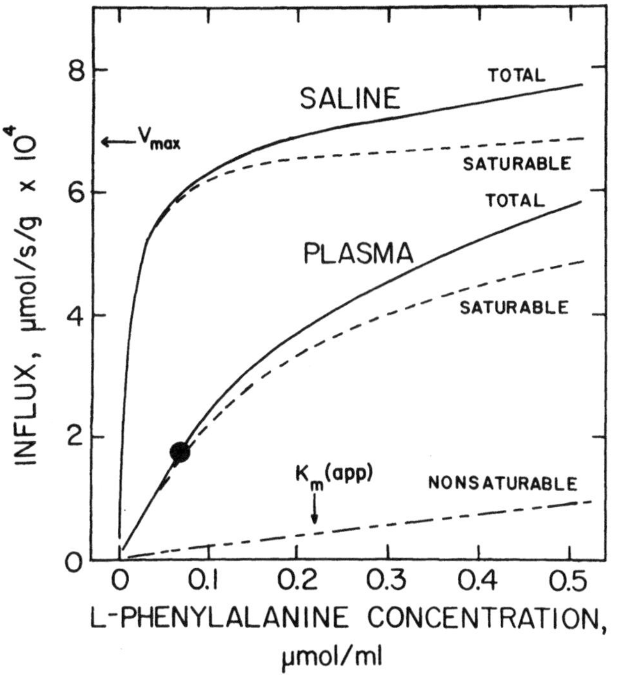

FIGURE 3
Concentration dependence of phenylalanine influx into rat cerebral cortex from saline and plasma. The curves represent total, saturable and nonsaturable influxes as predicted using Eq. 2. Values of Vmax, Km, and Kd are from Table 1. Km(app) was determined by direct measurement and equaled 0.21 ± 0.01 μmol/ml (15). Saline contains no competing NAAs. The point on the plasma curve represents the measured influx ($1.8 \pm 0.1 \times 10^{-4}$ μmol/s/g) at the plasma concentration of 0.069 ± 0.002 μmol/ml (15).

From inspection of Eq. 2 it can be seen that the influx of a NAA into brain depends primarily on three factors; the plasma concentration of the amino acid, the plasma concentration of competitors and the values of the kinetic constants. At normal plasma concentrations, Jin for a NAA will be essentially proportional to the plasma concentration of the NAA, since Km(app) ≫ C, and inversely proportional to the total plasma concentration of competing amino acids. This relation was summarized in 1972 by Fernstrom and Wurtman as the "plasma ratio", which was defined as $C/\sum Ci$ (28). They found in a number of nutritional, physiological and pharmacological experiments that the plasma ratio for tryptophan was an excellent predictor of brain tryptophan concentration. Later, Fernstrom and Faller extended the use of the plasma ratio for other essential NAAs, including PHE, TYR, LEY, ILE and VAL (29).

TABLE 2
Plasma concentration, apparent Km and influx into rat cerebral cortex

Amino Acid	Plasma Concentration (μmol/ml)	Km(app) (μmol/ml)	Unidirectional Influx (μ/s//g × 10^4)
Phenylalanine	0.081	0.17	2.33
Tryptophan	0.080	0.33	0.44
Leucine	0.175	0.50	2.67
Methionine	0.064	0.86	0.32
Isoleucine	0.087	1.21	0.67
Tyrosine	0.063	1.42	0.68
Histidine	0.095	2.22	0.47
Valine	0.181	4.89	0.45
Threonine	0.237	4.86	0.27
Glutamine	0.485	19.9	0.37

From Smith et al. (16).

The plasma ratio assumes that all NAAs have equal affinity for the cerebrovascular transport system. Since transport affinities differ widely, it would be expected based on theoretical grounds that a normalized ratio, $(C/Km)/[\sum Ci/Ki]$, would provide more precise estimates of brain NAA transport and concentration. In the normalized ratio, the plasma concentration of each NAA is weighted with the affinity of that amino acid for the cerebrovascular transport system.

Cerebrovascular NAA influx is determined not only by plasma NAA concentrations, but also by the transport constants, Vmax, Km and Kd, which vary regionally and may change under different physiologic and pathologic conditions. Hawkins *et al.* (13) demonstrated using quantitative autoradiography that phenylalanine influx from plasma varies approximately three fold among different brain regions and that regional influx rates correlate with regional capillary density. We found similar regional differences using the brain perfusion technique and demonstrated that the differences arise from variation in Vmax, not Km (15, 16). These results suggest that the affinity of the carrier is similar through out the brain and that the number of carriers varies regionally depending on the local number of capillaries. It is not known whether there are regional differences in brain NAA concentration that correlate with regional Vmax.

The values of the transport constants, especially Vmax, may change under differing conditions due to recruitment of unperfused capillaries or to synthesis on new carriers. Recent studies indicate that only ~ 50 % of brain capillaries are perfused in awake animals and that the percentage may vary with brain metabolic activity (30, 31) . If all capillaries contain similar numbers of carriers, then the measured Vmax for a region will depend on the number of perfused capillaries in the region. Two studies have presented preliminary evidence which suggests that the blood-brain barrier Vmax for glucose may vary 2—4 fold under different metabolic states (31, 32). Similar studies have not been conducted on brain NAA transport.

Cerebrovascular NAA transport activity has been reported to increase in various pathologic conditions, such as hyperammonemia and liver failure (33, 34). In addition, NAA transport activity is reported to decrease during development (35). In a recent study we showed that brain NAA transport activity does not change in chronic hyperaminoacidemia. Apparently, the blood-brain barrier may not have the capacity to modulate transport activity in the long term to correct amino acid imbalances in supply to the central nervous system (36).

Role of the Blood-brain Barrier in Brain NAA Regulation

The cerebrovascular NAA carrier is a passive exchange mechanism and would not be expected to homeostatically maintain brain NAA concentrations in the face of fluctuating plasma concentrations. Indeed, it has long been known that marked changes in brain concentrations of NAAs can be produced by peripheral administration of a large dose of a single NAA (7, 8). However, in many physiological situations brain NAA concentrations remain surprisingly constant. This regulation occurs in part because of transport saturation and competition at the blood-brain barrier.

Plasma NAA concentrations show considerable fluctuations that are related to food consumption and peripheral metabolism (5). Changes on the order of 3—10 fold are produced in rats fed diets containing 0—55 % protein. However, after a protein containing meal, plasma concentrations tend to change in the same direction as a group. When

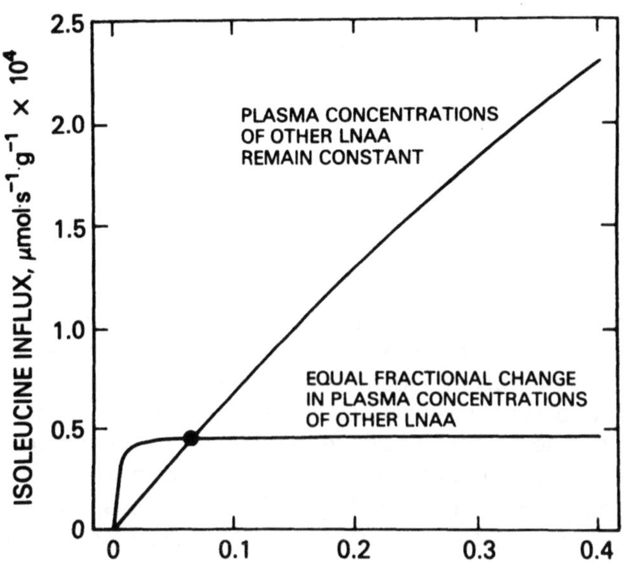

FIGURE 4

Isoleucine influx into rat cerebral cortex in relation to plasma isoleucine concentration. In the top curve, the plasma concentrations of all other NAAs remain constant, whereas in the bottom curve, there is an equal fractional change in the plasma concentrations of all NAAs. The point on the curve represents the measured isoleucine influx ($0.45 \pm 0.04 \times 10^{-4}$ μmol/s/g) at the plasma concentration of 0.064 μmol/ml (40).

plasma NAA concentrations change in the same direction and to a similar extent, alterations in brain NAA influx are minimized because of transport saturation and competition at the blood-brain barrier. Figure 4 illustrates that a five-fold increase or decrease in the plasma concentrations of all NAAs will change isoleucine influx by only ~ 12 %.

This relative protection of the brain from changes in plasma concentrations has been shown repeatedly in dietary studies (29, 37, 38). Table 3 summarizes results from three studies and shows that, on average, changes in brain concentrations are only 14—19 % of those in plasma in rats fed diets containing 0—55 % protein. Thus, even though the cerebrovascular NAA transporter is a passive, facilitated system, it does help minimize changes in brain NAA concentrations in many physiological situations. This regulatory role is consistent with the general homeostatic function of the blood-brain barrier to maintain constancy of the brain internal environment (39).

TABLE 3

Percent change in brain amino acid concentration relative to that in plasma in rats after ingestion of meals containing 0—55 % protein

Amino Acid	Fernstrom & Faller	Glaeser et al.	Peters & Harper
Phenylalanine	23.6 %	23.6 %	34.0 %
Tryptophan	6.2 %	14.3 %	16.1 %
Leucine	9.1 %	14.8 %	9.0 %
Methionine	19.9 %	16.0 %	—
Isoleucine	11.5 %	11.6 %	10.1 %
Tyrosine	20.9 %	29.2 %	32.4 %
Valine	6.8 %	8.3 %	10.7 %
Average	14.0 %	16.8 %	18.7 %
± SD	± 7.3	± 7.2	± 11.5

Data are from references 29, 37 and 38

Although the cerebrovascular NAA transport system will help maintain constancy of influx after a balanced protein meal, the transport system is susceptible to large imbalances in plasma NAA concentrations. Figure 4 shows that isoleucine influx is an approximately linear function of plasma isoleucine concentration up to 0.4 μmol/ml when plasma concentrations of competing amino acids remain constant. Furthermore, because the transport system is saturated, an increase in the concentration and thus influx of one NAA will decrease the influxes of all competing NAAs. Imbalances in plasma NAA concentrations occur in several pathologic conditions, such as diabetes, hepatic failure and the hyperaminoacidemias (9).

REFERENCES

1) REESE, T. S., KARNOVSKY, M. J. (1967). Fine structural localization of a blood-brain barrier to exogenous peroxidase. *J. Cell Biol.* **34:** 207—217.

2) OHNO, K., PETTIGREW, K. D., RAPOPORT, S. I. (978).Lower limits of cerebrovascular permeability to nonelectrolytes in the conscious rat. *Am. J. Physiol.* **235:** H299—H307.

3) RICHTER, J. J., WAINER, A. (1971). Evidence for separate systems for the transport of neutral and basic amino acids across the blood-brain barrier. *J. Neurochem.* **18:** 613—620.

4) OLDENDORF, W. H. (1971). Brain uptake of radiolabeled amino acids, amines and hexoses after arterial injection. *Am. J. Physiol.* **221:** 1629—1639.

5) PARDRIDGE, W. M. (1977). Regulation of brain amino acid availability. In: *Nutrition and the Brain* (Wurtman, R. J., Wurtman, J. J., eds.), **Vol. 1,** Raven Press, New York, pp. 142—205.

6) WURTMAN, R. J., HEFTI, F., MELAMED, E. (1981). Precursor control of neurotransmitter synthesis. *Pharmacol. Rev.* **32:** 315—335.

7) CHIRIGOS, M. A., GREENGARD, P., UDENFRIEND, S. (1960). Uptake of tyrosine by rat brain in vivo. *J. Biol. Chem.* **235:**: 2075—2079.

8) LAJTHA, A., TOTH, J. (1961). The brain barrier system — II: Uptake and transport of amino acids by the brain. *J. Neurochem.* **8:** 216—225.

9) PARDRIOGE, W. M. (1983). Brain metabolism: A perspective from the blood-brain barrier. *Physiol. Rev.* **63:** 1481—1535.

10) PRATT, O. E. (1980). The transport of nutrients into the brain: The effect of alcohol on their supply and utilisation. In: *Addiction and Brain Damage* (Richet, D., ed.), University Park Press, Baltimore, pp. 94—128.

11) SMITH, Q. R., TAKASATO, Y., RAPOPORT, S. I. (1984). Kinetic analysis of L-leucine transport across the blood-brain barrier. *Brain Res.* **311:** 167—170.

12) PARDRIDGE, W. M., LANDAW, E. M., MILLER, L. P., BRAUN, L. D., OLDENDORF, W. M. (1985). Carotid artery injection technique: Bounds for bolus mixing by plasma and by brain. *J. Cereb. Blood Flow Metabol.* **5:** 576—583.

13) HAWKINS, R. A., MANS, A. M., BIEBUYCK, J. F. (1982). Amino acid supply to individual cerebral structures in awake and anesthetized rats. *Am. J. Physiol.* **242:** E1—E11.

14) TAKASATO, Y., RAPOPORT, S. I., SMITH, Q. R. (1984). An in situ brain perfusion technique to study cerebrovascular transport in the rat. *Am. J. Physiol.* **247:** H484—H493.

15) MOMMA, S., AOYAGI, M., RAPOPORT, S. I., SMITH, Q. R. (1987). Phenylalanine transport across the blood-brain barrier as studied with the in situ brain perfusion technique. *J. Neurochem.* **48:** 1291—1300.

16) SMITH, Q. R., MOMMA, S., AOYAGI, M., RAPOPORT (1987). Kinetics of neutral amino acid transport across the blood-brain barrier. *J. Neurochem.* **49:** 1651—1658.

17) SHOTWELL, M. A., KILBERG, M. S., OXENDER, D. L. (1983). The regulation of neutral amino acid transport in mammalian cells. *Biochim. Biophys. Acta* **737:** 267—284.

18) CHRISTENSEN, H. N. (1985). On the strategy of kinetic discrimination of amino acid transport systems. *J. Membr. Biol. 84:* 97—103.

19) WADE, L. A., KATZMAN, R. (1975). Synthetic amino acids and the nature of L-dopa transport at the blood-brain barrier. *J. Neurochem.* **25:** 837—842.

20) BETZ, A. L., GOLDSTEIN, G. W. (1978). Polarity of the blood-brain barrier: Neutral amino acid transport into isolated brain capillaries. *Science* **202:** 225—226.

21) CHOI, T. B., PARDRIDGE, M. (1986). Phenylalanine transport at the human blood-brain barrier: Studies with isolated human brain capillaries. *J. Biol. Chem.* **261:** 6536—6541.

22) TAYARANI, I., LEFAUCONNIER, J. M., ROUX, F., BOURRE, J. M. (1987).Evidence for an alanine, serine, and cysteine system of transport in isolated brain capillaries. *J. Cereb. Blood Flow Metabol.* **7:** 585—591.

23) OLDENDORF, W. H., SZABO, J. (1976). Amino acid assignment to one of three blood-brain barrier amino acid carrier. *Am. J. Physiol.* **230:** 94—98.

24) AOYAGI, M., AGRANOFF, B. W., WASHBURN, L. C., SMITH, Q. R. (1988). Blood-brain barrier transport of 1-aminocyclohexanecarboxylic acid, a nonmetabolizable amino acid for in vivo brain transport. *J. Neurochem.*, in press.

25) WEISSBACH, L., HANDLOGTEN, M. E., CHRISTENSEN, H. N., KILBERG, M. S. (1982). Evidence for two Na-independent neutral amino acid transport systems in primary cultures of rat hepatocytes. *J. Biol. Chem.* **257:** 12006—12011.

26) LERNER, J., LARIMORE, D. L. (1986). Comparative aspects of the apparent Michaelis constant for neutral amino acid transport in several animal tissues. *Comp. Biochem. Physiol.* **84B:** 235—248.

27) SALTER, M., KNOWLES, R. G., POGSON, C. I. (1986).Transport of the aromatic amino acids in isolated rat liver cells. *Biochem. J.* **233:** 499—506.

28) FERNSTROM, J. D., WURTMAN, R. J. (1972). Brain serotonin content: Physiological regulation by plasma neutral amino acids. *Science* **178:** 414—416.

29) FERNSTROM, J. D., FALLER, D. V. (1978). Neutral amino acids in the brain: changes in response to food ingestion. *J.Neurochem.* **30:** 1531—1538.

30) WEISS, H. R., BUCHWEITZ, E., MURTHA, T. J., AULETTA, M. (1982). Quantitative regional determination of morphometric indices of the total and perfused capillary network in rat brain. *Circ. Res.* **51:** 494—503.

31) CREMER, J. E., CUNNINGHAM, V. J., SEVILLE, M. P. (1983). Relationships between extraction and metabolism of glucose blood flow, and tissue blood volume in regions of rat brain. *J. Cereb.Blood Flow Metabol.* **3:** 291—302.

32) GJEDDE, A., RASMUSSEN, M. (1980). Pentobarbital anesthesia reduces blood-brain glucose transfer in the rat. *J. Neurochem.* **35:** 1382—1387.

33) MANS, A. M., BIEBUYCK, J. F., SAUNDERS, S. J., KIRSCH, R. E., HAWKINS, R. A. (1979). Tryptophan transport across the blood-brain barrier during acute hepatic failure. *J. Neurochem.* **33:** 409—418.

34) JONUNG, T., RIGOTTI, P., JAMES, J. H., BRACKETT, K., FISCHER, J. E. (1985). Effect of hyperammonemia and methionine sulfoxime on the transport parameters of blood-brain transport of leucine and phenylalanine. *J. Neurochem.* **45:** 308—318.

35) NAGASHIMA, T., LEFAUCONNIER, J. M., SMITH, Q. R. (1987). Developmental changes in neutral amino acid transport across the blood-brain barrier. *J. Cereb. Blood Flow Metabol.* **7 (Suppl 1):** S524.

36) SMITH, Q. R., MOMMA, S., SWEENEY, D. J., RAPOPORT, S. I. (1987). Amino acid transport into brain during chronic hyperaminoacidemia. *J. Cereb. Blood Flow Metabol.* **7 (Suppl. 1):** 500.

37) GLAESER, B. S., MAHER, T. J., WURTMAN, R. J. (1983).Changes in brain levels of acidic, basic, and neutral amino acids after consumption of single meals containing various proportions of proteins. *J. Neurochem.* **41:** 1016—1021.

38) PETERS, J. C., HARPER, A. E. (1987). Acute effects of dietary protein on food intake, tissue amino acids, and brain serotonin. *Am. J. Physiol.* **252:**R902—R914.

39) BRADBURY, M. (1979).*The Concept of a Blood-Brain Barrier.* Wiley, Chichester, pp. 383—407.

40) SMITH, Q. R., TAKASATO, Y. (1986). Kinetics of amino acid transport at the blood-brain barrier studied using an in situ brain perfusion technique. *Ann. N. Y. Acad. Sci.* **481:** 186—201.

ENZYMES OF THE CEREBROVASCULAR ENDOTHELIUM; POSSIBLE RELEVANCE TO THE INTERPRETATION OF PRECURSOR-LOADING EXPERIMENTS

F. V. DeFeudis

Université Louis Pasteur
Faculté de Médecine
67084 Strasbourg Cedex
France

Present address:
Institute for BioScience
Grafton, MA, USA

INTRODUCTION

This article is focussed on the major component of the "blood-brain barrier" (BBB), the cerebrovascular endothelial cell (CVEC). CVECs serve not only to restrict solute entry into the brain, but more generally as a regulatory system consisting of transport processes, enzymatic mechanisms and a receptive surface to neurotransmitters (NTs) and neurohormones which limits the actions of blood-borne substances on brain function. However, BBB function does not depend solely upon the CVEC, but also upon the activities of glial cells, the basement membrane and pericytes.

Transport systems for large neutral amino acids (LNAAs) exist in lumenal and ablumenal membranes of CVECs and are involved in regulating LNAA availability (1). Some LNAAs serve as precursors for brain NTS. Choline, which is used for acetylcholine (ACh) synthesis, might be similarly regulated (1).

The metabolic component of CVEC function will be emphasized herein. NT metabolism by the BBB was first demonstrated by Bertler et al. (2) who found, using histochemical methods, that L-3,4-dihydroxyphenylalanine-decarboxylase (L-DOPA-decarboxylase; aromatic-L-amino acid decarboxylase; L-AAAD) is present in brain capillaries and pericytes. This finding led to the concept of the "enzymatic BBB", which limits the entry of circulating NTs and their precursors into the brain's interstitial fluid.

Why has NT-precursor loading (except for loading with L-DOPA) been of limited value in treating neurologic/psychiatric disorders? Several functional aspects of CVECs will be discussed in an effort to answer this question: (a) "completeness" of the enzymatic BBB; (b) "capacity" of the enzymatic BBB; (c) endothelium-dependent vasorelaxation.

NATO ASI Series, Vol. H20
Amino Acid Availability and Brain Function in Health and Disease. Edited by G. Huether

Endothelial Enzymes Involved in Handling Blood-Borne Precursors for Neurotransmitters or Neurohormones

Some enzymes of CVECs that are involved in handling NT-precursors, NTs or neurohormones are listed in Table 1.

TABLE 1
Some Enzymes of the Cerebrovascular Endothelium and Representative Substrates

Enzyme	Possible Substrate(s)
Aromatic-L-Amino Acid Decarboxylase	L-DOPA, TRP, TYR, PHE, HIS, L-5-Hydroxytryptophan
Monoamine Oxidase	DA, NE, Epinephrine, 5-HT, Tyramine, Phenylethylamine
Choline Acetyltransferase	Choline
Enzymes of Phospholipid Turnover	Choline
Phospholipases A-2 and C	1,2-*sn*-diacylphosphoglycerides (e. g., lecithin)
Phenylethanolamine-N-Methyltransferase	NE
Butyrylcholinesterase	ACh
Enzymes of Protein Synthesis	TYR, TRP, PHE, HIS
Peptidyl Dipeptidase	Angiotensin-I, Bradykinin, Enkephalins
5′-Nucleotidase	Adenosine-5′-Monophosphate

A. Aromatic-L-Amino Acid Decarboxylase (L-AAAD)
The histochemical demonstration of L-AAAD in brain capillaries (see above) was confirmed using isolated brain capillaries. L-AAAD is present in CVECs of microvessel walls and impedes the entry of aromatic amino acids into the brain parenchyma. As characterized in mouse and rat, L-AAAD activity is unique to CVECs of brain parenchymal microvessels (3). The small amount of L-AAAD activity found in other blood vessels of the brain or peripheral tissues is almost exclusively localized to perivascular sympathetic nerves (3, 4).

Hardebo *et al.* (3) showed that L-AAAD activity in rat whole brain is about 5.07 nmol/mg protein/hr. In pial vessels it was 2.68 nmol/mg protein/hr and was decreased to 0.35 nmol/mg protein/hr after sympathectomy. However, brain microvessels had L-AAAD activity of 4.85 nmol/mg protein/hr which was not reduced by sympathectomy, indicating that the enzyme was not localized in sympathetic nerve endings of the vessel wall (3).

It is because of this efficient mechanism for restricting L-DOPA's passage into the brain parenchyma as well as because of the high L-AAAD acitivity in peripheral tissues that such high doses of L-DOPA must be used in therapy for Parkinson's disease. DA, formed from L-DOPA in CVECs, could diffuse into the blood and/or into the brain's interstitial fluid, and such a mechanism could contribute to the therapeutic action of L-DOPA.

B. Enzymes of Protein Synthesis
The LNAAs phenylalanine (PHE) and tryptophan (TRP) that serve as precursors to catecholamine-NTs and serotonin (5-HT), respectively, like other LNAAs (e. g., leucine (LEU), valine (VAL)), are incorporated into protein in CVECs. A major portion of the protein syn-

thesis that occurs in the telencephalon of the rat is associated with the microvascular network (5). Audus and Borchardt (6) found that LEU transport in primary cultures of CVECs prepared from bovine brain microvessels had characteristics similar to those of the LNAA carrier of the BBB *in vivo* (cf. 7). Pre-incubation of CVEC monolayers for 20 minutes with cycloheximide (2.5 mg/ml; an inhibitor of protein synthesis) caused a 148 % increase in the amount of LEU transported across the CVEC monolayer. Thus, some of the LEU that was taken up by CVECs had been incorporated into protein during the transport process. The protein synthesis that occurs in CVECs could also restrict the passage of NT-precursor LNAAs across the BBB.

C. Choline Acetyltransferase (ChAT)

Recent immunocytochemical experiments have shown that ChAT is present in some CVECs of small blood vessels of rat visual cortex (8). Thus, CVECs might synthesize and store ACh. This ACh might be released following damage to CVECs and might contribute to a protective mechanism of vasodilatation (8; see below). ChAT activity has also been detected biochemically in capillaries of rat cerebral cortex (9). Maximum enzymatic activity occurred in 12-month-old rats (55 pmol/mg protein/min) and this decreased to a minimum (34 pmol/mg protein/min) in 24-month-old rats. The Km for ChAT activity of the capillary fraction was similar to those of synaptosomes and whole cortex homogenate (i. e., about 14—17 μM; acetyl-CoA as substrate). However, the Vmax of capillaries was only about 5 % of the Vmax values for synaptosomes or homogenate in 3-month-old rats. Gonzalez and Santos-Benito (10) showed further that ACh is synthesized in CVECs isolated from capillaries of rat cerebral cortex. The specific activity of ChAT in CVECs (0.264 nmol/mg protein/min) was about one-third of that present in total homogenate of cerebral cortex. Thus, ChAT seems to be intrinsically localized in CVECs, and it might restrict ACh-precursors from entering the interstitial fluid.

D. Enzymes of Phospholipid Synthesis and Degradation

Phosphatidylcholine (commercially available as "lecithin", but generally not pure) has been used as a potential ACh-precursor. However, if phosphatidylcholine or lysophosphatidylcholine enters CVECs, it could (like choline) undergo metabolic transformation therein, thus limiting its (or choline's) entry into the interstitial fluid. If phosphatidylcholine is catabolized peripherally so that choline enters the CVECs, then phospholipid synthesis and turnover, as well as ChAT could handle the choline. Hydrolysis of phosphatidylcholine in CVECs could liberate not only choline (phospholipase-D action) or phosphorylcholine (phospholipase-C action), but also unsaturated fatty acids like arachidonate (phospholipase-A2 or A1 action); lysophosphatidylcholine would also be formed. Metabolism of arachidonate in CVECs could lead to the formation of several potent vasoactive metabolites, such as prostacyclin (PGI-2), PAF-acether, and possibly "endothelium-derived relaxing factors" (EDRF), and the potent detergent action of lysophosphatidylcholine could alter membrane fluidity.

Lecithin, at doses used in loading studies or in human clinical trials (e. g., 50 g/day for 14 days as a treatment for tardive dyskinesia; Ref. 11)), could augment arachidonate levels in CVECs and lead to increased synthesis of the above-mentioned substances, and these in turn could alter transendothelial permeability and regional cerebral blood flow. Thus, lecithin-loading could influence brain function and behaviour by these mechanisms rather than by increasing the availability of choline for ACh synthesis. Even though the kinetics of BBB choline transport favor the passage of choline into the brain's interstitial space (i. e., the choline carrier is normally unsaturated with choline since its Km ≅ 440 μM and since the choline concentration in plasma is only about 10—30 μM; see e. g. Ref. 12), this metabolic system of CVECs could be of significance in interpreting studies of ACh-precursor loading.

E. Peptidyl Dipeptidase (Angiotensin-Converting Enzyme; ACE)
Angiotensin-I and bradykinin are the two most important naturally-occurring substrates for ACE. Angiotensin-I is converted to angiotensin-II, which is a potent vasoconstrictor, and bradykinin is catabolized, thereby limiting its vasodilatatory activity. ACE is an important functional component of CVECs *in situ,* and appears to contribute to the regulation of blood pressure. ACE activity in rabbit brain microvessels is about 20 times that of whole brain homogenate (13).

F. 5′-Nucleotidase
5′-Nucleotidase activity has been associated with myelin sheaths, plasmalemmae of astrocytes and with microvessel walls (14, 15). Both 5′-nucleotidase and unspecific "neutral" phosphatases are present in CVECs. Thes enzymes can hydrolyze adenosine-5′-monophosphate to yield adenosine, which appears to be involved in regulating cerebral blood flow via its vasodilatatory action (15).

G. Glutamate-α-Decarboxylase (GAD)
The presence of GAD in CVECs is of interest not only because it catalyzes the formation of γ-aminobytric acid (GABA), but also because it could serve to inactivate glutamate, minute amounts of which could enter CVECs from the blood, especially if vascular lesions exist. A non-neuronal form of GAD, having higher activity than GAD of peripheral blood vessels, has been found in pial arteries of dog and rabbit (16, 17). A microvessel preparation of rat forebrain had GAD activity with higher specific activity than was observed for pial vessels; GAD activity of microvessels was about half of that of brain parenchyma (18). GAD activity of microvessels is probably associated with CVECs.

H. Other Enzymes Possibly Present in CVECs
Maruki *et al.* (19) found that cultured CVECs (second to fourth generation) prepared from an isolated microvessel fraction of rat brain can take up and degrade 5-HT and synthesize 5-HT from L-TRP. Although production of 5-HT was only 900 fmol/mg protein/overnight when the cells were incubated in the presence of L-TRP and pargyline (to block MAO activity), sister cultures exposed to TRP-free medium had much lower 5-HT content. Addition of p-chlorophenylalanine to the feeding medium caused a 60 % decrease in 5-HT content and a 30 % decrease in 5-hydroxyindoleacetic acid content of these cells in 17 hours.

Some quantitative evaluation seems necessary. If one assumes that these cultured CVECs, like brain tissue, contain about 110 mg protein/g wet weight and converts the value of 900 fmol/mg protein/overnight (estimated as 12 hrs), the calculated rate of synthesis is about 1.45 ng 5-HT/g cultured CVECs/hr. This value is only about 0.35—0.75 % of the *in vivo* 5-HT synthesis rate measured in rat whole brain after MAO blockade (20, 21). However, as cultured CVECs are not fully functional, an accurate rate of 5-HT synthesis cannot be obtained from these data. Cultured CVECs lose their characteristic markers, especially after several passages, and therefore the *in vivo* rate of 5-HT synthesis in CVECs might be greater than that which was measured by Maruki *et al.* (19).

γ-Glutamyltranspeptidase (γ-GT) is present in cultured CVECs and in brain microvessels (22). γ-GT and the γ-glutamyl cycle could participate in regulating the transfer of certain NT-precursor amino acids across the BBB. Tyrosine (TYR) hydroxylase seems to be localized primarily in perivascular sympathetic nerves in pial and parenchymal blood vessels and in non-sympathetic nerve terminals that contaminate microvessel fractions rather than in CVECs (3).

Endothelial Enzymes Involved in Handling Blood-Borne Neurotransmitters or Neurohormones

A. Monoamine Oxidase (MAO)
MAO is present in isolated brain microvessels and in cultured CVECs (4, 22). After NT-precursor amino acids are converted to their respective monoamine-NTs by the action of L-AAAD, these monoamines are likely to be rapidly deaminated by the MAO of CVECs and pericytes (4). MAO of CVECs could also prevent blood-borne NE and/or 5-HT from entering the interstitial fluid (4), and would be expected to inactivate monoamine-NTs or other biologically-active amines that might be formed in excess following precursor-loading. As monoamines are actively taken up into microvessel walls ablumenally (4, 23), endothelial MAO could also inactivate excess monoamine-NTs released by neuronal activity that might be present in the interstitial fluid (4, 24). Remarkable species differences exist in MAO activity of brain microvessels, rat microvessels having about 6—46 times greater activity than those of human (25). MAO-B appears to be the predominant form of the enzyme in brain microvessels of the rat and several other species (25).

B. Catechol-O-Methyltransferase (COMT)
Although COMT activity of the microvessel wall has not been definitely associated with CVECs, it could prevent circulating monoamines from entering the interstitial fluid and/or could inactivate excess monoamines that might be released to the interstitial fluid. COMT is probably present in the smooth muscle cells of pial and brain parenchymal blood vessels (4). Hardebo *et al.* (3) considered that some COMT of rat brain microvessels might be present in CVECs. Recent studies which have shown that O-methylation of isoproterenol in rabbit isolated aorta was decreased by 30 % after de-endothelialization (26) add some support to this contention.

C. Phenylethanolamine-N-Methyltransferase (PNMT)
PNMT is present in cultured CVECs and in cerebral microvessels (22). This enzyme, which is probably also present in other elements of the capillary wall, including nerve-endings, provides mechanisms for inactivating NE that might enter CVECs from the blood or interstitial fluid and for the synthesis of epinephrine which is involved in regulating cerebral vascular permeability and blood flow.

D. Butyrylcholinesterase (BuChE) and Acetylcholinesterase (AChE)
Blood-borne ACh probably does not enter CVECs due to plasma cholinesterases. However, an enzymatic mechanism exists for its inactivation. BuChE activity is largely specific to CVECs, and is present in capillaries of most brain regions as well as in cultured CVECs (5, 22). In contrast, AChE is present mainly in the basement membranes of capillaries of brain regions possessing neurones with high AChE activity (27). In light of the finding that

ChAT is present in CVECs of rat cerebral cortex (8—10), one might speculate that BuChE, and possibly AChE, regulate the action of ACh that is synthesized in CVECs.

E. GABA-α-Oxoglutarate Transaminase (GABA-T) and Succinic Semialdehyde Dehydrogenase (SSDH)

The activities of GABA-T and SSDH, like that of BuChE, are much higher in cerebral blood vessels than in blood vessels of other tissues (5, 28, 29). High GABA-T activity also exists in intraparenchymal microvessel preparations (18, 30). Endothelial and smooth muscle layers of the vessel wall contain more GABA-T activity than the adventitial layer. GABA-T could inactivate GABA that might enter CVECs from the blood.

F. Peptidyl Dipeptidase (Enkephalinase)

Isolated brain capillaries contain enkephalinase activity which could be due, at least in part, to the presence of peptidyl dipeptidase (31). As blood-borne enkephalins would not be expected to enter CVECs to any great extent, this enzyme might act only when CVECs are injured, or in degrading enkephalins of the interstitial fluid.

Enzymes of Pericytes, Vascular Smooth Muscle Cells, Glia and Mast Cells that are Involved in Metabolizing NT-Precursors or NTs

Some of the enzymes mentioned above (e. g., L-AAAD, MAO) are present in pericytes (4). Thus, pericytes might be considered to constitute another barrier to the entry of certain NT-precursors and NTs into the interstitial fluid. Blood vessels larger than capillaries possess COMT, MAO, and L-AAAD activities within their smooth muscle cells (4). These represent other barrier mechanisms to circulating monoamines and/or their precursors.

Glial end-feet, which ensheath brain capillaries, also play a role in impeding the passage of blood-borne substances into the interstitial fluid. Histamine-methyltransferase activity has been found in glial cell cultures (32). Hence, conversion of histamine to *tele*-methylhistamine might represent a second barrier to circulating histamine. COMT, MAO (both A and B forms), and 5′-nucleotidase have also been detected in astrocytes (33; see above). Monoamine-NTs that are synthesized in CVECs and released ablumenally, or those that are released from perivascular nerves or mast cells, could be taken up by glia and metabolized.

Mast cells are present in the CNS of the rat, cat, hamster, gerbil, and several other species, but appear to be absent from the brains of humans (34). Diencephalic areas of adult brains contain the greatest numbers of mast cells (34). The histamine content of mast cells is relatively high, accounting for 10—100 % of total brain histamine content (34). Histamine, presumably within mast cells, has been found in isolated brain capillaries (35).

Endothelium-Dependent Vasorelaxation

Endothelium-dependent vasorelaxation (EDV) involves not only cerebral blood vessels, but also blood vessels of peripheral tissues. In capillaries (where no smooth muscle layers are present) this mechanism probably alters membrane permeability. EDV is mediated either by PGI-2 (and/or other vasodilator prostaglandins) or by EDRF, the exact mediator depending upon the blood vessel, the species, and the agent that evokes EDV (36, 37).

Some of the enzymes of CVECs, discussed above, are involved in handling NTs or neurohormones (e. g., ACh, 5-HT, NE, bradykinin) that can activate EDV or increase BBB permeability. Substances that activate EDV do not have to pass through the lumenal CVEC membrane; they interact with receptors present on the surfaces of CVEC membranes, or in some (possibly exceptional) cases, with receptors of vascular smooth muscle cells which in turn elicit endothelial synthesis and release of a secondary mediator (see below). Many agents can elicit EDV (see Table 2). Some of these are active on cerebral blood vessels. Differences among blood vessels and among species exist with regard to substances that trigger EDV.

TABLE 2
Agents That Produce Endothelium-Dependent Vasorelaxation

Agent	Species and Blood Vessel
ACh	Dog carotid artery, rabbit aorta
5-HT (S-1)	Dog coronary artery, pig renal artery
NE (α-2)	Dog coronary artery
Histamine (H-1)	Rat aorta, rabbit aorta
Histamine (H-2)	Rabbit middle cerebral artery
Substance P	Rabbit aorta, dog femoral artery
Bradykinin	Human mesenteric bed, dog renal artery
Cholecystokinin	Rabbit aorta
Neurotensin	Dog carotid artery
Bombesin	Dog carotid artery
Arachidonate	Rabbit aorta, rat aorta
Thrombin	Dog femoral artery
ATP, ADP	Rabbit aorta

For original references, see 36 and 37

Bradykinin represents a good example of a substance whose actions are mediated by EDV. Bradykinin would not be expected to pass the BBB, and therefore blood-borne bradykinin probably does not directly affect CNS neurones. However, bradykinin relaxes all arteries tested from dogs and humans via release of EDRF, and causes relaxation by releasing vasodilator prostaglandins from endothelial and other cell types in arteries of cats and rabbits (38). Bradykinin is about ten times more potent than ACh in eliciting EDRF-mediated EDV in dog arteries; threshold concentration ≅ 0.1—1.0 nM (Ref. 36), and it might increase capillary permeability by contracting endothelial cells (39).

Certain receptors involved in EDV (e. g., histaminergic, α-adrenergic) have been detected in brain capillaries, whereas others (e. g., serotoninergic, ACh muscarinic) do not appear to be present (40). However, such findings do not apply to blood vessels that have smooth muscle layers. Larger blood vessels might possess some of these other receptors on their endothelial cells. In any case, EDV can be indirectly triggered. Endothelial cells do not appear to have muscarinic ACh receptors, so that ACh-elicited EDV probably depends upon activation of the muscarinic receptors of smooth muscle cells (41). In contrast, it seems clear that receptors for substance P are located on endothelial cells of blood vessels larger than capillaries, so that EDV, in this case, probably occurs by direct activation of endothelial cells (41).

CONCLUDING COMMENTS

Investigators conducting precursor-loading studies should perhaps consider the effects that blood-borne NT-precursors, and/or their associated NTs or other active metabolites, might exert on endothelial cell metabolism and function and on vasomotor activity. Of course, this concept has already been considered with respect to L-DOPA therapy (see above). Some amplification will be given here.

How does L-DOPA act as a therapy for Parkinson's disease? An answer to this question might be: L-DOPA is given in a high dose that elevates plasma L-DOPA concentration so that some L-DOPA enters the brain and serves as a precursor for DA synthesis in remaining "hyperactive" DA-ergic neurones; i. e., L-DOPA compensates for the striatal DA deficiency by being converted to DA in the remaining nigral neurones (e. g., 42). Supersensitivity of the intact, but denervated, postsynaptic striatal DA-receptors might also be considered to play a role (e. g., 43). Although such mechanisms could be involved, the presence of L-AAAD in CVECs and in other non-neuronal cells of the brain indicates that surviving nigrostriatal DA-ergic neurones certainly do not represent the only site for conversion of systemically-administered L-DOPA to DA. In therapy for Parkinson's disease, the massive doses of L-DOPA (e. g., 3—10 g/day) required to saturate the L-AAAD of CVECs (in the absence of L-AAAD-inhibitors) would lead to circulating L-DOPA concentrations up to 10 μM (44). A 100 mg/kg (i. p.) dose of L-DOPA in the rat (which would be equal to a 6-gram dose in a 60-kg human, assuming equivalence of oral and i. p. routes of administration) leads to DA formation at a rate of 150 ng DA/g tissue/minute in regions of the CNS (e. g., cerebellum) that are devoid of DA-ergic neurones, thus providing strong evidence that this DA is formed in microvessel walls (4).

What would occur when an animal is "loaded" with very high doses of TRP (the 5-HT-precursor)? As mentioned above, TRP could be handled to some extent by the L-AAAD and enzymes of protein synthesis of CVECs. Also, if some of the excess 5-HT present in blood after TRP-loading escapes removal by non-aggregating platelets and peripheral endothelial cells, it could possibly influence brain function by activating EDV or by acting on intracerebral perivascular innervation.

One might argue that the capacity of these CVEC enzyme systems represents only a very small percentage of the whole brain's enzymatic capacity since the cytoplasmic volume of the CVEC compartment likely accounts for less than 1 % of the total brain water (45, 46), and therefore that changes in metabolite concentrations that occur in the CVEC compartment would not significantly alter metabolite levels determined in whole brain homogenates, or that there exists only a negligible capacity of these systems to inactivate NT-precursors or NTs. However, based on data obtained with L-DOPA-loading (see above), massive doses of NT-precursors might elevate the concentrations of newly-formed endothelial NTs enough to significantly alter total brain NT concentration, and especially NT concentrations in certain brain regions.

It might also be noted that even if such a large amount of NT-precursor is administered so that it is present in blood in such excess that it does traverse the CVECs, escapes inactivation by other elements of the BBB, and does enter the brain's neuronal pool, other mechanisms (e. g., intraneuronal catabolism of unreleased or non-releasable NT) could prevent the NT-precursor or its associated NT(s) from exerting neurophysiological effects (see e. g., 47—49).

Interpretations of experiments conducted with isolated brain microvessels or cultured CVECs must be made with caution since the functions of these preparations do not accurately represent the *in vivo* function of the BBB. Isolated capillaries have low ATP content, inadequate ion regulation and low internal potassium concentration, as compared with capillaries *in vivo* (e. g., 1, 50), and cultured CVECs have increased membrane permeability and altered enzymatic activities, especially after repeated passages (e. g., 51). However, even with these serious limitations, it appears evident from *in vitro* studies on CVECs that the *in vivo* BBB likely contains efficient enzyme systems for inactivating excess circulating NT-precursors, NTs, and neurohormones. Thus, further studies should perhaps be aimed at determining the effects of precursor-loading on NT synthesis and catabolism in CVECs and on EDV. Only after such information becomes available could one attempt to explain the mechanisms that underlie any changes in CNS neuronal activity or any behavioural or neurologic effects that might follow precursor-loading with L-DOPA or with any other substance.

REFERENCES

1) PARDRIDGE, W. M. (1983). Brain Metabolism: A perspective from the blood-brain barrier. *Physiol. Rev.* **63:** 1481—1535.

2) BERTLER, A., FALCK, B., OWMAN, C., ROSENGREN, E. (1966). The localization of monoaminergic blood-brain barrier mechanisms. *Pharmacol. Rev.* **18:** 369—385.

3) HARDEBO, J. E., EMSON, P. C., FALCK, B., OWMAN, C., ROSENGREN, E. (1980). Enzymes related to monoamine transmitter metabolism in brain microvessels. *J. Neurochem.* **35:** 1388—1393.

4) HARDEBO, J. E., OWMAN, C. (1980). Barrier mechanisms for neurotransmitter monoamines and their precursors at the blood-brain interface. *Ann. Neurol.* **8:** 1—11.

5) JOÓ, F. (1985). The blood-brain barrier in vitro: Ten years of research on microvessels isolated from the brain. *Neurochem. Int.* **7:** 1—25.

6) AUDUS, K. L., BORCHARDT, R. T.(1986). Characteristics of the large neutral amino acid transport system of bovine brain microvessel endothelial cell monolayers. *J. Neurochem.* **47:** 484—488.

7) PARDRIDGE, W. M., OLENDORF, W. H. (1975). Kinetic analysis of blood-brain barrier transport of amino acids. *Biochim. Biophys. Acta* **401:** 128—136.

8) PARNAVELAS, J. G., KELLY, W., BURNSTOCK, G. (1985). Ultrastructural localization of choline acetyltransferase in vascular endothelial cells in rat brain. *Nature* **316:** 724—725.

9) SANTOS-BENITO, F. F., GONZALEZ, J. L. (1985). Decrease of choline acetyltransferase activity of rat cortex capillaries with aging. *J. Neurochem.* **45:** 633—636.

10) GONZALEZ, J. L., SANTOS-BENITO, F. F. (1987). Synthesis of acetylcholine by endothelial cells isolated from rat brain cortex capillaries. *Brain Res.* **412:** 148—150.

11) JACKSON, I. V., DAVIS, L. G., COHEN, R. K., NUTTALL, E. A. (1981). Lecithin administration in tardive dyskinesia: Clinical and biomedical correlates. *Biol. Psychiat.* **16:** 85—90.

12) ZEISEL, S. H. (1986). Dietary influences on neurotransmission. *Adv. Pediatr.* **33:** 23—48.

13) BRECHER, P., TERCYAK, A., GAVRAS, H., CHOBANIAN, A. V. (1978). Peptidyl dipeptidase in rabbit microvessels. *Biochim. Biophys. Acta* **526:** 537—546.

14) KREUTZBERG, G. W., BARRON, K. D., SCHUBERT, P. (1987). Cytochemical localization of 5′-nucleotidase in glial plasma membranes. *Brain Res.* **158:** 247—257.

15) VORBRODT, A. W., LOSSINSKY, A. S., WISNIEWSKI, H. M. (1983). Enzyme cytochemistry of blood-brain barrier (BBB) disturbances. *Acta Neuropathol. (Berl.)* **Suppl. VIII:** 43—57.

16) KURIYAMA, K., HABER, B., ROBERTS, E. (1970). Occurrence of a new L-glutamic acid decarboxylase in several blood vessels of the rabbit. *Brain Res.* **23:** 121—123.

17) MIRZOYAN, S. A., KAZARAN, V. A., AKOPYAN, V. P. (1970). The glutamic decarboxylase activity in blood vessels of the brain. *Dokl. Akad. Nauk SSSR* **190:** 1241—1243.

18) DJURIČIĆ, B. M., ROGAČ, LJ., SPATZ, M., RAKIĆ, LJ. M., MRSULJA, B. B. 1978). Brain microvessels. I. Enzymatic activities. *Adv. Neurol.* **20:** 197—205.

19) MARUKI, C., SPATZ, M., UEKI, Y., NAGATSU, I., BEMBRY, J. (1984). Cerebrovascular endothelial cell culture: Metabolism and synthesis of 5-hydroxytryptamine. *J. Neurochem.* **43:** 316—319.

20) FERNSTROM, J. D., HIRSCH, M. J. (1977). Brain serotonin synthesis: Reduction in corn-malnourished rats. *J. Neurochem.* **28:** 877—879.

21) GAL, E. M., SHERMAN, A. D. (1980). L-kynurenine: Its synthesis and possible regulatory function in brain. *Neurochem. Res.* **5:** 223—239.

22) SPATZ, M., KARNUSHINA, I., NAGATSU, I., BEMBRY, J. (1984). Endothelial cell cultures: A new model for the study of cerebral vascular endothelium. In: *Recent Progress in the Study and Therapy of Brain Edema* (Go, K. G., Baethmann, A., eds.), pp. 373—380, Plenum Press, New York.

23) HAMBERGER, B. (1967). Reserpine-resistant uptake of catecholamines in isolated tissues of the rat. A histochemical study. *Acta Physiol. Scand. (Suppl.)* **295:** 1—56.

24) OLENDORF, W. H. (1984). The blood-brain barrier. In: *Handbook of Neurochemistry* (Lajtha, A., ed.), **Vol. 7,** pp. 485—499. Plenum Press, New York.

25) KALARIA, R. N., HARIK, S. I. (1987). Blood-brain barrier monoamine oxidase: Enzyme characterization in cerebral microvessels and other tissues from six mammalian species, including human. *J. Neurochem.* **49:** 856—864.

26) HEAD, R. J., PANEK, R., REID, J., STITZEL, R. E., BARONE, S. (1986). O-Methylation of isoproterenol by the endothelium of the rabbit thoracic aorta. *Blood Vess.* **23:** 279—287.

27) KREUTZBERG, G. W., TOTH, L. (1983). Enzyme cytochemistry of the cerebral microvessel wall. *Acta Neuropathol.* **8** (Suppl.): 35—41.

28) VAN GELDER, N. M. (1968). A possible enzyme barrier for γ-aminobutyric acid in the central nervous system. *Prog. Brain Res.* **29:** 259—268.

29) WAKSMAN, A., RUBINSTEIN, M. K., KURIYAMA, K., ROBERTS, E. (1968). Localization of γ-aminobutyric-αoxoglutaric acid transaminase in mouse brain. *J. Neurochem.* **15:** 351—357.

30) KRAUSE, D. N., ROBERTS, E., WONG, E., DEGENER, P., ROGERS, K. (1980). Specific cerebrovascular localization of GABA-related receptors and enzymes. *Brain Res. Bull* **5 (Suppl. 2):** 173—177.

31) PARDRIDGE, W. M., MIETUS, L. J. (1981). Enkephalin and blood-brain barrier: Studies of binding and degradation in isolated brain microvessels. *Endocrinology* **109:** 1138—1143.

32) GARBARG, M., BAUDRY, M., BRENDA, P., SCHWARTZ, J. C. (1975). *Brain Res.* **83:** 583—591 (Cited by Hough and Green, 1984).

33) KIMELBERG, H. K. (1986). Occurrence and functional significance of serotonin and catecholamine uptake by astrocytes. *Biochem. Pharmacol.* **35:** 2273—2281.

34) HOUGH, L. B., GREEN, J. P. (1979). Histamine and its receptors in the nervous system. In: *Handbook of Neurochemistry* (Lajtha, A. ed) **Vol. 6,** pp. 145—211. Plenum Press, New York.

35) JARROT, B., HJELLE, J. T., SPECTOR, S. (1984). Association of histamine with cerebral microvessels in regions of bovine brain. *Brain Res.* **168:** 323—330.

36) FURCHGOTT, R. F. (1984). Role of endothelium in responese of vascular smooth muscle to drugs. *Annu. Rev. Pharmacol. Toxicol.* **24:** 175—197.

37) DeFEUDIS, F. V. (1987). Endothelium dependent vasorelaxion — A new basis for developing cardiovascular drugs. *Drugs of Today* **23:** in press.

38) FURCHGOTT, R. F., MARTIN, W. (1985). Interactions of endothelial cells and smooth muscle cells of arteries. *Chest* **88** (Suppl.): 210S—213S.

39) REGOLI, D. (1984). Neurohumoral regulation of precapillary vessels: The kallikrein-kinin system. *J. Cardiovasc. Pharmacol.* **6:** S401—S412.

40) BETZ, A. L., GOLDSTEIN, G. W. (1984). Brain capillaries: Structure and function. In: *Handbook of Neurochemistry* (Lajtha, A., ed.), **Vol. 7,** pp. 465—484, Plenum Press, New York.

41) SUMMERS, R. J., MOLENAAR, P., STEPHENSON, J. A. (1987). Autoradiographic localization of receptors in the cardiovascular system. *Trends Pharmacol. Sci.* **8:** 272—276.

42) AGID, Y., JAVOY, F., GLOWINSKI, J. (1973). Hyperactivity of remaining dopaminergic neurons after destruction of the nigrostriatal dopaminergic system in the rat. *Nature* **245:** 150—151.

43) UNGERSTEDT, U. (1971). Postsynaptic supersensitivity after 6-hydroxy-dopamine induced degeneration of the nigrostriatal dopamine system. *Acta Physiol. Scand.* **367** (Suppl.): 69—93.

44) RINNE, U. K., SONNINEN, V., SIIRTOLA, T. (1973). Plasma concentration of levodopa in patients with Parkinson's disease. Response to administration of levodopa alone or combined with a decarboxylase inhibitor and clinical correlations. *Eur. Neurol.* **10:** 301—310.

45) PEROUTKA, S. J., MOSKOWITZ, M. A., REINHARD J. F., SNYDER, S. H. (1980). Neurotransmitter receptor binding in bovine cerebral microvessels. *Science* **208:** 610—612.

46) PARDRIDGE, W. M. (1986). Discussion. In: *Amino Acids in Health and Disease: New Perspectives* (Kaufman, S., ed.), p. 136. Alan R. Liss, New York.

47) COMMISSIONG, J. (1985). Monoamine metabolites: Their relationship to monoaminergic neuronal activity. *Biochem. Pharmacol.* **34:** 1127—1131.

48) KUHN, D. M., WOLF, W. A., YOUDIM, M. B. A. (1986). Serotonin neurochemistry revisited: A new look at some old axioms. *Neurochem. Int.* **8:** 141—154.

49) DeFEUDIS, F. V. (1987). The brain is protected from nutrient excess. *Life Sci.* **40:** 1—9.

50) GOLDSTEIN, G. W., BETZ, A. L., BOWMAN, P. D. (1986). In vitro studies of the blood-brain barrier using isolated brain capillaries and cultured endothelial cells. *Ann. N. Y. Acad. Sci.* **481:** 202—213.

51) GOLDSTEIN, G. W. (1986). In vitro studies of glial-endothelial cell interactions. *Fourth Colloq. Biol. Sci., Blood Brain Transfer, New York Acad. Sci.,* Nov. 3, 1986, **Abstract P-4.**

DYNAMICS OF THE CELLULAR AND EXTRACELLULAR COMPARTMENT OF BRAIN AMINO ACIDS — VISIONS AND REALITY OF THE DIALYSIS APPROACH

A. Hamberger, I. Jacobson, A. Lehmann, M. Sandberg

Institute of Neurobiology
University of Göteborg
P. O. Box 33031
S-40033 GÖTEBORG
Sweden

INTRODUCTION

Dialysis perfusion has in a couple of years attained the position of a most promising approach to the biochemistry of cell to cell signalling in the brain. Still, sampling of fluid which has equilibrated with the tissue is an old tradition in neurochemical work. The dialysis concept advances the investigator one step further away from squirting Ringer's medium on the brain. However, the *in vivo* monitoring of compounds in the extracellular space necessitates new considerations. Factors such as turnover and drainage of compounds form the extracellular fluid, sources of recruitment to the latter and thereby possible artifactual results, diffusion coefficients and changes in the extracellular volume become critical.

Perfusion Techniques

By perfusing the brain surface with artificial CSF via a small cup inserted in the skull bone, MacIntosh and Obarin (1) monitored acetylcholine release from the cerebral cortex. Some years later, Bhattacharya and Feldeberg (2) obtained acetylcholine by perfusion of the cat ventricular system. In order to monitor a distinct brain region, Gaddum (3) introduced the push-pull cannula which is implanted into the brain and consists of 2 concentric steel tubes. Medium is perfused via a small diameter inner tube and back through an outer tube after equilibrating with the surrounding tissue. Delgado and coworkers (4) modified this design by covering the tip with a semipermeable membrane to obtain a "dialytrode" which eliminates the erosive effects of the fluid on the tissue (5). The separation of the tissue from the medium with a membrane also excludes, if so desired, macro-molecules from the dialysis-perfusate which in turn, reduces the need for clean-up steps in the analysis procedure. Further refinements were introduced by Ungerstedt and Pycock (6), who introduced small diameter semipermeable tubings.

NATO ASI Series, Vol. H20
Amino Acid Availability and Brain Function in Health and Disease. Edited by G. Huether

Microdialysis

The dialysis tubing can be taken from almost any of the clinically used dialysers in which case the supply for a life time costs less than 50 USD. In our laboratory, the Cuprophan type B4AH (o. d. 0.3 mm) has been used almost exclusively. A 5—10 mm segment is glued between pieces of polyethylene tubing which serve for transport of the perfusion medium (7). This technique is suitable for the hippocampus and the striatum and has the advantage that the tubings follow the movements of the brain and that the dialysis membrane "integrates" extremely well with the tissue. The system is suitable for experiments lasting over several days (7). However, it requires an aligning system for a guide to pull the tubes through the brain via cannulated screws in holes drilled on each side of the

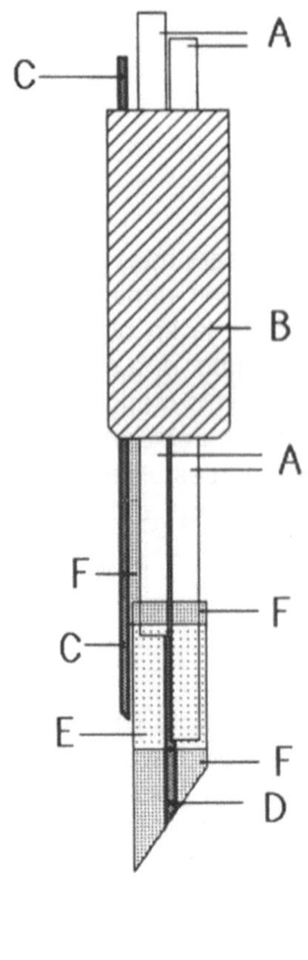

FIGURE 1

The dialysis-probe (8) consists of plastic coated glass capillaries as inlet and outlet (A), a supporting steel-cannula (B), wolfram-electrodes for recording (C) and supporting purposes and the dialysis-membrane (E). The assembly is fixed with cyano-acrylate based glue (F). The device is drawn to scale horizontally but not vertically. The diameter of the dialysis tubing is approximately 0.3 mm, but swells somewhat in wet conditions.

skull. Electrode-type dialytrodes which are introduced with stereotaxic instrumentation have by now been constructed in a number of laboratories and are as well commercially available (Carnegie Medicin, Stockholm, Sweden). The latest type developed in our laboratory consists of a 3—6 mm piece of dialysis tubing which is closed in its distal end (Fig. 1, (8)). In- and outlet of the perfusion medium takes place via very thin plastic coated glass capillaries (VS 170/100, Scientific Glass Engineering), both running parallel and inside the dialysis tube with their openings 0.5—3 mm apart depending on the structure studied. The glass capillaries offer the advantage of a small dead volume. Electrodes are placed outside and inside the dialysis-tubing for recording and as guides.

Implantation

As judged from the biochemical analysis of the perfusate, there is a stabilization of the tissue surrounding the probe 1—2 h after implantation (9, 10). The relative concentrations of amino acids in the perfusate (7, 11, 12), and the characteristics of evoked field potentials measured in parallel (12, 13) indicate a fairly good tissue preservation. Radiochemical tracer studies speak against severe damage to the blood-brain barrier (BBB) some hours after probe implantation (14, 15). However, implantation causes early, but within 24 h reversible disturbances in blood-flow and glucose metabolism (16). Ultrastructurally, an approximately 50 μm wide zone of intra- and extracellular oedema is seen around the probe during the first two days after implantation (17, 7).

Amino Acid Pools

The concentration of amino acids in the microenvironment of the CNS neurons is controlled by "barriers" against the blood and against the interior of the cells. The concentration ratios for most of the neuroactive amino acids in the extracellular fluid (or the CSF): blood: intracellular fluid are approximately 1:10:1000. Even a short perturbance of these gradients, as for example in ischemia, leads to functional chaos and ultimately neuronal death. Less than one per cent of total amino acids in the brain are present in the extracellular fluid and the amount in the CSF is of the same magnitude. Perfusion of dialysis tubings implanted in the brain gives access to the chemical composition of the extracellular compartment which, ideally can be estimated under controlled experimental conditions.

Diffusion

The working principle of dialysis probes is that solutes in the environment enter the perfusing medium due to concentration driving forces. Since the semipermeable membrane induces a diffusive resistance, it attenuates the mass transfer in the system to a certain degree. However, sampling and sensitivity of analysis give a time resolution in the order of minutes, thereby making these effects less worrying (18, 19, 20).

A simple diffusion model provides the basis for calculation of extracellular concentrations (18, 12). Mathematically, diffusion in the extracellular space (21), is determined by the formula $\underline{a}D^*dC/dX$ where $\underline{a}$ is the extracellular volume fraction, D^* the apparent diffusion coefficient and dC/dX the concentration gradient. Thus, the diffusion coefficient in the living brain with a 15 % extracellular space is only 30—40 % of that in a calibration solution in a test tube. The extracellular oedema after implantation of a probe may contribute to the small difference between estimates of extracellular concentrations of amino acids calculated with recovery factors (i. e., concentration ratio perfusate/medium outside the dialysis tube) from in vitro calibrations and estimates from computerized analysis based on the diffusional model (18). However, in dialysis experiments extracellular Ca^{2+} was calculated to 0.7 mM when recovery factors from in vitro calibrations were used (22), contrasting with 1.2—1.5 mM as measured with ion-selective microelectrodes (23, 24). Van Wylen and coworkers (10), using *in vitro* recovery factors, (10) estimated the extracellular concentrations of purine catabolites and found a 50 percent lower value at high perfusion rates (2.0 μl/min) than at low flow rates (0.1 μl/min). Further complications arise for the calculation of a stimulus-coupled response a dialysis perfusion experiment. It is known that for example epilepsy, hypoxia and spreading depression reduce the volume of the extracellular space considerably (25). If nothing else changes, the concentration of extracellular compounds consequently decreases in the dialysate. On the other hand, a stimulus coupled release from an intracellular compartment will be overestimated with dialysis perfusion, since the dilution will be smaller in a reduced extracellular fluid compartment (25). The effects of reduced diffusion and attenuated dilution may, theoretically,

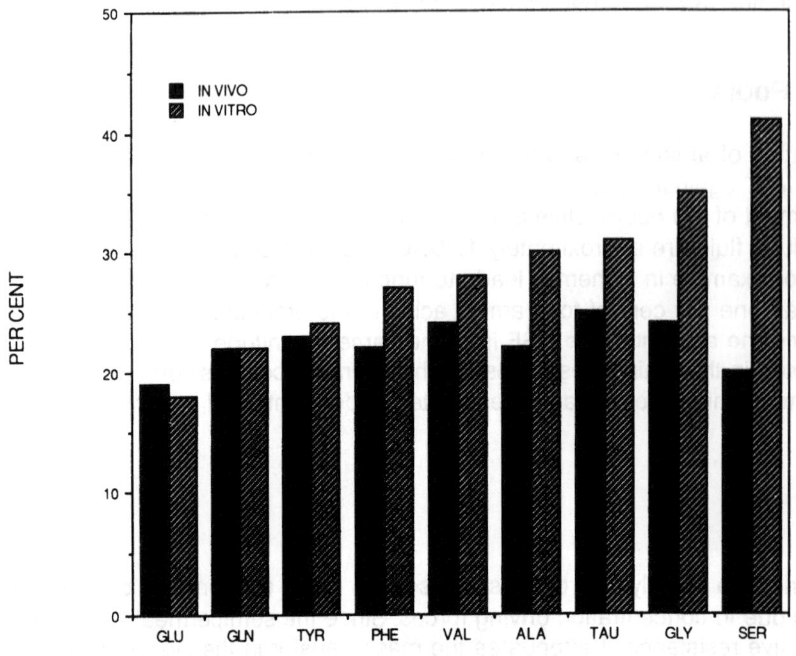

FIGURE 2
In vivo and in vitro "recovery" for amino acids. The in vivo value is the concentration in a medium which has passed once through a dialysis tubing in per cent of that which has recirculated 9—10 times, i. e., being in equilibrium with the surroundings. The in vitro value is the concentration in the perfusate in per cent of that in the outer medium.

influence equally and thus completely obscure an effect. Thus, extracellular markers (26, 27) or simultaneous measurement of several extracellular compounds is necessary. Since the diffusion which is relevant to dialysis experiments takes place mainly in the extracellular space, it is worth noticing that different substances have their own characteristics: when followed into the tissue by autoradiography after pulsing via the tubing, it is seen that the radius for glucose diffusion is larger than that for leucine (7).

Draining of Pools

In their evaluation of extracellular glucose concentration in the human abdominal subcutaneous region, Lönnroth and coworkers (28) stress the necessity of proper in vivo calibration of the dialysis probe. Recovery of glucose is smaller in vivo than in vitro. Furthermore, the recovery of glucose is a function of the concentration gradient across the dialysis membrane. The authors solved the problem by adding glucose to the medium.

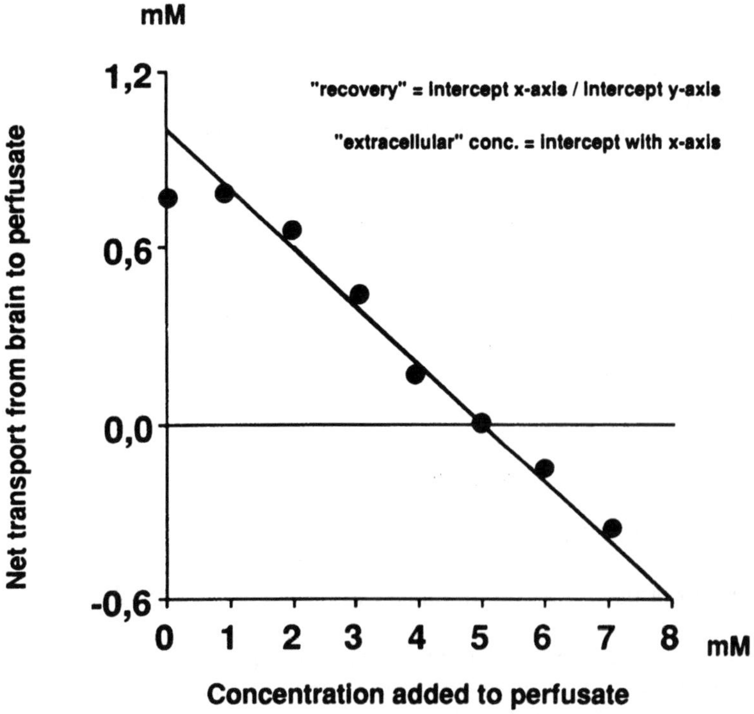

FIGURE 3
"In vivo recovery" and to the "true" concentration of glucose outside the dialysis tubing (28). Each point represents one concentration (0—7 mM) in the medium before the passage. No net transport occurs when the concentration in the medium is the same as in the surrounding tissue. A glucose-free medium is unsuitable for calculation, since "recovery" is too low possibly due to draining.

When the concentration gradient is kept small, changes in extracellular concentrations can be measured properly. The results may be due to the limited availability of glucose as compared to amino acids, as the intracellular glucose pool is relatively small. However, recent experiments in our laboratory suggest similar limitations, i. e., an actual drainage of the extracellular compartment, for some amino acids such as glycine and serine (Figs. 2 and 3). Fluid sampling, by dialysis perfusion thus induces a zone in which the concentration of some soluble compounds will be artifactually lowered. The situation may be critical in long time experiments.

Characteristics of the Extracellular Compartment

Since the relative amino acid concentrations in the dialysis-perfusates differ considerably from those inside the cells (7, 11, 12, 29), the extracellular compartment as measured with dialysis-perfusion is able to retain characteristics distinct from those of the intracellular pool (11, 12). There is good correlation between the magnitude of "basal" amino-acid efflux *in vivo* and *in vitro* in tissue slice experiments (30). Furthermore, the calculated concentrations for the neuroactive amino-acids are in the range of the K_m values for high affinity uptake from in vitro studies (31, 32, 33). Indeed, in dialysis-perfusion studies, the extracellular concentration of asp, glu and GABA increase upon addition of uptake inhibitors (30, 34, 35, 12). The relative importance of various cell compartments for the steady state concentration in the extracellular fluid is difficult to assess even under normal conditions. The extracellular space is largely bounded by glial membranes (36), which are likely to play an important role. However, compounds which affect depolarization-induced glu release (L-proline, 2-chloroadenosine, (37, 38)) reduce extracellular asp and glu in the striatum (39). General anesthetics and tetrodotoxin also depress the glu and GABA concentrations in push-pull and cortical cup experiments (40, 41). The extracellular concentration of the acidic amino acids is reduced in the striatum after decortication but not after a kainic acid induced lesion (42, 39). Such changes are not observed with the push-pull cannula (40) a discrepancy which is not easily understood. Consequently, there are indications that the composition of the extracellular fluid, at least as measured with the dialysis probe, is influenced by presynaptic pools and/or result from neuronal activity.

Dialysis Perfusion in Neurochemistry

Technically, dialysis perfusion has been applied to problems which can be categorized as follows:

a) Local administration i. e. via the dialysis tubing of drugs or other compounds which activate, depolarize or otherwise affect the surrounding tissue.
b) Induction of systemic of behavioural changes in the animal and recording of local CNS effects.
c) Physiological or electrical stimulation of nerve tracts which terminate in the region of the dialysis probe.
d) Monitoring of entrance of substances to the CNS from non CNS pools such as blood or CSF.
e) Clinical studies.

In the first type of studies (a), the system is used as a tissue slice perfusion experiment. In addition of elevated K^+ concentrations, veratrine, ouabain, transmitters and transmitter analogues, uptake inhibitors etc. have been employed (42, 43, 34). The results agree with slice work in some aspects but not in other. As a rule, the effect on the transmitter candidates, glu, asp and GABA are much smaller in dialysis experiments. Compounds which have gained interest in view of their marked increase in concentration in response to various stimuli delivered via the perfusion medium are taurine, glutamine, phosphoethanolamine and ethanolamine. Also alanine can be included in this group during certain conditions.

The second group (b) of studies includes the work on ischemia, epilepsy, hypoglycemia, hepatic coma etc. Most dramatic elevations of transmitter amino acids are seen in ischemia. The changes are rapidly reversible upon recirculation of blood (14). Taurine and phosphoethanolamine are also affected in ischemia, however, their peak concentration is usually delayed (31). When hypoglycemia has gone as far as isoelectricity of the EEG, transmitter amino acids appear at high concentrations in the dialysate (8, 44). However, the pattern differs from that in ischemia and peak levels are not so high. Contrasting to ischemia and hypoglycemia, epilepsy has small effects on amino acids in the dialysate. There is evidence that the increase in taurine in epilepsy occurs only after local application of a convulsive agent and is not part of the picture otherwise (45, 46). Alanine and phosphoethanolamine appear to be the only amino acids which change dramatically in epilepsy as measured with the dialysis technique.

The influence of experimental liver damage on CNS amino acids has been investigated with dialysis-perfusion (47, 48). The type of changes are less acute than for example in ischemia and there is probably not so much more to be learned with the dialysis approach as compared to with a CSF-drainage. The extracellular pool in ischemia shows an increase mainly because of a flow of amino acids from the interior of cells while in liver damage, on the other hand, high extracellular glutamine reflects high intracellular glutamine.

The potential of the dialysis probe as a tool to disclose the transmitter in various parts of the CNS (c) has not stood up to expectations. In spite of a large number of trials in different regions carried out in our laboratory, it has been very difficult to obtain a clear stimulus evoked response upcn tract stimulation. A possible explanation is that in the more intact *in vivo* system, the transmitter may not travel far from the synapse to obtain a successful recording with a dialysis probe. Results have been more promising with other transmitters and the area is apparently in rapid development.

The study of entrance from non-CNS pools (d) is also very much at its beginning. Data from several laboratories indicate that the BBB is not severely damaged, not even acutely after probe implantation. It is for example possible to monitor the BBB continuously by sampling from the dialysate various exogenous blood-born markers, i. e., ^{125}I-albumin (49).

Finally, (e) dialysis probes have recently entered into clinical research and may be used routinely within the next few years. Presently, the technique is employed in neurosurgery, general surgery and internal medicine in university hospitals in Sweden.

REFERENCES

1) MacINTOSH, F. C., OBARIN, P. E. (1953). Release of acetylcholine from intact cerebral cortex. *Proc. XIX Int. Cong. Physiol.*, pp. 580—581.

2) BHATTACHARYA, B. K., FELDBERG, W. (1958). Perfusion of cerebral ventricles: assay of pharmacologically active substances in the effluent from the cisterna and the aqueduct. *Br. J. Pharmacol.* **13:** 163—174.

3) GADDUM, J. H. (1961). Push-pull cannulae. *J. Physiol. (Lond.)* **155:** 1 P.

4) DELGADO, J. M. R., DeFEUDIS, F. V., ROTH, R. H., RYUGO, D. K., MITRUKA, B. M. (1972). Dialytrode for longterm intracerebral perfusion in awake monkeys. *Arch. Int. de Pharmacodyn.* **198:** 7—21.

5) NIEOULLON, A., CHERAMY, A., LEVIEL, V., GLOWINSKY, J. (1977). An adaptation of the push-pull cannula method to study the *in vivo* release of 3H-dopamine synthesized from 3H-tyrosine in the rat caudate nucleus: Effects of various tyrosine physical and pharmacological treatments. *J. Neurochem.* **28:** 819—828.

6) UNGERSTEDT, U., PYCOCK, C. (1974). Functional correlates of dopamine neurotransmission. *Bull. Schweiz. Akad. Med. Wiss.* **1278:** 1—5.

7) HAMBERGER, A., NYSTRÖM, B., BERTHOLD, C.-H., KARLSSON, B., LEHMANN, A. (1983). Extracellular GABA glutamate and glutamine *in vivo* — perfusion-dialysis of the rabbit hippocampus. In: *Glutamine, Glutamate and GABA in the Central Nervous System* (Hertz, L., Kvamme, E., McGeer, E. G., Schousboe, A., eds.), pp. 473—492. Alan R. Liss, New York.

8) SANDBERG, M., BUTCHER, S. B., HAGBERG, H. (1986). Extracellular overflow of neuroactive amino acids during severe insulin-induced hypoglycemia: In vivo dialysis of the rat hippocampus. *J. of Neurochem.* **47:** 178—184.

9) IMPERATO, A., DiCHIARA, G. (1984). Trans-striatal dialysis coupled to reverse phase high performance liquid chromatography with electrochemical detection: A new method for the study of the in vivo release of endogenous dopamine and metabolites. *J. Neurosci.* **4:** 966—977.

10) VAN WYLEN, D. G. L., PARK, T. S., RUBIO, R., BERNE, R. M. (1986). Increases in cerebral interstitial fluid adenosine concentration during hypoxia local potassium infusion and ischemia. *J. Cerebr. Blood Flow Metab.* **6:** 522—528.

11) JACOBSON, I. HAMBERGER, A. (1984). Veratridine-induced release in vivo and in vitro of amino acids in rabbit olfactory bulb. *Brain Res.* **299:** 145—155.

12) LERMA, J., HERRANZ, A. S., HERRERAS, O., ABRAIRA, V., MARTIN DEL RIO (1986). In vivo determination of extracellular concentration of amino acids in the rat hippocampus. A method based on brain dialysis and computerized analysis. *Brain Res.* **384:** 145—155.

13) SANDBERG, M., LINDSTRÖM, S. (1983). Amino acids in the dorsal lateral geniculate nucleus of the cat-collection *in vivo. J. Neurosci. Meth.* **9:** 65—74.

14) BENVENISTE, H., DREJER, J., SCHOUSBOE, A., DIEMER, N. H. (1984). Elevation of the extracellular concentrations of glutamate and aspartate in rat hippocampus during transient cerebral ischemia monitored by intracerebral microdialysis. *J. Neurochem.* **43:** 1369—1374.

15) TOSSMAN, U., UNGERSTEDT, U. (1986). Microdialysis in the study of extracellular levels of amino acids in the rat brain. *Acta Physiol. Scand.* **128:** 9—14.

16) BENVENISTE, H., DREJER, J., SCHOUSBOE, A., DIEMER, N. H. (1987). Regional cerebral glucose phosphorylation and blood flow after insertion of a microdialysis fiber through the dorsal hippocampus in the rat. *J. Neurochem.* **49:** 729—734.

17) BENVENISTE, H., DIEMER, N. H. (1987). Cellular reactions to implantation of a microdialysis tube in the rat hippocampus. *Acta Neuropathol.* **74:** 234—238.

18) JACOBSON, I., SANDBERG, M., HAMBERGER, A. (1985). Mass transfer in brain dialysis devices — a new method for the estimation of extracellular amino acids concentration. *J. Neurosci. Meth.* **15:** 263—268.

19) KORF, J., VENEMA, K. (1985). Amino acids in rat striatal dialysates: Methodological aspects and changes after electroconvulsive shock. *J. Neurochem.* **45:** 1341—1348.

20) SHARP, T., MAIDMENT, N. T., BRAZELL, M. P., ZETTERSTRÖM, T., UNGERSTEDT, U., BENNET, G. W., MARSDEN, C. A. (1985). Changes in monoamine metabolites measured by simultaneous in vivo differential pulse voltammetry and intracerebral dialysis. *J. Neurosci.* **12:** 1213—1221.

21) NICHOLSON, C., PHILLIPS, J. M. (1981). Ion diffusion modified by tortuosity and volume fraction in the extracellular microenvironment of the rat cerebellum. *J. Physiol. (Lond.)* **321:** 225—257.

22) LAZAREWICZ, J. W., HAGBERG, H., HAMBERGER, A. (1986). Extracellular calcium in the hippocampus of unanesthetized rabbits monitored with dialysis-perfusion. *J. Neurosci. Meth.* **15:** 317—328.

23) HEINEMANN, U., PUMAIN, R. (1980). Extracellular calcium activity changes in cat sensimotor cortex induced by iontophoretic application of amino acids. *Exp. Brain Res.* **40:** 247.

24) KRNJEVIC, K., MORRIS, M. E., REIFFENSTEIN, R. R., ROPERT, N. (1982). Depth distribution and mechanism of changes in extracellular K and Ca concentrations in the hippocampus. *Can. J. Physiol. Pharmacol.* **60:** 1658—1671.

25) LUX, H. D., HEINEMANN, U., DIETZEL, I. (1986). Ionic changes and alterations in the size of the extracellular space during epileptic activity. *Adv. Neurol.* **44:** 619—639.

26) CHASE, T. N., KOPIN, I. J. (1968). Stimulus-induced release of substances from olfactory bulb using the push-pull cannula. *Nature* **217:** 466—467.

27) DOLPHIN, A. C., ERRINGTON, M. L., BLISS, T. V. P. (1982). Longterm potentiation of the perforant path in vivo is associated with increased glutamate release. *Nature* **297:** 496—498.

28) LÖNNROTH, P., JANSSON, P. A., SMITH, U. (1987). A microdialysis method allowing characterization of the intercellular water space in man. *Am. J. Physiol.* (in press).

29) TOSSMAN, U., JONSSON, G., UNGERSTEDT, U. (1986). Regional distribution and extracellular levels of amino acids in rat central nervous system. *Acta Physiol. Scand.* **127:** 533—545.

30) JACOBSSON, I., HAMBERGER, A. (1985). Kainic acid-induced changes of extracellular amino acids, evoked potentials and EEG in the rabbit olfactory bulb. *Brain Res.* **348:** 289—296.

31) HAGBERG, H., LEHMANN, A., SANDBERG, M., NYSTRÖM, B., JACOBSON, I., HAMBERGER, A. (1985). Ischemia-induced shift of inhibitory and excitatory amino acids from intra- to extracellular compartments. *J. Cereb. Blood Flow and Metab.* **5.** 413—419.

32) HERTZ, L. (1979). Functional interactions between neurons and astrocytes. I. Turnover and metabolism of putative amino acid transmitters. *Prog. Neurobiol.* **13:** 277—323.

33) WOOD, J. D., SIDHU, H. S. (1986). Uptake of γ-aminobutyric acid by brain tissue preparations: a reevaluation. *J. Neurochem.* **46:** 739—744.

34) LEHMANN, A., ISACSSON, H., HAMBERGER, A. (1983). Effects of *in vivo* administration of kainic acid on the extracellular amino acid pool inthe rabbit hippocampus. *J. Neurochem.* **40:** 1314—1320.

35) LEHMANN, A., HAMBERGER, A. (1984). Dihydrokainic acid affects extracellular taurine and phosphoethanolamine levels in the hippocampus. *Neurosc. Lett.* **38.** 67—72.

36) WOLFF, J. R. (1979). Quantitative aspects of astroglia. In: *Proc. VI Int. Congr. Neuropath*, Macon & Cie, pp. 327—336.

37) DOLPHIN, A. C., ARCHER, E. R. (1983). An adenosine agonist inhibits and a cyclic AMP analogue enhances the release of glutamate but not GABA from slices of rat dentate gyrus. *Neurosci. Lett.* **43:** 49—54.

38) KELLER, E., DAVIS, J. L., TACHIKI, K. H., CUMMINS, J. T., BAXTER, C. F. (1981). L-Proline inhibition of glutamate release. *J. Neurochem.* **37:** 1335—1337.

39) YOUNG, A. M. J., BRADFORD, A. F. (1986). Excitatory amino acid neurotransmitters in the cortico-striate pathway-studies using intracerebral microdialysis in vivo. *J. Neurochem.* **47:** 1399—1404.

40) GIRAULT, J. A., BARBEITO, L., SPAMPINATO, V., GOZLAN, H., GLOWINSKI, J., BESSON, M.-J. (1986). In vivo release of endogenous amino acids from the rat striatum: Further evidence for a role glutamate and aspartate in corticostriatal neurotransmission . *J. Neurochem.* **47:** 98—106.

41) MORONI, F., PEPEU, G. (1984). The cortical cup technique. In: *Measurement of Neurotransmitter Release in vivo* (Marsden, C. A., ed.), pp. 63—79. John Wiley & Sons.

42) BUTCHER, S. P., HAMBERGER, A. (1987). *In vivo* studies on the extracellular, and veratrine releasable, pools of amino acids in the rat striatum: Effects of corticostriatal deafferentation and kainic acid lesion. *J. Neurochem.* **47.** 713—721.

43) JACOBSON, I., HAGBERG, H., SANDBERG, M., HAMBERGER, A. (1986). Ouabain-induced changes in extracellular aspartate, glutamate and GABA levels in the rabbit olfactory bulb *in vivo. Neurosci. Lett.* **64:** 211—215.

44) TOSSMAN, U., WIELOCH, T., UNGERSTEDT, U. (1985). Gamma-aminobutyric acid and taurine release in the striatum of the rat during hypoglycemic coma studied by microdialysis. *Neurosci. Lett.* **62:** 231—235.

45) LEHMANN, A. (1987). Alterations in hippocampal extracellular amino acids and purine catabolites during limbic seizures induced by folate injections into the rabbit amygdala. *Neuroscience* **22:** 573—578.

46) VEZZANI, A., UNGERSTEDT, U., FRENCH, E. D., SCWARCZ, R. (1985). In vivo brain dialysis of amino acids and simultaneous EEG measurements following intrahippocampal quinolinic acid injection: Evidence for a dissociation between neurochemical changes and seizures. *J. Neurochem.* **45.** 335—344.

47) HAMBERGER, A., NYSTRÖM, B. (1984). Extra- and intracellular amino acids, in the hippocampus during development of hepatic encephalopathy. *Neurochem. Res.* **9:** 1181—1192.

48) TOSSMAN, U., ERIKSSON, S., DELIN, A., HAGENFELDT, L., LAW, D., UNGERSTEDT, U. (1983). Brain amino acids measured by intracerebral dialysis in portacaval shuntet rats. *J. Neurochem.* **41:** 1046—1051.

49) NYSTRÖM, B., HAMBERGER, A., KARLSSON, J.-O. (1985). Changes of extracellular proteins in hippocampus during depolarization. *Neurochem. Int.* **9:** 55—59.

GLIAL CELLS AS METABOLIC REGULATORS OF NEURONS

Antonia Vernadakis, Nikos Sakellaridis, Dimitra Mangoura

Departments of Psychiatry and Pharmacology
University of Colorado School of Medicine
Denver, Colorado 80262
USA

INTRODUCTION

The brain is a complex organ composed of neuronal "wiring" and the more supportive glial and connective tissue cells. Neuronal function is highly dependent upon the interrelationships among these cells and their interactions with the microenvironment. A schematic representation of the various intercellular interactions is illustrated in Figure 1. It has been shown that neurons, neuroglia cells and connective tissue cells secrete into the microenvironment factors which in turn influence the function of other cells, i. e., neurotransmitter and other neurohormones, hypothalamic releasing factors, nucleotide messengers, etc. Exchange of intracellular molecules can occur via specialized intercellular junctions, i. e., ions, nutrients, regulatory molecules. Fixed insoluble molecules can form an extracellular matrix capable of interacting with other cells (the "microexudate" or

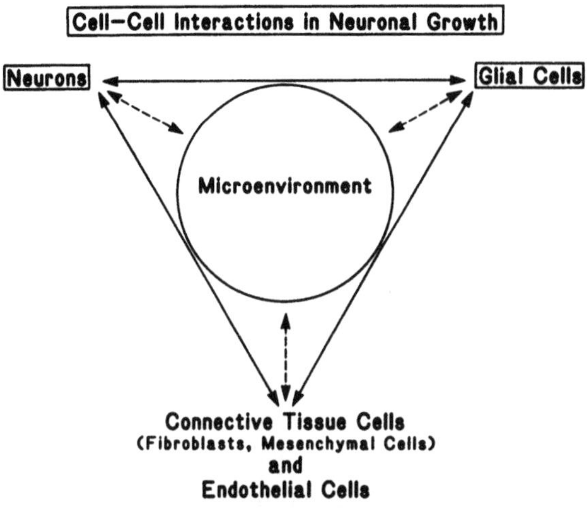

FIGURE 1
Schematic representation of cell-cell interactions in the Central Nervous System (From Ref. 48).

NATO ASI Series, Vol. H20
Amino Acid Availability and Brain Function in Health and Disease. Edited by G. Huether

basement membrane substance). Cell surface and intracellular molecules relay specific signal during early growth and differentiation of the neuron.

As also discussed in other chapters of this book, tissue culture techniques are offering important advantages in studying microenvironment-mediated effects on cell growth and differentiation. These techniques eliminate some of the *in vivo* complexities and expose the cells to a culture medium which functions as a completely manageable microenvironment: a "milieu interieur" that can be altered at will by dissecting out individual factors that possibly interfere with processes of cell development.

Changes in Glial Cell Population During Development and Aging

Considerable studies have been focusing on growth and differentiation of glial cells and the reader is referred to recent reviews by Fedoroff (1), Roots (2), Sturrock (3), Privat and Rataboul (4), Seeds *et al.* (5). In addition, the role of glial cells in neuronal growth and differentiation has been intensively investigated *in vivo* and with in culture models: glial cells have been shown to provide both cell-cell contact and secrete factors important for neuronal guidance and migration (6—9), growth and survival (review 10). We have used the chick embryo as an experimental animal to study both, *in vivo* and in culture, neural growth and differentiation. We have examined changes in the activities of two glial cell enzymes, glutamine synthetase (GS) and 2',' 3'-cyclic nucleotide 3'-phosphohydrolase (CNP) in chick embryonic brain and in cultures drived from chick embryonic brain (11). Glutamine synthetase has been used as a marker for astrocytes (12, 13) and CNP has been used as a marker for oligodendrocytes (14, 15). We found that the developmental profile of the two glial cell enzymes assayed in the chick embryo brain follows a similar

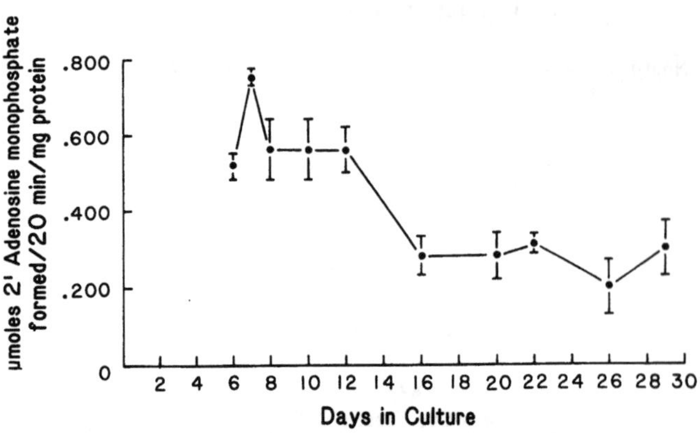

FIGURE 2

Changes with days in culture in 2'-3'-cyclic nucleotide 3'-phosphohydrolase activity in glial-enriched cultures dissociated from cerebral hemispheres of 15-day-old chick embryos. Activity is expressed in µmol of 2'-adenosine monophosphate formed in 20 min per mg protein and plotted *vs* days in culture. Points represent means ± S. E. of 3—4 separate culture dishes. (From Ref. 11).

pattern, increasing with age. In culture, both CNP and GS exhibit profiles different from those *in vivo*. CNP activity decreases in glial-enriched cultures derived from 15-day-old chick embryo cerebral hemispheres (Fig. 2). These findings are in contrast to those reported by Bansal and Pfeiffer (16) and Wernicke and Volpe (17) where CNP activity increased in cultures derived from rat brain. Whether the oligodendrocytes from the chick brain are more sensitive to the lack of neuronal input than those from rat is a possible consideration. For example, in another study when glial-enriched cultures were exposed to medium conditioned from neuron-enriched cultures, CNP activity increases (18). In contrast to the lack of growth of oligodendrocytes in culture, astrocyte growth in culture parallels the *in vivo* growth as shown by a progressive increase in GS activity (Fig. 3). It appears that astrocyte growth may not be as dependent on neuronal input but rather on information astrocytes carry in their own nuclei.

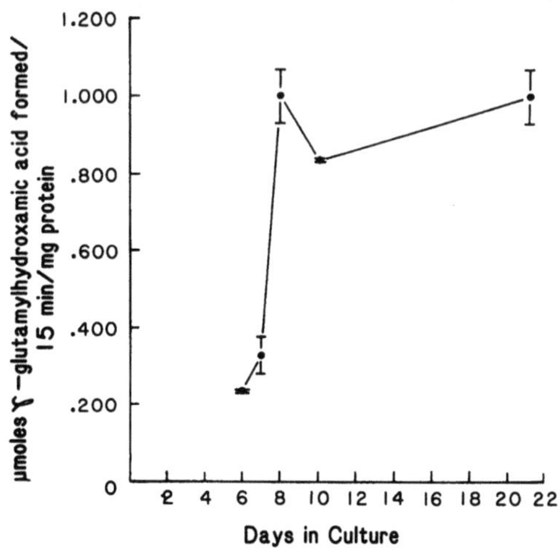

FIGURE 3
Changes with days in culture in glutamine synthetase activity in glial-enriched cultures dissociated from cerebral hemispheres of 15-day-old chick embryos. Activity is expressed in µmol of γ-glutamylhydroxamic acid formed in 15 min/mg protein and plotted *vs* days in culture. Points represent means ± S. E. of 3—5 separate culture dishes. (From Ref. 11).

Changes in the number and volume of glial cells in the aging brain have been reported *in vivo* (see reviews 19, 20). More recently, we have been studying aging in glial cells using both a glioma cell line (C-6 glia, 2B clone given to us by Dr. Jean de Vellis from University of California, Los Angeles) and dissociated cultures from newborn and aged, 18-month-old, mice (21—23). We found that C-6 glia cells at early passages (20—30) are predominantly oligodendrocytic using CNP as a marker, whereas cells at later passages (80—90) are predominantly astrocytic using GS as a marker. We have proposed that this shifting in cell types with cell passage may reflect an *in vitro* aging phenomenon, i. e., astrogliosis *in vitro*. This shift in glia expression is also noted in the cultures derived from aged mouse brain. We found that with subculturing CNP activity decreases (Fig. 4) whereas GS activity markedly increases (Fig. 5). Moreover, the marked GS activity coin-

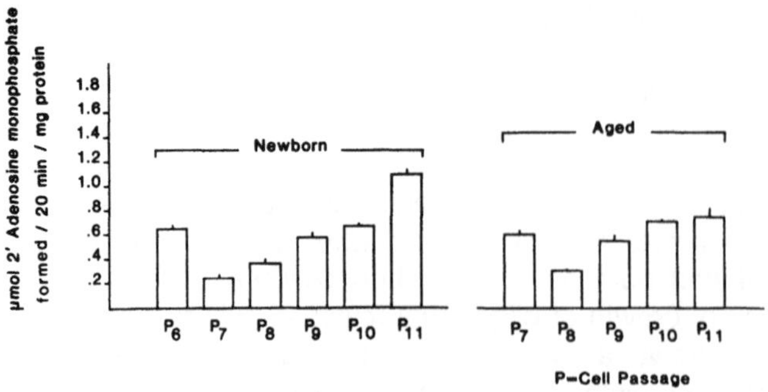

FIGURE 4
Cyclic nucleotide phosphohydrolase activity in glial cell cultures prepared from newborn and aged mouse cerebral hemispheres. Cells were from passage 6 (P6) to 11 (P11). Points with lines represent means ± S. E. for three to five cultures. For passages 0—5 see Ref. (22). (From Ref. 23).

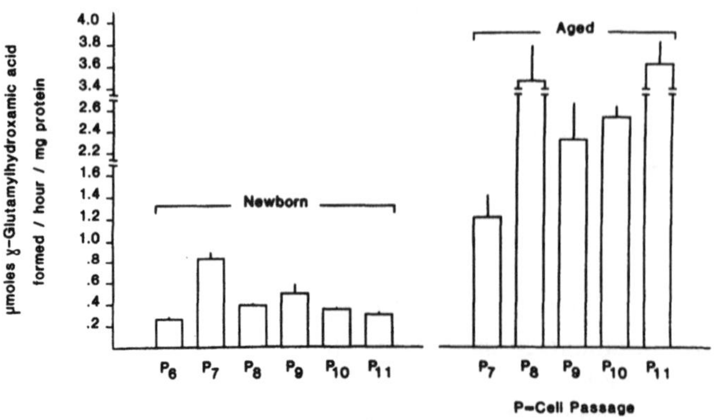

FIGURE 5
Glutamine synthetase activity. Details as in Figure 4.

cides temporally with the appearance of large polyhedral multinucleated cells characteristic of glia in the aging brain. Of interest was the finding that with progressive cell passage, the number of GFA-positive cells declined significantly as did also the intensity and pattern of immunoreactivity. This finding is in agreement with those reported by Lindsay *et al.* (24) in cultures derived from adult rat callosum; a 90 % decline in GFAP stained cells was noted after four subcultures and 8—10 weeks of cultivation. Based on our glioma studies (21) and also studies by Raff *et al.* (25) and Hallermayer and Hamprecht (26) showing shifting of glial types in culture, it is conceivable that in our study the decline in GFA-positive glial cells and the shift toward GS-containing glia reflects a phenomenon of cellular aging.

Functions of Glial Cells

Historically, the function of the brain was thought to be carried out entirely by the neurons. Virchow (27) described neuroglia, or "nerve glue" as the substance which held the neurons together. This notion has been subsequently challenged by showing that glial cells are involved in myelin formation. Myelin production is currently believed to be the primary function of the oligodendroglial cells. The other main class of glial cells, the astroglia, has been extensively studied with numerous functions ascribed to these nonneuronal cells (see review 28).

Although glial cells do much more than "glue the neurons together", they do serve as a limiting tissue, defining many synaptic clefts and providing the localized environment available for the process of synaptic modulation. Included in this capacity is the known function of oligodendroglia: production of myelin for insulation and allowing saltatory conduction along axons. The astrocyte is important for the structural integrity of the blood brain barrier, with the astrocytic "endfeet" covering the capillaries. The neurons are effectively protected from direct contact with the peripheral circulation by the astrocytic processes which interpose between neuron and capillary. This effective barrier may be a combination of passive and active intervention in the flow of materials and nutrients between the central nervous system and the peripheral circulation.

Glial cells have been postulated to influence the concentration of neurotransmitters in the synaptic cleft by a variety of mechanisms. Evidence that astrocytes may be intimately involved in neurotransmission processes has been provided by studies on electrophysical properties of astrocytes (review 29), catecholamine and serotonin uptake (review 30), amino acid uptake (reviews 31, 32), choline uptake (review 33), receptors (reviews 34, 35). The reader, therefore, is referred to these reviews for details and relevant references.

In this chapter, we will present two examples which illustrate the possible modulation of glial cells of the microenvironment as it relates to neurotransmission processes. In an early study, using the cerebellum from a 16-day-old chick embryo and maintained as an organotypic culture, we found a low affinity, 10^{-6} M, norepinephrine (NE) uptake which could be inhibited by cortisol, when present 2.76×10^{-5} M, in the culture medium for 24 h (36). In 1970, Iversen and Salt (37) reported that certain steroids are particularly powerful inhibitors of extraneuronal uptake, uptake$_2$. Based on this evidence, we proposed that the low affinity NE uptake in our cultures was reflecting uptake of NE in glia. In Figure 6 (36), it is illustrated schematically the fate of NE in the central nervous system. The importance of glial uptake of NE in the CNS can be speculated. If the role of glial cells in neurotransmission is to provide a safety ratio, i. e., to limit possible build-up of neurotransmitter substances extracellularly, then inhibition of glial cell uptake could lead to an intracellular-extracellular imbalance and result in deleterious cellular effects. For example, excessive amounts of NE in the synaptic cleft (Fig. 6) would make more NE available to stimulate the CNS and would result in CNS hyperexcitability known to occur with cortisol treatment (38). Moreover, since NE uptake in astrocytes is also sensitive to clinically effective antidepressants, as shown by Kimelberg and associates (review 30), the therapeutic effects of such agents may be partially mediated by their action on astrocytes.

The second example of a neuromodulatory role of glial cells is the evidence reported on the compartmentation of glutamate and GABA in neurons and glial cells (Fig. 7 from Ref. 39). As indicated by the width of the arrows in Figure 7, most of the GABA released

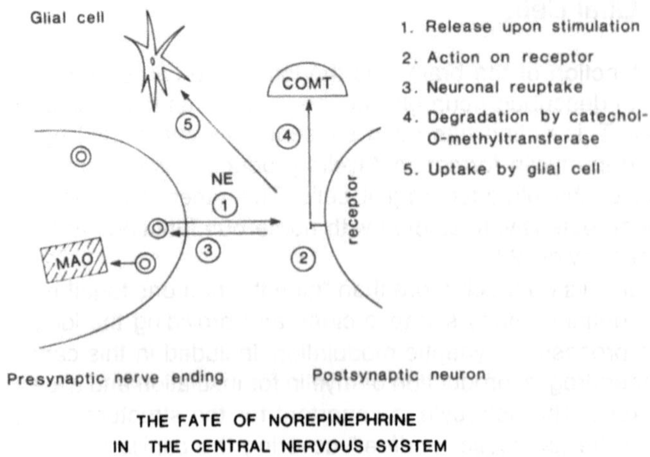

FIGURE 6
A diagrammatic representation of the fate of norepinephrine in the central nervous system. (From Ref. 36).

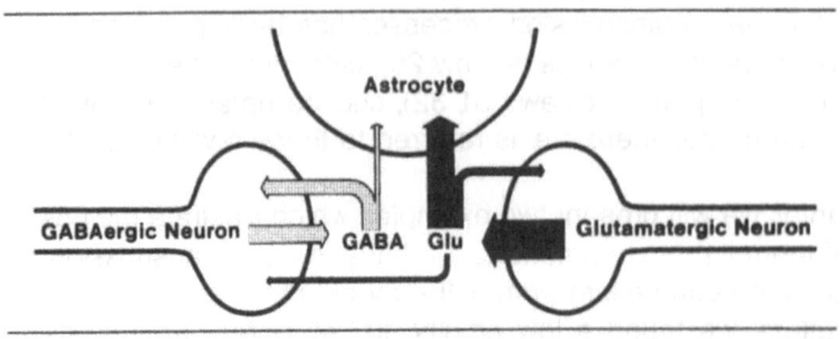

FIGURE 7
Schematic drawing of evoked release and uptake of glutamate and GABA in GABAergic or glutamatergic neurons and in astrocytes. The sizes of the *arrows* give an estimate of the relative magnitudes of the respective fluxes. It can be seen that neuronally released glutamamte to a major extent is accumulated into astrocytes, whereas most of the released GABA is reaccumulated into neurons. (From Ref. 39).

from cerebral cortical neurons is reaccumulated into these cells. Consequently, there is a modest drain of GABA from neurons to astrocytes. In contrast, the release of glutamate from glutamatergic neurons and its uptake by astrocytes is higher than that of GABA.

Hertz (40) has also described a glutamate-glutamine cycle involving a neuron-astrocyte interrelationship. Figure 8 (modified from Ref. 40) illustrates this interrelationship: glutamate released from neurons is taken up into astrocytes where it is converted to glutamine by glutamine synthetase. Glutamine released from astrocytes can be taken up into the neurons where it is hydrolysed by glutaminase and glutamate. As Hertz also discusses, the glutamate uptake into astrocytes is of sufficient intensity to be of major importance for termination of transmitter activity.

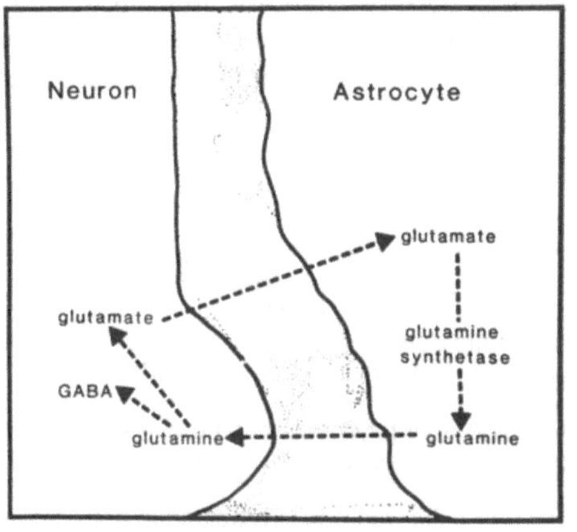

FIGURE 8
Schematic representation of uptake and release for glutamate and glutamine and their metabolic interconversion in neurons and astrocytes. (Modified from Ref. 40).

DISCUSSION AND SPECULATIONS

The high GS activity observed in both the glioma cell cultures with cell passage and in cultures dissociated from aged mouse brain leads us to propose that the marked GS activity representing GS-containing astrocytes may reflect a) increased cellular activity as a compensatory mechanism, mirroring *in vivo* "activated gliosis" and b) a shift in glia phenotypes possibly a cellular aging process observed *in vitro.* Increased cellular activity could be a response to continuous and accumulative exposure to environmental factors such as factors in the serum including hormones, neuropeptides and nucleotides, and substances contained in the medium, such as amino acids, ions, vitamins, and glucose. There is extensive evidence that several of these substances increase GS activity in glial cells (41, 42, 1, 43).

Glia compensatory responses have been observed in various situations in which neuronal injury or degeneration has occurred (see reviews 19, 20). Gliosis appears to represent a major alternate of injury to the CNS and has been viewed as being an important determinant of the extent and quality of neural repair in the mammal. "Activated gliosis" has been described as a beneficial and constructive process for astrocytes to provide cell-cell contacts and factors for neuronal repair. The signals which trigger this astrocyte response are not understood and probably include both neuronal signals and other signals from the microenvironment. We propose that the increase in GS activity (Fig. 5) may represent an activated state of astrocytes.

Glutamic acid is a major excitatory neurotransmitter substance in the CNS (44, 45, 32). Thus, the conversion of glutamate to glutamine by astrocytes (Fig. 8) is an important component of metabolic homeostasis. According to Hertz (40) the low affinity uptake of glutamine to neurons cannot be expected to transfer glutamine from astrocytes to neurons. Also, *in vivo* studies have shown a more pronounced transfer of glutamine than of glutamate from the brain to plasma (46). Johnson (47) has proposed that glutamine formation primarily serves to detoxify ammonia and to protect from excessive levels of glutamate. Based on these proposals, the functional state of astrocytes appears to be of paramount importance for brain metabolic adaptation in health and disease.

REFERENCES

1) FEDOROFF, S. (1986). Prenatal ontogenesis of astrocytes. In: *Astrocytes* (Fedoroff, S., Vernadakis, A., eds.), **Vol. 1,** pp. 35—74, Academic Press, New York.

2) ROOTS, B. I. (1986). Phylogenetic development of astrocytes. In: *Astrocytes* (Fedoroff, S., Vernadakis, A., eds.), **Vol. 1,** pp. 1—34, Academic Press, New York.

3) STURROCK, R. R. (1986). Postnatal ontogenesis of astrocytes. In: *Astrocytes* (Fedoroff, S., Vernadakis, A., eds.), **Vol. 1,** pp. 75—103, Academic Press, New York.

4) PRIVAT, A., RATABOUL, P. (1986). Fibrous and protoplasmic astrocytes. In: *Astrocytes* (Fedoroff, S., Vernadakis, A., eds.), **Vol. 1,** pp. 105—129, Academic Press, New York.

5) SEEDS, N. W., COLLINS, J. M., HAFFKE, S. C., FAFOUTIS, D. (1987). Oligodendrocyte development: Immunological studies. In: *Model Systems of Development and Aging of the Nervous System* (Vernadakis, A., Privat, A., Lauder, J. M., Timiras, P. S., Giacobini, E., eds.), pp. 101—103, Martinus Nijhoff, Boston.

6) HATTEN, M. E. (1984). Embryonic cerebellar astroglia in vitro. *Devel Brain Res.* **13:** 309—313.

7) HATTEN, M. E., LIEM, R. K. H., MASON, C. A. (1984a). Two forms of cerebellar glial cells interact differently with neurons in vitro. *J. Cell Biol.* **98:** 193—204.

8) HATTEN, M. E., LIEM, R. K. H., MASON, C. A. (1984b). Defects in specific associations between astroglia and neurons occur in microcultures of Weaver cerebellar cells. *J. Neurosci.* **4:** 1163—1172.

9) RAKIC, P. (1972). Mode of cell migration to the superficial layers of fetal monkey neocortex. *J. Comp. neurol.* **145:** 61—84.

10) MANTHORPE, MARSTON, RUDGE, J. S., VARON, S. (1986). Astroglial cell contributions to neuronal survival and neuritic growth. In: *Astrocytes* (Fedoroff, S., Vernadakis, A., eds.), **Vol. 2,** pp. 315—376, Academic Press, New York.

11) SAKELLARIDIS, N., BAU, D., MANGOURA, D., VERNADAKIS, A. (1983). Developmental profiles of glial enzymes in the chick embryo: in vivo and in culture. *Neurochem. Int.* **5:** 685—689.

12) NORENBERG, M. D., MARTINEZ-HERNANDEZ, A. (1979). Fine structural localization of glutamine synthetase in astrocytes of rat brain. *Brain Res.* **161.** 303—310.

13) RIEPE, R. E., NORENBERG, M. D. (1977). Muller cell localization of glutamine synthetase in rat retina — an immunohistochemical study. *Nature* **268:** 654—655.

14) PODUSLO, S. E., NORTON, W. T. (1972). Isolation and some chemical properties of oligodendroglia from calf brain. *J. Neurochem.* **19:** 727—736.

15) PODUSLO, S. E. (1975). The isolation and characterization of the plasma membrane and myelin fraction derived from oligodendroglia of calf brain. *J. Neurochem.* **24:** 647—664.

16) BANSAL, K., PFEIFFER, S. E. (1985). Developmental expression of 2', 3'-cyclic nucleotide 3'-phosphoydrolase in dissociated fetal rat brain cultures and rat brain. *J. Neurosci. Res.* **14:** 21—34.

17) WERNICKE, J. F., VOLPE, J. J. (1986). Glial differentiation in dissociated cell cultures on neonatal rat brain: noncoordinate and density-dependent regulation of oligodendroglial enzymes. *J. Neurosci. Res.* **15:** 39—47.

18) SAKELLARIDIS, N., MANGOURA, D., VERNADAKIS, A. (1986). Effects of neuron-conditioned medium and fetal calf serum content on glial growth in dissociated cultures. *Devel. Brain Res.* **27.** 31—41.

19) VERNADAKIS, A. (1985). Aging brain. In: *Symposium on Aging Process* (Gookas, M., ed.), *Clinics in Geriatrics Medicine,* **Vol. 1,** pp. 61—94, Saunders Philadelphia.

20) VERNADAKIS, A. (19869. Changes in astrocytes with aging. In: *Astrocytes* (Fedoroff, S., Vernadakis, A., eds.), **Vol. 2,** pp. 377—407, Academic Press, New York.

21) PARKER, K. P., NORENBERG, M. D., VERNADAKIS, A. (1980). Transdifferentiation of CG glial cells in culture. *Science* **208:** 179—181.

22) VERNADAKIS, A., DAVIES, D., SAKELLARIDIS, N., MANGOURA, D. (1984). Growth patterns of glial cells dissociated from newborn and aged mouse brain. *J. Nurosci. Res.* **11:** 253—262.

23) VERNADAKIS, A., DAVIES, D., SAKELLARIDIS, N., MANGOURA, D. (1986). Growth patterns of glial cells dissociated from newborn and aged mouse brain with cell passage. *J. Neurosci. Res.* **15:** 79—85.

24) LINDSAY, R. M., BARBER, P. C., SHERWOOD, M. R. C., ZIMMER, J., RAISMAN, G. (1982). Astrocyte cultures from adult rat brain. Derivation, characterization and neurotrophic properties of pure astroglial cells from corpus callosum. *Brain Res.* **243:** 329—343.

25) RAFF, M. C., MILLER, R. H., NOBLE, M. (1983). A glial progenitor cell that develops *in vitro* into an astrocyte or an oligodendrocyte dependig on culture media. *Nature* **303:** 390—399.

26) HALLERMAYER, K., HAMPRECHT, B. (1984). Cellular heterogenicity in primary cultures of brain cells revealed by immunochemical localization of glutamine synthetase. *Brain Res.* **295:** 1—11.

27) VIRCHOW, A. (1846). Uber das granulinte ansehan det wunderungen des gehirn-rentridal. *Allerg. Zschr. Psychiat.* **3:** 424—450.

28) VERNADAKIS, A. (1988). Neuron-glia interrelations. *Int. Neurobiol. Rev.* (in press).

29) RANSOM, B. R., CARLINI, W. G. (1986). Electrophysiological properties of astrocytes. In: *Astrocytes* (Fedoroff, S., Vernadakis, A., eds.), **Vol. 2,**, pp. 1—49, Academic Press, New York.

30) KIMELBERG, H. R. (1986). Catecholamine and serotonin uptake in astrocytes. In: *Astrocytes* (Fedoroff, S., Vernadakis, A., eds.), **Vol. 2,** pp. 107—131, Academic Press, New York.

31) HERTZ, L. SCHOUSBOE, A. (1986). Role of astrocytes in compartmentation of amino acid and energy metabolism. In: *Astrocytes* (Fedoroff, S., Vernadakis, A., eds.), **Vol. 2,** pp. 179—208, Academic Press, New York.

32) HOSLI, E., HOSLI, L, SCHOUSBOE, A. (1986). Amino acid uptake. In: *Astrocytes* (Fedoroff, S., Vernadakis, A., eds.), **Vol. 2,** pp. 133—153, Academic Press, New York.

33) MASSARELLI, K., MYKITA, S., SORRENTINO, G. (1986). The supply of choline to glial cells. In: *Astrocytes* (Fedoroff, S., Vernadakis, A., eds.), **Vol. 2,** pp. 155—178, Academic Press, New York.

34) HAMPRECHT, B. (1986). Astroglia cells in culture: receptors and cyclic nucleotides. In: *Astrocytes* (Fedoroff, S., Vernadakis, A., eds.), **Vol. 2,** pp. 77—106, Academic Press, New York.

35) LAUDER, J. McCARTHY, K. (1986). Neuronal-glial interactions. In: *Astrocytes* (Fedoroff, S., Vernadakis, A., eds.), **Vol. 2,** pp. 295—314, Academic Press, New York.

36) VERNADAKIS, A. (1974). Neurotransmission: a proposed mechanism of steroid hormones in the regulation of brain function. In: *Psychoneuroendocrinology Workshop* (Hatotani, ed.), pp. 251—258, Karger, Basel.

37) IVERSEN, L. L., SALT, P. J. (1970). Inhibition of catecholamine uptake $_2$ by steroids in the isolated heart. *Brit. J. Pharmacol.* **40:** 528—530.

38) VERNADAKIS, A., WOODBURY, D. M. (1963). Effects of cortisol on the electroshock seizure threshold in developing rats. *J. Pharmacol. Exp. Ther.* **139:** 110—113.

39) HERTZ, L., SCHOUSBOE, A. (1987). Primary cultures of gabaergic and glutamatergic neurons as model systems to study neurotransmitter functions in differentiated cells. In: *Model Systems of Development and Aging of the Nervous System* (Vernadakis, A., Privat, A., Lauder, J. M., Timiras, P. S., Giacobini, E., eds.), pp. 19—31. Martinus Nijhoff, Boston.

40) HERTZ, L. (1978). Functional interactions between neurons and astrocytes I. Turnover and metabolism of putative amino acid transmitters. *Prog. Neurobiol.* **13:** 277—323.

41) DAVIES, D. L., VERNADAKIS, A. (1986). Responses in astrocytic C6 glioma cells to ethanol and dibutyryl cyclic AMP. *Devel. Brain Res.* **24:** 153—260.

42) DE VELLIS, J., WU, D. K., KUMAR, S. (1986). Enzyme induction and regulation of protein synthesis. In: *Astrocytes* (Fedoroff, S., Vernadakis, A., eds.), **Vol. 2,** pp. 209—237, Academic Press, New York.

43) VERNADAKIS, A., PARKER, K., ARNOLD, E. B., NORENBERG, M. (1982). Role of glial cells in CNS aging. In: *The Aging Brain: Cellular and Molecular Mechanisms of Aging in the Nervous System* (Giacobini, E., Filogamo, G., Giacobini, G., Vernadakis, A., eds.), pp. 57—68, Raven, New York.

44) CURTIS, D. R., JOHNSTON, G. A. R. (1974). Amino acid transmitters in the mammalian central nervous system. *Ergb. Physiol. Biol. Chem. Exp. Pharmacol.* **69:** 97—188.

45) FONNUM, F. (1984). Glutamate: a neurotransmitter in mammalian brain. *J. Neurochem.* **42:** 1—11.

46) ABDUL-GHANI, A. S., MARTON, M., DOBKIN, J. (1978). Studies on the transport of glutamine in vivo between the brain and the blood in the resting state during afferent electrical stimulation. *J. Neurochem.* **31:** 541—546.

47) JOHNSON, J. L. (1978). The excitant amino acids glutamic and aspartic acid as transmitter candidates in the vertebrate central nervous system. *Prog. Neurobiol.* **10:** 155—202.

48) VERNADAKIS, A., SAKELLARIDIS, N. (1985). Role of glial cells in neurotransmission mechanisms. In: *Progress in Neuroendocrinology* (Parvez, H., Parvez, S., Gupta, V., eds.), **Vol. 1,** pp. 17—44, VNU, Science, The Netherlands.

UPTAKE AND RELEASE OF AMINO ACIDS FROM SYNAPTOSOMES

Elling Kvamme, Bjørg Roberg and Liv Johansen

Neurochemical Laboratory
University of Oslo
P.O.Box 1115 — Blindern
0317 Oslo 3, Norway

GENERAL

Synaptosomes are isolated, sealed synaptic nerve endings with intact membrane structure and therefore well suited for uptake and release studies of neurotransmitters (1).

Amino acids are transported through all membranes by diffusion and carrier-mediated transport. The former process is non-saturable and concentration-dependent, the latter is saturable using low and/or high affinity systems, and is either energy-dependent or non-energy-dependent facilitated. Glutamine at extracellular concentrations largely enter synaptosomes by diffusion, whereas amino acid transmitter candidates possess both low and high affinity uptake systems. Certain ions have important functions in transport processes, such as sodium which is required for the uptake of many transmitter amino acids, and chloride for that of GABA.

Transport has traditionally been studied with labelled amino acids by measuring the accumulation of isotopes. Transport studies using this technique overestimate the net accumulation by overlooking the counter transport. Thus the V_{max}'s for net accumulation of GABA and for H^3-GABA accumulation in synaptosomes were found to be 6 and 134 pmol/mg w. w. min, respectively (2).

Presynaptic reuptake of released transmitter is generally believed to be the most effective means of inactivating monoamine and amino acid transmitters in the CNS. An efficient mechanism of transmitter inactivation requires concentrative transport of high-affinity (low K_m) and high capacity (high V_{max}), as shown for dopamine, noradrenaline, glutamate, GABA and glycine.

The high-affinity uptake of neurotransmitters is reduced when the integrity of the nerve terminal plasma membrane is altered, e. g. by exposing the presynaptic nerve terminal to various active oxygen species (3, 4).

Transmitter release may be depolarization-induced or non-induced. Synaptosomal depolarization-induced release is usually Ca^{2+}-dependent in contrast to that of glial release.

In addition, a spontaneous release has been observed, probably caused by leakage from damaged synaptosomes. The extracellular $[Ca^{2+}]$ in mammals is about 10^{-3} M whereas in resting nerve cells $[Ca^{2+}]$ is of the order of 10^{-7} M. Transmitter release is probably activated when the cytosolic $[Ca^{2+}]$ increases above 10^{-6} M. The mechanism of the

NATO ASI Series, Vol. H20
Amino Acid Availability and Brain Function in Health and Disease. Edited by G. Huether

Ca^{2+}-stimulated release of amino acid transmitters is a matter of controversy. According to the calcium hypothesis, calcium ions directly trigger a transient release of the neurotransmitter, whereas the voltage hypothesis presumes that this release is triggered by the depolarization itself and that calcium, although needed, is of secondary importance. Recent results from electrophysiological experiments are consistent with the calcium hypothesis. Voltage may modulate, but not elicit transmitter release (5).

Glycine, GABA, glutamate and aspartate are likely to function as neurotransmitters in the CNS. It is uncertain whether tyrosine, taurine, tryptophane, histidine and other amino acids are neurotransmitters, but they may modulate synaptic transmission.

Glycine is assumed to be an inhibitory transmitter preferentially in the spinal cord. A high- and low-affinity Na^+-dependent uptake mechanism has been described in synaptosomes from spinal cord and cerebral cortex (6—8). It has been reported that glycine shares a Na^+-dependent brain synaptosomal uptake system with proline and pipecolic acid, and that this uptake system is inhibited by Leu- and Met-enkephalin (9). Moreover, the uptake and release process of glycine as well as of GABA and β-alanine is Cl^--dependent (10—12).

Three uptake systems have been found for β-alanine and nipecotic acid which are inhibitors of GABA uptake, namely a high-affinity, medium-affinity and low-affinity uptake system, respectively (13).

It is of interest that synaptosomes accumulate cysteic acid by a high-affinity transport system and that cysteate is likely to be transported via the same transport system as aspartate and glutamate (14).

In mammalian brain, glutamate, glutamine, GABA and aspartate account for more than 60 % of the free amino acid nitrogen (15) and these amino acids will be particularly dealt with here.

GAMMA-AMINOBUTYRIC ACID (GABA)

Uptake

Gamma-aminobutyric acid (GABA) is believed to be a major inhibitory transmitter in the CNS. GABA may be transmitter for up to 50 % of the brain synapses, depending on the region (16, 17). A high affinity and a low affinity uptake system have been reported with K_m values of 2—10 μM and approximately 1 mM, respectively (18, 19), However, Wood and Sidhu (20) presented evidence for three GABA uptake systems in synaptosomes with K_m values of 1.1 μM, 43 μM and 3.9 mM, respectively, and Debler and Lajtha (8) found no low-, only one high-affinity uptake system. The high-affinity uptake system is almost exclusively neuronal.

The uptake of GABA has specific requirements for Cl^-, but requires also Na^+. The uptake is electrogenic, and both a Na^+ and a Cl^- gradient are required to obtain maximal steady-state levels.

N-ethylmaleimide and p-chloromercuribenzoate inhibit GABA transport. The inhibitor profile for GABA uptake is about the same as that for its release (19), suggesting that there is an external, reactable SH-group on the transporter. The GABA transporter apparently binds 2 Na^+ and 1 Cl^- for each GABA being transported (21). The transporter is possibly asymmetric since the binding order of these molecules is different on the outside from that on the inside. The GABA transporter has been solubilized and reconstituted into liposomes, having the same binding characteristics as in the native membrane vesicles. Using the reconstitution assay as a tool, the transporter protein has been highly purified (21).

Release

The physiological release mechanism is less clear than that of the uptake. A membrane potential-dependent Ca^{2+} channel apparently limits the GABA release (22).

Whether Ca^{2+} induced GABA release occurs from synaptic vesicles or from the cytosol, is a matter of different opinions. There is a poor correlation between the Ca^{2+} dependent GABA release and the number of synaptic vesicles in the amacrine cells of 3—10 day old rabbits, favoring a Ca^{2+} dependent non-vesicular process for the release (23).

Besides the Ca^{2+}-dependent release there is a release which depends on external Na^{+} and is independent of Ca^{2+} in the medium (24—26). A widely accepted explanation for Ca^{2+} independent neuronal GABA release is that it represents a reversal of the ordinary GABA carrier in the cell membrane (24).

There is no evidence for a tight coupling of the synaptosomal uptake and release mechanisms for GABA. Thus Na^{+} free medium inhibits uptake, but has little effect on synaptosomal release (27), and the presence of Na^{+} is not required for release from previously loaded pools (7, 28). However, extracellular Na^{+} is also not required to support Ca^{2+} dependent release from brain slices, but Na^{+} causes a decrease in the Ca^{2+} independent releasing action of elevated K^{+} (29). It should be noted that commonly used substitutes for Na^{+}, such as Li^{+} and choline, are reported to have some inhibitory effect on the uptake of monoamine and amino acid transmitters (30). Ouabain increases both the synaptosomal release of [^{3}H]-GABA (31), as well as that of [^{3}H]-GABA, β-alanine, glycine and aspartate from brain prisms (32).

Glutamine derived glutamate is generally believed to be the precursor for transmitter GABA, but evidence has been produced that GABA can be formed from putrescine in the caudate-putamen, and released in a Ca^{2+} dependent manner (33).

GLUTAMATE AND ASPARTATE

Uptake

L-glutamate is probably one of the major excitatory neurotransmitters in the CNS, and the excitatory action is mainly terminated by the uptake process (34). Both high- and low-affinity uptake are dependent on the Na^{+}-concentration. Debler and Lajtha (8) report that there exist 4 distinct synaptosomal transport systems, transporting glycine, taurine, GABA and L-glutamate in addition to L-aspartate, respectively, and that only high affinity uptake systems could be selected for GABA, glutamate and aspartate. It has been suggested that a single Na^{+}-dependent uptake system exists for L-glutamate, L-aspartate and D-aspartate since the uptake of each of these amino acids is inhibited by the others (35, 36). However, Ferkany and Coyle (37) report that the Na^{+}-dependent transport mechanisms for L-glutamate and L- and D-aspartate are dissimilar in striatal synaptosomes. Moreover, ATP has a non-competitive inhibitory action on the high-affinity uptake of L-glutamate, but has no effect on that of L-aspartate (38, 39).

Based on studies using membrane vesicle preparations Kanner and Radian (40) conclude that glutamate uptake is electrogenic; it requires both external Na^+ and internal K^+, and the requirement for K^+ is specific. Na^+ is cotransported with glutamate and the mechanism of transport is chemiosmotic, similar to that of GABA. The transport also requires a coupling ion which is K^+ for glutamate, instead of Cl^- for GABA (40).

K^+ as well as Na^+ has been reported to be a competitive inhibitor for binding of L-glutamate or L-aspartate in brain membrane preparations (41). Danbolt and Storm-Mathisen (42) have shown that inhibition of Na^+ dependent uptake by K^+ also occurs in membrane-bounded saccules in the preparation and thus in the absence of transmembrane gradients of Na^+ and K^+.

Cl^- enhances L-glutamate binding to synaptic membranes (43) and it has been suggested that this in part reflects a sequestration process driven by glutamate exchange (44).

In contrast to previous reports (45) it has been demonstrated that L-glutamate is specifically taken up by synaptic vesicles in an ATP- and temperature-dependent, but low-affinity and Na^+ independent manner. The uptake is stimulated by Cl^- and is possibly driven by electrochemical proton gradients generated by Mg-ATPase (46, 47). These findings are supported by the immunocytochemical evidence that glutamate is concentrated in synaptic vesicles (48).

Release

The release of glutamate is countertransported with K^+ from the outside to allow for return of the unloaded transporter. Binding order is assymetric and internal Na^+ is required (40). In addition there is an exchange of glutamate which occurs in the presence or absence of external Na^+ and glutamate.

Kainic acid inhibits reuptake of released glutamate and aspartate from synaptosomes, and does also stimulate in a calcium-dependent way the basal release of endogenous glutamate and aspartate (49). Kainic acid stimulates both the release of newly synthesized L-glutamate and L-aspartate and the fraction of these amino acids accumulated through high-affinity mechanisms (49). However, kainic acid inhibits the basal as well as the high K^+-stimulated release of endogenously accumulated [D-^{3}H] aspartate, demonstrating that [D-^{3}H] aspartate is not a reliable substitute for L-glutamate and L-aspartate in such studies.

Synaptosomes have been shown to contain high concentrations of Cl^- (about 91 mM) (50), and a sudden reduction of [Cl^-] in the incubation medium stimulates the release of glutamate and aspartate from synaptosomes (51).

Whether transmitter glutamate is released from a cytoplasmic or vesicular pool, has been a matter of debate. Using a continuous enzymatic assay Nicholls and Sihra (52) have been able to distinguish between glutamate release from non-cytoplasmic and cytoplasmic pools in cerebrocortical synaptosomes. They have demonstrated that 15—20 % of the total glutamate can be rapidly released following depolarization in a Ca^{2+}-dependent manner from a non-cytoplasmic pool, whereas D-aspartate can be used to release glutamate from a cytoplasmic pool.

Long-term potentiation is associated with an increase in release of prelabelled, newly synthesized glutamate, as well as an increase in Ca^{2+} dependent K^+ induced release of [^{14}C]glutamate from hippocampal synapto-somes (53, 54). Lynch and Bliss (54) report that calmodulin and oleoyl-acetyl-glycerol, a synthetic analogue of diacylglycerol, enhance the K^+-induced release of radiolabelled glutamate from the synaptosomes. The authors suggest that calmodulin and endogenous diacylglycerol are involved in control of glutamate release and that the increased release is associated with long-term potentiation. In this context is is of interest that Hu *et al.* (55) recently have obtained evidence that injection of calcium/diacyl-glycerol-dependent protein kinase into hippocampal pyramidal cells mimics long-term potentiation.

UPTAKE AND RELEASE OF GLUTAMINE

The mechanism for synaptosomal uptake and release of glutamine is less well understood than that of glutamate, and conflicting reports have been published. Since the extracellular glutamine concentration is rather high 0.5—0.6 mM (56), most workers feel that glutamine generally enters mammalian cells by low affinity mechanisms (K_m 0.15—3.3 mM) which has been described for brain slices, neuroblastomas, glioblastomas and synaptosomes (57).

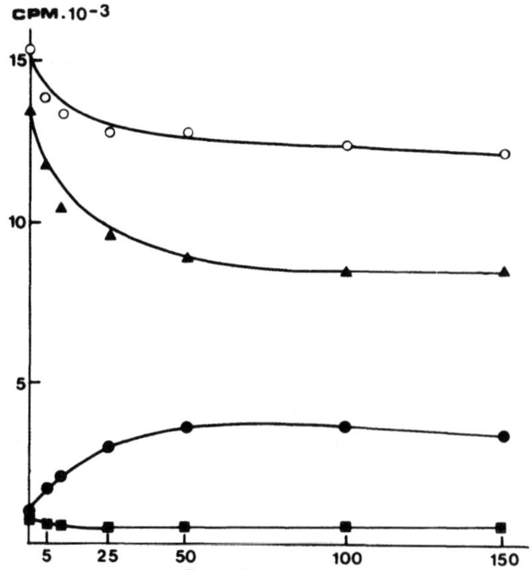

FIGURE 1
Effect of Na^+ on the total [^{14}C]-glutamine uptake and on the [^{14}C]glutamine and [^{14}C]glutamate accumulation and release. The incubation of a crude synaptosomal fraction (P2) prepared from rat brain was performed in a Ringer buffer pH 7.4 with no Ca^{2+} at pH 7.4. Total [^{14}C]glutamine uptake (○) is the calculated sum of [^{14}C]glutamine in the synaptosomes (■), [^{14}C]glutamine-derived glutamate in the synaptosomes (●) and [^{14}C]glutamine-derived glutamate released to the incubation medium (▲).

However, high-affinity Na^+-dependent uptake of glutamine has been reported for brain prisms, dorsal roots, dorsal ganglia (57) and also in rat brain synaptosomes (58). Some workers have not taken into account that glutamine is partially hydrolyzed by phosphate activated glutaminase (PAG) to form glutamate which is rapidly released to the incubation medium. Therefore, when incubating with radiolabelled glutamine, the number of counts found in the cells do not give a correct estimate of the glutamine taken up. The true net uptake of [^{14}C]-glutamine represents the sum of [^{14}C]glutamine and glutamate accumulated in the cells and the [^{14}C]glutamate released to the incubation medium (Fig. 1). Measured in this way we found that glutamine uptake is Na^+ independent and unaffected by the Na^+K^+-ATPase inhibitor ouabain. Since it has been suggested that synaptosomal glutamine uptake is regulated by the gamma-glutamyl cycle (58), and the development of the endogenous glutamate pool in cultured neurons correlates with that of solubilized gamma-glutamyl transferase, (59) we also investigated the effect of Acivicin, a specific inhibitor of this enzyme on glutamine uptake. However, we found no effect of Acivicin neither on glutamine nor on glutamate uptake (60). Our experiments gave no support to the notion that glutamine uptake, as measured under our conditions, is of the high-affinity Na^+-dependent type. However, the higher neuronal and synaptosomal glutamine concentration than that of the extracellular fluid still remains to be explained, but may be caused by restriction on glutamine release.

It is indicated that newly synthesized glutamine derived glutamate as well as glutamate taken up from the extrasynaptosomal space, are compartment-mentalized from endogenous glutamate (61). Moreover, when incubating or superfusing polarized (60) or depolarized (62) synaptosomes with [^{14}C]glutamine, the specific radioactivity of released glutamine-derived glutamate is much higher than that of the remaining glutamate in the synaptosomes, indicating that newly synthesized glutamate is preferentially released (Table 1).

TABLE 1

Specific activity of the [^{14}C]glutamine-derived glutamate in the P2 fraction and the incubation medium. Conditions as in Fig. 1.

	SPECIFIC ACTIVITY	
NaCl mM	Synaptosomes cpm/nmol	Medium cpm/nmol
0[a]	8.3 ± 1.6	130.0 ± 16.7
125	30.8 ± 5.0	79.2 ± 6.5

n = 6 (Mean ± S.D)
[a] NaCl substituted with choline-Cl

A model of the glutamine and glutamate pools is illustrated in Figure 2. PAG which is a mitochondrial enzyme, localized in the inner membrane and probably also in the matrix region, hydrolyzes external glutamine to glutamate that is largely excreted to the incubation medium. External glutamine and glutamate give rise to cytosolic pools of these amino acids and external glutamate is a potent inhibitor of PAG. However, the endogenous glutamate pool is sufficiently large to inhibit PAG profoundly, if available to the enzyme. This is apparently not the case (61), and PAG will hydrolyze any glutamine that is not compartmentalized. In addition, when considering the higher specific activity of released glutamine-derived glutamate than that of the glutamate remaining in the synaptosomes, it is indicated that there exists distinct cytosolic and endogenous pools of glutamine as well as of glutamate. Whether the cytosolic or the compartmentalized, endogenous glutamate pool contain the transmitter glutamate is still a matter of controversy.

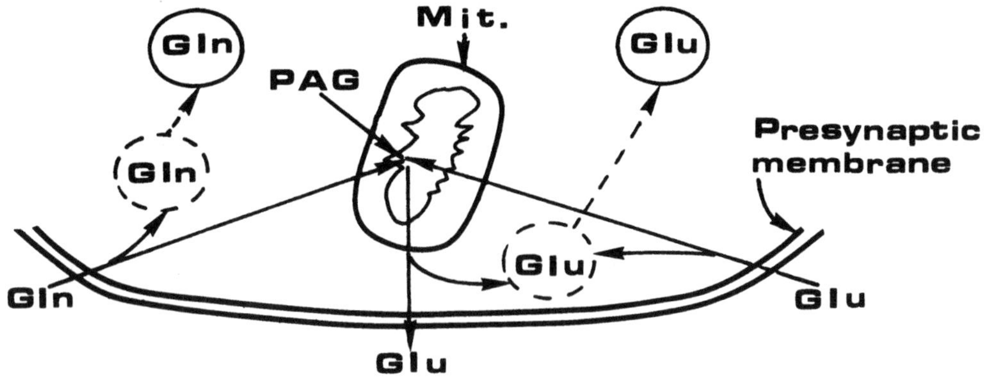

FIGURE 2

Pools of glutamine (Gln) and glutamate (Glu) in synaptosomes. Mit: Mitochondrion.

CONCLUSION

The main synaptosomal uptake and release processes for amino acids and in particular for putative transmitter amino acids, are briefly reviewed. Furthermore, the uptake and release process for each individual amino acid is modulated by other amino acids, peptides, monoamines, etc., and the coexistence of several transmitters within the same neuron opens up great perspectives for interrelated regulatory mechanisms.

REFERENCES

1) WHITTAKER, V. P. (1972).The subcellular fractionation of brain tissue with special reference to the preparation of synaptosomes and their component organelles. In: *Methods of Neurochemistry* (Fried, R., ed), **Vol. 2,** Marcel Dekker, Inc., New York, pp. 1—52.

2) HALVARSSON, G. B., KARLSSON, I., SELLSTRØM, Å. (1985). The use of ^{3}H-gamma-aminobutyric acid for transport studies with isolated nerve-terminals from rat brain. *Life Sci.* **37:** 209—216.

3) CHAN, P. H., KERLAN, R., FISHMAN, R. A. (1983). Reductions of gamma-aminobutyric acid and glutamate uptake and (Na^{+} + K^{+})-ATPase activity in brain slices and synaptosomes by arachidonic acid. *J. Neurochem.* **40:** 309—316.

4) DEBLER, E. A., SERSHEN, H., LAJTHA, A., GENNARO, J. F. Jr. (1986). Superoxide radical-mediated alteration of synaptosome membrane structure and high-affinity gamma-[^{14}C]-aminobutyric acid uptake. *J. Neurochem.* **47:** 1804—1813.

5) ZUCKER, R. S., LANDO, L. (1986). Mechanism of transmitter release: Voltage hypothesis and calcium hypothesis. *Science* **231:** 574—579.

6) LOGAN, W. J., SNYDER, S. H. (1971). Unique high affinity uptake systems for glycine, glutamic and aspartic acids in central nervous tissue of the rat. *Nature* **234:** 297—299.

7) SIMON, J. R., MARTIN, D. L., KROLL, M. (1974). Sodium-dependent efflux and exchange of GABA in synaptosomes. *J. Neurochem.* **23:** 981—991.

8) DEBLER, E. A., LAJTHA, A. (1987). High-affinity transport of gamma-aminobutyric acid, glycine, taurine, L-aspartic acid, and L-glutamic acid in synaptosomal (P_2) tissue: A kinetic and substrate specificity analysis. *J. Neurochem.* **48:** 1851—1856.

9) RHOADS, D. E., PETERSON, N. A., RAGHUPATHY, E. (1984). Iminoglycine transport system in synaptosomes and its interaction with enkephalins. *Biochemistry* **23:** 117—121.

10) KANNER, B. I. (1978). Active transport of gamma-aminobutyric acid by membrane vesicles isolated from rat brain. *Biochemistry* **17:** 1207—1211.

11) ZAFRA, F., ARAGÓN, M. C., VALDIVIESO, F., GIMÉNEZ, C. (1984). β-Alanine transport into plasma membrane vesicles derived from rat brain synaptosomes. *Neurochem. Res.* **9:** 695—707.

12) ARAǴON, M. C., GIMÉNEZ, C. (1986). Efflux and exchange of glycine by synaptic plasma membrane vesicles derived from rat brain. *Biochim. Biophys. Acta* **855:** 257—264.

13) SIDHU, H. S., WOOD, J. D. (1986). Three uptake systems in synaptosomes for nipe cotic acid and beta-alanine. *Neuropharm.* **25:** 555—558.

14) WILSON, D. F., PASTUSZKO, A. (1986). Transport of cysteate by synaptosomes iso lated from rat brain: Evidence that it utilizes the same transporter as aspartate, glutamate, and cysteine sulfinate. *J. Neurochem.* **47:** 1091—1097.

15) TALLAN, H. H., MOORE, S., STEIN, W. H. (1954). Studies of the free amino acids and related compounds in the tissues of the cat. *J. Biol. Chem.* **211:** 927—939.

16) IVERSEN, L. L., BLOOM, F. E. (1972). Studies of the uptake of [^{3}H]-GABA and [^{3}H]-glycine in slices and homogenates of rat brain and spinal cord by electron microscopic autoradiography. *Brain Res.* **41:** 131—143.

17) TURNER, A. J., WHITTLE, S. R. (1983). Biochemical dissection of the gamma-amino butyrate synapse. *Biochem. J.* **209:** 29—41.

18) SNYDER, S. H., YOUNG, A. B., BENNETT, J. P., MULDER, A. H. (1973). Synaptic biochemistry of amino acids. *Fed. Proc.* **32:** 2039—2047.

19) TROEGER, M. B., WILSON, D. F., ERECINSKA, M. (1984). The effect of thiol reagents on GABA transport in rat brain synaptosomes. *FEBS Lett.* **171:** 303—308.

20) WOOD, J. D., SIDHU, H. S. (1986). Uptake of gamma-aminobutyric acid by brain tissue preparations: A reevaluation. *J. Neurochem.* **46:** 739—744.

21) KANNER, B. I., RADIAN, R. (1985). Ion-coupled neurotransmitter transport across the synaptic plasma membrane. *Ann. N.Y. Acad. Sci.* **456:** 153—161.

22) ASAKURA, T., HOSHINO, M., KOBAYASHI, T. (1982). Effect of calcium ion on the release of gamma-aminobutyric acid from synaptosomal fraction. *J. Biochem.* **92:** 1919—1923.

23) JØNSSON, U., LUNDSTRØM, M., SELLSTRØM, Å, EHINGER, B. (1986). Calcium-independent release of gamma-aminobutyrate from nerve processes in the developing rabbit retina. *Neurosci.* **17:** 1235—1241.

24) MARTIN, D. L. (1976). Carrier-mediated transport and removal of GABA from synaptic regions. In: *GABA in Nervous System Function* (Roberts, E., Chase, T. N., Tower, D. B., eds.), Raven Press, New York, pp. 347—386.

25) LEVI, G., BANAY-SCHWARTZ, M., RAITERI, M. (1978).Uptake, exchange and release of GABA in isolated nerve endings. In: *Amino Acids as Chemical Transmitters* (Fonnum, F., ed.), Plenum Press, New York, pp. 327—350.

26) ARIAS, C., SITGES, M., TAPIA, R. (1984). Stimulation of [^{3}H]gamma-aminobutyric acid release by calcium chelators in synaptosomes. *J. Neurochem.* **42:** 1507—1514.

27) REDBURN, D. (1978). Relationship between synaptosomal uptake and release of [^{14}C]GABA, [^{14}C]diaminobutyric acid and [^{14}C]β-alanine. *J. Neurochem.* **31:** 939—945.

28) COTMAN, C. W., HAYCOCK, J. W., WHITE, W. F. (1976). Stimulus-secretion coupling processes in brain: Analysis of noradrenaline and gamma-aminobutyric acid release. *J. Physiol.* **254:** 475—505.

29) NADLER, J. V., WHITE, W. F., VACA, K. W., REDBURN, D. A., COTMAN, C. W. (1977). Characterization of putative amino acid transmitter release from slices of rat dentate gyrus. *J. Neurochem.* **29:** 279—290.

30) SHANK, R. P., SCHNEIDER, C. R., TIGHE, J. J. (1987). Ion dependence of neurotransmitter uptake: Inhibitory effects of ion substitutes. *J. Neurochem.* **49:** 381—388.

31) RAITERI, M., FEDERICO, R., COLETTI, A., LEVI, G. (1975).Release and exchange studies relating to the synaptosomal uptake of GABA. *J. Neurochem.* **24:** 1243—1250.

32) O'FALLON, F. V., BROSEMER, R. W., HARDING, J. W. (1981). The Na^+, K^+-ATPase: A plausible trigger for voltage-independent release of cytoplasmic neuro-transmitters. *J. Neurochem.* **36:** 369—378.

33) NOTO, T, HASHIMOTO, H., NAKAO, J., KAMIMURA, H., NAKAJIMA, T. (1986). Spontaneous release of gamma-aminobutyric acid formed from putrescine and its enhanced Ca^{2+}-dependent release by high K^+ stimulation in the brains of freely moving rats. *J. Neurochem.* **46:** 1877—1880.

34) FONNUM, F. (1984). Glutamate: A neurotransmitter in mammalian brain. *J. Neurochem.* **42:** 1—11.

35) TAKAGAKI, G., KONAGAYA, H. (1985). Properties of the uptake and release of neurotransmitter glutamate in cerebral cortical tissue of guinea pigs. *Neurochem. Res.* **10:** 1059—1069.

36) SCHOUSBOE, A., DREJER, J., HERTZ, L. Uptake and release of glutamate and glutamine in neurons and astrocytes in primary cultures. In: *Glutamine and Glutamate in Mammals* (Kvamme, E., ed.), CRC Press, Boca Raton, Florida, in press.

37) FERKANY, J., COYLE, J. T. (1986). Heterogeneity of sodium-dependent excitatory amino acid uptake mechanisms in rat brain. *J. Neurosci. Res.* **16:** 491—503.

38) WARNER, S. J. C., CANTRILL, R. C., BRENNAN, M. J. W. (1981). Differential action of adenosine triphosphate on high affinity transport of L-glutamate and L-aspartate in rat brain synaptosomes. *Life Sci.* **28:** 163—165.

39) ARBUTHNOT, P. B., CANTRILL, R. C. (1985). Purine nucleoside and nucleotide regulation of high affinity [^{3}H]glutamate and [^{3}H]aspartate uptake into rat brain synaptosomes. *Int. J. Biochem.* **17:** 753—755.

40) KANNER, B. I., RADIAN, R. (1986). Mechanisms of reuptake of neurotransmitters from the synaptic cleft. In: *Exitarory Amino Acids* (Roberts, P. J., Storm-Mathisen, J., Bradford, H. F., eds.), Macmillan Press, London, pp. 159—172.

41) KRAMER, K., BAUDRY, M. (1984). Low concentrations of potassium inhibit the Na-dependent [^{3}H]glutamate binding to rat hippocampal membranes. *Eur. J. Pharmacol.* **102:** 155—158.

42) DANBOLT, N. C., STORM-MATHISEN, J. (1986). Inhibition by K^+ of Na^+-dependent D-aspartate uptake into brain membrane saccules. *J. Neurochem.* **47:** 825—830.

43) MENA, E. E., FAGG, G. E., COTMAN, C. W. (1982). Chloride ions enhance L-glutamate binding to rat brain synaptic membranes. *Brain Res.* **243:** 378—381.

44) ZACZEK, R., ARLIS, S., MARKE, A., MURPHY, T., DRUCKER, H., COYLE, J. T. (1987). Characteristics of chloride-dependent incorporation of glutamate into brain membranes argue against a receptor binding site. *Neuropharmacol.* **26:** 281—287.

45) MANGAN, J. L, WHITTAKER, V. P. (1966). The distribution of free amino acids in subcellular fractions of guinea-pig brain. *Biochem. J.* **98:** 128—137.

46) DISBROW, J. K., GERSHTEN, M. J., RUTH, J. A. (1982). Uptake of L-[^{3}H] glutamic acid by crude and purified synaptic vesicles from rat brain. *Biochem. Biophys. Res. Commun.* **108:** 1221—1227.

47) NAITO, S. UEDA, T. (1985). Characterization of glutamate uptake into synaptic vesicles. *J. Neurochem.* **44:** 99—109.

48) STORM-MATHISEN, J., LEKNES, A. K., BORE, A. T., VAALAND, J. L., EDMINSON, P., HAUG, F. M. S., OTTERSEN, O. P. (1983).First visualization of glutamate and GABA in neurones by immunocytochemistry. *Nature* **301:** 517—520.

49) VIRGILI, M., POLI, A., CONTESTABILE, A., MIGANI, P., BARNABEI, D. (1986). Synaptosomal release of newly-synthetized or recently accumulated amino acids. Differential effects of kainic acid on naturally occurring excitatory amino acids and on [D-^{3}H]aspartate. *Neurochem. Int.* **9:** 29—33.

50) MARCHBANKS, R. M. (1975). The chloride content, anion deficit and volume of synaptosomes. *J. Neurochem.* **25:** 463—470.

51) HARDY, J. A., BOAKES, R. J., THOMAS, D. J., E., KIDD, A. M., EDWARDSON, J. A., VIRMANI, M., TURNER, J., DODD, P. R. (1984). Release of aspartate and glutamate caused by chloride reduction in synaptosomal incubation media. *J. Neurochem.* **42:** 875—877.

52) NICHOLLS, D. G., SIHRA, T. S. (1986). Synaptosomes possess an exocytotic pool of glutamate. *Nature* **321:** 772—773.

53) FEASEY, K. J., LYNCH, M. A., BLISS, T. V. P. (1986). Long-term potentiation is associated with an increase in calcium-dependent potassium release of [^{14}C]glutamate from hippocampal slices: an ex vivo study in the rat. *Brain Res.* **364:** 39—44.

54) LYNCH, M. A., BLISS, T. V. P. (1986). Long-term potentiation of synaptic transmission in the hippocampus of the rat; Effect of calmodulin and oleoyl-acetyl-glycerol on release of [^{3}H]glutamate. *Neurosci Lett.* **65:** 171—176.

55) HU, G. Y., HVALBY, Ø., WALAAS, S. I., ALBERT, K. A., SKJEFLO, P., ANDERSEN, P., GREENGARD, P. (1987). Protein kinase C injection into hippocampal pyramidal cells elicits features of long term potentiation. *Nature* **328:** 426—429.

56) PERRY, T. L. Glutamine, glutamate and GABA in human diseases. In: *Glutamine and Glutamate in Mammals* (Kvamme, E., ed.), CRC Press, Boca Raton, Florida, in press.

57) KVAMME, E. (1983). Glutamine. In: *Handbook of Neurochemistry* (Lajtha, A., ed.), **Vol. 3,** 2nd ed., Plenum Press, New York, pp. 405—422.

58) MINN, A., BESAGNI, D. (1983). Uptake of L-glutamine into synaptosomes. Is the gamma-glutamyl cycle involved? *Life Sci.* **33:** 225—232.

59) KVAMME, E., SCHOUSBOE, A., HERTZ, L, TORGNER, I. A., SVENNEBY, G. (1985). Developmental change of endogenous glutamate and gamma-glutamyl transferase in cultured cerebral cortical interneurons and cerebellar granule cells, and in mouse cerebral cortex and cerebellum in vivo . *Neurochem. Res.* **10:** 993—1008.

60) JOHANSEN, L, ROBERG, B., KVAMME, E. (1987). Uptake and release for glutamine and glutamate in a crude synaptosomal fraction from rat brain. *Neurochem. Res.* **12:** 135—140.

61) KVAMME, E., LENDA, K. (1981). Evidence for compartmentalization of glutamate in rat brain synaptosomes using the glutamate sensitivity of phosphate-activated glutaminase as a functional test. *Neurosci. Lett.* **25:** 193—198.

62) BRADFORD, H. J., WARD, H. K., THOMAS, A. J. (1978). Glutamine — a major substrate for nerve endings. *J. Neurochem.* **30:** 1453—1459.

SUMMARY AND DISCUSSION-REPORT OF CHAPTER II

Oliver E. Pratt

SUMMARY

With regard to the supply of amino acids to the brain, we have to keep always in mind that the passive, reversible character of the transport system taking amino acids across the brain capillary endothelium means that the net flux (whatever the magnitude of the unidirectional influx) is driven by the concentration gradient across the capillary endothelium, that is the ratio between the concentration in the blood plasma and that in the extracellular space. This concentration gradient will be maintained by the net uptake of amino acids into brain cells. This latter process is normally energy-dependent, sodium-pump driven and (in contrast to the capillary endothelial transport) amino acids can be taken into the cells against a concentration gradient. The reason that the concentrations of most amino acids in the extracellular space (or in the CSF) are below those in the blood plasma is due to the large V_{max} for the uptake into brain cells. It must be remembered that the surface area of the brain cells greatly exceeds that of the cerebral capillaries. A second corollary of the model is that if transport across the capillary endothelium becomes limiting (e. g., for tryptophan due to high levels of competing large neutral amino acids), then the concentration gradient of that amino acid will fall to a low value in the extracellular space of the brain. We should be able to detect such an imbalance by the new perfusion sampling techniques such as that used by Hamberger and Lehmann (this volume).

Measurement of flux across the brain capillary barrier needs to be comprehensive. It is not enough to measure unidirectional flux by the admittedly quick and convenient carotid bolus, reference tracer method (Oldendorf, 1971, Pardridge and Oldendorf, 1975) or by one of the various steady state methods (Pratt, 1985, Smith, this volume). Unidirectional influx needs to be supplemented either by the measurement of efflux (usually difficult), extracellular space sampling, or net flux. The latter can be measured in various ways, e. g., by arterial-venous differences of carrier amino acids. Steady state measurement methods, though less convenient are much more powerful. Thus their sensitivity to low fluxes is some orders of magnitude greater. The time course of flux changes may provide compartment space information. Efflux measurements are more likely to be possible by perfusion methods (Grennwood et al., 1985, Smith, this volume).

NATO ASI Series, Vol. H20
Amino Acid Availability and Brain Function in Health and Disease. Edited by G. Huether

DISCUSSION-REPORT

How significant is efflux of amino acids from the brain in maintaining the stability of brain amino acids pools and in achieving new steady-states of brain amino acids concentrations?

Harper: Concentrations of branched-chain amino acids increase in brains of animals consuming high protein diets. Might the increased concentrations of branched-chain amino acids contribute to the stability of brain concentrations of other amino acids under these conditions by competing with the other amino acids for exit and reducing their rates of efflux?

In this context, I would like to draw your attention to the general importance of efflux and the significance of efflux in relation to the achievement of steady states, with stability of the amino acid concentrations, or the achievement of changes which disturb this steady state and result in substantial fluctuations. I wonder, if somebody can give us some information on the potential of this very significant factor.

Qu. Smith: At present we know very little about the mechanisms that mediate amino acid efflux from brain. It is likely that neutral and basic amino acids are transported out of the central nervous system in part by the same brain capillary carriers that mediate their influx. These carriers (tentatively identified as the L1 and y + systems) are passive, bidirectional and function primarily to facilitate amino acid exchange between plasma and brain extracellular fluid. In addition to these exchange carriers, there may be other transport systems that mediate active amino acid efflux from brain. Recent studies have identified two active transport carriers, the A and the ASC systems, in brain capillary endothelial cells and have suggested that these carriers function to transport amino acid out of brain (Betz and Goldstein, 1975, Science 202: 225—226; Tayarani et al., 1987, J Cereb Blood Flow Metabol 7: 585—591). Although it is difficult at this time to speculate on the role of efflux in maintaining stability in brain amino acid pools, it seems likely that efflux, through active transport, has a critical role in producing the large concentration gradient that exists for most amino acids between plasma and brain extracellular fluid. For most amino acids, concentrations in brain extracellular fluid (as estimated from cerebrospinal fluid) are only 10—30 % of those in plasma (Snodgrass et al., 1969, Am J Physiol 221: 214—217).

Bradberry: I would like to draw your attention to the question, what are the parameters that govern the diffusion of organic molecules through the extracellular compartment? It is not simple linear diffusion which one sees in solutions. It is tortuous diffusion around neurons. This results in apparent diffusion coefficients much lower than are determined in a beaker. It has also been reported that aromatic amines diffuse much more slowly than aromatic acids. It was originally thought this was a charge effect, but it is now thought that neuronal and extraneuronal uptake processes may contribute to this difference. The relevant papers here are: Rice, et al., Neurosci., 15, 891—902 (1985); and Nicholson and Rice, Can. J. Physiol. Pharmacol., 65, 1086—1091 (1987).

Does neuronal activity affect the precursor availability independently from precursor supply:

Huether: Most data on brain amino acid uptake have been obtained using anaesthetized animals. It may not only be asked whether the kinetic models are adequate, but also how far they are applicable to stimulated states of brain activity. Does neuronal stimulation affect regional blood flow and how would such changes alter the regional amino acid uptake? This is not unimportant since serotoninergic fibres do release serotonin at blood vessels and serotonin is known to alter tone and permeability of the vasculature. Does the stimulation of serotoninergic systems affect brain amino acid uptake and precursor-availability?

Bruinvels: With regard to tyrosine uptake and tyrosine availability for catecholamine synthesis, one observation which may be of interest, was made by us several years ago, when we studied synaptosomal tyrosine uptake under conditions of sodium depletion. We found an increased tyrosine uptake when we used a sodium deficient buffer. The question is, is sodium deficiency near the postsynaptic membrane physiologic or not? During the depolarization of the neuron we find a removal of sodium ions from the membrane surface. So you will have during depolarization a small localized area in the membrane where sodium is removed and where tyrosine is taken up more rapidly. If you use a label to monitor the fate of labelled tyrosine which is taken up more rapidly under sodium deficient conditions you will find an increased production of labelled CO_2, indicating an increased formation of catecholamines. Interestingly enough, if you do the same thing with tryptophan you find the same increased uptake, but you do not find an increased formation of serotonin under these conditions.

Is the free or the bound tryptophan the crucial determinant of tryptophan availability for serotonin synthesis?

Wurtman: For several years controversy has ensued concerning whether the availability of tryptophan to the brain depends on its binding to circulating albumin (which presumably restricts its passage across the blood-brain barrier) or its competition with other large neutral amino acids (LNAA) for brain uptake. I think we can now resolve this scientific disagreement: Both of these factors have been shown sometimes to influence brain tryptophan levels. Most of the time, the competition among LNAA is most important. But situations do exist — when plasma free fatty acid levels have been highly elevated (e. g., with sympathetic stimulation) — when the binding of tryptophan to albumin (or its lack) becomes an important determinant of brain tryptophan.

Curzon: It is a pity that the word "control" has been so often used when discussing this question as it tends to be interpreted in an "either/or" way. "Influences" is better as it implies the existence of more than one variable that affects brain tryptophan concentration. As for whether competition is operative most of the time .... well, yes, I would agree insofar as if the competers were not present in plasma then the availability of tryptophan to the brain would be much greater. How important are feeding-provoked changes of competition? Physiologically, probably rather slight, pathologically, perhaps more important.

Chapter III.

INFLUENCE OF ALTERED PRECURSOR AMINO ACID SUPPLY ON TRANSMITTER METABOLISM AND BRAIN FUNCTION

PPROCESSES THAT COUPLE AMINO ACID AVAILABILITY TO NEUROTRANSMITTER SYNTHESIS AND RELEASE

Ian N. Acworth, Matthew J. During and Richard J. Wurtman

Department of Brain and Cognitive Science,
Massachusetts Institute of Technology,
Cambridge, Massachusetts, 02139, USA

INTRODUCTION

Certain neurotransmitters are synthesized from precursor amino acids (e. g. catecholamines from tyrosine; serotonin [5-hydroxytryptamine or 5HT] from tryptophan) which must be obtained from the circulation. The concentrations of tyrosine and tryptophan, as well as of the other large neutral amino acids (LNAA) in plasma, are subject to wide variations (1, 2), changing not only when people or animals receive various drugs (3, 4) or the amino acids themselves (5, 6), but also in disease states affecting amino acid metabolism (7, 8), and, of major physiological relevance, in association with eating (1, 2) or with strenuous exercise (9, 10, 11). As will be described below, these variations often cause parallel changes in the levels of these amino acids within the monoaminergic neurons, and thereby may influence both neurotransmitter synthesis and even central neurotransmission (12).

The effects of particular macronutrients, particularly dietary proteins and the insulin-releasing carbohydrates, on the blood and brain have been best studied: Carbohydrates tend to decrease plasma levels of the LNAA, whereas dietary proteins contribute these compounds to the blood stream, thereby increasing their plasma concentrations (1, 2). Individual LNAA vary not only in their responses to insulin but also in their abundance within dietary proteins. Hence, food-induced changes in their plasma levels are not uniform: For example, a high-protein meal can almost double plasma isoleucine concentrations while increasing plasma tryptophan and phenylalanine levels by only 30—50 % (1). Moreover, as detailed below, each of the circulating LNAA competes with the others for uptake into the brain, so that a large increase in some of LNAA (leucine; isoleucine; valine) as occurs after protein consumption, coupled with a smaller increase in the others (tryptophan or phenylalanine), can actually *reduce* brain levels of the latter compounds (13, 14). Fortunately, the relationships between plasma amino acid patterns and brain LNAA levels are predictable, if complex (12), and there are only a very limited number of ways in which real foods can change these patterns. [The brain, depending on whether its tryptophan levels do or do not rise after a meal, tends to "read" the meal as either high-carbohydrate-low-protein or other (15)].

In order for changes in brain tryptophan or tyrosine levels to affect serotonin or catecholamine synthesis, their concentrations within the monoaminergic terminals (the locus

NATO ASI Series, Vol. H20
Amino Acid Availability and Brain Function in Health and Disease. Edited by G. Huether

at which most of the neurotransmitter molecules are synthesized) must be allowed to vary. This requires both the *lack* of a regulatory mechanism (for keeping intraneuronal levels constant) and the presence of a transport system (for moving the precursor from the brain's extracellular space into the nerve terminal) which is not saturated under normal circumstances. Additionally, the enzyme within the nerve terminal which initiates the precursor's conversion to its neurotransmitter product (tryptophan hydroxylase for serotoninergic neurons; tyrosine hydroxylase for those releasing catecholamines) must also be unsaturated with the precursor amino acids (i. e. it must have a relatively high K_m for its substrate). Although these two requirements may seem obvious or even simplistic, they are worth pointing out because of the number of neurotransmitter candidates for which they are *not* satisfied. For example, the transmitter gamma-aminobutyric acid (GABA) is synthesized from a brain amino acid, glutamate, however brain glutamate levels are regulated within narrow limits by a glutamate efflux system at the blood-brain barrier (16). One large family of neurotransmitters, the neuropeptides, almost certainly is not subject to direct precursor control, inasmuch as brain polysomal activity and thus protein synthesis apparently are not influenced by physiological variations in plasma amino acid composition (at least in the developed brain) (17).

It seems contrary to homeostatic principles for variations in nutrient intake to be able to dictate brain composition and even function. One is left pondering the evolutionary advantages in having such an open-loop system, and one for which there are no obvious endocrine counterparts. As discussed below this system may be important in regulating the relative intakes of macronutrients.

Major aberrations in the plasma amino acid pattern may underlie the neurological disturbances of a number of disease states (e. g. phenylketonuria, maple syrup disease, tyrosinemia, hepatic encephalopathy) (18). Disturbances in plasma amino acid levels may also contribute to abnormal eating behaviors (19).

Macronutrient Intake and Brain Serotonin

The ability of nutrients to affect brain composition was first demonstrated in 1971, in experiments in which rats consuming meals containing carbohydrate and fat (i. e. lacking protein) were found to have, soon thereafter, increased brain levels of the essential (and scarce) amino acid tryptophan (20). They also exhibited increased brain levels of serotonin and its metabolite 5-hydroxyindole acetic acid (5HIAA) (20) — findings compatible with the view that the rise in brain tryptophan had increased the substrate-saturation of tryptophan hydroxylase, the initial and rate-limiting enzyme in serotonin synthesis (21). (The change in 5HIAA was taken as indirect evidence that serotonin's release had increased in parallel.) The increase in brain tryptophan was initially thought to have resulted from the small increase in plasma tryptophan concentration which also occurs after carbohydrate intake (in rats but not in humans). It was unclear at that time why both plasma and brain tryptophan levels should *rise* following ingestion of a meal *deficient* in tryptophan, especially as insulin release (in response to carbohydrate) was known to *lower* the plasma levels of most other LNAA (22). This unusual response to carbohydrate intake became clear several years later when tryptophan's binding to serum albumin was taken into account (23, 24, 25). Insulin, the major antilipolytic hormone, increases the transfer of free fatty acids from serum albumin to adipocytes. This, in turn, diminishes the free fatty acid content of the albumin, *increasing* its affinity for tryptophan (24). Plasma levels of free

(i. e. non-albumin bound) tryptophan actually fall in response to insulin, but these are compensated by the rise in albumin-bound tryptophan — a moiety that is almost as accessable to brain as the "free" fraction (26).

When rats were fed protein-rich meals, there appeared at first to be no relationship between plasma and brain tryptophan levels: Although plasma tryptophan levels rose (derived from tryptophan in the meal's protein), brain trypthophan and serotonin either failed to rise or, if the meal contained sufficient protein, actually fell (13). It was suggested that this phenomenon reflected competition for brain uptake between tryptophan and the other LNAA — which are more abundant in protein, an explanation supported by Pardridge and Oldendorf's subsequent characterization (27) of the mechanism that transports the LNAA across the blood-brain barrier (28). Tryptophan shares the LNAA carrier — a facilitated diffusion system — with other LNAAs (tyrosine, phenylalanine, histidine, leucine, isoleucine, valine, threonine, methionine); its affinity for tryptophan (0.052 mM) is less than that of phenylalanine (0.032 mM), but in the same order as those of tyrosine, leucine and methionine (28). The carrier's net affinity for tryptophan; the amino acid's net flux into the brain; and, after sufficient time, even brain tryptophan levels can all be estimated by calculating a "plasma tryptophan ratio": The ratio of the plasma tryptophan concentration to the summed concentrations of the other LNAA which bind to the carrier with a reasonably high affinity (see above) (13). Since all dietary proteins are considerably richer in the competing LNAA than in tryptophan (which generally comprises only 1.0—1.5 % of protein), consumption of a meal that is rich in protein (e. g. 30 to 40 % of calories) can cause the plasma tryptophan *ratio* to fall, even as plasma tryptophan *levels* rise. Similar competitive mechanisms mediate the fluxes of tryptophan and other LNAAs between the brain's extracellular space and individual neurons, however this competition is not limiting because the V_{max} of transport at this locus is tenfold that at the blood-brain barrier (28). Moreover, similar plasma ratios predict brain levels of each of the other LNAAs after treatments that modify plasma amino acid patterns (14). There is some evidence that blood-brain barrier transport macromolecules for LNAA and other ligands might be subject to up- or down-regulation after prolonged changes in plasma amino acid patterns — as might occur with inborn errors of amino acid metabolism, or in experimental animals given a single type of meal (high- or low-protein) chronically (28): However, the usual eating patterns of rats, men, and other omnivores, with the varieties of macronutrient composition represented in the foods and the frequent variation in meal protein contents, would not be expected to trigger up- or down-regulation.

It seems counterintuitive that meals which most effectively raise brain tryptophan are those lacking tryptophan entirely (i. e. containing carbohydrates but no proteins), while protein-rich meals, which elevate plasma tryptophan concentrations substantially, have opposite effects on the brain. Recently, Yokogoshi et al. (15) demonstrated that addition of *as little as 5 %* of a protein, casein, to a 70 % carbohydrate diet was sufficient to *block the increase* in both the plasma tryptophan ratio and brain tryptophan observed with carbohydrate alone (Table 1). It thus appears that the rat's brain "recognizes" — by whether or not they increase its tryptophan and serotonin levels — just two main kinds of meals, i. e. those rich in carbohydrate and lacking in protein, or those either containing 5 % or more protein, or lacking carbohydrate, or both. (These two types of meals can produce, in non-fasting humans, plasma tryptophan ratios that differ by about 50 % (29) (Fig. 1)). This ability to distinguish between the two kinds of meals coincides with the omnivore's need to divide its food intake between foods which are almost protein-free (fruits, honey) or low in protein (grains); and those which are rich in protein (meats, eggs). The om-

TABLE 1

Effect of dietary protein content on plasma tryptophan levels and Ratios

DIET		TRYPTOPHAN LEVEL (nmol/ml)	TRYPTOPHAN RATIO (TP/LNAA)
Fasting		53	.197
Casein	0%	82*	.475*
Casein	2.5%	75	.330*
Casein	5%	74	.244
Casein	10%	90*	.185
Lactalbumin	10%	165*	.369*
Egg White	10%	110*	.278
Peanut Meal	10%	60	.225
Gelatin	10%	40	.143

Rats trained to eat between 9 AM and 2 PM daily received the indicated food at 9 AM and were killed two hours later. All diets contained 70% dextrose. * $p<0.05$ (fasting). From Yokogoshi & Wurtman, *Metabolism,* 1986. (Ref. 15).

nivore's problem is not distinguishing foods with 20 % protein from those containing 40 % protein; rather it is telling either type from foods containing little or no protein at all. For the human, this translates into distinguishing a breakfast of eggs, bacon, and milk (ca. 30 % protein) from one of toast, jelly and juice (ca. 5 % protein). For the bear it seems knowing when to stop eating the honey and go catch a fish.

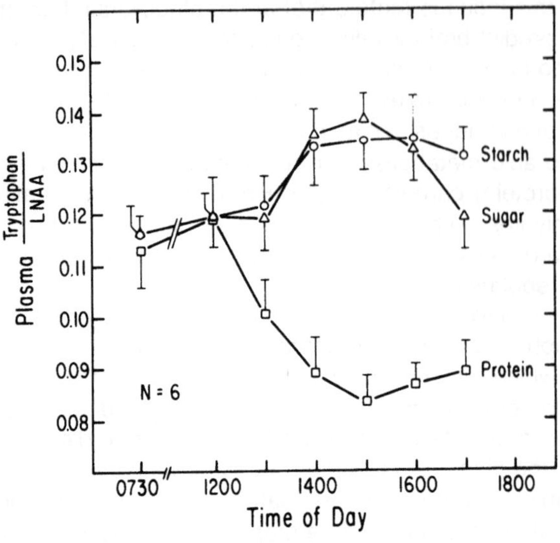

FIGURE 1

The effects of approximately-isocaloric lunch meals on tryptophan/LNAA ratios (mean±SEM). Six male subjects consumed a standard breakfast meal (220 calories; 40 g carbohydrate, 4 g protein) at 8 AM. At noon on three days, subjects a+e, in random order, one of three lunches (612—672 calories) consisting mainly of starches, sucrose, or proteins. (Ref. 29).

In normal humans, plasma tryptophan ratios can vary between 0.065 and 0.160 (1), depending largely on the composition of the last meal (or snack) eaten, and the interval that has passed since its ingestion. Such variations are sufficient, in the rat, to cause sizable differences in brain tryptophan levels and serotonin synthesis (20). Subnormal plasma tryptophan ratios are seen in obese people, reflecting elevated plasma levels of the branched-chain amino acids (30, 31), perhaps caused by insulin resistance; these ratios are further reduced if the subjects are put on a high-protein, low carbohydrate diet (32). In contrast, the plasma tryptophan ratio tends to be elevated (33) in liver failure, reflecting both decreased hepatic catabolism of tryptophan and decreased plasma levels of the branched-chain amino acids, probably secondary to the hyperinsulinemia associated with this condition (34) (which would facilitate uptake of these amino acids into muscle). Brain levels of tryptophan and its metabolites are increased in parallel with the plasma tryptophan ratio among rats with liver failure (35).

It had been known since 1965 that administration of pure tryptophan could increase brain serotonin synthesis (36) and thereby affect a number of serotonin-dependent brain functions (e. g. alertness; mood state (37)). What was novel about the above findings was the demonstration that brain tryptophan levels — and serotonin synthesis — are influenced by *normal* meal-to-meal variations in what one eats. Although administration of a carbohydrate meal or of tryptophan itself could clearly be shown to accelerate serotonin *synthesis* (e. g. the rate at which brain accumulated an intermediate, 5-hydroxytryptophan, in animals also given a decarboxylase inhibitor (38)), there was debate as to whether serotonin *release* was enhanced accordingly. It was known that in animals given drugs which increase intraneuronal serotonin levels (monoamine oxidase inhibitors) or act intrasynaptically to enhance serotonin-mediated neurotransmission (reuptake blockers), one or more feedback systems, based on slowing raphe firing or activating serotonin autoreceptors, could operate to diminish subsequent serotonin release (39, 40). A similar effect could be seen in rats given very large doses of tryptophan (41); this was taken as evidence of a homeostatic mechanism designed to keep serotonin release (and thus serotonin-mediated neurotransmission) within the physiological range. However, when rats were given *small* doses of tryptophan, which increased brain tryptophan but kept it within the peak levels seen under normal physiological conditions, *no* decrease in raphe firing was observed (42). Hence it seems that food-induced changes in serotonin synthesis are, in fact, "allowed" to modulate the amounts of serotonin released from raphe neurons per firing without activating a compensatory mechanism to slow the neurons' firing frequencies. How the brain might take advantage of this property is discussed below.

More recently two techniques have been used to assess changes in serotonin release, i. e. *in vivo* voltammetry, which measures the levels of serotonin's metabolite 5HIAA in the brain's extracellular fluid (e. g. after animals receive tryptophan) and superfused electrically stimulated brain slices. In the former experiments, tryptophan loading did in fact increase 5HIAA levels (43). In the latter, addition of tryptophan to the superfusion medium in concentrations that caused twofold variations in its levels within the slices, (i. e. a range that occurs normally in brains of rats consuming carbohydrate-rich vs. protein-rich meals) did cause dose-related increases in serotonin release, in proportion to the increases in serotonin levels within the slices (44) (Fig. 2).

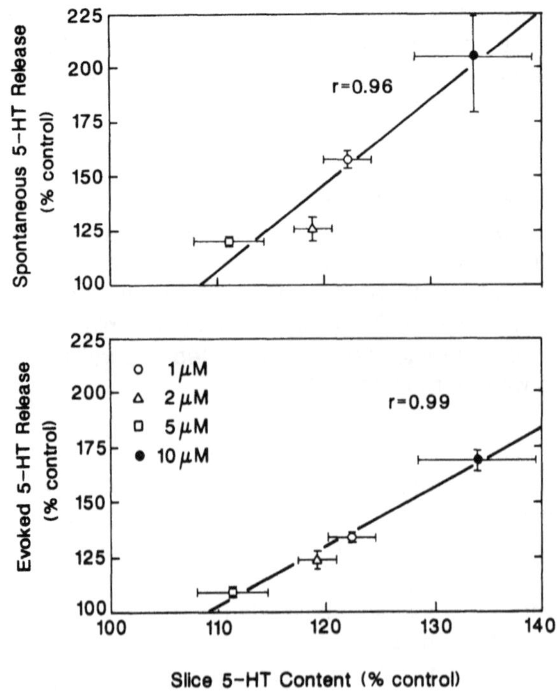

FIGURE 2
Spontaneous and electrically-evoked serotonin release were measured over an eighty minute period, fifty minutes after the start of the experiment. Slices were electrically stimulated (1400 bipolar pulses, 5 Hz, 2 ms, 100 mA/cm^2) at three time points during the experiment (at 60, 85 and 110 minutes). After superfusion, individual slices were assayed for serotonin (5 HT), and these levels were correlated with serotonin release.

Tryptophan, Dietary Carbohydrate, and the Human Brain

Numerous behavioral and neurological effects have been associated with tryptophan administration, starting with Smith and Prockop's original observation that it causes drowsiness and euphoria (45). Since then many clinical reports have described effects of tryptophan, in normal subjects and in those with, for example, depression (46) or agitation (47). Tryptophan administration can also affect neuroendocrine secretion, its i. v. administration increasing plasma prolactin and cortisol levels (48). [There is still debate regarding the mechanism of these responses, however the prolactin response probably is mediated by increasing serotoninergic neurotransmission, inasmuch as methysergide, a serotonin antagonist, inhibited the prolactin response to i. v. tryptophan (48)].

Apparently no data are available concerning neurochemical responses of the human brain to carbohydrate intake. There are, however, some data that allow comparisons of *behavioral* responses to dietary carbohydrates (49) or to tryptophan itself. Thus, sugar administration to children with attention deficit disorder tended to *reduce* motor activity, (similar to tryptophan's effect in normal individuals), and a high-carbohydrate lunch increased *sleepiness* in women, calmness in men, and, in subjects over the age of forty, tended to increase *errors* in a standardized test of performance (50). Hyperactive children reportedly consume larger quantities of sugar than normal children in an uncontrolled environment (51). This might reflect a greater need for energy, or, alternatively an unwitting attempt at self-medication. A generalized increase in carbohydrate consumption (non-sweet starches as often as sweets) is usually seen in patients suffering from SADS (seasonal affective disorder syndrome) (52), and episodic carbohydrate cravings characterise many people with obesity (53). Such cravings are also described by girls with normal-weight bulimia, and women with premenstrual syndrome (54). We have hypothesized that this overconsumption of carbohydrates (without protein) might constitute an attempt to overcome a dysphoria perhaps attributable to the "inadequate" release of serotonin from central raphe neurones. Some support for this hypothesis is derived from the evidence that levels of serotonin and 5HIAA often are subnormal in brains of people dying from suicide (55). Moreover, virtually all anti-depressant drugs are thought to enhance serotoninergic neurotransmission.

Brain Serotonin and Macronutrient Selection

If rats, an omnivorous species, are presented with two diets, differing in their proportions of protein and carbohydrate, the animals tend to choose among the two so as to obtain fairly constant (for each animal) amounts of these macronutrients (56, 57). The protein/carbohydrate ratio tends to be low (58) soon after weaning; thereafter some animals increase it by selecting a higher protein intake, while others maintain juvenile nutrient-choice patterns. The long-term implications of this difference between animals — e. g. for vulnerability to development of obesity — await characterization. Other physiological states — like lactation or ovulation — can also affect the dietary protein/carbohydrate ratio (59). If however, the animals consume a high carbohydrate "snack" (60) or are given a drug that facilitates serotoninergic neurotransmission (58) prior to having access to the two test diets, they subsequently modify their food intake so as to consume less carbohydrate. Such observations suggest that the brain may normally use food-induced changes in serotonin release in order to maintain macronutrient balance, and that disturbances in this mechanism can contribute to diseases of appetite. A recent study examined the effects on elective protein and carbohydrate intakes of three serotoninergic drugs, i. e. d-fenfluramine, which both releases serotonin and blocks its reuptake, and fluoxetine or CGS 10686B, which are reuptake blockers (61). All three drugs selectively suppressed carbohydrate intake, but only if the composition of test diets was such that one of the diets, eaten alone, would have increased brain tryptophan and serotonin levels and the other diet would not. (The drugs selectively suppressed consumption of the first of the diets.) Experiments designed to study mechanisms of macronutrient selectivity would thus do well to include one meal containing 5 % or less protein, and the other rich in protein. The sweetness of the carbohydrate does not seem to matter.

In humans, similar nutrient-choice mechanisms appear to operate and, as is often the case, these mechanisms are easiest to demonstrate in subjects — carbohydrate-cravers — in which they are disturbed. Subjects at MIT's Clinical Research Center were allowed to choose from among six different isocaloric foods (containing varying amounts of protein and carbohydrate but constant fat) at each meal; between meals they had unlimited access to a computerized vending machine stocked with isocaloric and iso-fat mixed snacks (either high in carbohydrate or high in protein). It was observed that the basic parameters of each person's food intake, i. e. the total number of calories; grams of carbohydrate and protein; number and composition of snacks; tended to vary within only a narrow range from day to day and was unaffected by a cellulose placebo (53, 62).

To assess the possible involvement of brain serotonin in this regulation, meal and snack intakes were measured in obese "carbohydrate cravers" (previously shown to consume large quantities of carbohydrate-rich snacks, usually at a characteristic time of day, and often in association with dysphoric feelings) given placebo or a low dose (15 mg b.i.d.) of d-fenfluramine. The d-fenfluramine caused a major reduction in snack carbohydrate intake (62); a smaller but significant reduction in mealtime carbohydrate; and no significant changes in mealtime protein nor fat intake. (Too few protein-rich snacks were consumed to allow assessment of the drug effects on such snacks).

Severe carbohydrate-craving is also a symptom of SADS, a variant of bipolar depressive disorder associated with a late autumn onset; a higher frequency in populations that are far from the equator; and concurrent hypersomnia and weight-gain (52). Since serotoninergic neurons appear to be involved in both appetite control and mood state they may constitute the link between a patient's eating behavior and his or her affective symptoms: People with inadequate serotonin-mediated neurotransmission, for whatever reason, might present to their physicians complaining of obesity, reflecting their "overuse" of dietary carbohydrates to treat their dysphoria, or, alternatively of seasonal depression (found, on questioning, to be associated with concurrent carbohydrate craving and weight gain). Alternatively they might seek treatment for normal-weight bulimia (and describe, on questioning, signs of depression and the desire to binge on carbohydrates), or even the premenstrual syndrome. Other syndromes presenting with disturbances of meal and appetite — and, perhaps also, of other serotonin-dependent behaviors like sleep and ag-

TABLE 2

Evidence that carbohydrate consumption is regulated

1. A rat choosing food from various diet pairs tends to consume constant proportions of CHO (and of protein).
2. Protein intake by rats increase during pregnancy and lactation; the proportion of CHO thus declines.
3. Consumption by rats of a CHO (or protein) "pre-meal" decreases subsequent mealtime CHO (or protein) intake.
4. Rats deprived of CHO for several days exhibit "CHO-craving" thereafter (even for non-sweet CHO).
5. Serotoninergic drugs selectively diminish CHO intake by normal rats; non-serotoninergic anorexics (like d-amphetamine) lack this effect.
6. Placement of serotonin or d-norfenfluramine within the rat's hypothalamus selectively decreases CHO intake.
7. "CHO-craving" is a characteristic finding in several human disorders (e.g., SADS; PMS; CHO-craving obesity; normal-weight bulimia) and in patients treated with non-serotoninergic antidepressant drugs.
8. Administration of d-fenfluramine has been shown to suppress the CHO-craving in obesity and in SADS.

gression — might also be anticipated. The participation of serotoninergic neurons in a large number of brain functions might have the effect of making these functions "hostages to eating" (seen, for example, in the sleepiness that can follow carbohydrate intake), just as it could cause mood-disturbed individuals to consume large amounts of carbohydrates for reasons related neither to nutritional value nor the taste of these foods.

Evidence that carbohydrate consumption normally is *regulated* is summarized in Table 2.

Exercise, Precursor Availability and Monoaminergic Neurotransmission

Plasma concentrations of the precursor amino acids are also influenced by prolonged exercise and stress, and perhaps by additional physiological conditions as well. Although preliminary observations of the effects of exercise on some amino acids were published three decades ago, only recently have well-controlled studies examined the consequences of prolonged exercise on plasma levels of the precursor amino acids. Endurance exercise (greater than sixty minutes) in rats has been shown to increase total (9) and free (63) tryptophan, as well as their corresponding plasma LNAA ratios. Consistent with these findings, both brain tryptophan and 5HIAA levels also increased. In humans, Decombaz *et al.* showed that following a 100 kilometer run, plasma amino acid concentrations were dramatically altered (11): In general, aromatic amino acid levels increased, while those of the branched chain amino acids decreased. Unfortunately tryptophan was not measured. More recently we have shown that levels of both plasma total and free tryptophan levels (and their LNAA ratios) were significantly increased following the Boston Marathon in fit, well trained runners (10). Although no indices of brain serotonin release were measured, plasma prolactin levels (an *indirect* index of such release (48)) also rose, correlating with the plasma tryptophan ratio and perhaps brain tryptophan levels.

Tyrosine and phenylalanine availability to the human brain also increase post exercise (10, 11). The increase in dopamine turnover (64, 65, 66) seen following increased activity or exercise in rats may, in part, be secondary to increased precursor availability.

The changes in central monoaminergic neurotransmission following prolonged exercise may be partly responsible for changes in mood and behavior reported in both normal and mildly depressed individuals (67, 68). A parallel may be drawn between the reported euphoria with extremes of exercise and the mood-elevating effects of carbohydrates, acting through a serotoninergic mechanism as discussed above. (Although endogenous opiates have been proposed as causing the mood elevations that can follow exercise, this elevation is not blocked by naloxone, an opiate antagonist; moreover circulating endorphins apparently are not able to cross the blood-brain barrier).

When Will Nutrient Intake Affect Neurotransmission?

On the basis of the tryptophan-serotonin relationship, one can formulate a sequence (12) of biochemical processes that would have to occur in order for any nutrient to affect the release of its neurotransmitter product:

1). Plasma levels of precursor (as well as any competitors) must *not* be under tight homeostatic control, and must be allowed to vary after its administration or its ingestion as a constituent of foods. Plasma levels of LNAA do, in fact, vary several-fold after consumption of normal foods (1, 2).

2). The brain level of the precursor must depend on its plasma level; a carrier must exist to transport the precursor across the blood-brain barrier and this carrier must not be fully saturated with its ligand. Such carriers do, in fact, exist (16).

3). The enzyme catalyzing the first (and rate-limiting) step in the conversion of precursor to its neurotransmitter product must be unsaturated with its substrate, and must not be susceptible to end-product inhibition. Both criteria are satisfied for tryptophan hydroxylase, also for tyrosine hydroxylase when the enzyme has been activated (see below).

Available evidence suggests that only a few neurotransmitters are subject to precursor control — principally the monoamines discussed above (serotonin and the catecholamines), acetylcholine (69), histamine (70), and glycine (71). Pharmacological doses of histidine or threonine can elevate brain concentrations of their respective neurotransmitter products, histamine (70) and glycine (71). As discussed by J. Growdon elsewhere, threonine administration also elevates CSF glycine concentrations in humans (72), possibly with useful clinical consequences.

It cannot be stated for certain whether or not the excitatory amino acid neurotransmitters (aspartate, glutamate) are or are not under precursor control, because the precursors for these compounds within *glutamatergic* or *aspartatergic* neurons await possible discovery. Similarly, experiments have not yet been done to assess effects on GABA synthesis caused by raising brain levels of GABA's precursor, glutamate, because exogenous glutamate fails to cross the blood-brain barrier.

In order for a precursor-mediated increase in neurotransmitter synthesis to affect its release the neuron involved must continue to fire at its normal frequency. Feedback processes operating both via short presynaptic (autoreceptor) or long post-synaptic loops may prevent this from happening. The absence of overt physiologic effects after normal humans consuming tyrosine (5) or choline (as lecithin) (73) suggests that these controlling mechanisms are operating. However, circumstances *do* exist in which receptor-mediated feedback mechanisms apparently are *not* activated by increases in transmitter release, thus allowing precursor administration to amplify neurotransmission. Such circumstances include, among others, *neurodegenerative disorders,* in which surviving neurons will continue to exhibit accelerated firing frequencies so long as the total quantity of neurotransmitter released from the afflicted tract or nerve is less than adequate; *physiological states,* in which a prolonged increase in neurotransmitter is required (e. g. to sustain blood pressure in hypovolemia, or blood glucose after insulin administration); and *neurons not "hard-wired" into multisynaptic feedback loops,* or those serving as components of *positive* feedback loops.

Precursors also are "allowed" to amplify transmitter release after drug treatments which cause release to be inadequate (e. g. tyrosine after reserpine treatment; tryptophan after treatment with drugs which activate raphe neurons).

Tyrosine Effects on Dopamine and Norepinephrine Synthesis

Because tyrosine administration generally does not increase brain dopamine or norepinephrine levels among otherwise-untreated animals, most investigators assumed that catecholamine synthesis was not under precursor control. This apparent lack of precursor-responsiveness was perhaps surprising, in view of the facts that plasma tyrosine levels were known to increase severalfold after protein intake (1, 2) or tyrosine administration (5); that the LNAA transport system was as capable of ferrying tyrosine as tryptophan across the blood-brain barrier (16); and that tyrosine hydroxylase, which catalyses the rate-limiting step in catecholamine synthesis was, like tryptophan hydroxylase, likely to be unsaturated with its amino acid substrate *in vivo* (74). The possibility was considered that a small and rapidly-turning-over catecholamine pool did exist which was tyrosine-responsive, but that this pool was *too* small, in relation to total catecholamine stores, to allow detection of an increase in its size after tyrosine administration. Initial studies were therefore conducted to determine whether catecholamine *synthesis* or *release,* assessed independently of brain catecholamine *levels,* might be affected by changes in brain tyrosine concentrations. At first, catechol synthesis was estimated by following the rate at which L-dopa accumulated in brains of animals treated acutely with an inhibitor or aromatic acid decarboxylase: Tyrosine administration did, in fact, accelerate L-dopa's accumulation, while administration of the other LNAAs decreased both it and brain tyrosine levels (75). Catecholamine *release* was then estimated by measuring brain levels of the major metabolites of dopamine (DOPAC; HVA) or norepinephrine (MHPG-SO_4) in animals given the amino acid. The administration of even large doses of tyrosine had no consistent effects on these metabolites. However, if the experimental animals were given an *additional* treatment designed to *accelerate* the firing of dopaminergic of noradrenergic tracts [e. g. dopamine antagonists (76); cold exposure (77); 6-hydroxydopamine lesions (78); reserpine (79)] the supplemental tyrosine now markedly enhanced the accumulation of the catecholamine metabolites. These initial observations formed the basis for the hypothesis that catecholaminergic neurons *become* tyrosine-sensitive when they are physiologically active, and lose this capacity when quiescent.

The biological mechanism that couples a neuron's firing frequency to its ability to respond to supplemental tyrosine involves phosphorylation of the enzyme tyrosine hydroxylase, a process that occurs when the firing frequency of a particular catecholaminergic neuron increases (80, 81). This phosphorylation, which is short-lived, enhances the enzyme's affinity for its cofactor (tetrahydrobiopterin) and makes it *in*sensitive to end-product inhibition (by norepinephrine and other catechols): These changes allow its net activity to depend on the extent to which it is saturated with tyrosine. An additional mechanism underlying this coupling may be an actual depletion of tyrosine within nerve terminals, as a consequence of its accelerated conversion to catecholamines: If slices of rat caudate nucleus superfused with a Krebs-Ringer solution (which lacks tyrosine or other amino acids) are depolarized repeatedly, they are unable to sustain their release of dopamine (82); concurrently their contents of tyrosine — but not of other LNAA — decline markedly. The addition of tyrosine to the superfusion solution enables the tissue to continue releasing dopamine at the initial rates, and also protects it against depletion of its tyrosine. The concentrations of tyrosine needed for these effects are proportional to the number of times that the neurons are depolarized. [This ability of supplemental tyrosine to maintain catecholamine release in the face of repeated firing apparently is not true for all catecholaminergic neurons: Norepinephrine release from superfused hypothalamic slices was not sustained by adding tyrosine to the medium (83)].

The coupling of tyrosine-responsiveness to neuronal firing probably explains tyrosine's paradoxical effects on blood pressure; the amino acid *elevates* blood pressure (and sympatho-adrenal catecholamine release) in hypotensive animals (84), but *lowers* blood pressure (without effecting sympatho-adrenal catecholamine release) in hypertensive animals (85). This latter effect probably results from its conversion to norepinephrine in brain stem neurons active in the depressor pathway; in support of this hypothesis, tyrosine increases brainstem MHPG-SO_4 levels in spontaneously hypertensive rats (85). [Tyrosine fails to affect blood pressure at all in normotensive humans and animals (85, 5)].

Clinically, supplemental tyrosine may be useful to treat some patients with early Parkinson's Disease (86), although its effect apparently is short-lived. It may also have some use in depression, given with or without 5-hydroxytryptophan or tryptophan (56, 86). Its utility in treating hypertension or other cardiovascular diseases (e. g. cardiac arrhythmias) awaits evaluation. The amino acid may also have some value in prophylaxis or treatment of stress responses: Rats subjected to tail-shock stress were found, immediately thereafter, to have depressed brain norepinephrine levels (particularly in the locus coeruleus and the hypothalamus), probably reflecting the inability of norepinephrine synthesis to keep up with its release (87); the animals also showed norepinephrine-related behavioral abnormalities and elevated plasma corticosterone levels (88). All of these changes, including the adrenocortical response, were suppressed by supplemental oral tyrosine (87, 88), but not if the tyrosine was co-administered with another large neutral amino acid (valine) that blocked its brain uptake.

In a preliminary study Bandaret *et al.* (89) investigated the effect of tyrosine on performance, symptoms and mood, of US Air Force pilot volunteers under situations of stress (a cold environment at low barometric pressure). Tyrosine enhanced performance, including reaction time, vigilance and complex information processing, and reduced subjective symptoms of coldness, muscle discomfort and headache. Mood states (e. g. anxiety and tension) were also improved.

Plasma tyrosine is derived from both the tyrosine and the phenylalanine in dietary proteins, since the latter amino acid is metabolized by being hydroxylated to tyrosine in the liver. Phenylalanine itself can apparently serve as a substrate for tyrosine hydroxylase, low concentrations partially sustaining dopamine release (in the brain slice preparation discussed above) when tyrosine is lacking. However in higher concentrations (200 micromolar) and in the presence of tyrosine, phenylalanine *inhibited* tyrosine's hydroxylation and suppressed dopamine efflux from the brain slice preparation (90).

We have recently started to explore the effects of the precursor amino acids, tyrosine and phenylalanine, on the *in vivo* release of brain dopamine, using the novel technique of intracerebral microdialysis (91). Tyrosine, in doses of 50—200 mg/kg i. p., increased striatal dopamine release, as did phenylalanine when given in doses (e. g. 200 mg/kg i. p) which are rapidly converted in the liver to tyrosine. Larger doses of phenylalanine (>500 mg/kg) significantly *inhibited* dopamine release, probably by competing with tyrosine for transport into the brain and by directly inhibiting tyrosine hydroxylase activity (92) (Fig. 3). Tyrosine's short-lived enhancement of striatal dopamine release is prolonged if the frequency of neuronal firing (and presumably tyrosine hydroxylase activity) is increased (by giving drugs like haloperidol or partially destroying catecholaminergic tracts by 6-hydroxydopamine). There appear to be regional variations in the sensitivity of groups of dopaminergic neurons to changes in precursor availability (93): Those in the nucleus accumbens are more responsive to precursor loading than those located in the striatum, a finding compatible with its the higher dopamine synthesis rate constant (94) and faster dopamine turnover of the former group (95) (Fig. 4). We did not observe a clear direct or reciprocal relationship, after tyrosine or phenylalanine administration, between dopamine release and levels of the dopamine metabolites DOPAC and HVA in extracellular brain fluid. This suggests that the widespread use of whole-brain metabolite levels to estimate dopamine release may not be warranted.

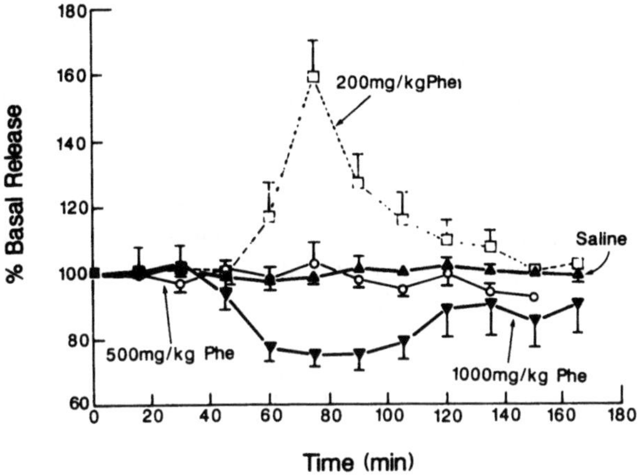

FIGURE 3
Effects of phenylalanine (200, 500 or 1000 mg/kg, i. p.) or saline on release of dopamine into striatal dialysate. Groups of rats (n = 4) were anaesthetized with alpha-chloralose/urethane (0.05/0.5 g per kg i. p.). Animals received the amino acid in saline at zero time; dialysates (1.5 μl/min) were collected at 15 min intervals and essayed for dopamine, DOPAC and HVA. No changes were observed in dialysate DOPAC and HVA levels.

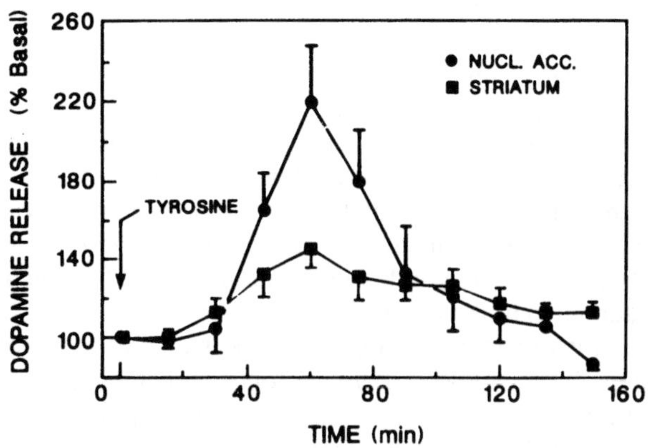

FIGURE 4
Effect of tyrosine (200 mg/kg i. p.) on release of dopamine into dialysates from stratium (n = 4) and nucleus accumbens (n = 4). Rats were anaesthetized with alpha-chloralose/urethane (0.05/0.5 g per kg i. p.) and received tyrosine (in saline) at zero time; dialysates were collected at 15 min intervals and assayed for dopamine, DOPAC and HVA. Control animals received saline at time zero. Saline administration failed to effect dopamine release in these areas.

The inhibition of dopamine release seen with high doses of phenylalanine might become clinically significant among people who consume very large quantities of the dipeptide sweetener aspartame (56 % phenylalanine). When aspartame is ingested alone (e. g. in beverages), or, especially, in combination with high-carbohydrate foods (which, by releasing insulin, deplete the plasma of competing LNAAs), the plasma ratio of phenylalanine to LNAA is significantly increased (96). This effect is *not* seen after protein consumption (1, 2) because as described above, although dietary protein elevates plasma phenylalanine levels, it causes equal or greater increases in plasma levels of the other LNAAs which are more abundant than phenylalanine. Hence the plasma phenylalanine ratio, and presumably, brain phenylalanine concentration are not increased. If the release of striatal dopamine (and perhaps of catecholamines elsewhere) is decreased in humans consuming large amounts of aspartame (as our preliminary dialysis data suggest), this effect could contribute to the neurological symptoms that sometimes appear to be temporally associated with aspartame consumption [e. g. seizures (97); precipitation of headaches (98)].

The competitive nature of brain LNAA uptake may underlie variations in the therapeutic effect of L-dopa, itself an LNAA. The "On-Off" effect in Parkinsonian patients receiving L-dopa is significantly worsened by ingestion of a high protein meal (99).

CONCLUSIONS

The ingestion of certain foods and pure nutrients may have major physiological effects on the central nervous system by modulating neurotransmission through the monoamines (and acetylcholine). Moreover, the macronutrient composition of the foods consumed electively appears to be regulated by central serotonin pathways, responding to the changes that foods bring about in plasma large neutral amino acid patterns. Diet, by altering these patterns, dictates the rate at which tryptophan enters the brain; saturates tryptophan hydroxylase; and, thereby enhances serotonin synthesis and release. Recent studies using *in vivo* voltammetry and superfused-stimulated brain slices provide indirect evidence that precursor-induced increases in brain serotonin levels do in fact result in increased serotonin *release*. A number of pathological conditions may be associated with impairment of the "macronutrient-plasma tryptophan-brain serotonin" axis; these include carbohydrate-craving obese patients, those with SADS, and some PMS sufferers. These patients may all overconsume carbohydrates in an attempt to self-medicate, i. e. to raise brain serotonin, which seems to make them feel better.

Brain catecholamine metabolism also appears to be sensitive to variations in precursor availability, at least in physiologically-active neurons. The catecholamines may be derived not only from dietary tyrosine, but also from dietary phenylalanine (which has been metabolized to tyrosine in the liver). Like tryptophan, tyrosine's entry into the brain is dependent on competition at the blood-brain barrier. Using *in vivo* dialysis, we have recently shown that increasing the availability of tyrosine to the brain increases the *release* of dopamine from both the striatum and nucleus accumbens. Moreover, exogenous phenylalanine inhibits dopamine release (by inhibiting dopamine synthesis).

Endurance exercise also alters the availability of tryptophan and tyrosine to the brain, and has been shown to change monoamine synthesis. These changes may participate in alterations in mood and behavior often found to accompany extremes of exercise.

ACKNOWLEDGEMENTS

These studies were supported in part by grants from The United States Air Force, The National Aeronautics and Space Administration, and The Center for Brain Sciences and Metabolism Charitable Trust.

REFERENCES

1) FERNSTROM, J. D., WURTMAN, R. J., HAMMARSTROM-WIKLUND, B., RAND, W. M., MUNRO, H. N., and DAVIDSON, C. S. (1979). Diurnal variations in the plasma concentrations of tryptophan, tyrosine and other neutral amino acids: Effect of dietary protein intake. *Am. J. Clin. Nutr.* **32:** 1912—1922.

2) MAHER, T. J., GLAESER, B. S., and WURTMAN, R. J. (1984). Diurnal variations in the plasma concentrations of basic and neutral amino acids in red cell concentrations of aspartate and glutamate: Effects of dietary protein intake. *Am. J. Clin. Nutr.* **39:** 722—729.

3) BADWAY, A. A.-B., and EVANS, M. (1981). Inhibition of rat liver pyrrolase activity and elevation of brain tryptophan by administration of antidepressants. *Biochem. Pharmacol.* **30:** 1211—1216.

4) VALZELLI, L., BERNASCONI, S., COEN, E., and PETKOV, V. V. (1980). Effect of different psychoactive drugs on serum and brain tryptophan levels. *Neuropsychobiol.* **6:** 224—229.

5) GLAESER, B., MELAMED, E., GROWDON, J. H., and WURTMAN, R. J. (1979). Elevation of plasma tyrosine after a single oral dose of L-tyrosine. *Life Sci.* **25:** 265—272.

6) FERNSTROM, J. D., and WURTMAN, R. J. (1971). Brain serotonin content: Physiological dependence on plasma tryptophan levels. *Science* **173:** 149—152.

7) CURZON, G. (1980). Transmitter metabolism and behavioral abnormalities in liver failure. In: *The Biochemistry of Psychiatric Disturbances.* (Curzon, G. ed.), pp. 89—111. Wiley, Chichester.

8) CURZON, G., KANTAMANENI, B. D., CALLIGHAN, N., and SULLIVAN, P. A. (1982). Brain transmitter precursors and metabolites in diabetic ketoacidosis. *J. Neurol. Neurosurg. Psychiat.* **45:** 489—493.

9) ACWORTH, I. N., NICHOLASS, J., MORGAN, B., and NEWSHOLME, E. A. (1986). Effect of sustained exercise on concentrations of plasma aromatic and branched-chain amino acids and brain amines. *Biochem. Biophys. Res. Commun.* **137**: 149—163.

10) ACWORTH, I. N., DURING, M. J., WATKINS, C. J., KREMPF, M. ELSENBERG, M., and WURTMAN, R. J. (1987). The effects of marathon running on plasma free and total large neutral amino acids and free fatty acids. In preparation.

11) DECOMBAZ, J., REINHARDT, P., ANANTHARAMAN, K., von GLUTZ, G., and POORTMANS, J. R. (1979). Biochemical changes in a 100 km run: Free amino acids, urea, and creatine. *Eur. J. Appl. Physiol.* **41:** 61—72.

12) WURTMAN, R. J., HEFTI, F., and MELAMED, E. (1980). Precursor control of neurotransmitter synthesis. *Pharmacol. Rev.* **32:** 315—335.

13) FERNSTROM, J. D., and WURTMAN, R. J. (1972). Brain serotonin content: Physiological regulation by plasma neutral amino acids. *Science* **178:** 414—416.

14) FERNSTROM, J. D., and FALLER, D. V. (1978). Neutral amino acids in the brain: Changes in response to food ingestion. *J. Neurochem.* **30:** 1531—1538.

15) YOKOGOSHI, H., and WURTMAN, R. J. (1986). Meal compositions and plasma amino acid ratios: Effect of various proteins and carbohydrates, and of various protein concentrations. *Metabolism* **35:** 837—842.

16) PARDRIDGE, W. M. (1979). Regulation of amino acid availability to the brain: Selective control mechanisms for glutamate. In: *Glutamic Acid: Advances in Biochemistry and Physiology* (Filer, L. J., Kare, M. R., Garattini, S., and Wurtman, R. J., eds.), pp. 125—137. Raven Press, New York.

17) BARRA, H. S., UNATES, L. E., SAAVERDRA, M. S., and CAPPUTO, R. (1972). Capacities of binding amino acids by t-RNAs from rat brain and their changes during development. *J. Neurochem.* **19:** 2287—2297.

18) STANBURY, J. B., WYNGAARDAN, J. B., and FREDRICKSON, D. S. (1966). The Metabolic Basis for Inherited Disease. McGraw-Hill Book Co., New York.

19) KAYE, W. H., GWIRTSMAN, H. E., BREWERTON, T. D., GEORGE, T. D., and WURTMAN, R. J. (1987). Bingeing behavior and plasma amino acids: A possible involvement of brain serotonin in bulimia nervosa. *Psychiatr. Res. (submitted).*

20) FERNSTROM, J. D., and WURTMAN, R. J. (1971). Brain serotonin content: Increase following ingestion of carbohydrate diet. *Science* **174:** 1023—1025.

21) BRADFORD, H. F. (1986). Chemical Neurobiology. W. H. Freeman and Co., New York.

22) FERNSTROM, J. D., and WURTMAN, J. R. (1972). Elevation of plasma tryptophan by insulin in the rat. *Metabolism* **21:** 337—342.

23) McMENAMY, R. H., and ONCLEY, J. L. (1958). The specific binding of L-tryptophan to serum albumin. *J. Biol. Chem.* **233:** 1436—1440.

24) MADRAS, B. K., COHEN, E. L., MESSING, R., MUNRO, H. N., and WURTMAN, R. J. (1974). Relevance of serum-free tryptophan to tissue tryptophan concentrations. *Metabolism* **23:** 1107—1116.

25) KNOTT, P. J., and CURZON, G. (1972). Free tryptophan in plasma and brain tryptophan metabolism. *Nature* **239:** 452—453.

26) YUWILER, A., OLDENDORF, W. H., GELLER, E., and BRAUN, L. (1977). The effect of albumin binding and amino acid competition on tryptophan uptake into brain. *J. Neurochem.* **28:** 1015—1023.

27) PARDRIDGE, W. M., and OLDENDORF, W. H. (1975). Kinetic analysis of blood-brain barrier transport of amino acids. *Biochem. Biophys. Acta.* **401:** 128—136.

28) PARDRIDGE, W. M. (1986). Potential effects of the dipeptide sweetener aspartame on the brain. In: *Nutrition and the Brain* (Wurtman, R. J., and Wurtman, J. J., eds.) **7,** pp. 199—241. Raven Press, New York.

29) LEIBERMAN, H., CABALLERO, B., and FINER, N. (1986). The composition of lunch determines afternoon plasma tryptophan ratios in humans. *J. Neural. Transm.* **65:** 211—217.

30) ASHLEY, D. V. M., FLEURY, M. O., GOLAY, A., MAEDER, E., and LEATHWOOD, P. D. (1985). Evidence for deminished brain 5HT biosynthesis in obese diabetic and non-diabetic humans. *Am. J. Clin. Nutr.* **42:** 1240—1245.

31) CABALLERO, B., FINER, N., and WURTMAN, R. J. (1987). Plasma amino acid levels in obesity: Effects of insulin resistance. In: *Amino Acids in Health and Disease: New Prospectives.* (Kaufman, S., ed.), pp. 369—382. Alan R. Liss Inc., New York.

32) HERAIEF, E., BUCKHARDT, P., MAURON, C., WURTMAN, J. J., and WURTMAN, R. J. (1983). The treatment of obesity by carbohydrate deprivation suppresses plasma tryptophan and its ratio to other neutral amino acids. *J. Neural. Transm.* **57:** 187—195.

33) FERNSTROM, J. D., WURTMAN, R. J., HAMMARSTROM-WIKLUND, B., RAND, W. M., MUNRO, H. N., and DAVIDSON, C. S. (1979). Diurnal variations in plasma amino acid concentrations in patients with cirrhosis: Effect of dietary protein. *Am. J. Clin. Nutr.* **32:** 1923—1933.

34) FERNSTROM, J. D., ARNOLD, M. A., WURTMAN, R. J., HAMMARSTROM-WIKLUND, B., MUNRO, H. N., and DAVIDSON, C. S. (1978). Diurnal variations in plasma insulin concentrations in normal and cirrhotic subjects: Effect of dietary protein. *J. Neural. Transm.* **14:** 133—142.

35) BLOXAM, D. L., and CURZON, G. (1978). A study of proposed determinants of brain tryptophan concentration in rats after portocaval anastomosis or sham operation. *J. Neurochem.* **31:** 1255—1263.

36) ASHCROFT, G. W., ECCLESTON, D., and CRAWFORD, T. B. B. (1965). 5-hydroxyindole metabolism in rat brain: A study of intermediate metabolism using the technique of tryptophan loading. *J. Neurochem.* **12:** 483—492.

37) YOUNG, S. (1985). The clinical psychopharmacology of tryptophan. In: *Nutrition and the Brain.* (Wurtman, R. J., and Wurtman, J. J., eds.) **7,** pp. 49—88. Raven Press, New York.

38) COLMENARES, J. L., WURTMAN, R. J., and FERNSTROM, J. D. (1975). Effect of ingesting a carbohydrate-fat meal on the levels and synthesis of 5-hydroxyindoles in various regions of the rat central nervous system. *J. Neurochem.* **25:** 825—829.

39) MORET, C. (1985). Pharmacology of the serotonin autoreceptor. In: *Neuropharmacology of Serotonin.* (Green, A. R., ed.), pp. 21—49. Oxford University Press, New York.

40) COOPER, J. R., BLOOM, F. E., and ROTH, R. H. (1982). The Biochemical Basis of Neuropharmacology. Oxford University Press, New York.

41) GALLAGER, D. W., and AGHAJANIAN, G. K. (1976). Inhibition of firing of raphe neurons by tryptophan and 5-hydroxy-tryptophan: Blockage by inhibiting serotonin synthesis with R04-4602. *Neuropharmacology* **14:** 149—158.

42) BRAMWELL, G. (1974). Factors affecting the activity of 5HT containing neurons. *Brain Res.* **79:** 515—519.

43) DE SIMONI, M. G., SOKOLA, A., FODRITTO, F., TOSO, G. D., and ALGERI, S. (1987). Functional meaning of tryptophan-induced increase of 5-HT metabolism as clarified by in vivo voltammetry. *Brain Res.* **411:** 89—94.

44) SCHAECHTER, J., and WURTMAN, R. J. (1987). Effect of tryptophan availability on release of endogenous serotonin from rat hypothalamic slices. *Society for Neuroscience Abstracts Volume* **13,** New Orleans.

45) SMITH, B., and PROCKUP, D. J. (1962). Central-nervous-system effects of ingestion of L-tryptophan by normal subjects. *N. Engl. J. Med.* **267:** 1338—1341.

46) MOLLER, S. E., KIRK, L., and HONORE, P. (1980). Relationship between plasma ratio of tryptophan to competing amino acids and the response to L-tryptophan treatment in endogenously depressed patients. *J. Affective Disord.* **2:** 47—49.

47) WILCOCK, G. K., STEVENS, J., and PERKINS, A. (1987). Trazadone/tryptophan for aggressive behavior. *Lancet* **1:** 929—930.

48) MacINDOE, J. H., and TURKINGTON, R. W. (1973). Stimulation of human prolactin secretion by intravenous infusion of L-tryptophan. *J. Clin. Investig.* **52:** 1972—1978.

49) SPRING, B. (1985). Effects of foods and nutrients on the behavior of normal individuals. In: *Nutrition and the Brain.* (Wurtman, R. J., and Wurtman, J. J., eds.) **7,** pp. 1—47. Raven Press, New York.

50) SPRING, B., MALLER, O., WURTMAN, J. J., DIGMAN, L., and COZOLINO, L. (1983). Effects of protein and carbohydrate meals on mood and performance. *J. Psychiatr. Res.* **17:** 155—167.

51) PRINZ, R. J., ROBERTS, W. A., and HARTMAN, E. (1980). Dietary correlates of hyperactive behavior in children. *J. Consult. Clin. Psychol.* **48:** 760—769.

52) ROSENTHAL, N. E., and HEFFERMAN, M.M. (1985). Bulimia, carbohydrate craving, and depression: A central connection. In: *Nutrition and the Brain.* (Wurtman, R. J., and Wurtman J. J., eds.) **7,** pp. 139—166. Raven Press, New York.

53) WURTMAN, J. J., WURTMAN, R. J., GROWDON, J. H., LIPSCOMB, H. P., and ZEISEL, S. A. (1981). Carbohydrate craving in obese people: Supression by treatments affecting serotoninergic transmission. *Int. J. Eating Disord.* **1:** 2—15.

54) REID, R. (1985). Premenstrual syndrome. *Current Problems in Obstetrics* **8:** 1—57.

55) VAN PRAAG, H. M., and LEMUS, C. (1985). Monoamine precursors in the treatment of psychiatric disorders. In: *Nutrition and the Brain.* (Wurtman, R. J., and Wurtman, J. J., eds.) **7,** pp. 89—138. Raven Press, New York.

56) ANDERSON, G. H. (1977). Regulation of protein intake by plasma amino acids. In: *Advances in Nutritional Research.* (Draper, H. H., ed.) **1,** pp. 145—166. Plenum Press, New York.

57) WURTMAN, J. J., and WURTMAN, R. J. (1979). Drugs that enhance central serotoninergic transmission diminish elective carbohydrate consumption by rats. *Life Sci.* **24:** 895—904.

58) YOKOGOSHI, H., THEALL, C. L., and WURTMAN, R. J. (1986). Selection of dietary protein and carbohydrate by rats: Changes with maturation. *Physiol. and Behavior* **36:** 972—982.

59) WURTMAN, J. J., and BAUM, M. (1980). Estrogen reduces total food and carbohydrate intake in female rats. *Physiol. and Behavior* **24:** 823—827.

60) WURTMAN, J. J., MOSES, P. L., and WURTMAN, R. J. (1983). Prior carbohydrate consumption affects the amount of carbohydrate that rats choose to eat. *J. Nutr.* **113:** 70—78.

61) KIM, S.-H., and WURTMAN, R. J. (1987). Selective effects of CGS 10686B, DL-fenfluramine or fluoxetine on nutrient selection. *Physiol. and Behavior* (Submitted).

62) WURTMAN, J. J., WURTMAN, R. J., MARK, S., TSAY, R., GILBERT, W., and GROWDON, J. H. (1985). D-fenfluramine selectively suppresses carbohydrate snacking by obese subjects. *Int. J. Eating Disorders.* **4:** 89—99.

63) CHAOULOFF, F., KENNET, G. A., SERRURRIER, B., MERINO, D., and CURZON, G. (1986). Amino acid analysis demonstrates that increased plasma free tryptophan causes the increase in brain tryptophan during exercise in the rat. *J. Neurochem.* **46:** 1647—1650.

64) ACWORTH, I. N., MORGAN, B., and NEWSHOLME, E. A. (1987). Dopamine turnover is increased in the brains of untrained and trained rats after endurance exercise. *In preparation.*

65) CHAOULOFF, F., LAUDE, D., GUEZENNEC, Y., and ELGHOZI, J. L. (1986). Motor activity increases tryptophan, 5HIAA and HVA in ventricular CSF of the conscious rat. *J. Neurochem.* **46:** 1313—1316.

66) BLISS, E. L., and AILION, J. (1971). Relationship between stress and activity to brain dopamine and homovanillic acid. *Life Sci.* **10:** 1161—1169.

67) RANSFORD, C. P. (1982). A role for amines in the antidepressant effect of exercise: A review. *Med. Sci. Sports Exer.* **14:** 1—10.

68) BROWN, B. S., RAMIREZ, D. E., and TAUB, J. M. (1978). The prescription of exercise for depression. *Phys. Sportsmed.* **6:** 34—45.

69) COHEN, E., and WURTMAN, R. J. (1976). Brain acetylcholine synthesis: Control by dietary choline. *Science* **191:** 561—562.

70) SCHWARZ, J. C., LAMPART, C., and ROSE, C. (1972). Histamine formation in rat brain in vivo: Effects of histidine loads. *J. Neurochem.* **19:** 801—810.

71) MAHER, T. J., and WURTMAN, R. J. (1980). L-threonine administration increases glycine concentrations in the rat central nervous system. *Life Sci.* **26:** 1283—1286.

72) NADER, T. M. A., GROWDON, J. H., MAHER, T. J., and WURTMAN R. J. (1986). L-threonine administration increases CSF glycine levels and suppresses spasticity. *American Academy of Neurology Meeting.* New York.

73) WOOD, J. L., and ALLISON, R. G. (1981). Effects of consumption of choline and lecithin on neurological and cardiovascular systems. In: *FASEB: Technical Report,* pp. 1—105. SRO Press, Washington.

74) SVED, A. F. (1983). Precursor control of the function of monoaminergic neurons. In: *Nutrition and the Brain* (Wurtman, R. J., and Wurtman, J. J., eds.) **6,** pp. 223—275. Raven Press, New York.

75) WURTMAN, R. J., LARIN, F., MOSTAFAPOUR, S., and FERNSTROM, J. D. (1974). Brain catechol synthesis: Control of brain tyrosine concentration. *Science* **185:** 183—184.

76) SCALLY, M. C., ULUS, I., and WURTMAN, R. J. (1977). Brain tyrosine levels control striatal dopamine synthesis in haloperidol-treated rats. *J. Neural. Transm.* **41:** 1—6.

77) GIBSON, C. J., and WURTMAN, R. J. (1978). Physiological control of brain norepinephrine synthesis by brain tyrosine concentration. *Life Sci.* **22:** 1399—1406.

78) MELAMED, E., HEFTI, F., and WURTMAN, R. J. (1980). Tyrosine administration increases striatal dopamine release in rats with partial nigrostriatal lesions. *Proc. Natl. Acad. Sci. USA* **464:** 4305—4309.

79) SVED, A. F., FERNSTROM, J. D., and WURTMAN, R. J. (1979). Tyrosine administration decreases serum prolactin levels in chronically reserpinized rats. *Life Sci.* **25:** 1293—1300.

80) EL MESTIKAWAY, S., GLOWINSKI, J., and HAMON, M. (1983). Tyrosine hydroxylase activation in depolarized dopaminergic terminals: Involvement of calcium. *Nature* **302:** 830—832.

81) LOVENBERG, W., AMES, M. M., and LERNER, P. (1978). Mechanisms of short-term regulation of tyrosine hydroxylase. In: *Psychopharmacology: A Generation of Progress* (Lipton, M. A., DiMascio, A., and Killam, K. F., eds.), pp. 247—259. Raven Press, New York.

82) MILNER, J. D., and WURTMAN, R. J. (1986). Catecholamine synthesis: Physiological coupling to precursor supply. *Biochem. Pharmacol.* **35:** 875—881.

83) IRIE, K., and WURTMAN, R. J. (1987). Release of norepinephrine from rat hypothalamic slices: Effects of desipramine and tyrosine. *Brain Res.* In press.

84) CONLAY, L. A., MAHER, T. J., and WURTMAN, R. J. (1981). Tyrosine increases blood pressure in hemorrhagic shock. *Science* **212:** 559—560.

85) SVED, A. F., FERNSTROM, J. D., and WURTMAN, R. J. (1979). Tyrosine administration reduces blood pressure and enhances brain norepinephrine release in spontaneously-hypertensive rats. *Proc. Natl. Acad. Sci. USA* **76:** 3511—3514.

86) GROWDON, J. H. (1979). Neurotransmitter precursors in the diet: Their use in the treatment of disease. In: *Nutrition and the Brain* (Wurtman, R. J., and Wurtman, J. J., eds.) **3,** pp. 117—181. Raven Press, New York.

87) LEHNERT, H., REINSTEIN, D. K., STROWBRIDGE, B. W., and WURTMAN, R. J. (1984). Neurochemical and behavioral consequences of acute, uncontrollable stress: Effects of dietary tyrosine. *Brain Res.* **303:** 215—223.

88) REINSTEIN, D. K., LEHNERT, H., and WURTMAN, R. J. (1985). Dietary tyrosine suppresses the rise in plasma corticosterone following acute stress in rats. *Life Sci.* **37:** 2157—2163.

89) BANDERET, L. E., LIEBERMAN, H. R., FRANCESCONI, R. P., GOLDMAN, R. F., SCHNAKENBERG, D. D., RAUCH, T. M., ROCK, P. B., and MEADORS, G. F. (1987). Development of a paradigm to assess nutritive and biochemical substances in humans: A preliminary report on the effects of tyrosine upon altitude- and cold-induced stress responses. In: *Biochemical Enhancement of Performance. The "AGARD" Conference Proceedings. No. 415.* Specialized Printing Services Lt., Loughton, Essex.

90) MILNER, J. D., IRIE, K., and WURTMAN, R. J. (1986). Phenylalanine inhibition and enhancement of endogenous dopamine release from rat striatal slices. *J. Neurochem.* **47:** 1444—1448.

91) DURING, M. J., ACWORTH, I. N., and WURTMAN, R. J. (1987). An in vivo study of dopamine release in striatum: The effects of phenylalanine. *In press.* In: *Dietary Phenylalanine and brain function* (Wurtman, R. J., and Ritter-Walker, E., eds.). Birkhauser, Boston.

92) IKEDA, M., LEVITT, M., and UDENFRIEND, S. (1967). Phenylalanine as substrate and inhibitor of tyrosine hydroxylase. *Arch. Biochem. Biophys.* **120:** 420—427.

93) DURING, M. J., ACWORTH, I. N., and WURTMAN, R. J. (1987). Precursor influence on in vivo dopamine release: Regional effects of tyrosine. *Brain Res.* Submitted.

94) ANDEN, N. E., GRABOWSKA-ANDEN, M., LINDGREN, S., and THORNSTROM, U. (1983). Synthesis rate of dopamine: Difference between corpus striatum and limbic system as a possible explanation of variations in reactions to drugs. *Nauyn-Schmiedeberg's Arch. Pharmacol.* **323:** 193—198.

95) BANNON, M. J., and ROTH, R. H. (1983). Pharmacology of mesocortical dopamine neurons. *Pharmacol. Rev.* **35:** 53—68.

96) YOKOGOSHI, H., ROBERTS, C. H., CABALLERO, B., and WURTMAN, R. J. (1984). Effects of aspartame and glucose administration on brain and plasma levels of large neutral amino acids and brain 5-hydroxyindoles. *Am. J. Clin. Nutr.* **40:** 1—7.

97) MAHER, T. J., and PINTO, J. M. B. (1987). Aspartame, phenylalanine, and seizures in experimental animals. *In press.* In: *Dietary Phenylalanine and Brain Function* (Wurtman, R. J., and Ritter-Walker, E., eds.). Birkhauser, Boston.

98) JOHNS, D. R. (1987). Aspartame and headache. *In press.* In: *Dietary Phenylalanine and Brain Function* (Wurtman, R. J., and Ritter-Walker, E., eds.). Birkhauser, Boston.

99) NUTT, J. G., WOODWARD, W. R., HAMMARSTAD, J. P., and CARTER, J. H. (1984). The "on-off" phenomenon in Parkinson's disease: Relation to levodopa absorption and transport. *N. Eng. J. Med.* **310:** 483—488.

TRYPTOPHAN AVAILABILITY AND SEROTONIN SYNTHESIS IN BRAIN

John D. Fernstrom

Departments of Psychiatry and Behavioral Neuroscience
University of Pittsburgh School of Medicine
Western Psychiatric Institute and Clinic
3811 O'Hara Street
Pittsburgh PA 15213

INTRODUCTION

One of the most important factors governing the rate of serotonin (5HT) synthesis in the mammalian brain is the local concentration of the substrate amino acid, L-tryptophan (see (1)). As a result, a variety of factors that influence brain tryptophan levels are also found to modify 5HT synthesis. These include such phenomenon as the ingestion of single meals, stress, the development of diabetes, and certain forms of malnutrition (see (1)). Where studied, it has been possible to show that when 5HT synthesis has been modified by one of the above phenomenon, 5HT release is also changed, as evidenced by effects on particular brain functions thought to be regulated in part by 5HT neurons. This article reviews data that have established connections between brain tryptophan levels and 5HT synthesis, and between tryptophan-related changes in 5HT synthesis and 5HT release at the synapse. It also considers evidence linking meal ingestion to alterations in brain tryptophan levels and 5HT synthesis, and food-induced changes in 5HT synthesis to appetites for a specific macronutrient (carbohydrate). It should be evident from the discussion that while a substantial amount of evidence can now be assembled to support many of these relationships, several require additional experimental validation before they can be fully evaluated or accepted.

Control of Brain 5HT Synthesis Rate by Tryptophan Levels

Serotonin is synthesized in the brain from L-tryptophan. The initial biosynthetic step involves the hydroxylation of the amino acid to form 5-hydroxytryptophan. This reaction is catalyzed by tryptophan hydroxylase (2). 5-Hydroxytryptophan is converted to 5HT by aromatic-L-amino acid decarboxylase (3). The activity of this enzyme is much greater than that of tryptophan hydroxylase (4), and thus the hydroxylase is generally regarded as catalyzing the rate-limiting step in 5HT synthesis (2, 5). The main 5HT metabolite in brain is 5-hydroxyindoleacetic acid (5HIAA), formed in a two-step reaction catalyzed sequentially by monoamine oxidase and aldehyde dehydrogenase.

NATO ASI Series, Vol. H20
Amino Acid Availability and Brain Function in Health and Disease. Edited by G. Huether

Because tryptophan hydroxylase catalyzes the rate-limiting step in 5HT synthesis, it has been studied as the locus for controlling the rate of 5HT formation. Central to the present discussion is the repeated finding that changes in brain tryptophan level rapidly influence the rates of tryptophan hydroxylation and 5HT formation (6—9). The biochemical basis for this relationship involves the degree of substrate saturation of tryptophan hydroxylase *in vivo.* The K_m of this enzyme for tryptophan is about 2×10^{-5} M (when biopterin is the cofactor), which approximates normal brain tryptophan concentrations (about $1—5 \times 10^{-5}$ M in rats) (7). This implies that the enzyme is normally not saturated *in vivo* with its substrate. Hence, variations in the local tryptophan concentration can directly affect the degree of enzyme saturation, and thus the rate of tryptophan hydroxylation. Parenthetically, the importance of this control mechanism is underscored by the fact that tryptophan hydroxylase lacks another form of control previously thought to be typical of monoamine-synthesizing neurons; i. e., direct end-product inhibition (10—12).

Precursor availability also appears to be important *chronically* in the control of brain 5HT synthesis, as evidenced by the results of studies in which tryptophan supply to the body has been increased or decreased substantially for long periods. In such studies, the removal of tryptophan from the diet, either by feeding a diet containing a casein hydrolysate (which lacks tryptophan) or a protein-source naturally low in tryptophan (corn), or a mixture of synthetic amino acids devoid of tryptophan, has been shown to reduce brain 5HT levels and synthesis over a several week period (13—16). Such effects were eliminated if normal amounts of tryptophan were added to the diet (about 0.2 % of the amino acid content of casein, on a weight basis). Additionally, if supranormal amounts of tryptophan were added to the diet, increments in brain 5HT levels were observed (17, 18). Such results, though originally designed to make a nutritional point, also allow an important biochemical observation about the regulation of neuronal 5HT synthesis: apparently, the 5HT neuron does not attempt to modulate the rate of 5HT synthesis (or does not succeed in doing so) via changes in the tryptophan hydroxylating capacity of the neuron, even in situations in which elevations or reductions in brain tryptophan level are chronically driving or restraining 5HT production, respectively.

Meal-induced Changes in Brain Tryptophan Level and 5HT Synthesis

The ability of the local tryptophan level to influence 5HT synthesis became physiologically interesting when 5HT synthesis rate was found to be sensitive to even a *modest* change in brain tryptophan level. Acute increases in brain tryptophan level that are smaller than those seen normally in untreated animals readily stimulate 5HT synthesis (7). As a result of this finding, considerable energy was devoted to ascertaining what physiologic phenomenon might normally influence brain tryptophan levels, and subsequently the rate of 5HT production. Early on, eating was identified as one such phenomenon. Subsequent studies indeed showed this to be the case, and the results brought to light the fact that the ingestion by fasted rats of a carbohydrate meal induces effects on brain tryptophan level and 5HT synthesis (it increases them) quite different from those produced by the ingestion of a protein-containing meal (it does not alter them) (19). The biochemical basis for this difference follows from the ability of the meal to change the uptake of tryptophan into brain: Brain tryptophan levels are remarkably sensitive to the supply of the amino acid from the circulation. The major factor governing this uptake process is a transport carrier located at the blood-brain barrier (20). This carrier transports not only tryptophan, however, but also several other large, neutral amino acid (LNAA), including tyrosine, phenylalanine, leucine, isoleucine, and valine. The transport carrier is saturable and competitive (21), and thus the uptake of tryptophan depends not only on its own blood level, but also on the blood concentrations of the other LNAA. Hence, one could raise brain tryptophan uptake and level either by raising blood tryptophan concentrations, *or* by lowering the blood levels of any or all of the other LNAA. One could lower brain tryptophan uptake either by reducing blood tryptophan *or* by increasing the blood concentrations of any or all of the other LNAA.

This competitive transport notion is central to the effects of food ingestion on brain tryptophan levels and 5HT synthesis. As indicated above, food intake studies revealed that the ingestion of a carbohydrate meal by fasting rats rapidly elevates brain tryptophan level and stimulates the rate of 5HT synthesis, while the consumption of a protein-containing meal does not (19, 22). These results are readily explained by the effects of each meal on the blood levels of tryptophan and the other LNAA, when placed into the context of competitive transport across the blood-brain barrier. In the fasting rat, the ingestion of a carbohydrate meal induces insulin secretion, which raises blood tryptophan levels and lowers the blood concentrations of the other LNAA. The result is to increase greatly tryptophan's competitive advantage for uptake into brain, and as a consequence, brain tryptophan level. The ingestion of a protein-containing meal (e. g., 18—40 % protein, dry weight) also induces insulin secretion, which tends to raise blood tryptophan and lower

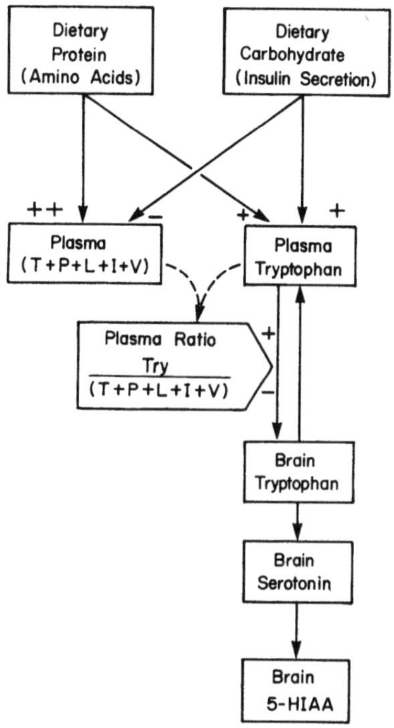

FIGURE 1

Model summarizing meal-induce changes in the serum tryptophan/LNAA ratio, brain tryptophan levels and brain serotonin synthesis. T = tyrosine, P = phenylalanine, L = leucine, I = isoleucine, V = valine. See text for details.

the blood levels of the other LNAA. But this meal also provides amino acids to the circulation from the digested dietary proteins, which has the effect of raising the blood levels of all the LNAA. The net result of consuming the protein-containing meal is that the blood level of tryptophan and the combined blood levels of the other LNAA rise proportionally by about the same amount. To the competitive transport carriers, such changes signal no net alteration in competition for transport sites. Consequently, uptake into brain is not changed, and brain tryptophan levels do not change following ingestion of this meal. These competitive effects are summarized in the model shown in Figure 1, the essential feature of which is the serum tryptophan/LNAA ratio. This ratio summarizes the competitive relationship between tryptophan and the other LNAA following ingestion of a single meal, and predicts the meal's effects on brain tryptophan levels (and 5HT synthesis): when a carbohydrate meal is consumed, the ratio rises (blood tryptophan increases; the blood levels of the other LNAA fall), and tryptophan uptake into brain increases. Brain tryptophan levels accordingly rise, and 5HT synthesis is stimulated. When a protein-containing meal is ingested, the ratio does not change (blood tryptophan rises, but the blood levels of the other LNAA rise by a proportionally similar amount), predicting no net change in brain tryptophan uptake. Brain tryptophan levels are thus unaltered by this meal, as is the rate of 5HT synthesis (1, 19).

Precursor-induced Changes in Brain 5HT Synthesis: Effect on 5HT Release

Since increases in brain tryptophan level, induced by a physiologic phenomena (e. g., eating a single meal) can stimulate 5HT production in brain neurons, it has been of interest to assess if such increments in 5HT synthesis also cause more transmitter to be released, when 5HT neurons depolarize. If so, then brain functions and behaviors controlled by neuronal circuits that include 5HT neurons may be influenced by tryptophan- and meal-induced changes in 5HT synthesis.

One indirect approach to answering this question has been to explore for changes in particular brain functions following tryptophan administration. An example of this approach has been the study of blood pressure effects of L-tryptophan. When administered to spontaneously-hypertensive rats, L-tryptophan produces a modest, dose-related decline in blood pressure (23). This effect can be blocked by the co-administration of other LNAA, suggesting that in order to lower blood pressure, the amino acid must gain access to brain (the locus of the 5HT neurons involved in the control of blood pressure (24)). Moreover, the effect of L-tryptophan can be blocked by pretreatment with an inhibitor of tryptophan hydroxylase (Table 1), indicating that the tryptophan must be hydroxylated to lower blood pressure (i. e., it must be converted to 5HT; it is not acting as untransformed L-tryptophan or as decarboxylated tryptophan (tryptamine)). And, coadministration of a 5HT receptor antagonist will also block tryptophan's antihypertensive effects, further focusing the effect on the 5HT receptor. Finally, tryptophan's efficacy can be enhanced by administering it with an inhibitor of presynaptic 5HT reuptake (fluoxetine) (23) (Table 1). This latter result clearly implies that the administqation of L-tryptophan leads to an enhancement of 5HT synthesis *and* release. The only way tryptophan's blood pressure-lowering effects could be enhanced by fluoxetine, given that these effects of L-tryptophan follow from its stimulation of 5HT synthesis, is if L-tryptophan administration also enhanced 5HT release from the nerve terminal.

TABLE 1
Effects of serotonergic drugs on the tryptophan-induced decline in blood pressure in spontaneously hypertensive rats

Treatment	Blood Pressure Before	Blood Pressure After	Change in Blood Pressure
Experiment 1:			
PCPA + Saline	182 ± 5	177 ± 5	- 5 ± 5
PCPA + Tryptophan	180 ± 4	174 ± 6	- 6 ± 4
Experiment 2:			
Vehicle + Vehicle	194 ± 3	192 ± 2	- 2 ± 3
Metergoline + Vehicle	199 ± 3	197 ± 2	- 2 ± 1
Vehicle + Tryptophan	194 ± 2	173 ± 4*	-21 ± 1*
Metergoline + Tryptophan	198 ± 5	197 ± 4	0 ± 4
Experiment 3:			
Vehivle + Vehicle	196 ± 2	197 ± 3[a]	+ 2 ± 2[a]
Tryptophan + Vehicle	196 ± 4	186 ± 5[b]	-10 ± 4[b]
Vehicle + Fluoxetine	195 ± 6	174 ± 8[b]	-21 ± 5[b]
Tryptophan + Fluoxetine	195 ± 2	149 ± 4[c]	-46 ± 5[c]

In *experiment 1,* spontaneously hypertensive rats [SHR] (n = 5/group) received saline or PCPA (100 mg/kg ip) once a day for 3 days. On the fourth day, base-line blood pressures were taken, after which saline or tryptophan (100 mg/kg) was injected ip. Blood pressures were taken 2 hr later. In *experiment 2,* SHR (n = 5/group) were injected with metergoline (2 mg/kg ip) or vehicle after base-line blood pressure determination; 30 min later, they received tryptophan (100 mg/kg) or vehicle and blood pressures were measured 2 hr later. In *experiment 3,* SHR (n = 5/group) were injected ip with tryptophan (50 mg/kg), fluoxetine (10 mg/kg), both compounds together, of vehicle after determination of base-line blood pressures.Blood pressures were again measured 2 hr later. In experiment 2, * indicates group differs from all other groups, $P < 0.01$; in experiment 3, values followed by different symbols are significantly different, $P < 0.05$ (analysis of variance, Newman-Keuls test). Adapted from (23).

However, it should be noted that it is by no means an accepted fact that L-tryptophan-induced increases in 5HT synthesis enhance 5HT release. For example, L-tryptophan administration was reported to slow the spontaneous firing rate of raphe neurons (i. e., 5HT neurons) in a dose-related manner (25, 25). These results were interpreted as showing that the 5HT neuron quickly reduces its firing frequency when 5HT synthesis has been stimulated by L-tryptophan. Hence, 5HT release per unit time presumably remains constant despite different rates of 5HT production. Recently, this study was repeated (27), with the L-tryptophan administered as a part of the diet. Surprisingly, no effect of the amino acid was noted on raphe unit firing, despite the presence of the expected differences in brain 5HT and 5HIAA levels in the different treatment groups. This finding thus diverges from the earlier results. If true, this latter result would support the notion that when more 5HT is produced following L-tryptophan ingestion, more is released (i. e., if there is more 5HT in the nerve terminal, and it fires at the same frequency, more 5HT will be released per unit time). However, this implication was recognized by the investigator, and parallel biochemical studies were thus performed to quantitate 5HT release. To this end, labeled 5HT release into CSF was measured following the administration of labeled tryptophan into the lateral ventricles. Labeled 5HT release in the CSF was found to be unaffected by L-tryptophan ingestion, despite increments in the brain levels of (unlabeled) 5HT. Since labeled 5HT release was unaffected by brain tryptophan level, the conclusion was drawn that 5HT release is not influenced by tryptophan-induced (and diet-induced) vagaries in 5HT synthesis in brain (27).

There is a more likely interpretation, however. In this study, precursor (i. e., tryptophan) specific activity was not measured. Undoubtedly, the animals with higher brain tryptophan levels (due to its ingestion) experienced a greater dilution of label by the cold brain tryptophan pool than did those animals with lower brain tryptophan levels. As a consequence, even if 5HT synthesis rate were higher in the animals with high brain tryptophan levels, the absolute number of labeled tryptophan molecules converted to labeled 5HT in animals with elevated brain tryptophan levels would not by themselves reflect this increased synthesis (e. g., see Table 1 in (28) for a specific example of this phenomenon). Hence, had the two groups been releasing 5HT at comparable rates, the amount of labeled 5HT appearing in the CSF should have been *less* in the animals consuming *excess* L-tryptophan than in those that did not. However, since the actual counts of 5HT were found to be comparable between the high and normal brain tryptophan groups, the correct conclusion is likely to be that the brain of animals with high tryptophan levels synthesized *and* released 5HT at a faster rate than those of animals with normal brain tryptophan levels. (A similar argument can also be made for the third diet group in this study, in which brain tryptophan level was lowered by feeding excess LNAA in the diet (27)).

In summary, though there continues to be controversy, available evidence continues to support the likelihood that changes in 5HT synthesis induced by alterations in neuronal tryptophan level do cause like changes in 5HT release.

Diet, Brain Serotonin and Appetite For Carbohydrates

The biochemical findings showing effects of single meals on brain tryptophan levels and 5HT synthesis (and presumably release) have also stimulated a line of behavioral investigation to ascertain if food-induced changes in brain tryptophan level and 5HT synthesis provide the brain with chemical information it uses to control the ingestion of macronutrients. One hypothesis of recent interest, for example, a model for the meal-to-meal regulation of carbohydrate intake, includes as a biochemical "sensing" mechanism in brain the changes in brain tryptophan level and %HT synthesis induced by carbohydrate ingestion (29). The model is constructed from three experimental observations: first, the report that over time, rats seem to maintain their carbohydrate intake at a surprisingly constant level (29); second, the observation that the ingestion of a carbohydrate meal increases brain 5HT synthesis; and third, the results of a series of pharmacologic studies purporting to show that the administration of drugs that enhance transmission across 5HT synapses (e. g., fluoxetine, fenfluramine) selectively suppresses carbohydrate intake (29). Together, these findings have been woven into the following model to describe the meal-to-meal regulation of carbohydrate intake in mammals: When a carbohydrate meal is ingested, the blood levels of tryptophan rise and those of the other LNAA fall, brain tryptophan level rises, and 5HT synthesis is stimulated. This leads to enhanced release of 5HT from brain neurons, which in turn reduces the intake of carbohydrates. As carbohydrate intake subsequently falls, the serum level of tryptophan falls while those of the other LNAA rise, and thus brain tryptophan uptake and levels, and serotonin synthesis and release begin to decline. As a result, the inhibition of carbohydrate intake (appetite) moderates, and the animal resumes eating carbohydrates (see Figure 2).

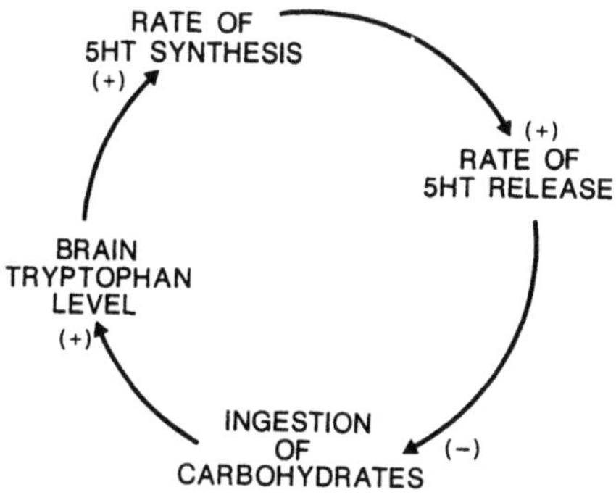

FIGURE 2
Proposed model of carbohydrate intake regulation. The ingestion of carbohydrates is known to raise brain tryptophan levels and stimulate the rate of serotonin synthesis. The increase in synthesis may lead to increased 5HT release from nerve terminals, though this is at present unproven. Based on results of pharmacologic studies, the increased release of 5HT is then said to inhibit specifically subsequent appetite for carbohydrates. The proposed model is thus an inhibitory feedback control system.

Though this model is appealing, it is seriously flawed. First, and most important, there is ample evidence that the level of carbohydrate intake consumed over extended time-periods by individual animals is actually *not* tightly controlled. There appears to be great variability in the amount of carbohydrate individual rats elect to eat over extended periods (e. g., (30)), suggesting the absence of a regulatory system. Second, animals will not maintain an elected level of carbohydrate intake when they must press a bar an ever increasing number of times to obtain carbohydrate food pellets. In contrast, they *will* defend their calorie and protein intakes, when such bar-pressing demands are made (31). If carbohydrate intake were a regulated commodity, the animals *should* defend their level of carbohydrate intake (i. e., maintain the "set point" for carbohydrate intake), but they do *not.* Third, there is no known nutritional requirement for carbohydrate in rats or man, as there is for total calories and for protein (32, 33). In the absence of such a requirement, there is no reason to have a control system for regulating carbohydrate intake. And fourth, the pharmacologic studies in rats and humans purporting to show selective suppression of carbohydrate intake by drugs that stimulate 5HT transmission actually appear to show the opposite: viz., that these drugs do *not* produce a selective suppression of carbohydrate intake. For example, in the studies published by Wurtman and Wurtman (29, 34), effects of fenfluramine that were interpreted as showing a selective suppression of carbohydrate intake actually show a suppression of fat intake as well (see also (35)). This nonspecific suppression has also been seen by other investigators: fenfluramine administration reduced the ingestion of *all* three macronutrients (36, 37). Consequently, one is left with the impression that fenfluramine does *not* selectively suppress appetite for or intake of carbohydrates. This conclusion for fenfluramine is very important, since (a) the fenfluramine findings are the principal database cited in support of a selective inhibitory effect of 5HT release on carbohydrate intake (29, 34, 38—40); and (b) the selective inhibitory effect of fenfluramine (i. e., of enhanced 5HT release) on carbohydrate intake is required in order for the regulatory loop (Figure 2) to exist. For if fenfluramine (5HT release) suppresses carbohydrate *and* protein (and fat) intakes, as appears to be the case, then the result of administering this drug would be to lower plasma levels of *both* tryptophan and the other LNAA together as total macronutrient intake fell, leading to no net change in the competitive uptake of tryptophan into brain, and hence no net change in 5HT synthesis (and thus release). Consequently, there would be no modified signal in brain to change the ingestive behavior induced initially by the drug (i. e., increased 5HT release), one way or the other. Therefore, in the absence of a *selective* inhibitory effect of 5HT release on carbohydrate intake, the carbohydrate regulatory loop cannot exist. Together, all of the above considerations suggest that the proposed model (Figure 2) for regulating carbohydrate intake cannot be valid.

SUMMARY AND CONCLUSIONS

This brief review has attempted to focus on three specific issues related to the biochemical finding that 5HT synthesis in brain is tied to local tryptophan concentrations. First, evidence has been summarized supporting the sensitivity of 5HT synthesis to small changes in brain tryptophan levels produced by physiologic phenomenon (in this case, the ingestion of single meals). Second, the key question of whether tryptophan-induced alterations in 5HT synthesis cause parallel changes in 5HT release at the nerve terminal has been considered. The conclusion drawn is that 5HT release *does* reflect changes in 5HT synthesis secondary to modifications in brain tryptophan supply. And third, the hypothesized connection of diet-induced changes in brain tryptophan levels and 5HT synthesis to appetite for carbohydrates has been reviewed. The conclusion drawn from the analysis is that available data do *not* provide convincing support for the putative control mechanism for regulating carbohydrate appetite. This latter hypothesis had provided a most appealing explanation for why 5HT synthesis in brain neurons might be tied to the diet via food-induced changes in the blood LNAA pattern. The search for the ultimate significance of linking 5HT synthesis rate to the supply of tryptophan to the brain must therefore continue in new and/or renewed directions.

ACKNOWLEDGEMENT

The work described in this article was supported in part by a grant from the US Public Health Service (MH38178). The author is also the recipient of a Research Scientist Development Award (Level II) from the US National Institute of Mental Health (MH00254).

REFERENCES

1) FERNSTROM, J. D. (1983). Role of precursor availability in the control of monoamine biosynthesis in brain. *Physiol. Rev.* **63:** 484—546.

2) LOVENBERG, W., JEQUIER, E., SJOERDSMA, A. (1968). Tryptophan hydroxylation in mammalian systems. In: *Advances in Pharmacology* **Vol. 6A** (Garattini, S., Shore, P. A., eds.). Academic Press New York, pp. 21—36.

3) LOVENBERG, W., WEISSBACH, H., UDENFRIEND, S. (1962). Aromatic L-amino acid decarboxylase. *J. Biol. Chem.* **237:** 89—92.

4) ICHIYAMA, A., NAKAMURA, S., NISHIZUKA, Y., HAYAISHI, O. (1968). Tryptophan 5-hydroxylase in mammalian brain. In: *Advances in Pharmacology,* **Vol. 6A** (Garattini, S., Shore, P. A., eds.). Academic Press New York, pp. 5—17.

5) KAUFMAN, S. (1974). Properties of pterin-dependent aromatic amino acid hydroxylases. In: *Aromatic Amino Acids in the Brain* (Wolstenholme, G. E. W., FitsSimons, D. W., eds.). Elsevier North Holland, pp. 85—108.

6) ASHCROFT, G. W., ECCLESTON, D., CRAWFORD, T.B. B. (1965). 5-Hydroxyindole metabolism in rat brain. A study of intermediate metabolism using the technique of tryptophan loading. *J. Neurochem.* **12:** 483—492.

7) FERNSTROM, J. D., WURTMAN, R. J. (1971). Brain serotonin content: physiological dependence on plasma tryptophan levels. *Science* **173:** 149—152.

8) HESS, S. M., DOEPFNER, W. (1961). Behavioral effects and brain amine content in rats. *Arch. Intl. Pharmacodyn.* **134:** 89—99.

9) WEBER, L. J., HORITA, A. (1965). A study of 5-hydroxytryptamine formation from L-tryptophan in the brain and other tissues. *Biochem. Pharmacol.* **14:** 1141—1149.

10) LIN, R. C., NEFF, N. H., NGAI, S. H., COSTA, E. (1969). Turnover rates of serotonin and norepinephrine in brain of normal and pargyline-treated rats. *Life Sci.* **8:** 1077—1084.

11) MEEK, J. L., FUXE, K. (1971). Serotonin accumulation after monoamine oxidase inhibition. Effects of decreased flow and of some antidepressants and hallucinogens. *Biochem. Pharmacol.* **20:** 653—706.

12) MILLARD, S. A., COSTA, E., GAL, E. M. (1972). On the control of brain serotonin turnover rate by end product inhibition. *Brain Res.* **40:** 545—551.

13) CULLEY, W. J., SAUNDERS, R. N., MERTZ, E. T., JOLLY, E. T. (1963). Effect of a tryptophan-deficient diet on brain serotonin and plasma tryptophan level. *Proc. Soc. Exptl. Biol. Med.* **113:** 645—648.

14) GAL, E. M., DREWES, P. A. (1962). Studies on the metabolism of 5-hydroxytryptamine (serotonin) II. Effect of tryptophan deficiency in rat. *Proc. Soc. Exptl. Biol. Med.* **110:** 368—371.

15) ZBINDEN, G., PLETSCHER, A., STUDER, A. (1958). Alimentäre Beeinflussung der enterochromafinen Zellen und des 5-Hydroxytryptamin-Gehaltes von Gehirn und Darm. *Z. Ges. Exptl. Med.* **129:** 615—620.

16) FERNSTROM, J. D., HIRSCH, M. J. (1977). Brain serotonin synthesis: reduction in cornmalnourished rats. *J. Neurochem.* **28:** 877—879.

17) GREEN, H., GREENBERG, S. M., ERICKSON, R. W., SAWYER, J. L., ELLISON, T. (1962). Effect of dietary phenylalanine and tryptophan upon rat brain amine levels. *J. Pharmacol. Exp. Ther.* **136:** 174—178.

18) WANG, H. L., HARWALKAR, V. H., WAISMAN, H. A. (1962). Effect of dietary phenylalanine and tryptophan on brain serotonin. *Arch. Biochem. Biophys.* **96:** 181—184.

19) FERNSTROM, J. D., WURTMAN, R. J. (1972). Brain serotonin content: physiological regulation by plasma neutral amino acids. *Science* **178:** 414—416.

20) OLDENDORF, W. H. (1971). Brain uptake of radiolabeled amino acids, amines, and hexoses after arterial injection. *Amer. J. Physiol.* **221:** 1629—1639.

21) PARDRIDGE, W. M., OLDENDORF, W. H. (1975). Kinetic analysis of blood brain barrier transport of amino acids. *Biochim. Biophys. Acta* **401:** 128—136.

22) FERNSTROM, J. D., WURTMAN, R. J. (1971). Brain serotonin content: increase following ingestion of carbohydrate diet. *Science* **174:** 1023—1025.

23) SVED, A. F. Van ITALIE, C. M., FERNSTROM, J. D. (1982). Studies of the antihypertensive action of L-tryptophan. *J. Pharmacol. Esp. Ther.* **221:** 329—333.

24) KUHN, D. M., WOLF, W. A., LOVENBERG, W. (1980). Review of the role of the central serotonergic neuronal system in blood pressure regulation. *Hypertension* **2:** 243—255.

25) GALLAGER, D. W., AGHAJANIAN, G. K. (1976). Inhibition of firing of raphe neurons by tryptophan and 5-hydroxytryptophan: blockade by inhibiting serotonin synthesis with RO-4-4602. *Neuropharmacology* **15:** 149—156.

26) TRULSON, M. E., JACOBS, B. L. (1976). Dose-response relationship between systemically administered L-tryptophan or L-5-hydroxytryptophan and raphe unit activity in the rat. *Neuropharmacology* **15:** 339—344.

27) TRULSON, M. E. (1985). Dietary tryptophan does not alter the function of brain serotonin neurons. *Life Sci.* **37:** 1067—1072.

28) FERNSTROM, M. H., BAZIL, C. W., FERNSTROM, J. D. (1984). Caffeine injection raises brain tryptophan level, but does not stimulate the rate of serotonin synthesis in rat brain. *Life Sci.* **35:** 1241—1247.

29) WURTMAN, J. J., WURTMAN, R. J. (1979). Drugs that enhance central serotoninergic transmission diminish elective carbohydrate consumption by rats. *Life Sci.* **24:** 895—904.

30) LEATHWOOD, P. D., ASHLEY, D. V. M. (1983). Strategies of protein selection by weanling and adult rats. *Appetite* **4:** 97—112.

31) ASHLEY, D. V. M. (1985). Factors affecting the selection of protein and carbohydrate from a dietary choice. *Nutrition Res.* **5:** 555—571.

32) NATIONAL ACADEMY OF SCIENCES (1978). *Nutrient requirements of domestic animals,* **Number 10:** Nutrient requirements of laboratory animals. Third revised edition. National Academy of Sciences, Washington, pp. 7—37.

33) NATIONAL ACADEMY OF SCIENCES (1980). *Recommended Dietary Allowances,* Ninth Revised Edition. Committee on Dietary Allowances, Food and Nutrition Board, National Research Council. Washington: National Academy of Sciences, p. 33.

34) WURTMAN, J. J., WURTMAN, R. J. (1984). D-fenfluramine selectively decreases carbohydrate but not protein intake in obese subjects. *Intl. J. Obesity* **8 (Suppl. 1):** 79—84.

35) FERNSTROM, J. D. (1987). Food-induced changes in brain serotonin synthesis: is there a relationship to appetite for specific macronutrients? *Appetite* **8:** 163—182.

36) ORTHEN-GAMBILL, N., KANAREK, R. B. (1982). Differential effects of amphetamine and fenfluramine of dietary self-selection in rats. *Pharmacol. Biochem. Behav.* **16:** 303—309.

37) McARTHUR, R. A., BLUNDELL, J. E. (1983). Protein and carbohydrate self-selection: modification of the effects of fenfluramine and amphetamine by age and feeding regimen. *Appetite* **4:** 113—124.

38) WURTMAN, J. J., WURTMAN, R. J. (1982/1983). Studies on the appetite for carbohydrates in rats and humans. *J. Psychiat. Res.* **17:** 213—221.

39) WURTMAN, J. J., WURTMAN, R. J., GROWDON, J H., HENRY, P., LIPSCOMB, A., ZEISEL, S. H. (1981). Carbohydrate craving in obese people: suppression by treatments affecting serotoninergic transmission. *Intl. J. Eat. Disord.* **1:** 2—15.

40) WURTMAN, J. J., MOSES, P. L., WURTMAN, R. J. (1983). Prior carbohydrate consumption affects the amount of carbohydrates that rats choose to eat. *J. Nutr.* **113:** 70—78.

DISPROPORTIONATE AMINO ACID DIETS AND ANOREXIC RESPONSES IN RATS: THE ROLE(S) OF LIMBIC BRAIN AREAS AND NORADRENERGIC AND SEROTONINERGIC SYSTEMS

D. W. Gietzen, Q. R. Rogers, and P. M. B. Leung

Department of Physiological Sciences
School of Veterinary Medicine
University of California, Davis DA 95616
USA

INTRODUCTION

The role of protein and amino acids in the control of food intake has long been of interest. As early as 1931, Rose (1) stated that an amino acid deficiency was synonymous with a reduction in food intake. Indeed, along with energy, each of the macronutrients in turn has been considered to have the primary role in hypotheses about the control of feeding behavior. The "Aminostatic" theory has gained few adherents, and it is acknowledged that diets with balanced protein content between approximately 15 and 30 or 35 % have little or no effect on feeding. However, reductions in food intake and dietary avoidance are reliably found with diets that: 1, have a protein content outside the 15—35 % range, 2, contain an excess of one or more amino acids, 3, are devoid of one or more amino acids, or 4, induce an amino acid imbalance. The food intake depression that occurs with amino acid imbalance will be the focus of the studies to be discussed here.

Imbalanced amino acid diets are prepared by the addition of an excess of all but one of the essential amino acids to a low protein basal diet limiting in that essential amino acid. The response of rats to an imbalanced or devoid diet is a reliable and rapid reduction in food intake, i. e., an anorexic response to the dietary disproportion of amino acids. The nutritional, biochemical, and behavioral responses of rats to imbalanced amino acid diets have been well studied (2, 3, 4). Therefore, this model provides a robust set of behavioral and biochemical responses to a well defined dietary paradigm, and provides a convenient model in which to study one of the influences of amino acids on the neural control of feeding.

An emphasis on neural control of amino acid intake is based on several lines of evidence, each of which points to the brain as having a major role in the anorexic responses of rats to imbalanced amino acid diets: 1, the limiting amino acid is more rapidly depleted from brain than from plasma after a single meal of an imbalanced amino acid diet (5); 2, infusion of the limiting amino acid into the carotid artery blocks the anorexic response at only 4 mg/day, whereas infusions into the jugular vein require much greater concentrations to block the response (6); and 3, lesions of certain limbic brain areas attenuate or block the responses to these diets (7). In addition, neither olfactory bulbectomy, complete gastric vagotomy, nor hypophysectomy has reversed the feeding depression.

NATO ASI Series, Vol. H20
Amino Acid Availability and Brain Function in Health and Disease. Edited by G. Huether

Lesion Studies

Electrolytic lesions of various brain areas have been used to study the role of each of these areas in the rat's response to disproportionate amino acid diets. Of the more than 15 studies that have been reported over the last several years by our laboratory and those of others, only lesions of the anterior prepyriform cortex (PPC) and the medial basal amygdala (AMYG) have been shown to attenuate or reverse the initial feeding depression to amino acid imbalances in the rat (7).Of these, only the PPC lesion also blunts the feeding depression seen with a diet completely devoid of an essential amino acid. To re-emphasize this point, although certain hypothalamic nuclei, e. g., the ventromedial hypothalamus (VMH) and lateral hypothalamus (LH) have been included in reports from several laboratories (7, 8, 9), and the paraventricular nucleus (PVN) has been studied by us, only extrahypothalamic, rhinencephalic (limbic) structures have been implicated in the feeding response to amino acid imbalance by lesion studies.

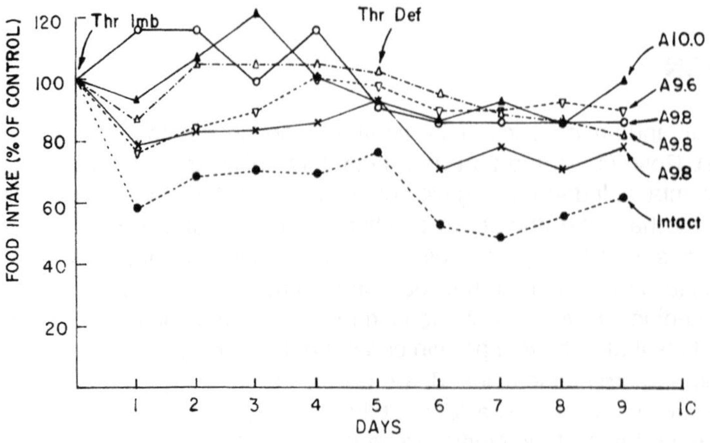

FIGURE 1

Mean food intake of rats with bilateral lesions of the prepyriform cortex (PPC) at anterior (A) coordinates indicated on the right. Values are percent of the animal's basal diet intake for the 3 days prior to introduction of the imbalanced diet. N = 6—15 per group. Taken from Leung and Rogers (10) with permission.

The responses of PPC-lesioned animals are shown in Figure 1. As may be seen in the figure, the food intake of rats with bilateral lesions of the PPC from anterior (A) 9.6 to A 10.0, was 80—116 % of control, while intact rats reduced their intake to about 60 % of control. It should be noted that, although the PPC is an olfactory relay in the brain, olfactory bulbectomy did not alter the animals' normal rejection of the imbalanced diet, while it did reduce their ability to distinguish between control diets to which odors had been added (3).

Conversely, VMH lesioned animals, all of whom were hyperphagic, retained their capability to regulate amino acid intake. Figure 2 shows the mean food intake response of hyperphagic VMH lesioned rats, who still depressed their intake of a moderately or severely imbalanced or a devoid amino acid diet (11).

Although these animals did not depress their intake of a mild isoleucine imbalanced diet as much as did the sham-lesioned controls (Fig. 2, right, Ile Imb-1), the response to mild imbalances are usually blunted in older animals past the stage of rapid growth.

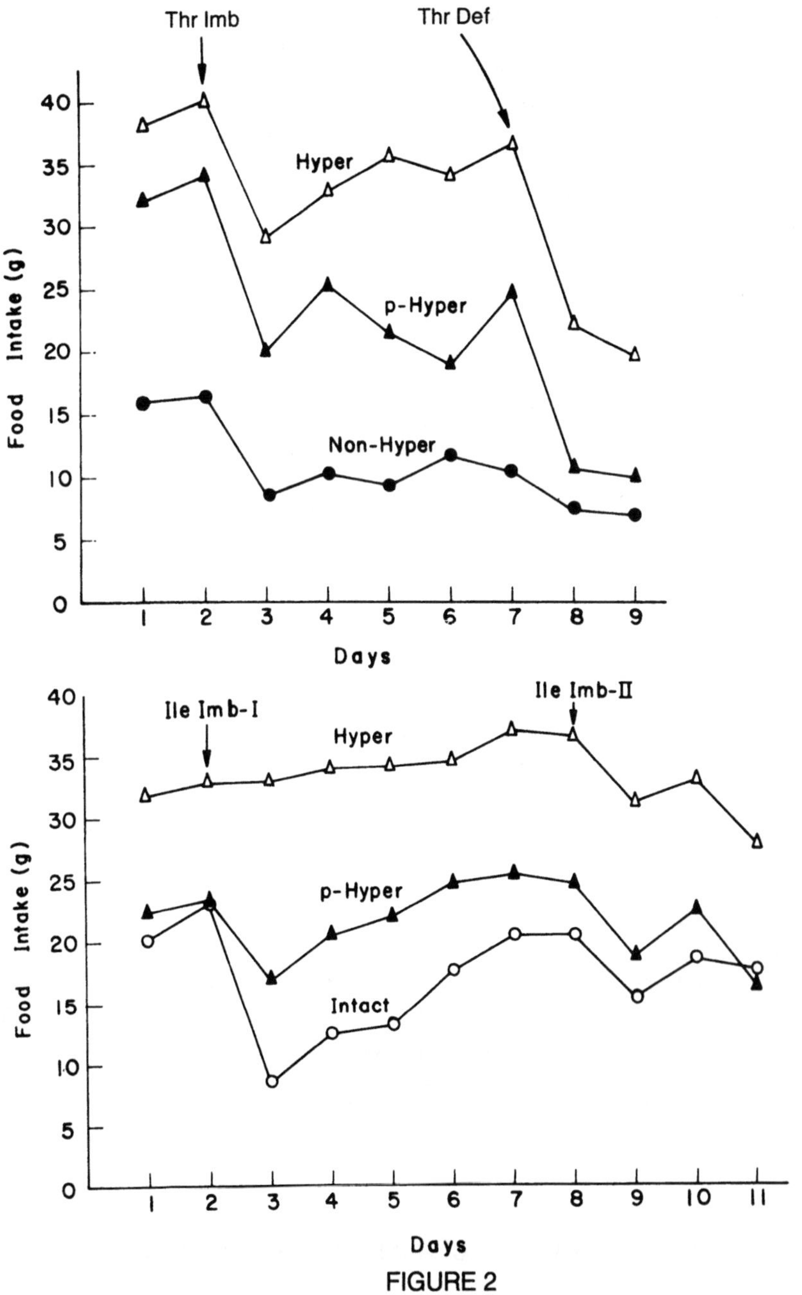

FIGURE 2

Mean food intake (grams/24 hr, threonine limited diets on the left, isoleucine limited diets on the right) of animals with ventromedial hypothalamic (VMH) lesions resulting in either partial (PHyper) or pronounced (Hyper) hyperphagia. Intact: control, Thr Imb: threonine imbalanced diet, Thr Def: threonine devoid diet, Ile Imb-I: mild isoleucine imbalanced diet, Ile Imb-II: severe isoleucine imbalanced diet. Diet presentation is indicated at the arrows. Taken from Leung and Rogers (11) with permission.

On the other hand, the initial response was exacerbated in animals with lesions of the ventral tegmental nuclei (TEG (12)) and in hypophysectomized (HYPOX (13)) rats. Interestingly, replacement with a crude pituitary extract did not reverse the depression in the HYPOX animals, but, in fact, increased the exacerbation (13). Several other brain areas, listed in Table 1, have been shown to be involved in the adaptive phase of the response, as well. These areas are also extrahypothalamic limbic structures (7). Of particular interest is the effect of a lesion in the area postrema, site of vagal and other peripheral integration, which increased intake of a threonine devoid diet, and accelerated adaptation of rats to a 75 % casein diet (Table 1). Of the several brain areas studied in this laboratory over the years, this is the first instance of a lesion that has affected intake of the high protein diet.

TABLE 1
Effect of lesions in specific neural areas on food intake responses of rats fed disproportionate amounts of dietary amino acids

	Food intake responses					
	Initial depression			Adaption		
Neural areas	Imb[a]	Devoid	Excess	Imb	Devoid	Excess
Response of intact rat	Yes[b]	Yes	Yes	Slow	No	Slow
Ventromedial hypothalamus	Yes[c]	Yes	Yes	Slow	No	Slow
Thalamic taste nuclei	Yes	Yes	Yes	Rapid	No	Slow
Anterior prepyriform cortex	No	No	Yes	Rapid	Rapid	Slow
Amygdala (medial)	No	Yes	Yes	Rapid	No	Slow
Hippocampus (dorso-lateral)	Yes	Yes	Yes	Rapid	No	Slow
Septum (lateral)	Yes	Yes	Yes	Rapid	No	Slow
Anterior cingulate cortex	Yes	Yes	Yes	Rapid	No	Slow
Ventral tegmental nuclei	Yes	Yes	Yes	Slow	No	Slow
Raphe nuclei (dorsal and median)	Yes	Yes	Yes	Slow	No	Slow
Area postrema	Yes	Yes	Yes	Slow	No[d]	Rapid
Nucleus of the solitary tract	Yes	Yes	Yes	Slow	No	Slow
Olfactory bulbectomy	Yes	Yes	Yes	Slow	No	Slow

[a] Imb: imbalanced diet, devoid: devoid diet, excess: high-protein diet. [b] Yes: present, No: absent or attenuated, Rapid: accelerated adaption, Slow: "normal" slow adaption. [c] However, no depression with mild isoleucine imbalanced diet. [d] Increased intake but not true adaption.

Regional Differences in Reduction of the Limiting Amino Acid in Brain

One of the most consistent biochemical changes, that has been observed with the feeding of imbalanced amino acid diets over the years, is the altered amino acid pattern in the plasma. The plasma amino acid profile includes a depression in the limiting amino acid and increases in all of the essential amino acids added to cause the imbalance (Fig. 3). The depression of the limiting amino acid is even more pronounced in whole brain than in plasma, while the increases of the essential amino acids, added to cause the imbalance, are blunted in whole brain homogenates (5), and in some brain areas, such as the PPC (Fig. 4).

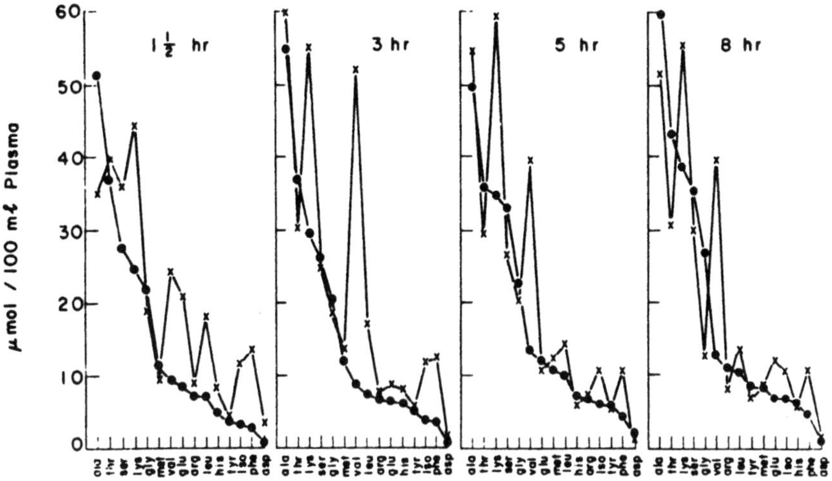

FIGURE 3
Plasma amino acid patterns at various times after ingestion of a threonine imbalanced (x) or control (o) diet. Aminograms have abbreviations for the several amino acids on the abscissa. Adapted from Leung, P. M. B., Rogers, Q. R., and Harper, A. E. (1968), *J. Nutr.* **96:** 303—318, with permission.

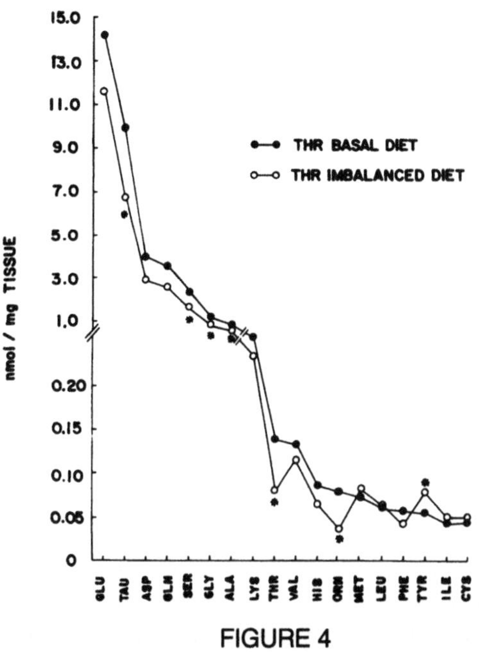

FIGURE 4
Aminogram for the prepyriform cortex from animals fed threonine diets for 2.5 hr prior to killing. *Sig p ≤ 0.05. Taken from (4) with permission.

However, not all brain areas show the depression of the limiting amino acid. Shown in Figure 5 are the concentrations of the limiting amino acid, threonine, in several brain areas from rats fed basal, imbalanced, or corrected diets with threonine as the growth-limiting amino acid. The expected increase in the concentrations of threonine after ingestion of the corrected diet was seen in all but one of these brain areas, but the depression of the limiting amino acid in the imbalanced diet group did not occur in VMH, LH, or hippocampus (HIP), while the anterior cingulate cortex (AC), PPC, and AMYG did show the expected decrease.

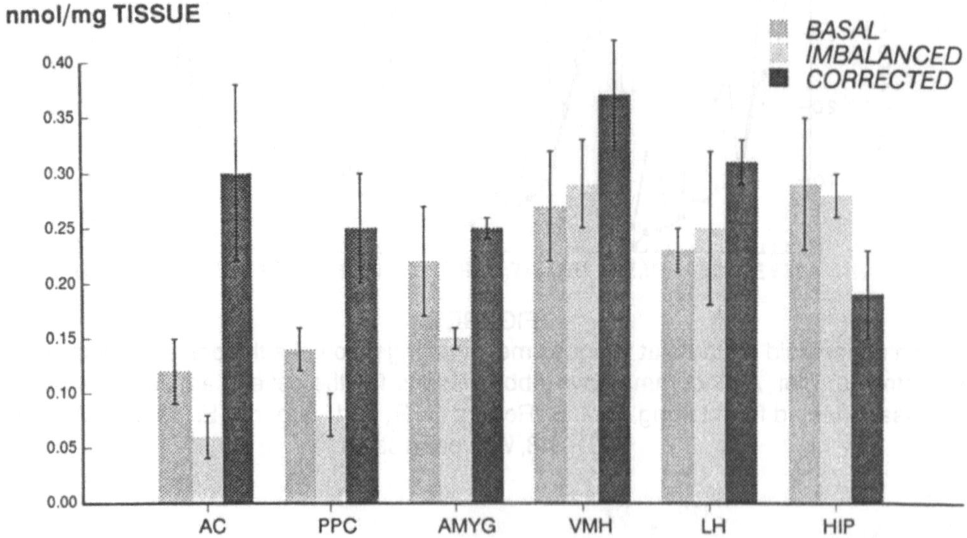

FIGURE 5

Concentrations of the limiting amino acid, threonine, in brain regons after 2.5 hr ingesting the threonine-basal, -imbalanced, or -corrected diet. Values are nmol/mg wet tissue weight. PPC: anteror prepyriform cortex, AC: anterior cingulate cortex, VMH: ventromedial hypothalamus, LH: lateral hypothalamus, HIP: hippocampus.

Therefore, we may not assume that there is a uniform distribution of the amino acid profile in the brain after feeding imbalanced amino acid diets. Since hypothalamic regions were among those areas that did not show a decrease in the concentration of the limiting amino acid, these findings, if extended to other amino acid imbalances, would confirm the lack of evidence to support an inhibitory effect of the hypothalamus on feeding responses to imbalanced diets.

Regional Concentrations of Monoamines in Brain

Since the monoamines have been shown to have important effects on feeding, we began our inquiry into the role of these amino acid metabolites by measuring their concentrations in specific brain areas from animals fed amino acid imbalanced diets and in controls given basal or corrected diets. Purified powdered diets using either L-threonine or L-isoleucine as the growth-limiting amino acid were used. As is customary in this model, the animals were pretreated with the appropriate low protein basal diet, adapted to a reversed light cycle with lights off at noon, and to manual measurements of food intake. For these experiments, to synchronize the first meal of the dark period, the food was removed at the beginning of the prior light period. The diets were presented at noon, the beginning of the dark cycle. At the selected time, just before the feeding depression with the mild isoleucine diet and just after a significant depression with the more severe threonine imbalanced diet, the animals were decapitated, and the brains frozen on dry ice. Concentrations of the monoamines were subsequently determined in 14 microdissected brain areas by high pressure liquid chromatography with electrochemical detection (HPLC-EC) as described previously (4).

Regional concentrations of neurotransmitters, in themselves, can only suggest alterations in the transmitter system if changes are found, since, for example, the lack of a change in concentration can occur in the face of great transmitter activity if synthesis is increased along with release. Nevertheless, changes in concentration could be the result of alterations in parameters such as release, synthesis, or uptake (see Commissiong, this volume) and are unlikely to be seen in the absence of some perturbation in the system. Selected results from our regional monoamine measurements are outlined in Table 2. The concentration of norepinephrine (NE) was decreased by 30 % in PPC (14) and increased to 115 % of control in VMH from animals fed both isoleucine and threonine imbalanced diets, when compared with their basal and corrected diet fed controls. Preliminary results of push-pull perfusion in the PPC suggest that the decreases in NE concentration in that area may be the result of an increase in turnover in the imbalanced group.

TABLE 2
Monoamines in Brain Areas of Animals Fed Imbalanced Amino Acid Diets

Brain Area	Limiting Amino Acid			
	Isoleucine		Threonine	
	NE	5HT	NE	5HT
Anterior cingulate cortex	—[a]	—	—	—
Anterior prepyriform cortex	↓	—	↓	—
Paraventricular nucleus	—	—	—	—
Lateral hypothalamus	—	↑	—	—
Ventromedial hypothalamus	↑	—	↑	—
Hippocampus	—	—		

[a]: unchanged from both basal and corrected diet controls, ↓: decrease relative to controls, ↑: increased relative to controls.

The concentration of serotonin (5HT) was increased only in LH and only from animals fed the isoleucine imbalanced diet. No changes were noted in the concentration of dopamine in any of the brain areas that have been analyzed.

Comparisons of metabolite (5-hydroxyindoleacetic acid, [5HIAA] to 5HT: 5HIAA/5HT) ratios in 14 brain areas showed significant increases in the isoleucine-imbalanced group in the hippocampus (HIP) and locus ceruleus (LC), relative to both basal and corrected groups. In the PPC, the ratio in the isoleucine-imbalanced group was less than that in the basal-diet group, and similar to the corrected-diet group. In the threonine-imbalanced group the 5HIAA/5HT ratio was increased in anterior cingulate cortex (AC), PPC, and LC relative to the basal-diet group, and also in PPC and LC relative to the corrected-diet group. The ratio was decreased in the threonine-imbalanced group relative to both control groups in the septum, and similar to basal but increased relative to the corrected-diet group in the parabrachial nucleus (PBP) and the nucleus of the solitary tract (NTS). In the isoleucine imbalanced group, 6 of 7 areas caudal to the hypothalamus showed at least some increase relative to the controls. Thus the 5HIAA/5HT ratio was altered in imbalanced diet-fed animals in both studies (increased in LC and PPC, and decreased in septum in the threonine study; increased in HIP and LC and decreased in PPC in the isoleucine study).

When the 5HIAA/5HT ratio was examined relative to the concentration of dietary protein, i. e., in the direction of the precursor hypothesis, which predicts an increase in 5HT activity in animals fed the lower protein diet, in only one of the 14 brain areas studied (PPC) was the expected change noted, i. e., a decrease after 2.5—3.5 hours ingestion of the higher protein (corrected) diet, relative to the lower protein (basal) diet. The basal diets include the equivalent of approximately 8—10 % crude protein, and the corrected diets contain the equivalent of approximately 17 % crude protein. The 5HIAA/5HT ratio was actually higher in rats fed the corrected diet than the basal diet (i. e., after 17 % vs 8—10 % protein) in 2 brain areas, the AC and NTS. Thus, of 14 brain areas in two dietary situations, intake of the higher protein diet was associated with an increased ratio in two areas (AC, NTS) with the threonine diets and a decreased ratio (the expected change, according to the "TRP/LNAA as a predictor of protein intake" hypothesis) in only one brain area, the PPC, in the isoleucine study. It should be noted that these circumstances may not fulfill the special requirements for alteration of 5HT function outlined by Fernstrom (this volume).

Pharmacological Studies

Since it has been suggested that 5HT induces a generalized inhibition of food intake, and the only consistent changes in the 5HIAA/5HT ratio were found in the mild isoleucine imbalanced diet group, some facet of the feeding depression in animals fed imbalanced amino acid diets could well involve activation of this inhibitory 5HT feeding system. Therefore, the pharmacological responsiveness of the 5HT system was investigated in animals fed diets with isoleucine as the limiting amino acid.

Treatments which may be analogous to an increase in 5HT activity exacerbated the feeding depression seen in animals ingesting the imbalanced diet. The four treatments were: 1, the addition of the precursor, L-tryptophan (TRP 1 % added to the diets) and 2, injection of the agonist quipazine (5 %mg/kg ip); two additional treatments that cause depletion of 5HT and subsequent receptor supersensitivity (15) also exacerbated the feeding depression: 3, injection of parachlorophenylalanine (PCPA 300 mg/kg in a divided dose, ip); and 4, intraventricular injection of 5,7-dihydroxytryptamine (DHT 200 μg/rat injected into the cisterna magna 30 min after desmethylimipramine, 25 mg/kg ip). The suggestion of receptor supersensitivity was supported by the feeding response to quipazine in previously DHT treated animals, and in previously PCPA treated animals given TRP. Both of these combined treatments resulted in a further exacerbation of the depression, which may be interpreted as agonist enhancement of supersensitivity.

A reduction in 5HT activity was induced by subcutaneous injection of the $5HT_{1A}$ (autoreceptor) agonist 8OH-DPAT (DPAT, 500 μg/kg sc given in the light period, 30 min before the onset of the dark cycle, at which time the test diets were introduced). The results of these pharmacological interventions are given in Fig. 6, along with the response of control animals, given saline injections, to the imbalanced diet. As may be seen in the figure, after PCPA or DHT, with presumed receptor supersensitivity, or injection of the agonist, quipazine, the initial 24-hr food intake was reduced to around 40 % of control, while in DHT treated animals, quipazine decreased intake even further, to about 27 % of control. In contrast, DPAT, used to inhibit 5HT activity by stimulating the autoreceptor, severely attenuated the feeding depression.

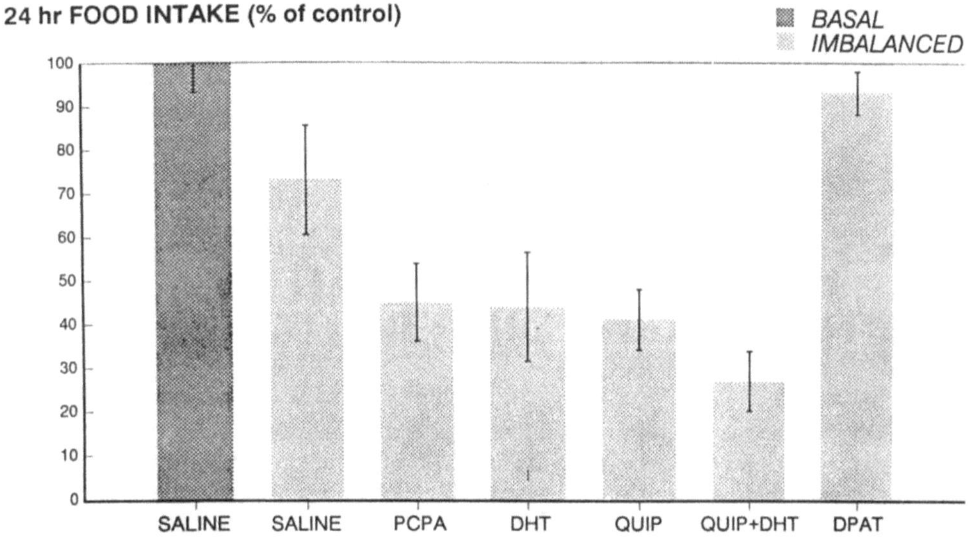

FIGURE 6

Food intake (percent of control as described in Fig. 1) of basal or mild isoleucine imbalanced diets by animals after various serotonergic treatments. Saline: 0.9 % NaCl 1 ml/kg ip, PCPA: parachlorophenylalanine 300 mg/kg ip in 3 doses, DHT: 5,7-dihydroxytryptamine 200 μg/rat intracisternally, QUIP: quipazine 5 mg/kg ip, DPAT: 8-OH DPAT 100 μg/kg sc. Bars: mean food intake as % of control for the first 24 hours of imbalanced diet feeding, vertical lines represent standard errors.

Taken together, we have interpreted these data as supportive of the suggestion that an increase in 5HT activity is associated with the depression in food intake of rats fed imbalanced amino acid diets. The location of this increased 5HT activity must involve a postsynaptic site, since DHT should have destroyed presynaptic elements. This postsynaptic site may belong to the $5HT_{1B}$ category, described as anorexigenic by Curzon (this volume), since treatment with pirenpirone, the $5HT_2$ antagonist (100 μg/kg sc), produced a pronounced exacerbation of the anorexic response, similar to that seen with quipazine plus DHT. Potentiation of an inhibitory 5HT effect by a $5HT_2$ antagonist has been suggested to act via modulation of a $5HT_1$ site (16). Quipazine also acts at the $5HT_{1B}$ site. Taken together, these observations strengthen our suggestion of a role for 5HT in the anorexic response of rats to imbalanced diets. Still, the full interpretation of these findings awaits further study; 5HT must be only one of several systems interacting in these responses.

With regard to the noradrenergic system, peripheral injection of agonists or antagonists has not produced any change in the feeding depression to date. However, we have reported a decrease in NE concentration in the PPC (14), and injection of the NE autoreceptor agonist clonidine (100 μg/0.5 μl) into the PPC attenuated the feeding depression with a threonine imbalanced diet (Table 3). Also, as noted above, preliminary results with push-pull perfusion indicate that NE release may be increased with imbalanced diet feeding. Further studies of both the 5HT and NE systems are in progress in our laboratory.

TABLE 3
Effect of clonidine injection into the anterior prepyriform cortex on consumption of a threonine imbalanced diet

Injection	% of Basal Diet Intake
Saline[a]	68.8 ± 2.4
Artificial CSF	69.6 ± 6.4
Clonidine	85.1 ± 7.8*

[a] Injection volume, 0.5 μl. Basal diet intake: 24 hr intake of the basal diet after similar injection. Values are imbalanced diet intake/basal diet intake × 100, ± SE. N = 8. * Significant at $p < 0.05$.

SUMMARY AND CONCLUSIONS

It is clear that dietary amino acids can affect food intake. Reliable depressions in feeding are seen with either a deficiency, an imbalance, or an excess of amino acids in the diet. When a diet containing an imbalanced proportion of amino acids is fed, the dietary amino acid pattern is reflected in plasma and peripheral tissues and the decrease in the limiting amino acid is also reflected in the CSF and much of the brain.

Several lines of evidence suggest that the brain is important in the feeding responses of rats to dietary amino acids. Infusion studies suggested that structures perfused by the carotid artery were involved in these responses. Lesion studies have implicated certain specific brain areas, particularly in the extrahypothalamic limbic system, as being important in the response, and vagotomy and olfactory bulbectomy have eliminated at least 2 sources of peripheral input as essential to the response.

Regional changes in amino acid patterns, norepinephrine, serotonin, and the ratios of the major metabolites to each of these neurotransmitters occur. Serotonin and norepinephrine appear to play important roles in these feeding responses; there is some evidence for increased activation of both systems after ingestion of imbalanced amino acid diets. Regional studies must be extended, in order to provide enough information for a proper understanding of the roles of the specific brain areas and the various neurotransmitter systems in the mediation of the animal's response to amino acids. Further experiments should clarify these roles and the probable interactions among the amino acids and neurotransmitters in the brains of animals ingesting and responding to diets containing disproportionate quantities of amino acids.

ACKNOWLEDGEMENTS

The expert secretarial help of Tracy Schuster is greatly appreciated. This work was supported by NIH grants # AM13252, AM07355, and AM07557.

REFERENCES

1) ROSE, W. C. (1931). Feeding experiments with mixtures of highly purified amino acids. I. The inadequacy of diets containing 19 amino acids. *J. Biol. Chem.* **94:** 155—165.

2) HARPER, A. E., BENEVENGA, N. J., WOHLHUETER, R. M. (1970). Effects of ingestion of disproportionate amounts of amino acids. *Physiol. Rev.* **50:** 428—558.

3) ROGERS, Q. R., LEUNG, P. M. B. (1977). The control of food intake: When and how are amino acids involved? In: *The Chemical Senses and Nutrition* (Kare, M. R., Maller, O., eds.). Academic Press, New York, pp. 213—249.

4) GIETZEN, D. W., LEUNG, P. M. B., CASTONGUAY, T. W., HARTMAN, W. J., ROGERS, Q. R. (1986). Time course of food intake and plasma and brain amino acid concentrations in rats fed amino acid-imbalanced or -deficient diets. In: *Interaction of the Chemical Senses with Nutrition* (Kare, M. R., Brand, J. G, eds.). Academic Press, New York, pp. 415—456.

5) PENG, Y., TEWS, J. K., HARPER, A. E. (1972). Amino acid imbalance, protein intake, and changes in rat brain and plasma amino acids. *Am. J. Physiol.* **222:** 314—321.

6) LEUNG, P. M. B., ROGERS, Q. R. (1969). Food intake: Regulation by plasma amino acid pattern. *Life Sci.* **8:** 1—9.

7) LEUNG, P. M. B., ROGERS, Q. R. (1987). The effect of amino acids and protein on dietary choice. In: *Umani: A Basic Taste* (Kawamura, Y., Kare, M. R., eds.). Marcell Dekker Inc., New York, pp. 565—610.

8) KRAUSS, P. M., MAYER, J. (1965). Influence of protein and amino acids on food intake in the rat. *Am. J. Physiol.* **209:** 479—483.

9) NASSET, E. S., RIDLEY, P. T., SCHENK, E. A. (1967). Hypothalamic lesions related to ingestion of an imbalanced amino acid diet. *Am. J. Physiol* **213:** 645—650.

10) LEUNG, P. M. B., ROGERS, Q. R. (1971). Importance of prepyriform cortex in food intake response of rats to amino acids. *Am. J. Physiol.* **221:** 929—935.

11) LEUNG, P. M. B., ROGERS, Q. R. (1970). Effect of amino acid imbalnce and deficiency on food intake of rats with hypothalamic lesions. *Nutr. Rep. Internat* **1:** 1—10.

12) LEUNG, P. M. B., ROGERS, Q. R. (1980). Hyperphagia after ventral tegmental lesions and food intake responses of rats fed disproportionate amounts of dietary amino acids. *Physiol. Behav.* **25:** 457—464.

13) LEUNG, P. M. B., ROGERS, Q. R. (1971). Effects of pituitary extract on food intake of intact and hypophysectomized rats fed imbalanced amino acid diets. *Nutr. Rep. Internat* **4:** 207—215.

14) GIETZEN, D. W., LEUNG, P. M. B., ROGERS, Q. R. (1986). Norepinephrine and amino acids in prepyriform cortex of rats fed imbalanced amino acid diets. *Physiol. Behav.* **36:** 1071—1080.

15) GREEN, A. R., HEAL, D. J. (1985). The effects of drugs on serotonin-mediated behavioural models. In: *Neuropharmacology of Serotonin* (Green, A. R., ed.). Oxford Univ. Press, Oxford, pp. 326—365.

16) LAKOSKI, J. M., AGHAJANIAN, G. K. (1985). Effects of ketanserin on neuronal responses to serotonin in the prefrontal cortex, lateral geniculate and dorsal raphe nucleus. *Neuropharmacology* **24:** 265—273.

EFFECTS OF TRYPTOPHAN ADMINISTRATION ON THE SYNTHESIS, STORAGE AND METABOLISM OF SEROTONIN IN THE HYPOTHALAMUS OF NORMAL AND RAPHE-STIMULATED RATS

K. J. Lookingland, N. J. Shannon and K. E. Moore

Department of Pharmacology and Toxicology
Michigan State University
East Lansing, MI 48824
USA

INTRODUCTION

The various steps in the synthesis and metabolism of serotonin (5-hydroxytryptamine; 5HT) in the brain have been well characterized (see reviews 1, 2). The initial and rate-limiting step in the synthesis of 5HT is the conversion of L-tryptophan to 5-hydroxytryptophan (5HTP), a reaction catalyzed by tryptophan hydroxylase. Newly formed 5HTP is rapidly decarboxylated by aromatic L-amino acid decarboxylase to 5HT which is either stored in vesicles for subsequent release, or is metabolized by mitochondrial monoamine oxidase to 5-hydroxyindoleacetic acid (5HIAA). The calculated K_m for tryptophan hydroxylase is higher than tryptophan concentrations measured in the brain, suggesting that under normal physiological conditions the activity of this enzyme is substrate limited. Accordingly, procedures that increase the concentrations of tryptophan within the brain produce corresponding changes in 5HT concentrations.

Neurochemical estimates of the activity of 5HT neurons in the bran are based upon the coupled relationship between neurotransmitter synthesis, release and metabolism. An increase in 5HT release in response to an electrical impulse is accompanied by an increase in the synthesis and metabolism of 5HT and this coupling maintains a "steady state" level of releasable neurotransmitter within the neuron. The accumulation of 5HTP following inhibition of decarboxylase activity has been shown to be an *in vivo* index of the rate of synthesis of 5HT in the brain (3), and 5HTP accumulation has been utilized as an index of 5HT neuronal activity (4, 5). In addition, concentrations of 5HIAA increase following activation of 5HT neurons (6—9), and concentrations of 5HIAA and the ratio of 5HIAA to 5HT in the brain have been shown to reflect the activity of 5HT neurons.

One potential problem with the use of neurochemical measures to determine the activity of 5HT neurons relates to the availability of substrate for tryptophan hydroxylase. Since an increase in tryptophan hydroxylase activity induced by an increase in 5HT release during neuronal activation is not accompanied by an increase in the intracellular availability of tryptophan (10), estimates of neuronal activity by measuring 5HT synthesis may be underestimated under normal *in vivo* conditions. Conversely, a decrease in neu-

NATO ASI Series, Vol. H20
Amino Acid Availability and Brain Function in Health and Disease. Edited by G. Huether

ronal activity may not result in a corresponding decrease in 5HT synthesis since tryptophan hydroxylase activity is already submaximal due to substrate limitations. Furthermore, drugs of experimental manipulations that alter the availability of tryptophan may confound interpretation of results from experiments utilizing 5HTP accumulation as an index of 5HT neuronal activity (e. g. 11, 12). It follows then that development of an *in vivo* model in which the availability of tryptophan is not a limiting factor would be advantageous in the study of 5HT neurons.

Effects of Tryptophan on Neurochemical Estimates of the Basal Activity of 5HT Neurons

As a first step towards determining the importance of substrate availability on the function of 5HT neurons, we characterized the effects of acute tryptophan administration on neurochemical indices of 5HT synthesis, storage and metabolism in the rat hypothalamus under basal conditions (13). In these studies, tryptophan administration produced a dose-dependent increase in the rate of 5HTP accumulation in the hypothalamus following inhibition of decarboxylase activity, but had no effect on the accumulation of the catecholamine precursor 3,4-dihydroxyphenylalanine. These results indicate that the intrinsic activity or tryptophan hydroxylase in hypothalamic 5HT neurons can be accelerated by increasing the availability of its substrate, and increasing tryptophan concentrations in the brain does not alter the synthesis of dopamine or norepinephrine in hypothalamic catecholamine neurons.

As a consequence of an accelerated rate of 5HT synthesis following tryptophan administration there is a concurrent increase in the storage and metabolism of 5HT, as indicated by an increase in brain concentrations of 5HT and 5HIAA, respectively. Since there is no change in the 5HIAA/5HT ratio following tryptophan despite marked elevations in both the amine and its deaminated metabolite, the storage capacity for 5HT is apparently less than the amount synthesized in response to exogenous tryptophan and significant amounts of cytoplasmic 5HT become available for metabolism by mitochondrial monoamine oxidase.

In order to determine the relative contributions of unreleased 5HT and released/recaptured 5HT to the total pool of 5HT metabolized to 5HIAA we examined the effects of the 5HT uptake inhibitors chlorimipramine and fluoxetine (14, 15) on 5HIAA concentrations in hypothalamic tissue of normal and tryptophan-treated rats. By inhibiting the reuptake of released 5HT into presynaptic terminals and glial cells (16), these drugs increase synaptic 5HT content and prolong the stimulation of both pre- and post-synaptic 5HT receptors and, through a mechanisms mediated by inhibitory neuronal feedback circuits and/or presynaptic 5HT autoreceptors, decrease the rates of firing of 5HT neurons (17, 18). This decrease in 5HT neuronal activity is associated with a decrease in 5HT synthesis (19, 20) and a decrease in spontaneous 5HT release (17). Since the inhibitory effects of uptake inhibitors on the activity of 5HT neurons is dependent upon the presence of released 5HT in the synapse, and tryptophan does not alter 5HT release unless metabolism of newly synthesized 5HT is blocked (10, 21), we reasoned that these drugs should have similar effects in normal and tryptophan-treated rats. These assumptions seem warranted since uptake inhibitors have been shown to produce similar decreases in 5HTP accumulation in normal and tryptophan-treated rats (19).

Both chlorimipramine and fluoxetine decrease 5HIAA concentrations in hypothalamic regions of normal rats by only 10—40 % (13) suggesting that the majority of 5HIAA measured in hypothalamic tissue reflects metabolism of intraneuronal 5HT rather than released 5HT. These results are in agreement with previous reports (22, 23) and imply that 5HIAA concentrations measured in brain tissue are more a reflection of 5HT metabolism than 5HT neuronal activity under non-stimulated basal conditions (also see 24). Although the possibility that an increase in intraneuronal metabolism of 5HT following application of uptake inhibitors cannot be ruled out, there is an accompanying decrease in 5HT synthesis following inhibition of 5HT reuptake (19, 20), and any elevation in intraneuronal metabolism of 5HT under these conditions would for the most part be negligible.

The administration of tryptophan to chlorimipramine- and fluoxetine-treated rats causes an increase in 5HIAA concentrations indicating that tryptophan increases intraneuronal metabolism of newly synthesized 5HT (13). Since inhibition of 5HT uptake in tryptophan-treated rats produces a similar decrease in 5HIAA concentrations as that observed in normal rats, tryptophan administration apparently does not produce a marked change in 5HT release (also see 25). On the other hand, if the metabolism of newly synthesized 5HT is blocked by the administration of a monoamine oxidase inhibitor, postsynaptic receptor-mediated effects are observed following tryptophan suggesting that 5HT release is enhanced (21, 18). Taken together these results demonstrate the importance of intraneuronal metabolism in eliminating excess nonvesicular 5HT from the cytoplasm and suggest that increasing 5HT synthesis does not alter spontaneous 5HT release in the hypothalamus.

Effects of Tryptophan on Neurochemical Estimates of the Activity of 5HT Neurons Following Raphe Stimulation

5HT neurons located within the hypothalamus originate primarily in the mesencephalic dorsal and median raphe nuclei (26, 27). 5HT neurons in raphe nuclei have relatively slow firing rates (28),and under basal conditions the relationship between neurotransmitter synthesis, release and metabolism may be uncoupled (24, 29, 20). In agreement, the results from studies described above suggest that synthesis of 5HT in quiescent neurons is in excess of functional needs and excess newly synthesized neurotransmitter is metabolized within neurons without being released (also see 31).

The purpose of the present study was to examine the effects of tryptophan administration on neurochemical estimates of 5HT neuronal activity following activation by electrical stimulation. In these studies the effects of stimulation of the dorsal raphe nucleus on the synthesis, storage and metabolism of 5HT in the suprachiasmatic nucleus were examined since 5HT terminals in this brain region have been implicated in the regulation of a variety of physiological processes.

MATERIALS AND METHODS

Male Long-Evans rats (225—250 g) were purchased from Charles River Breeding Laboratories and maintained in a temperature- (22°C) and light- (lights on between 0600 h and 1800 h) controlled environment with food (Wayne Lablox) and tap water provided *ad libitum.* On the day of the experiment animals were anesthetized with chloral hydrate

(400 mg/kg; i. p.), positioned in a stereotaxic apparatus, and implanted with a coaxial bipolar stainless steel electrode (NE-100, Rhodes Medical Instruments, Woodland Hills, CA) in the dorsal raphe nucleus (A 0 mm, L 0 mm, H −1.0 mm; 32). Brains were stimulated for either 15, 30 or 60 min with cathodal monophasic electrical pulses of 1 ms duration and 0.3 nA current at 10 Hz frequency as described previously (5). Sham stimulation consisted of placement of an electrode into the dorsal raphe nucleus without application of a current.

Following the appropriate treatments, animals were killed by decapitation and brains were removed and frozen on aluminum foil placed directly over dry ice. Frontal brain sections (600 μm) were prepared in a cryostat (−9 °C) and the suprachiasmatic nucleus dissected from the appropriate section according to the method of Palkovits (33). Samples were placed in 60 μl of 0.1 M phosphate-citrate buffer (pH 2.5) containing 15 % methanol and stored at − 20 °C until assayed.

5HTP, 5HT and 5HIAA contents of the tissue samples were determined by high performance liquid chromatography with electrochemical detection as described previously (34). Tissue pellets were dissolved in 1.0 N NaOH and assayed for protein (35).

Statistical analysis of differences between groups was determined by Bonferroni's t-test for multiple comparisons (36). Differences were considered significant if the probability of error was less than 5 %.

RESULTS AND DISCUSSION

As shown in Figure 1, in the absence of a decarboxylase inhibitor (time zero) there was very little 5HTP present within the suprachiasmatic nucleus, but by 15 and 30 min following the i. v. administration of 3-hydroxybenzylhydrazine (NSD 1015), 5HTP accumulated in sham-stimulated rats progressively with time reflecting the rate of 5HT synthesis. Electrical stimulation of the dorsal raphe nucleus for 15 and 30 min significantly increased the accumulation of 5HTP in the suprachiasmatic nucleus (Figure 1). These results are in agreement with previous reports from our laboratory (5, 8) and indicate that activation of 5HT neurons following stimulation of the dorsal raphe nucleus increases the rate of 5HT synthesis in the hypothalamus.

Concurrent with an increase in 5HT synthesis, electrical stimulation of the dorsal raphe nucleus caused a significant increase in the concentrations of 5HIAA in the suprachiasmatic nucleus, but had no effect on the levels of 5HT (Table 1; Figure 2). Consequently, there was an increase in the relative amount of 5HT metabolized in this brain region with respect to the amount of 5HT stored (as indicated by the ratio of 5HIAA/5HT concentrations). Taken together these results suggest that the stimulus-induced increase in 5HT release is coupled to an increase in the synthesis and metabolism of 5HT, and the maintenance of 5HT stores in hypothalamic 5HT neurons.

To determine the effects of tryptophan administration on the relative contribution of unreleased 5HT to the total pool of 5HT metabolized to 5HIAA we examined the effects of electrical stimulation of the dorsal raphe nucleus on 5HIAA concentrations in the suprachiasmatic nucleus of saline- and fluoxetine-treated rats. In saline-treated rats, stimulation of the dorsal raphe nucleus increased 5HIAA concentrations to approximately 125 % of that observed in sham-stimulated controls (Fig. 2). Similarly, raphe stimulation in tryptophan-treated rats produced an increase in 5HIAA concentrations, but this effect was greater (150 % of sham-stimulated controls) than that observed in vehicle-treated

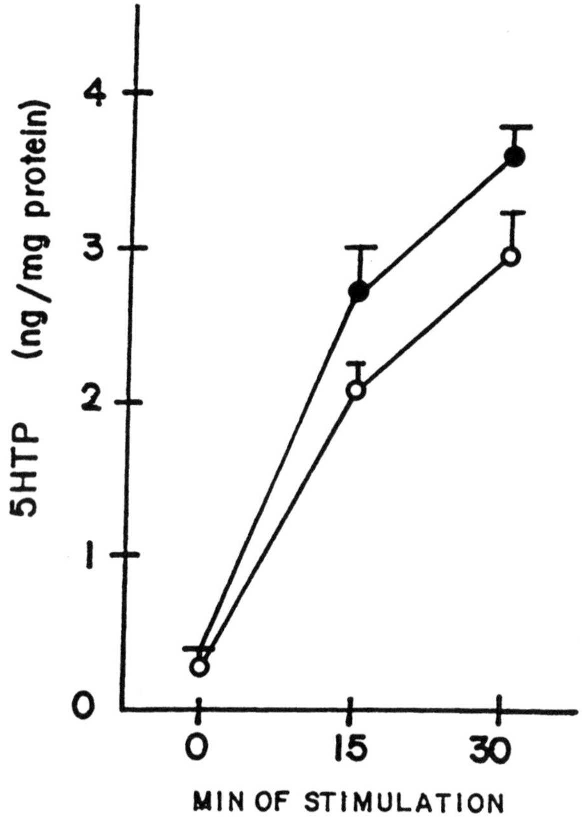

FIGURE 1

Time course of the effects of electrical stimulation of the dorsal raphe nucleus on the accumulation of 5HTP in the suprachiasmatic nucleus. Rats were killed either 15 or 30 min after sham (open symbols) or electrical stimulation (closed symbols) of the dorsal raphe nucleus. Symbols represent the means and vertical lines 1 SEM (n = 8—12) of 5HTP concentrations determined 15 or 30 min after NSD 1015 (25 mg/kg; i. v.) administration.

TABLE 1

Concentrations of 5HIAA and 5HT in the suprachiasmatic nucleus following electrical stimulation of the dorsal raphe nucleus

	Duration of Stimulus (min)		
	0 (Control)	15	30
5HIAA (ng/mg protein)	4.48 ± 0.34	5.44 ± 0.29*	5.51 ± 0.32*
5HT (ng/mg protein)	7.78 ± 0.72	6.80 ± 0.46	7.02 ± 0.48
5HIAA/5HT	0.59 ± 0.03	0.83 ± 0.08*	0.81 ± 0.05*

Values represent the means ± 1 SEM of 7—12 determinations. * indicates values that are significantly different ($P < 0.05$) from zero time controls.

animals. By contrast, tryptophan failed to alter the effects of raphe stimulation on 5HIAA concentrations in fluoxetine-treated rats. These results indicate that tryptophan administration increases the total pool of 5HT metabolized to 5HIAA. On the other hand, tryptophan failed to alter the intraneuronal metabolism of 5HT following stimulation of the raphe, suggesting that the stimulated increase in 5HIAA is derived from released 5HT. It would appear, therefore, that tryptophan increases the releasable pool of 5HT so that when the neurons are activated more transmitter can be released. This is in agreement with the previous reports (37, 38) and indicates that the tryptophan-induced increase in 5HT synthesis and storage augments evoked 5HT release.

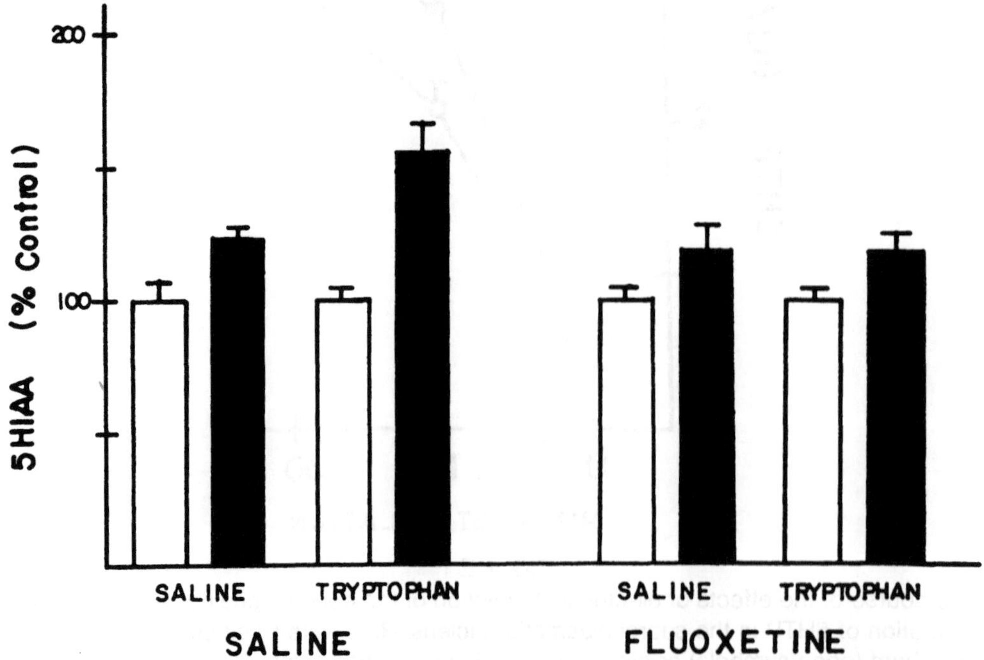

FIGURE 2
Comparison of the effects of tryptphan on 5HIAA concentrations in the suprachiasmatic nucleus following electrical stimulation of the dorsal raphe nucleus in control and fluoxetine-treated rats. Rats were injected with tryptophan methylester HCl (100 mg/kg; i. p.) or its saline vehicle (2 ml/kg; i. p.) and fluoxetine HCl (10 mg/kg; s. c.) or its saline vehicle (1 ml/kg; s. c.) 60 min prior to decapitation. Closed columns represent the means and vertical lines 1 SEM (n = 8—10) of 5HIAA concentratons in raphe-stimulated animals expressed as a percentage of levels in sham-stimulated controls (open columns set at 100 %).

ACKNOWLEDGEMENTS

The authors would like to thank Marty Burns for preparation of this manuscript. The experiments described in this study were supported by U. S. Public Health Service grant NS15911.

REFERENCES

1) FERNSTROM, J. D. (1983). Role of precursor availability in control of monoamine biosynthesis in brain. *Physiol. Rev.* **63:** 484—546.

2) LOVENBERG, W., KUHN, D. M. (1982). Substrate regulation of serotonin synthesis. In: *Serotonin in Biological Psychiatry* (Ho, B. T. et al., eds.). Raven Press, New York.

3) CARLSSON, A., DAVIS, J. N., KEHR, W., LINDQVIST, M., ATACK, C. V. (1972). Simultaneous measurement of tyrosine and tryptophan hydroxylase activities in brain in vivo using an inhibitor of the aromatic amino acid decarboxylase. *Arch. Pharmacol.* **275:** 153—168.

4) BOURGOIN, S., OLIVERAS, J. L., BRUXELLE, J., HAMON, M., BESSON, J. M. (1980). Electrical stimulation of the nucleus raphe magnus in the rat. Effects on 5-HT metabolism in the spinal cord. *Brain Res.* **194:** 377—389.

5) DUDA, N. J., MOORE, K. E. (1985). Simultaneous determination of 5-hydroxytryptophan and 3,4-dihydroxyphenylalanine in rat brain by HPLC with electrochemical detection following electrical stimulation of the dorsal raphe nucleus. *J. Neurochem.* **44.** 128—133.

6) AGHAJANIAN, G. K., ROSECRANS, J. A., SHEARD, M. H. (1967). Serotonin: release in the forebrain by stimulation of midbrain raphe. *Science* **156:** 402—403.

7) CURZON, G., FERNANDO, J. C. R., MARSDEN, C. A. (1978). 5-Hydroxytryptamine: the effects of impaired synthesis on its metabolism and release in rat. *Br. J. Pharmacol.* **63:** 627—634.

8) SHANNON, N. J., GUNNET, J. W., MOORE, K. E. (1986). A comparison of biochemical indices of 5-hydroxytryptaminergic neuronal activity following electrical stimulation of the dorsal raphe nucleus. *J. Neurochem.* **47:** 958—996.

9) SHEARD, M. H., AGHAJANIAN, G. K. (1968). Stimulation of the midbrain raphe: effect on serotonin metabolism. *J. Pharmacol. Exp. Ther.* **163:** 425—430.

10) ELKS, M. L., YOUNGBLOOD, W. W., KIZER, J. S. (1979). Serotonin synthesis and release in brain slices: independence of tryptophan. *Brain Res.* **172:** 471—486.

11) FULLER R. W., SNODDY, H. D., PERRY, K. W., ROUSH, B. W., MOLLOY, B. B., BYMASTER, F. P., WONG, D. T. (1976). The effects of quipazine on serotonin metabolism in rat brain. *Life Sci.* **18:** 925—934.

12) NECKERS, L. M., BIGGIO, G., MOJA, G., MEEK, J. L. (1977). Modulation of brain tryptophan hydroxylase activity by brain tryptophan content. *J. Pharmacol. Exp. Ther.* **201:** 110—116.

13) LOOKINGLAND, K. J., SHANNON, N. J., CHAPIN, D. S., MOORE, K. E. (1986). Exogenous tryptophan increases synthesis, storage, and intraneuronal metabolism of 5-hydroxytryptamine in the rat hypothalamus. *J. Neurochem.* **47:** 205—212.

14) LEMBERGER, L., ROWE, H., CARMICHAEL, R. (1978). Fluoxetine, a selective serotonin uptake inhibitor. *Clin. Pharmacol. Ther.* **23:** 421—429.

15) WONG, D. T., HORNG, J. S., BYMASTER, F. P., HAUSER, K. L., MOLLOY, B. B. (1974). A selective inhibitor of serotonin uptake: Lilly 110140, 3-(p-trifluoromethylphenoxy)-n-methyl-3-phenylpropylamine. *Life Sci.* **15:** 471—479.

16) KIMELBERG, H. K. (1986). Occurrence and functinal significance of serotonin and catecholamine uptake by astrocytes. *Biochem. Pharmacol.* **35.** 2273—2281.

17) GALLAGER, D. W., AGHAJANIAN, G. K. (1975). Effects of chlorimipramine and lysergic acid diethylamide on efflux of precursor-formed 3H-serotonin: correlations with serotonin impulse flow. *J. Pharmacol. Exp. Ther.* **193:** 785—795.

18) WILLNER, P. (1985). Antidepressants and serotonergic neurotransmission: an integrative review. *Psychopharmacol.* **85:** 387—404.

19) CARLSSON, A., LINDQVIST, M. (1978). Effects of antidepressant agents on the synthesis of brain monoamines. *J. Neural Trans.* **43:** 73—91.

20) MARCO, J. M., MEEK, J. L. (1979). The effects of antidepressants on serotonin turnover in discrete regions of rat brain. *Arch. Pharmacol.* **306:** 75—79.

21) GRAHAME-SMITH, D. G. (1971). Studies in vivo on the relationship between brain tryptophan, brain 5-HT synthesis and hyperactivity in rats treated with a monoamine oxidase inhibitor and L-tryptophan. *J. Neurochem.* **18:** 1053—1066.

22) WOLF, W. A., YOUDIM, M. B. H., KUHN, D. M. (1985). Does brain 5-GIAA indicate serotonin release or monoamine oxidase activity? *Eur. J. Pharmacol.* **109:** 381—387.

23) WOLF, W. A., KUHN, D. M. (1986). Uptake and rlease of tryptophan and serotonin: an HPLC method to study the flux of endogenous 5-hydroxyindoles through synaptosomes. *J. Neurochem.* **46:** 61—67.

24) COMMISSIONG, J. W. (1985). Monoamine metabolites: their relationship and lack of relationship to monoaminergic neuronal activity. *Biochem. Pharmacol.* **34.** 1127—1131.

25) TRULSON, M. E. (1985). Dietary tryptophan does not alter the function of brain serotonin neurons. *Life Sci.* **37:** 1067—1072.

26) STEINBUSCH, H. W. M., NIEUWENHUYS, R. (1981). Localization of serotonin-like immunoreactivity in the central nervous system and pituitary or the rat, with special reference to the innervation of the hypothalamus. In: *Serotonin: Current Aspects of Neurochemistry and Function* (Haber, B., et al., eds.). Plenum Publishing Corp., New York.

27) VAN DE KAR, L. D., LORENS, S. A. (1979). Differential serotonergic innervation of individual hypothalamic nuclei and other forebrain regions by the dorsal and median raphe nuclei. *Brain Res.* **162:** 45—54.

28) AGHAJANIAN, G. K. (1982). Regulation of serotonergic neuronal activity: autoreceptors and pacemaker potentials. In: *Serotonin in Biological Psychiatry* (Ho, B. T., et. al., eds.). Raven Press, New York.

29) KUHN, D. M., WOLF, W. A., YOUDIM, B. H. (1986). Serotonin neurochemistry revisited: a new look at some old axioms. *Neurochem. Int.* **8:** 141—154.

30) MURPHY, D. L. (1986). Serotonin neurochemistry: a commentary on some of its quandaries. *Neurochem. Int.* **8:** 161—163.

31) CURZON, G., MARSDEN, C. A. (1975). Metabolism of a tryptophan load in the hypothalamus and other brain regions. *J. Neurochem.* **25:** 251—256.

32) KÖNIG, J. F. R., KLIPPEL, R. A. (1963). The rat brain: A stereotaxic atlas of the forebrain and lower parts of the brain stem. Williams and Wilkins, Baltimore.

33) PALKOVITS, M. (1973). Isolated removal of hypothalamic or other hypothalamic brain nuclei of the rat. *Brain Res.* **5:** 459—450.

34) CHAPIN, D. S., LOOKINGLAND, K. J., MOORE, K. E. (1986). Effects of LC mobile phase compostion on retention times for biogenic amines, and their precursors and metabolites. *Current Sep.* **7:** 68—70.

35) LOWRY, O HL, ROSEBROUGH, N. J., FARR, A. L., RANDALL, R. J. (1951). Protein measurement with the Folin phenol reagent. *J. Biol. Chem.* **193:** 265—275.

36) MYERS, J. L. (1979). Fundamentals of Experimental Design. Allyn and Bacon Inc., Boston.

37) JONES, R. S. G., BROADBENT, J. (1982). Further studies on the role of indoleamines in the responses of cortical neurones to stimulation of nucleus raphe medianus: effects of indoleamine precursor loading. *Neuropharmacol.* **21:** 1273—1277.

38) SUTER, H. A., COLLARD, K. J. (1983). The regulation of 5-hydroxytryptamine release from superfused synaptosomes by 5-hydroxytryptamine and its immediate precursors. *Neurochem. Res.* **6:** 723—730.

AMINO ACID AVAILABILITY AND CONTROL OF HISTAMINERGIC SYSTEMS IN THE BRAIN

C. O. Enwonwu

Meharry Medical College
Center for Nutrition—A73
Nashville, Tennessee 37208
USA

INTRODUCTION

Histamine (Hm) in the mammalian brain is generally partitioned into a larger "neuronal" fraction with rapid turnover, and a much smaller pool with slow turnover in the mast cells. There are however, reports suggesting the absence of mast cells in the brains of the guinea pig, nonhuman and human primates (1). Synthesis of Hm in the mammalian brain involves the specific histidine decarboxylase (L-histidine carboxylase, EC 4.1.1.22) with a K_m (apparent) of approximately 3.10^{-5} M which is high relative to normal brain levels of histidine. Thus, histidine (His) is converted to Hm by a specific substrate-limited enzyme. Within the last two decades, several lines of evidence mainly neurochemical and neuropharmacological, strongly suggest that Hm endogenous to the central nervous system, functions as an important neuroregulatory substance (2, 3). Mammalian brain contains at least two kinds of Hm receptors (H1 and H2), and both receptor subtypes have been fairly well characterized (3, 4). The relatively recent immunocytochemical visualization of neurons containing Hm and histidine decarboxylase (HDC) in the mammalian brain (5, 6) has strengthened the claim of a neurotransmitter status for Hm.

As in the monoaminergic system, highest levels of Hm and HDC in the mammalian brain are found in the hypothalamus with marked variations in the various hypothalamic nuclei (1, 4, 7). Degradation of Hm in mammalian brain is through inactivation to telemethylhistamine by the specific histamine-N-methyltransferase, EC 2.1.1.8 (HNMT), and N^t-methylhistamine is oxidatively deaminated by monoamine oxidase (MAO-B type) to yield 1-methyl-4-imidazoleacetic acid (1, 3). HNMT utilizes S-adenosylmethionine (SAM) as the methyl donor, and is inhibited by the substrate Hm in concentrations exceeding 10^{-5} M, as well as by products of the reaction (7). Other substances which competitively inhibit the activity of HNMT are the aminoisoquinolines and the diaminopyrimidines (1, 7). The former group includes drugs such as chloroquine and amodiaquine which are extensively used as antimalarials in malnourished tropical communities.

The physiological roles of Hm in the CNS have continued to be a challenging puzzle. There are however suggestions that it participates in a wide range of important neuroendocrine and vegetative responses (1, 8—10). This report examines the availability of His to the brain in mammalian protein malnutrition and its functional relevance to the histaminergic system.

NATO ASI Series, Vol. H20
Amino Acid Availability and Brain Function in Health and Disease. Edited by G. Huether

Histidine Metabolism in Protein Malnutrition

As part of the physiological adaptation to protein-energy malnutrition (PEM) in children (11) and in experimental animals (8, 9, 12, 13), there is accelerated breakdown of muscle proteins with increased mobilization of amino acids to the liver. In protein depleted adult cockerels, for example, breast muscle carnosine (β-alanylhistidine) decreases to less than 5 per cent of control level while the pool of free His is markedly increased (13). The major degradative route of His in the liver is deamination catalyzed by the enzyme histidase (L-histidine ammonia-lyase, EC 4.3.1.3) via the urocanic acid pathway. Activity of this enzyme is severely impaired in the kwashiorkor clinical syndrome of human PEM (11, 14) as well as in experimental PEM in laboratory animals (15). Thus, unlike the other indispensable amino acids, His level in blood is either increased or relatively unaffected in protein malnutrition (11). In rats fed graded levels of casein, serum His concentration is highest in the zero per cent casein group and decreases gradually with increasing dietary casein level (16). In contrast, serum levels of the other essential amino acids, which include most of the neutral amino acids, increase with increasing dietary casein intake.

TABLE 1
Levels of free amino acids in Guinea pig plasma

Amino acid	(μmoles/100 ml plasma) Normal protein pair-fed n:7	Low protein (ad. libitum) n:6
Threonine	15.3 ± 0.9	5.7 ± 0.5 (37)[b]
Serine	16.8 ± 2.5	12.5 ± 2.4 (74)
Asparagineglutamine	43.3 ± 6.0	31.7 ± 4.3 (73)
Glutamic acid	8.6 ± 0.7	6.7 ± 0.8 (78)
Glycine	106.1 ± 11.1	129.3 ± 14.1 (122)
Alanine	22.2 ± 1.7	19.2 ± 2.0 (86)
Valine	20.4 ± 1.4	6.8 ± 1.3 (33)[b]
Methionine	3.1 ± 0.2	1.7 ± 0.3 (55)[b]
Isoleucine	10.3 ± 0.7	3.2 ± 0.5 (31)[b]
Leucine	14.3 ± 1.6	4.3 ± 0.5 (30)[b]
Tyrosine	5.1 ± 0.4	2.7 ± 0.5 (53)[b]
Phenylalanine	6.5 ± 0.5	3.0 ± 0.4 (46)[b]
Lysine	10.8 ± 0.6	5.2 ± 0.6 (48)[b]
Histidine	7.6 ± 0.7	7.1 ± 0.8 (93)
Arginine	8.8 ± 0.7	3.3 ± 1.1 (38)[b]
Essential amino acids (EA)[a]	80.7	29.9 (37)
Total amino acids	299.3	242.4 (81)
His in % of EA	9.4	23.7
His in % of NAA	10.1	25.9

The malnourished young guinea pigs were fed a 3 % casein for 3 weeks. Data expressed as means + S. E. Value in parenthesis represents percentage of control level.
a: Sum of essential amino acids including histidine, tyrosine and arginine.
b: P < 0.005 or 0.001 (Comparison with normal protein/pair-fed).
NAA: Met, Thr, Val, Leu, Ileu, Tyr, Phe.

As clearly indicated in Table 1, dietary protein deficiency in guinea pigs elicits a prominent decrease in plasma essential (EA)/nonessential amino acid ratio but a marked increase in His/neutral amino acid (NAA) ratio. Similar findings are documented in rats (16, 17, 18), as well as in nonhuman primates (12, 19) and children (11).

Histidine and Histamine Levels in Brain in Protein Malnutrition

Amino acids are transported into the brain by carrier mechanisms that are charge and size specific, and within each carrier group, individual amino acids compete with one another for uptake (20). Histidine is transported through the blood-brain barrier system (BBB) principally as a neutral amino acid. Thus, the crucial factor that determines His availability in the brain is not the plasma level of His per se but rather, the ratio of plasma His concentration to the sum of concentrations of competing neutral amino acids. Table 2 shows the marked changes in plasma His/NAA ratio with duration of feeding a 3 per casein diet to weanling guinea pigs and the equally prominent increase in brain/plasma His ratio. Thus brain His content, like the level in plasma, is inversely correlated with dietary protein intake but the response is more pronounced in brain than in plasma because of marked reduction in plasma levels of the competing NAA (16, 17, 21). Increased brain levels of His do not occur in animals fed adequate protein diets in restricted amounts (8, 19, 21), an observation consistent with the plasma amino acid ratio concept (22), and with the conclusion that influx of His into the brain in PEM is due primarily to protein and not energy deficiency. Changes observed in brain His level in protein deficient laboratory animals are likely to apply to PEM in children since K_m of BBB neutral amino acid transport in the human is low (22).

TABLE 2
Brain-plasma histidine ratio and ratio of plasma histidine to sum of plasma levels of competing NAA in protein-energy deficient Guinea pigs*

Dietary Group	Brain/Plasma Histidine %	Plasma Histidine/NAA %
Control	101.0	10.2
Low Protein, Day 7	143.5	17.6
Low Protein, Day 14	166.7	27.1
Low Protein, Day 18	207.1	24.5

Malnourished guinea pigs were fed a 3 % casein diet ad. libitum.
Data based on 6—8 animals per dietary group.
*NAA: (Met, Thr, Val, Ileu, Leu, Tyr, Phe).

Mammalian brain, unlike the liver, lacks the urocanic acid as well as the imidazolylpyruvic acid pathways for His metabolism and therefore disposes of this amino acid through utilization for protein synthesis, formation of peptides such as homocarnosine (γ-amino-butyryl-L-histidine) and decarboxylation to Hm by a specific HDC. Evaluation of the metabolic fate of His in brains of protein deficient rats, using tritium-labeled His, indicates accelerated uptake of the amino acid into brain without any evidence of increased utilization for cerebral protein synthesis (21). In contrast, marked increase in brain content of free His in protein malnutrition is accompanied by equally prominent elevations in the levels of Hm and homocarnosine (Table 3). Our reported findings of in-

TABLE 3
Whole brain histamine and regional distribution of histidine and homocarnosine in the Guinea pig[a]

	Control Diet (pair fed)			Low Protein (3 % Casein) (ad libitum)		
	Cerebrum	Cerebellum	Brain Stem	Cerebrum	Cerebellum	Brain Stem
Histidine	7.89	8.67	7.96	17.34 (220)	16.42 (189)	14.01 (176)
Homocarnosine	15.06	12.94	24.22	29.48 (196)	18.26 (141)	39.02 (161)
Histamine		0.49[b] ± 0.03			1.29 ± 0.06 (263)[b]	

Guinea pigs were fed their respective diets for 3 weeks.
a: Values expressed as μmol/100g wet wt of tissue; figures in parenthesis per cent of corresponding control value
b: $P<0.001$

creased brain Hm in protein malnutrition in virtually all animal species so far investigated (10, 19) have been confirmed and extended by Ramanamurthy (23) who has noted a 2—3 fold increase in Hm content of synaptosomes isolated from the brains of protein depleted rats in comparison with similar preparations from well fed control rats. These findings are consistent with the reported observations that if the body pool of free His in weanling rat is increased by changing dietary His content from 0.1 to 0.8 per cent, Hm level and HDC activity in brain increase 5- and 2-fold respectively over control values (24). At variance is the recent report (25) that His transport into rat brain synaptosomes has no role in the regulation or maintenance of neurotransmitter pools of Hm. No explanation can be offered for this apparent discrepancy. Table 4 shows that dietary rehabilitation of protein deficient rats restores brain His and Hm contents to normal levels within one week, an observation consistent with the conclusion that Hm synthesis is rate-affected by precursor supply as determined by BBB transport (20, 22).

TABLE 4
Effect of diet on levels of histidine and histamine in rat brain

	Control (n = 14)	Malnourished* (n = 10)		Rehabilitated** (n = 5)	
Brain wt (g)	1.58 ± 0.04	1.50 ± 0.05		1.62 ± 0.08	
Histidine (μmole/g)	0.06 ± 0.01	0.33 ± 0.03	(515)[a]	0.08 ± 0.03	(117)
Histamine (ng/g)	53.19 ± 277	168.92 ± 6.79	(318)[a]	50.69 ± 2.27	(110)

* Fed 0.5 % protein (lactalbumin diet) for 5—8 weeks.
** Malnourished rats refed 18 % lactalbumin diet for 8 days.
Value in parenthesis is per cent of control.
a: Significantly different from control value ($P < 0.001$).

DISCUSSION

The mammalian brain is a very complex organ and this makes it difficult to reach simple conclusions on the physiological functions of Hm in the central nervous system. The situation is even more complicated in protein malnutrition since the marked changes in brain His and Hm contents are accompanied by equally prominent alterations in the other neuronal systems (21, 23).

Several studies have, for example, demonstrated very significant reduction in brain levels of tryptophan and 5-hydroxytryptamine (5-HT) in protein malnutrition (16, 23). There are also suggestions that Hm interferes with release of norepinephrine (NE) and 5-HT by synaptosomes (26). Additionally, a study of discrete brain nuclei has demonstrated that protein malnutrition, which elicits only a slight reduction in the arcuate nucleus content of NE, produces a 75 per cent reduction in the level of the same amine in paraventricular nucleus/median eminence (27). Taylor (28) has summarized the points of interaction between Hm and several other biogenic amines. The pathophysiological implications of the markedly increased brain burden of Hm in PEM still remain largely speculative at this stage. What is however clear is that in view of the fact that normal and adaptive activities in the CNS are the net effects of coordinated dynamic interactions within and between different neuronal pathways, many functions usually attributed to other well studied neurotransmitters in chronic protein malnutrition are probably mediated in part through the histaminergic systems.

The reported roles of Hm in the CNS include modulation of neuroendocrine functions at the hypothalamic level, especially when the negative feedback mechanism of the corticotropinergic system is broken (29). There are several excellent reviews of the neuroendocrine and vegetative responses attributed to Hm in the brain (4, 10, 30). These include antidiuresis resulting from increased vasopressin secretion, regulation of adrenocorticotropin release, potentiating effect on thyrotropin-releasing hormone/luteinizing hormone-releasing hormone stimulated prolactin (PRL) and luteinizing hormone (LH) secretion, as well as emetic, cardiovascular and thermoregulatory responses. Histamine is also an extracellular messenger involved in such biological processes as allergy, host responses to severe stress, anaphylaxis and injury. It increases capillary permeability, constricts both vascular and non-vascular smooth muscle, stimulates gastric secretory activity, and exerts both inotropic and chronotropic actions on the heart (9, 10, 28).

SUMMARY AND CONCLUSION

Histidine in the brain is converted to the putative neurotransmitter Hm by a substrate-limited enzyme (HDC), and this pathway is rate-affected by substrate supply as determined by the important regulatory role of the BBB. In both human PEM and the experimental animal prototypes, there is increased circulating level of free His with concomitant severe reduction in blood levels of the NAAs valine, leucine, isoleucine, methionine, tryptophan, phenylalanine and tyrosine. The situation in the human is made worse in some malnourished communities who rely heavily on plantain (Musa sapientum var. paradisiaca) as a staple food. This food has a very high content of histidine (about 400 mg His per gm. nitrogen). The prominent increase in blood His/NAA ratio in protein malnutrition results in accelerated influx of His into the brain with marked stimulation of Hm formation. Although there are, to the best of my knowledge, no reported data on brain

His metabolism in human PEM, the severely impaired hepatic metabolism in such children as well as indirect evidence from various animal species, suggests increased brain burden of Hm in malnourished children. It is also conceivable that in such impoverished children especially in the tropical world, degradation of brain Hm is severely restricted as a result of extensive use of antimalarials which competitively inhibit the activity of HNMT. Many of the prominent pathophysiological features of PEM in children such as defective thermoregulatory homeostasis, increased circulating cortisol, fluid/electrolyte imbalance, impaired cell-mediated immunity, reduced cardiac output with prolongation of systemic recirculation time, and apathy bordering on a severe clinical state of depression, are consistent with the effects of Hm as determined mainly by neurochemical and neuropharmacological studies (10). Evaluation of histaminergic systems along with the other well-studied neuronal systems, may therefore help to clarify the complex pathophysiological features associated with the kwashiokor clinical syndrome of PEM in children.

REFERENCES

1) HOUGH, L. B., GREEN, H. P. (1984). Histamine and its receptors in the nervous system. In: *Handbook of Neurochemistry. Receptors in the Nervous System* (Lajtha, A., ed.), **Vol. 6,** pp. 145—211.

2) SCHWARTZ, J. C. (1977). Histaminergic mechanisms in brain. *Annu. Rev. Pharmacol. Toxicol.* **17:** 325—339.

3) TAYLOR, J. E. (1982). Neurochemical and neuropharmacological aspects of histamine receptors. *Neurochem. International* **4:** 89—96.

4) SCHWARTZ, J. C., GARBARG, M., LEBRECHT, U., NOWAK, J., POLLARD, H., RODERGAS, E., ROSE, C., QUACH, T., MORGAT, J L., FOY, J. (1982). Histaminergic systems in brain: studies on localisation and actions. In: *Advances in Histamine Research* (Uvnas, B., Tasaka, K., eds.). Pergamon Press, Oxford, pp. 71—80.

5) PANULA, P., YANG, H. Y. T., COSTA, E. (1984). Histamine containing neurons in the rat hypothalamus. *Proc. Natl. Acad. Sci. USA* **81:** 2572—2576.

6) WATANABE, T., TAGUCHI, Y., SHIOSAKA, S., TANAKA, J., KUBOTA, H., TERANO, Y., TOHYOMA, M., WADA, H. (1984). Distribution of the histaminergic neuron system in the central nervous system of rats: a fluorescent immunohistochemical analysis with histidine decarboxylase as a marker. *Brain Res.* **295:** 13—25.

7) GREEN, J. P., KHANDELWAL, J. K. (1985). Histamine turnover in regions of rat brain. In: *Frontiers In Histamine Research* (Ganellin, C. R., Schwartz, J. C., eds.). Pergamon Press, Oxford, pp. 185—195.

8) ENWONWU, CO. O., OKADIGBO, G. (1983). Rapid development of oedema and defective brain histidine metabolism in young guinea pigs fed a protein-energy deficient diet. *Br. J. Exp. Path.* **64:** 487—496.

9) ENWONWU, C. O. (1986). Potential relevance of impaired histidine metabolism to the immunodeficiency in human protein-energy malnutrition. *Nutr. Res.* **6:** 337—348.

10) ENWONWU, C. O. (1987). Pathophysiological implications of increased brain burden of histamine in protein malnutrition. *Medical Hypotheses* **22:** 1—13.

11) ALLEYNE, G. A. O., HAY, R. W., PICOU, D. I., STANFIELD, J. P., WHITEHEAD, R. G. (1977). Protein-Energy Malnutrition. Edward Arnold, London.

12) ENWONWU, C. O., WORTHINGTON, B. S. (1974). Regional distribution of homocarnosine and other ninhydrin positive substances in brains of malnourished monkeys. *J. Neurochem.* **22:** 1045—1052.

13) FISHER, H., KONLANDE, J., STRUMEYER, D. (1975). Levels of histidine and histidine derivatives in breast muscle of protein depleted and repleted adult cockerels. *Nutr. Metabol.* **18:** 120—126.

14) WHITEHEAD, R. G. (1969). The assessment of nutritional status in protein-malnourished children. *Proc. Nutr. Soc. England* **28:** 1—16.

15) RAO, D. R., DEODHAR, A. D., HARIHARAN, K. (1965). Histidine metabolism in experimental protein malnutrition in rats. *Biochem. J.* **97:** 311—317.

16) GUSTAFSON, J. M., DODDS, S. J., BURGUS, R. C., MERCER, L. P. (1986). Prediction of brain and serum free amino acid profiles in rats fed graded levels of protein. *J. Nutr.* **116:** 1667—1681.

17) ENWONWU, C. O. (1987). Differential effect of total food withdrawal and dietary protein restricition on brain content of free histidine in the rat. *Neurochem. Res.* **12:** 483—487.

18) GIFFORD, C. D., DODDS, S. J., JOHNSON, L. K., SMITH, D. L., MERCER, L. P. (1987). Metabolic adaptation to protein deficiency in rats: histidine. *Nutr. Res.* **7:** 617—627.

19) ENWONWU, C. O., OKOLIE, E. E. (1983). Differential effects of protein malnutrition and ascorbic acid deficiency on histidine metabolism in the brains of infant nonhuman primates. *J. Neurochem.* **41:** 230—238.

20) PARDRIDGE, W. M. (1983). Brain metabolism: a perspective from the blood-brain barrier. *Physiol. Rev.* **63:** 1481—1535.

21) PAO, S. K., DICKERSON, J. W. T. (1975). Effect of a low protein diet in the weanling rat on the free amino acids of the brain. *Nutr. Metabl.* **18:** 204—216.

22) PARDRIDGE, W. M., CHOI, T. B. (1986). Neutral amino acid transport at the human blood-brain barrier. *Federation Proc.* **45:** 2073—2078.

23) RAMANAMURTHY, P. S. V. (1977). Maternal and early postnatal malnutrition and transmitter amines in rat brain. *J. Neurochem.* **28:** 253—254.

24) LEE, N. S., FITZPATRICK, D., MEIER, E., FISHER, H. (1981). Influence of dietary histidine on tissue histamine concentration, histidine decarboxylase and histamine methyltransferase activity in the rat. *Agents Actions* **11:** 307—311.

25) HEGSTRAND, L. R., SIMON, J. R. (1985). Histidine transport into rat brain synaptosomes. *J. Neurochem.* **45:** 407—414.

26) ANONYMOUS (1977). Nutrition and transmitter amines in rat brain. *Nutr. Revs.* **35:** 283—285.

27) HAWRYLEWICZ, E. J., KISSANE, J. Q. (1980). The effect of protein restriction on brain biogenic amines. In: *Biogenic Amines in Development* (Parvez, H., Parvez, S., eds.). Elsevier/North Holland, Amsterdam, p. 493.

28) TAYLOR, K. M. (1975). Brain histamine. In: *Biochemistry of Biogenic Amines* (Iversen, L. L., Iversen, S. D., Snyder, S. H., eds.). **Vol. 3,** pp. 327—279, Plenum Press, New York.

29) WADA, H., WATANABE, T., YAMATODANI, A., MAEYAMA, K., ITOI, N., CACABELOS, R., SEO, M., KIYONO, S., NAGAI, K., NAKAGAWA, H. (1985). Physiological functions of histamine in the brain. In: *Frontiers in Histamine Research* (Ganellin, C. R., Schwartz, J. C., eds.). Pergamon Press, Oxford, pp. 225—235.

30) WEINER, R. I., GANONG, W. F. (1978). Role of brain monoamines and histamine in regulation of anterior pituitary secretion. *Physiol. Rev.* **58:** 905—976.

SUMMARY AND DISCUSSION-REPORT OF CHAPTER III

Moussa B. H. Youdim

SUMMARY

In this session the emphasis was placed on amino acid precursors availability, neurotransmitter metabolism and function. It is necessary to keep in mind that a number of important enzymatic and non-enzymatic processes will occur before a neurotransmitter is synthesized and released into function. In many instances, the kinetics of these processes cannot be estimated accurately *in vivo,* and erroneous conclusion are reached. Thus, for example, the measurement of 5-HIAA in the brain after treatment with various drugs or physiological changes and in certain pathological conditions is frequently used as an index of 5-HT release and functional activity. However, the use of 5-HIAA as an indirect measurement of the status of 5-HT neurotransmission is not certain, since this surely depends on the mechanism by which 5-HT is released from the presynaptic neurones. Therefore, the levels of 5-HIAA in brain must reflect very different neurochemical processes. If 5-HIAA reflects some measure of 5-HT utilization, several assumptions must be met before this conclusion can be considered valid. 1. 5-HT is deaminated by MAO only after release into synapse. 2. 5-HT is released by a mechanism which prevents monoamine oxidase from deamination 5-HT until after release, and 3. Reuptake occurs into a cellular (5-HT or non-5-HT) compartment where deamination can occur. If 5-HT release occurred solely by an exocytotic process, this would satisfy all the three above criteria, since any vesicle bound 5-HT would be protected from the action of MAO, which is located outside of vesicles in the mitochondrial outer membrane within the neurone. By contrast, if 5-HT is available in the neuronal cytoplasm it could be deaminated before or after release can occur. Any 5-HT deaminated, before it can be released, would indicate only the presence of MAO activity within the neurone. While 5-HT release is generally thought to be an exocytotic process, there is now compelling evidence form *in vitro* and *in vivo* studies for release involving a cytoplasmic pool or in a non-exocytotic manner. The high levels of brain 5-HIAA as compared to 5-HT after tryptophan loading could very well explain the deamination of 5-HT prior to release and after reuptake. It is rather interesting that reserpine which depletes the brain of its 5-HT and increases 5-HIAA does not cause increased 5-HT function. Rather, increased 5-HT functional activity is observed only when monoamine oxidase is inhibited prior to reserpinization. Therefore, it would be difficult to conclude whether 5-HIAA accurately reflects 5-HT utilization or simply provides an index of monoamine oxidase activity until the release process for 5-HT is more completely understood. The theories discussed in the study of 5-HT metabolism and release could very well be applied to dopamine metabolism and neurotransmission.

NATO ASI Series, Vol. H20
Amino Acid Availability and Brain Function in Health and Disease. Edited by G. Huether

DISCUSSION-REPORT

What is the actual contribution of the precursor-dependent synthesis step in the control of synaptic monoamine content?

Huether: We are talking about changes in neuronal transmitter content as if the changes caused by an altered precursor supply were the only ones to be expected in a living animal. Neuronal transmitter content is a steady state but not a fixed value. Nor is it like the water-level in a bath tube, which rises automatically if you open the water supply. Neuronal transmitter content is continuously filled up by synthesis and — if not stored in a protected compartment — continuously emptied by degradation. It is therefore subject to at least three equally important regulatory mechanisms: the rate of synthesis, the rate by which it can be protected against degradation after its synthesis or release and the rate of degradation.

You all know how many possibilities exist to affect the rate of degradation of monoamines by modulating monoamine oxidase activity, or by changing the reuptake and transport of monoamines into vesicular stores. The rate of synthesis of amines can be altered by changes in the availability of co-factors for the hydroxylases (BH_4, oxygen, divalent cations etc.) or by modifications which change the affinity of the enzyme to these co-factors (by e. g., phosphorylation). So, saturation of the hydroxylation step with substrate is only one out of the many possibilities to affect the content of monoamines within the neuron. But, according to my knowledge, even the actual K_m of tryptophan-hydroxylase, the degree to which this enzyme is saturated, is still a matter of debate.

Youdim: Many enzymes can operate well below their K_m . The idea that tryptophan hydroxylase would be saturated was born at a time when we used to measure tryptophan hydroxylase with the synthetic co-factor $DMPH_4$. In the meantime we learned that all aromatic amino acid hydroxylases use the natural co-factor, BH_4. In the presence of the synthetic co-factor you will find a K_m for tryptophan hydroxylase of about 10^{-4} M. If you, however, use the natural co-factor, the K_m will be two orders of magnitude lower, namely 10^{-6} M. According to my opinion the question whether the enzyme is saturated is a red herring. We all know that enzymes can operate at low substrate concentrations. The problem is the V_{max} not the K_m. For instance: monoamine oxidase is able to oxidize monoamines at extremely low concentrations in spite of the fact that its apparent K_m-values for monoamines are 200—400 μM, a concentration range of amines, which is never seen in a living organism, as far as I know.

Fernstrom: The issue has two parts. First, to what extent do local changes in precursor level influence the rate of transmitter *synthesis?* In the case of the tryptophan-serotonin pathway, it would appear that in virtually any situation studied, serotonin synthesis rate is quite responsive to changes in local tryptophan level. The importance of precursor level to synthesis rate in the serotonin pathway is further amplified by the fact that this pathway shows no regulation by end-product inhibition (Fernstrom, 1983, Physiol Rev 63: 484—546). Whether there are other factors controlling synthesis is a matter of current study, but there is little doubt that the serotonin pathway is very vulnerable to changes in precursor level. Paranthetically, this relationship holds because the substrate K_m of tryptophan hydroxylase (measured using the natural cofactor, BH_4) does approximate the normal brain tryptophan level *in vivo,* which is around 10^{-5} M (Fernstrom, 1983, Physiol Rev 63: 484—546). The tyrosine-catecholamine pathways are a somewhat *different* matter. Taking the tyrosine-dopamine pathway as an example, local tyrosine levels appear to be somewhat higher than the substrate K_m of tyrosine hydroxylase, but not so much higher as to saturate the enzyme (Fernstrom, 1983, Physiol Rev 63: 484—546). Nevertheless, injections of tyrosine have little impact on hydroxylation rate or dopamine synthesis rate unless the neurons under study are actively firing. Though there are various explanations for this need for high activity to see tyrosine effects, one that is appealing is that dopamine synthesis is so stimulated during active firing, that the immediate local pool of tyrosine is partly depleted (the supply cannot keep up with the demand). As a consequence, the saturation of the hydroxylase falls, and a subsequent injection of tyrosine can then sufficiently raise the local tyrosine pool to produce a clear increase in enzyme saturation and synthesis rate. Hence, the dopamine pathway responds to tyrosine supply, but only when the relevant neurons are firing rapidly (Fernstrom, 193, Physiol Rev 63: 484—546). In addition, and unlike the serotonin pathway, there is clear feedback inhibition of synthesis, mediated partly by direct end-product inhibition, and partly indirectly via pre-and post-synaptic dopamine receptors. Hence, overall, while dopamine synthesis rate is influenced by precursor supply, this represents only one of several identified factors governing this synthetic pathway. Second, to what extent do precursor-related changes in transmitter synthesis influence the amounts *released* into the synapse? There are experimental situations in which it can be demonstrated that increases in serotonin or catecholamine synthesis produced by the administration of their respective precursors lead to increases in transmitter release. Such studies include work both *in vitro* (Milner and Wurtman, 1984, Neurosci Lett 59: 215—220) and *in vivo* (Ternaux et al., 1976, 1977, Brain Res 101: 533—548). But the results of such studies are pharmacologic, and few data are available that convincing show physiologically-induced increases (or decreases) in the brain level of a precursor and the synthesis rate of its respective transmitter product alter the amount of the transmitter released. Further work is called for in this area.

Curzon: Moussa Youdim says that whether tryptophan hydroxylase is saturated or not is a red herring. Nevertheless, as giving tryptophan does alter 5-HT synthesis, its availability must be not entirely devoid of interest. Enzyme activity, transmitters release, receptor properties, interneurones, modulators and so on are also of interest as possible influences on the thing that matters most, i. e. the end expression of transmitter action. But none of them (including enzyme saturation) is the only thing that matters.

How is increased synthesis of a transmitter related to increased transmitter release and transmitter function?

van Gelder: Can we seriously say that depending on the tryptophan level the postsynaptic response of a neuron is going to change? Let us assume that if we have an action potential generated by the release of X molecules of transmitter released. If we change the presynaptic concentration of this transmitter, or to go even further away, if we change the concentration of the precursor by let's say 30 %, would this mean that this postsynaptic potential is going down or increasing by 30 % as well?

Wurtman: Perhaps. Our new data on serotonin release from electrically-stimulated hypothalamic slices do indicate that a physiologic increase in brain tryptophan levels does cause a parallel change in serotonin release, and this latter effect may very well influence signal transmission across the synapse: Only future studies will resolve this question. But I think it is terribly important for the neurophysiologists to perceive what the neurochemists want to say. What we are saying is, that the earlier quantal notion on the number of molecules that are being released and the postsynaptic effects of these molecules being fixed is wrong. We are saying that the amount of information that is transmitted is not necessarily of a quantitatively fixed size and may vary in proportion to the level of presynaptic material.

Is the "serotonin-syndrome" a challenge to the theory of how precursor supply affects brain function or is it a pharmacologic peculiarity?

Youdim: We know from our earlier studies that, if you give large doses of tryptophan you will not see functional changes including behaviour in these rats, at least with respect to the parameters that we used. But if you include a monoamine oxidase inhibitor you will produce an overt behavioural syndrome which is called the serotonin syndrome. We studied the question of what degree of monoamine oxidase inhibition is required in order to see this behavioural response after a tryptophan load? We found that one can inhibit the enzyme by about 85 % without seeing any behavioural changes although serotonin concentrations in the brain did rise significantly. If you went over the treshold of 85 % monoamine oxidase inhibition the full serotonin syndrome was observed. The rate of serotonin synthesis jumped to 0.52 μg per hour per gram of tissue when we gave 100 mg tryptophan/kg and this appeared to be the rate of serotonin synthesis required to achieve the functional changes. So both compartmentation and metabolism of serotonin are important, and clinically, if you combine tryptophan treatment with an monoamine oxidase inhibitor you get a definite clinical response which is absent with tryptophan alone.

Recently we have found a rather simple and elegant way in which we have demonstrated that not only metabolism by monoamine oxidase is important but also the uptake system. Serotonin is metabolized by monoamine oxidase A. We have used reserpinized animals where brain serotonin content is reduced by more than 90 %. If these animals are given a monoamine oxidase A inhibitor, which is selective for serotonin degradation, and additionally a selective serotonin re-uptake inhibitor such as fluoexetine, the serotonin syndrome can be produced in the absence giving any precursor. This clearly shows that what is synthesized can be released into function if you get rid of the two monoamine inactivating systems, namely metabolization by monoamine oxidase and re-uptake by nerve endings.

Wurtman: There are a number of intelligent people who think that the serotonin syndrome is in fact a tryptamine syndrome and that in fact it is not caused by the accumulation of serotonin at all, but that it is due to the accumulation of tryptamine which is normally not found in high concentrations in the brain.

Youdim: This was just the reason why I mentioned that we have used reserpinized animals that received a monoamine oxidase inhibitor (clorgyline) which is specific for the inhibition of serotonin degradation plus a selective serotonin uptake inhibitor (fluoxetine) to get the full syndrome. Under these conditions serotonin is elevated. Tryptamine is not a substrate for monoamine oxidase A but for monoamine oxidase B which in this case was not inhibited. Furthermore not very much tryptamine can be found in the brain in the absence of tryptophan loading. Therefore the 5-HT syndrome is a property of post synaptic 5-HT function

Curzon: It is fascinating that the serotonin syndrome in rats given p-chloroamphetamine is apparently unaltered by reserpine. It would be interesting to know whether the drug releases as much 5-HT as in normal animals and also whether it affects the responsiveness of the receptors mediating the syndrome. It is also important to remember that p-chloroamphetamine releases 5-HT by a non-physiological mechanism. Another question is to what extent does reserpine alter physiological release? It alters behaviour so it presumably affects transmitter release in some way. As for the serotonin syndrome and tryptophan; tryptamine (some of extracerebral origin) has a big role in the syndrome as provoked by tryptophan and tranylcypromine (Marsden and Curzon, 1979, Neuropharmacology 18: 159—164). Conceivably, even though normal brain tryptamine levels are very low it might play a part in the syndrome provoked by p-chloroamphetamine. It is relevant that synergistic effects of tryptamine with 5-HT have been described.

Young: As mentioned in my presentation, treatment of rats with single doses of pargyline or impramine caused three-fold increases of CSF serotonin without causing any gross behavioral change. On the other hand treatments which caused the serotonin syndrome raised CSF serotonin 16 to 20-fold. Thus, the serotonin syndrome is a peculiarity even compared with other pharmacologic treatments.

Youdim: It is obvious that both Gerald Curzon and Richard Wurtman have missed the point with regards to the studies using selective monoamine oxidase A and selective serotonin uptake inhibitors. The dosage used was just sufficient to selectively inactive the two processes. And in control as well as in reserpinized rats the combination of the drugs (in absence of precursor treatment) induced the full 5-HT syndrome similar to what is observed with 5-HT releaser, parachloroamphetamine. Both drugs have been used clinically as antidepressants with some success.

Chapter IV.

INFLUENCE OF ALTERED PRECURSOR AMINO ACID SUPPLY ON FUNCTIONAL NEUROTRANSMISSION

THE RELATIONSHIP OF THE SYNTHESIS AND METABOLISM OF CATECHOLAMINES TO BRAIN FUNCTION

John W. Commissiong

Department of Physiology
McGill University
3655 Drummond Street
Montreal, Quebec, Canada
H3G 1Y6

INTRODUCTION

The three principle monoamines dopamine (DA) noradrenaline (NA) and 5-hydroxytryptamine (5-HT) are widely distributed throughout the CNS. These neuromodulator substances are known to affect neuronal excitability at the single cell level (1), as well as to influence the output of large networks of neurons (2). At present, there is considerable uncertainty about how the processes of synthesis, release and metabolism are regulated in monoaminergic neurons. In the periphery, reuptake is the main mechanism of transmitter conservation (3, 4). In the CNS, similar reuptake mechanisms have been demonstrated for the monoamines (5, 6). However, in recent years, the concept that reuptake may play a primary role in transmitter conservation has been largely ignored. Instead we have come to focus on the idea that the released transmitter is largely catabolized. This hypothesis has, in turn, led to a number of corollary beliefs, most of which have been supported by published evidence. Among these beliefs are: 1) Levels of metabolities (DOPAC, HVA, MHPG, 5-HIAA) are indices of transmitter release, and by implication, neuronal firing. 2) Synthesis is activated during release. 3) Precursor availability may become a limiting factor in transmitter synthesis, during episodes of increased neuronal firing/release. 4) Differences in turnover rates (TR) of monoamines, are indicative of the rates of neuronal firing/transmitter release in those regions. For a further discussion of these problems, see (7). The purpose of this short review is to explore selective aspects of the questions raised above. Data published in recent years, roughly since 1985, suggest that the interpretation of results referred to above may have been too simplistic. A working model will be presented as a first step in the effort to arrive at a correct understanding of monamine neurotransmitter dynamics, within a functional context.

Turnover Rates

Data on the concentration, half-life (in hours), rate constant (hr^{-1}) and turnover rate (TR) of NA and DA in the dorsal horn of the cervical spinal cord of the rat are presented in Table 1.

NATO ASI Series, Vol. H20
Amino Acid Availability and Brain Function in Health and Disease. Edited by G. Huether

TABLE 1
Concentrations and Turnover Rates of Noradrenaline and Dopamine in the Dorsal Horn of the Cervical Spinal Cord of the Rat

Monamine	Steady-state concentration (pmol/mg protein)	Half-life (hr)	Rate Constant (hr^{-1})	Turnover Rate (pmol/mg protein/hr)
NA	40 ± 3	2.2	0.32	13 ± 1
DA	7.1 ± 0.9	0.25	2.7	19 ± 2

Data are expressed as $\bar{x}$ ± S.E.M. (N = 5). Modified from ref. no. 8, with permission.

The traditional interpretation of these data would be that dopaminergic neurotransmission in the cervical dorsal horn is more intense than noradrenergic transmission, despite the initial, misleading higher concentration of NA (8). However, since neither neuronal activity, nor release was measured, the conclusion may not be valid. The result may merely reflect a higher level of MAO activity in dopaminergic nerve terminals versus noradrenergic nerve terminals in this region of the spinal cord. This point has been forcefully argued by Kuhn et al. (9), with reference to the serotonergic neurons. The general functional significance of neurotransmitter turnover data has been discussed in a recent monograph (10). It might be that monoamine transmitter turnover data really do reflect real rates of neuronal firing and release. However, adequate data to support this radical conclusion are almost never presented.

Increased Dopamine Synthesis and Metabolism after Blockade of Neurotransmission

Tetrodotoxin (TTX), at a concentration of 3.0×10^{-7} M has been shown to cause a cessation of conducted action potentials in nerves (11). However, after the infusion of TTX into the medial forebrain bundle (MFB), which contains the nigrostriatal dopaminergic fibres, there is a marked increase in the synthesis and metabolism of striatal DA, Figure 1. Previously, the dominant idea had been that increased synthesis, secondary to activation of tyrosine hydroxylase (TH), was a homeostatic response to increased nerve activity in dopaminergic neurons (12, 13). This conclusion may still be valid. However, there is now the curious result that two extremes, increased nerve activity (usually induced by electrical stimulation), and cessation of activity (after TTX), produce the same end result. Therefore, the common sense view that increased synthesis, secondary to increased neuronal firing/release, is a correct, physiologic, homeostatic response, no longer seems assured. In effect, the regulation of TH is now thrust into a much broader perspective.

The Precursor Dependence Hypothesis

The finding that TTX, when infused into the MFB caused a robust increase in the synthesis and metabolism of DA, provided a method of testing the idea that during episodes of increased synthesis of catecholamines, precursor availability may become a limiting fac-

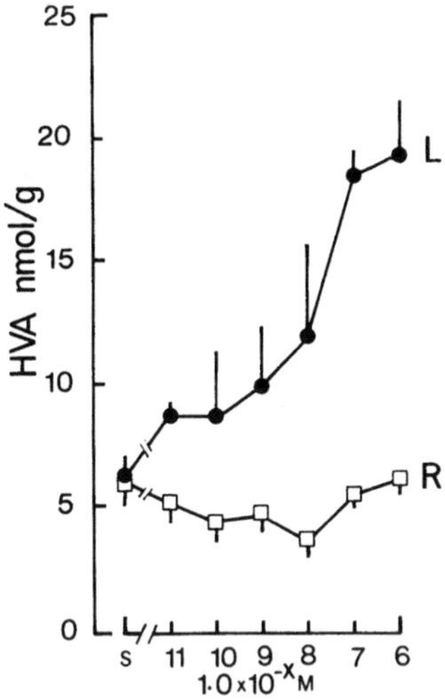

FIGURE 1
TTX (1.0 × 10^{-11} – 1.0 × 10^{-6} M; TTX 1.5 μl; 1.0 μl mm^{-1}) was infused into the left MFB, of the rat, anaesthetized with Brietal (50 mg/kg i. p.). DA, DOPAC and HVA were measured in the left (L) and right (R) striata 4 hrs later. There was a dose-related increase in all three compounds (data for HVA only illustrated) on the injected left side.

tor (14). The results illustrated in Table 2 suggest that this hypothesis, if true, may not always be valid.

More recent data for the prefrontal and cingulate cortices suggest that the turnover rate of DA is quite high, and that increased blood levels tyrosine do lead to an increase in the synthesis of DA (15). Presumably, therefore, the role that catecholamine precursors (phenylalanine and tyrosine) play in regulating the synthesis of DA and NA in different regions of the CNS, will have to be explored on a case by case basis.

TABLE 2
The Effect of Tyrosine on the TTX-activated Synthesis and Metabolism of Dopamine in the Striatum of the Rat

Treatment		DA	DOPAC nmol/g ± S.D. (N=5)	HVA
SAL-SAL	L	66 ± 3	12.4 ± 3.62	7.01 ± 1.62
	R	68 ± 2	11.7 ± 1.79	6.48 ± 1.76
TTX-SAL	L	91 ± 7	21 ± 6	17 ± 4
	R	66 ± 5	12.5 ± 2.26	6.52 ± 2.42
TTX-TYR	L	87 ± 11	32 ± 10	18 ± 6
	R	67 ± 5	10.0 ± 2.25	9.13 ± 1.56

Tyrosine (275 mg/kg) was suspended in saline and administered by gavage, immediately after the infusion of TTX (3.0 × 10^{-7} M; 1,5 μl; 1.0 μl min^{-1}). The results suggest that the administered tyrosine had no effect on the synthesis or metabolism of striatal DA (SAL: saline; TTX: TYR: tyrosine; TTX: tetrodotoxin).

The Synthesis and Metabolism of Dopamine Outside of Dopaminergic Neurons

After a chronic (100 days), mid-thoracic, complete transsection of the spinal cord in rat (16), and the subsequent degeneration of the descending dopaminergic nerve fibres, DA is synthesized and metabolized (to both DOPAC and HVA) in the denervated lumbar spinal cord as efficiently as in the intact spinal cord, Table 3. Exactly the same pattern of result has been obtained in the chronically denervated striatum of the rat, Table 4.

TABLE 3
The Synthesis and Metabolism of Dopamine
in the Chronically Transected Spinal Cord of the Rat

Group	DA	DOPAC nmol/g ± S.D. (N=5)	HVA
Unoperated Control	0.31 ± 0.08	0.18 ± 0.03	0.26 ± 0.03
Unoperated L-DOPA + NIAL:			
T6 Chronic Transection	10.7 ± 2.83	—	—
L-DOPA + NIAL	10.1 ± 1.05	—	—
Unoperated Control: L-DOPA	—	19 ± 3	14 ± 1
T6 Chronic Transection: L-DOPA	—	19 ± 2	14 ± 1

Animals were anaesthetized with Brietal (50 mg/kg i. p.), and a mid-thoracic laminectomy was done. The T6 region of the cord was totally transected. At 100 days later, the animals were treated with DOPA + NIAL or with DOPA alone; DA and DOPAC + HVA were measured 1 hr later in the different groups as indicated. L-DOPA: 100 mg/kg i. p.; NIAL (Nialamide, a MAO inhibitor): 200 mg/kg i. p. given 30 min before L-Dopa.

These two latter results suggest two points. The first is that provided the TH regulatory step is by-passed, DA can be very efficiently synthesized and metabolized in several regions of the CNS. This result is important for an understanding of the mechanism of ac-

TABLE 4
The Synthesis and Metabolism of Dopamine
in the Chronically-Denervated Stratium of the Rat

Group	Dopamine	DOPAC nmol/g ± S.D. (N=3)	HVA
Unlesioned: Saline	35 ± 4	21 ± 5	7.69 ± 1.58
Unlesioned: L-DOPA	58 ± 14	146 ± 22	59 ± 13
Lesioned: Saline	2.48 ± 0.84	DET	DET
Lesioned: L-DOPA	10.2 ± 3.74	218 ± 45	45 ± 14

Animals were anaesthetized with Brietal (50 mg/kg i. p.) and the MFB cut using a stereo taxically-placed brain knife. Eight days later the animals were given L-DOPA (100 mg/kg i. p.) and DA, DOPAC and HVA measured 1 hr later. DET: Detectable only.

tion, and sites of action of L-DOPA in Parkinson's disease. The second point is that in normally innervated tissues, the rate of metabolism of released DA may be a function of the level of MAO activity in elements other than dopaminergic nerves, e. g. glia (17). This possibility raises a further caution against the common tendency to interpret DOPAC and HVA levels as indices of functional dopaminergic neurotransmission.

SUMMARY AND CONCLUSIONS

It is almost certain that in the release /reuptake cycle of monoamines, certain fraction (Δf) of the released transmitter is lost by diffusion away from the synaptic cleft, and by catabolism. Therefore, since transmitter concentration remains relatively constant during release, new synthesis would depend upon the steepness of the diffusion gradient, as well as the site(s) of localization and levels of activity of the catabolic enzymes MAO and COMT. Following publication of the results illustrating a firm link between release, synthesis, metabolism and TH activity in the striatum (12, 13), it was accepted for many years that these results were a physiologic manifestation of the real workings of central monoaminergic neurons. However, it has now been demonstrated that increased metabolism is not always causally coupled to increased release (18). Furthermore, DA release may be reduced, at a time when the efflux of DOPAC and HVA is still increasing (19). It has also been shown that increased synthesis and metabolism of DA can occur without increased release (20). All of these results can be accommodated in the model suggested in Figure 2.

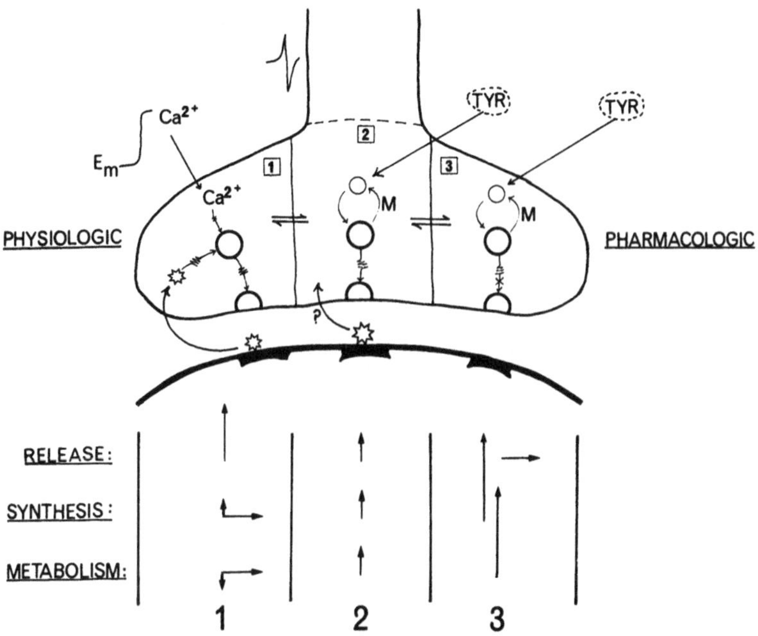

FIGURE 2
A model to account for the variety of processes related to release, synthesis and metabolism of DA in dopaminergic nerve terminals.

It is suggested that the dopaminergic nerve terminal may operate in different modes, depending on the functional state of the animal. Under normal, physiologic, drug-free and unanaesthetized states, reuptake and re-release, with minimal catabolism of the released transmitter may predominate (compartment 1). At the other extreme (compartment 3), drugs like gamma-butyrolactone (GBL) may cause substantial increases in synthesis and metabolism, with no, or very little increase in release. Between these two extremes, in compartment 2, lies the state in which all three processes, release, synthesis and metabolism are increased. Drugs like haloperidol induce this last effect. However, it is also known that this haloperidol-mediated release of DA habituates rapidly (21).

As stressed previously (7), a limiting factor in the study of monoamine dynamics had always been the lack of a reliable method to measure release directly. With the introduction of the technique of intracerebral dialysis (19, 22), a partial solution to this problem has been found. It may therefore be expected that a substantive reinterpretation of the monoamine literature will ensue as the dialysis technique becomes more widely used.

REFERENCES

1) MARSHALL, K. C., ENGBERG, I. (1979). Reversal potential for noradrenaline induced hyperpolarization of spinal motoneurons. *Science* **205:** 422—424.

2) COMMISSIONG, J. C. (1981). Spinal monoaminergic systems: an aspect of somatic motor function. *Fed. Proc.* **40:** 2771—2777.

3) IVERSEN, L. L. (1973). Catecholamine uptake processes. *Br. Med. Bull.* **29:** 130—135.

4) IVERSEN, L. L. (1971). Role of transmitter uptake mechanisms in synayptic neurotransmission: Third Gaddum Memorial Lecture. *Br. J. Pharmac.* **41:** 571—591.

5) COYLE, J. T., SNYDER, S. H. (1969). Catecholamine uptake by synaptosomes in homogenates of rat brain: stereospecificity in different areas. *J. Pharmacol. Exp. Ther.* **170:** 221—231.

6) SNYDER, S. H., COYLE, J. T. (1968). Regional differences in ^{3}H-norepinephrine and ^{3}H-dopamine uptake into rat brain homogenates. *J Pharmac. Exp. Ther.* **165:** 78—86.

7) COMMISSIONG, J. W. (1985). Monoamine metabolites: their relationship and lack of relationship to monoaminergic neuronal activity. *Biochem. Pharmac.* **34:** 1127—1131.

8) KAROUM, F., COMMISSIONG, J. W., NEFF, N. H., WYATT, R. J. (1981). Regional differences in catecholamine formation and metabolism in the rat spinal cord. *Brain Res.* **212:** 361—366.

9) KUHN, D. M., WOLF, W. A., YOUDIM, B. H. (1986). Sertonin neurochemistry revististed: a new look at some old axioms. *Neurochem. Int.* **8:** 141—154.

10) PYCOCK, C. V., TABERNER, P. V. (1981). *Central Neurotransmitter Turnover.* University Part Press, Baltimore, M. D.

11) NARAHASHI, T., ANDERSON, N., MOORE, J. (1967). Comparison of tetrodotoxin and procaine on internally perfused squid grand axon. *J. Gen. Physiol.* **50:** 1413—1428.

12) MURRIN, L. C., ROTH, R. H. (1976). Dopaminergic neurons: effects of electrical simulation on dopamine biosynthesis. *Mol. Pharmacol.* **12:** 463—475.

13) MURRIN, L. C., MORGENROTH, V. H., ROTH, R. H. (1976). Dopaminergic neurons: effects of electrical stimulation on tyrosine hydroxylase. *Mol. Pharmacol.* **12:** 1070—1081.

14) WURTMAN, R. J., HEFTI, F., MELAMED, E. (1980). Precursor control of neurotransmitter synthesis. *Pharm. Rev.* **32:** 315—335.

15) ROTH, R. H. (1984). CNS Dopamine autoreceptors: distribution, pharmacology and function. *Am. N. Y. Acad. Sci.* **430:** 27—53.

16) COMMISSIONG, J. W. (1985). The synthesis and metabolism of catecholamines in the spinal cord of rat after acute and chronic transections. *Brain Res.* **347:** 104—111.

17) KIMBELBERG, H. K. (1986). Occurrence and functional significance of serotonin and catecholamine uptake by astrocytes. *Biochem. Pharmacol.* **35:** 2273—2281.

18) COMMISSIONG, J. W. (1987). The role of precursors in the regulation of striatal dopamine synthesis. In: *Sweeteners: Health Effects* (Williams, G., ed.). Priceton Scientific Publishing Co. Inc., N. J. In press.

19) IMPERATO, A., DI CHIARA, G. (1985). Dopamine release and metabolism in awake rats after systemic neuroleptics as studied by trans-striatal dialysis. *J. Neurosci.* **5:** 297—306.

20) WALTERS, J. R., ROTH, R. H. (1976). Dopaminergic neurons: an in vivo system for measuring drug interactions with presynaptic receptors. *Naunyn-Schmiedeberg's Arch. Pharmacol.* **296:** 5—14.

21) DI CHIARA G., IMPERATO, A. (1985). Rapid tolerance to neuroleptic-induced stimulation of dopamine release in freely moving rats. *J. Pharmac. Exp. Ther.* **235:** 487—494.

22) ZETTERSTROM, T., SHARP, T., MARSDEN, C. A., UNGERSTEDT, U. (1983). In vivo measurement of dopamine and its metabolites by intracerebral dialysis: changes after d-amphetamine. *J. Neurochem.* **41:** 1769—1773.

PRECURSOR AVAILABILITY AND FUNCTION OF MIDBRAIN DOPAMINERGIC NEURONS

Robert H. Roth, See-Ying Tam, Charles W. Bradberry, Dan H. Karasic and Ariel Y. Deutch

Departments of Pharmacology and Psychiatry
Yale University School of Medicine
333 Cedar Street
New Haven, CT 06510
U.S.A.

INTRODUCTION

Midbrain dopamine (DA) neurons were once thought of as a very homogeneous population of neurons with similar biochemical, physiological, and pharmacological properties and regulatory controls. However, it is now known that these midbrain neurons are quite heterogeneous and are comprised of several distinct subsets. Biochemical and electrophysiological studies have demonstrated that certain mesocortical DA neurons, in contrast to the intensively-studied nigrostriatal and mesolimbic DA neurons, appear to lack impulse-regulating somatodendritic and synthesis-modulating nerve terminal autoreceptors (1). The absence or insensitivity of this important class of receptors on mesoprefrontal and mesocingulate DA neurons may, in part, explain some of the unique biochemical and electrophysiological properties of these two subpopulations of midbrain neurons (2). For example, the mesoprefrontal and mesocingulate DA neurons appear to have a faster firing rate, exhibit more bursting, and have a more rapid turnover of transmitter than those DA neurons innervating the striatum, nucleus accumbens, and pyriform cortex. The mesoprefrontal and mesocingulate DA neurons also exhibit diminished biochemical and electrophysiological responsiveness to DA agonists and antagonists. Chronic administration of antipsychotic drugs leads to the development of biochemical tolerance in those midbrain DA neurons possessing autoreceptors, but not in the systems lacking autoreceptors. Transmitter synthesis is also readily influenced by altered availability of the dopamine precursor tyrosine in those midbrain DA neurons lacking autoreceptors; this observation is perhaps related to the enhanced rate of physiological activity of this subpopulation of midbrain DA neurons.

The mesoprefrontal DA neurons have other unique properties which do not appear to be attributable completely to the absence of autoreceptors. For example, the mesoprefrontal DA neurons are very sensitive to stress, conditioned fear, and other environmental perturbations. The stress-induced activation of these neurons can be prevented by pretreatment with anxiolytic benzodiazepines. Recent studies have demonstrated that the mesoprefrontal neurons are also selectively activated by systemic administration of certain benzodiazepine inverse agonists (anxiogenic beta-carbolines) (3). This activation is reversed by pretreatment with central benzodiazepine receptor agonists and antagonists,

NATO ASI Series, Vol. H20
Amino Acid Availability and Brain Function in Health and Disease. Edited by G. Huether

suggesting that a benzodiazepine-GABA recognition site exerts a powerful modulatory control on this subset of VTA DA neurons (4). Since the mesoprefrontal DA system has been implicated in emotional responses to stress and is suspected to play a role in cognitive function (5), we were interested in pursuing our original findings suggesting that this subset of midbrain DA neurons may be more susceptible to precursor regulation of synthesis than other mesotelencephalic DA neurons.

It has been reported that the rate of catecholamine synthesis in the brain varies directly with local tyrosine supply (6). However, this relationship appears to hold only under conditions in which tyrosine hydroxylase is in an activated state (7). The activation of tyrosine hydroxylase in both noradrenergic and dopaminergic neurons appears to be associated with conditions which increase the firing rate of these neurons. Thus, in the majority of midbrain DA neurons most of the tyrosine hydroxylase remains in a non-activated state under basal conditions and is not very responsive to precursor regulation. Since the mesoprefrontal and mesocingulate DA neurons have a higher basal rate of firing and exhibit more bursting than other midbrain DA neurons under normal conditions (1), it is not surprising to find that the rate of tyrosine hydroxylation in these neurons is sensitive to tyrosine administration. A recent report indicates that a greater proportion of the tyrosine hydroxylase associated with the mesoprefrontal DA neurons is in an activated form than that found in other DA terminal fields, supporting the speculation that this may in part explain the enhanced responsiveness to tyrosine (8).

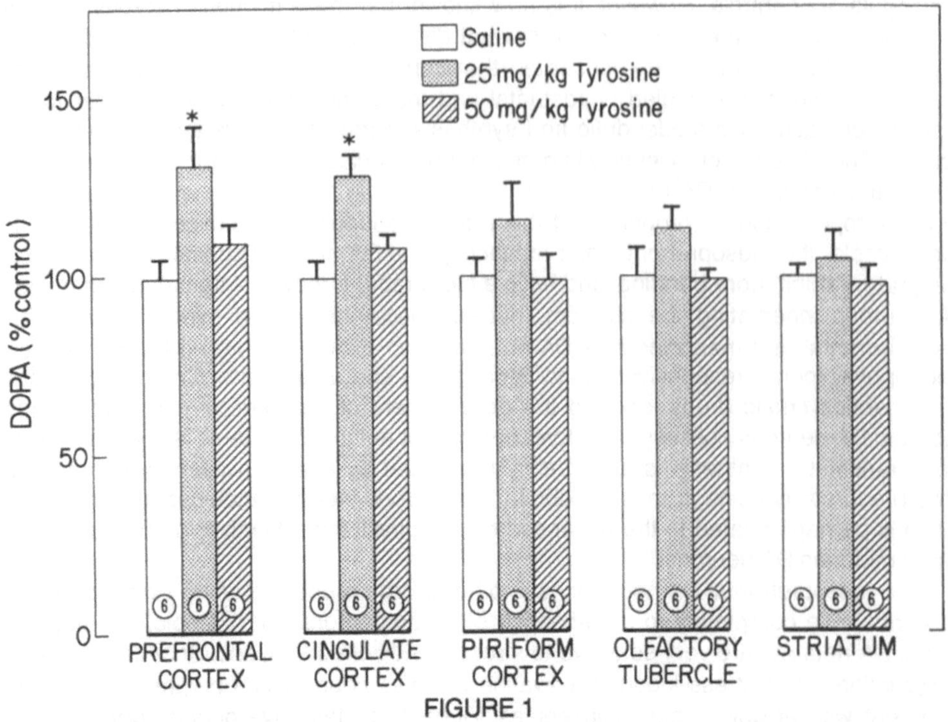

FIGURE 1

Rats were administered tyrosine, i.p., one hour before sacrifice. NSD-1015 (100 mg/kg, i.p.) was administered 30 min after tyrosine administration. Numbers in columns refer to the number of individual animals. Values are expressed as a percentage of control DOPA levels, mean ±S.E.M. * Differs significantly from saline controls ($p<0.05$).

EFFECTS OF TYROSINE ON IN VIVO TYROSINE HYDROXYLATION IN MIDBRAIN DA NEURONS

Our initial studies investigating the effects of systemic tyrosine on *in vivo* tyrosine hydroxylation in midbrain DA systems revealed that physiologically relevant doses of tyrosine (25 mg/kg) produced a significant increase in tyrosine hydroxylation only in the prefrontal and cingulate cortices (Fig. 1) (9, 10). Other midbrain DA systems were unaffected.

We were struck by the sharp dose response for this effect. Doses of tyrosine in excess of 25 mg/kg (50—200 mg/kg) were ineffective in enhancing tyrosine hydroxylation in the prefrontal cortex, despite the fact that this treatment resulted in a dose-dependent increase in cortical levels of tyrosine (Fig. 2). Since administration of 25 mg/kg, but not 50 mg/kg, tyrosine was found to enhance tyrosine hydroxylation in the prefrontal cortex, we sought to determine if prefrontal cortical DA levels could also be altered by tyrosine treatment. A significant increase in DA levels was observed in the prefrontal cortex following administration of 50 mg/kg but not 25 mg/kg of tyrosine both in theprescuse and absence of decarboxylase inhibition (see Table I).

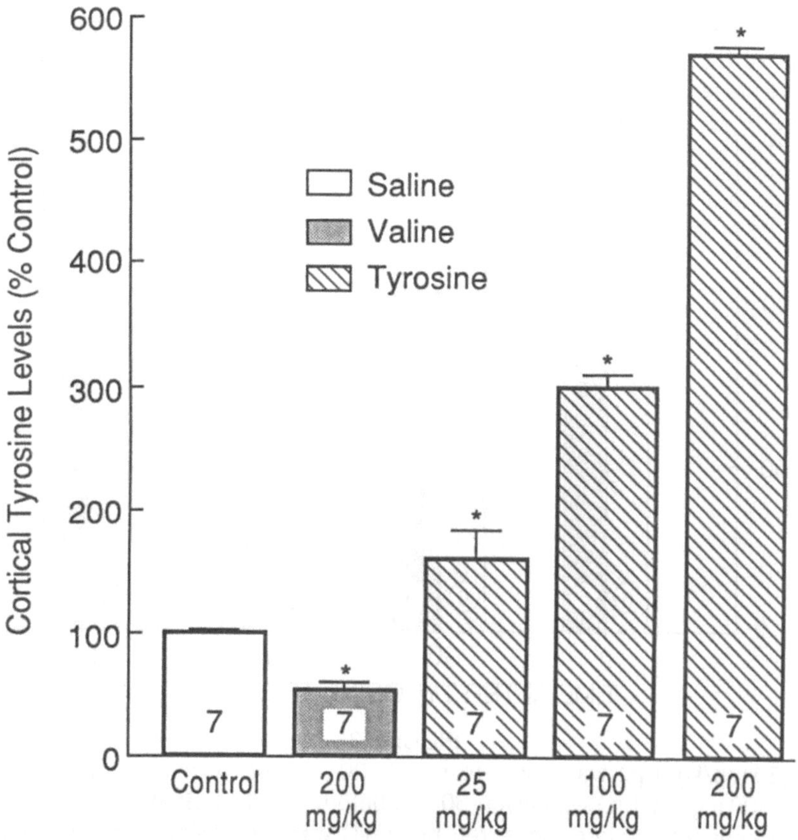

FIGURE 2

Rats were administered tyrosine (i.p.) in the form of the methyl ester, one hour before sacrifice. Values are expressed as percentage of control, mean ±S.E.M.

TABLE I

Effects of Tyrosine on Dopamine Levels in Prefrontal Cortex

Tyrosine was administered, i. p., one hour before sacrifice. In experiment II NSD-1015 (100 mg/kg, i. p.) was given 30 minutes after tyrosine administration. Data are expressed as the mean ± S. E. M. (n = 6 and 12 in experiments I and II respectively).

	DA (ng/g)	
	Exp. I	Exp. II
Saline	85.1 ± 3.2	28.2 ± 2.3
Tyrosine (25 mg/kg)	99.6 ± 4.7	32.6 ± 2.3
Tyrosine (50 mg/kg)	105.3 ± 5.4*	43.5 ± 3.4**

Differs significantly from saline controls * ($p<0.05$), ** (<0.01)

We were perplexed by these findings which indicated that 50 mg/kg tyrosine increased DA levels but did not enhance tyrosine hydroxylation in the prefrontal cortex. We reasoned that elevated levels of DA caused by an initial increase in tyrosine hydroxylation might inhibit subsequent tyrosine hydroxylase activity by end-product inhibition. Thus, we carried out additional studies to examine the time-course of the effects of tyrosine admin-

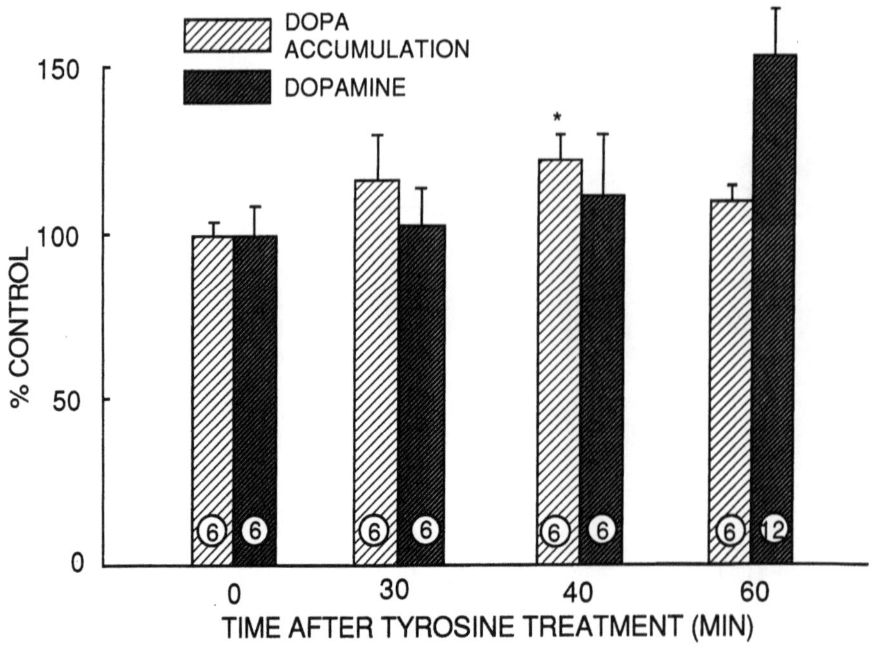

FIGURE 3

Tyrosine (50 mg/kg, i. p.) was given to rats 0, 30, 40 and 60 min before sacrifice, and NSD-1015 (100 mg/kg, i. p.) was given 30 min before sacrifice. Both DOPA and DA levels in the prefrontal cortex were measured by HPLC with electrochemical detection. Numbers in columns refer to the number of individual animals. Values are expressed as a percentage of the levels in saline treated controls at each respective time point. * Differs significantly from respective saline controls ($p < 0.05$)

istration (50 mg/kg) on *in vivo* tyrosine hydroxylase activity and DA levels in the prefrontal cortex. Rats were administered tyrosine 0, 30, 40 and 60 min before decapitation, and the decarboxylase inhibitor, NSD-1015, was given 30 min before sacrifice. DOPA accumulation (for the last 30 min of each time point) and DA levels in the prefrontal cortex were measured concurrently and compared with values from saline-treated controls. As illustrated in Figure 3, there were no significant alterations in DOPA accumulation and DA levels in the prefrontal cortex 30 min after tyrosine treatment (although there was a trend towards a slight increase in DOPA accumulation). However, 40 min after tyrosine treatment, DOPA accumulation in the prefrontal cortex was significantly elevated as compared with the saline controls, whereas the increase in the respective prefrontal DA levels was not significant at this time point. At 60 min after tyrosine administration, DOPA accumulation was not significantly altered by tyrosine treatment. However, the DA levels at this time were significantly elevated as compared to saline controls (54 %) (see Fig. 3). These data are thus consistent with the possibility that the higher dose of tyrosine (50 mg/kg) may have initially enhanced tyrosine hydroxylation in the prefrontal cortex. The subsequent increase in endogenous DA levels observed at 60 min may thereafter negatively feedback on tyrosine hydroxylase activity and normalize the rate of tyrosine hydroxylation in the prefrontal cortex. These biochemical changes are in contrast to the situation occurring after administration of 25 mg/kg of tyrosine, where a significant increase in DOPA accumulation is observed 60 min after tyrosine administration, without any change in DA levels.

Effects of Valine on In Vivo Tyrosine Hydroxylation in Prefrontal Cortex

Since tyrosine was found to enhance DA synthesis in mesoprefrontal DA neurons, experimental conditions which lowered brain tyrosine levels would be expected to reduce tyrosine hydroxylation in the prefrontal cortex. Administration of valine (200 mg/kg, i. p.), a large neutral amino acid which competes with tyrosine for uptake into the brain, resulted in a significant decrease in cortical tyrosine levels (Fig. 2). This treatment was associated with a modest but significant decrease in DOPA accumulation in the prefrontal cortex. A similar inhibitory effect of valine on tyrosine hydroxylation was observed in the striatum.

Effects of Tyrosine Availability on DA Metabolite Levels in Prefrontal Cortex

Previous studies (11) have shown that brain concentrations of DA metabolites such as DOPAC can be elevated by administration of a low dose of tyrosine (50 mg/kg). However, administration of either tyrosine or valine (at dosages which, in our laboratory, were shown to be effective in altering tyrosine hydroxylation in the prefrontal cortex) failed to change the levels of DA metabolites such as DOPAC and HVA (Table 2). Similarly, such tyrosine and valine treatments did not significantly affect DOPAC levels in the olfactory tubercle or striatum (data not shown). Thus, at least under basal conditions, the mesoprefrontal DA neurons appear to have sufficient intrinsic controls to minimize the influence of short-term alterations in precursor availability on transmitter output, despite the transient effects observed on synthesis.

TABLE II

Effect of Tyrosine on DA Metabolites in Prefrontal Cortex

Rats were administered tyrosine, i.p., one hour before sacrifice, or valine (200 mg/kg), i.p., 90 min before sacrifice. Data are expressed as the mean ±S.E.M.

	DOPAC (ng/g)	HVA (ng/g)
Saline	28 ± 2 (6)	44 ± 5 (6)
Tyrosine (25 mg/kg)	26 ± 2 (6)	46 ± 3 (6)
Tyrosine (50 mg/kg)	28 ± 2 (6)	N.D.
Valine (200 mg/kg)	26 ± 2 (7)	43 ± 3 (7)

Effects of Tyrosine on Activated Mesoprefrontal DA Neurons

The DA projection to the prefrontal cortex is very susceptible to various forms of stress and can be metabolically activated by a number of experimental paradigms (4, 12—14). Mild foot shock, restraint stress, conditioned fear or administration of anxiogenic beta carbolines all result in a selective activation of the mesoprefrontal DA system. In view of the susceptibility of this cortical system to environmental stress, it was of interest to deter-

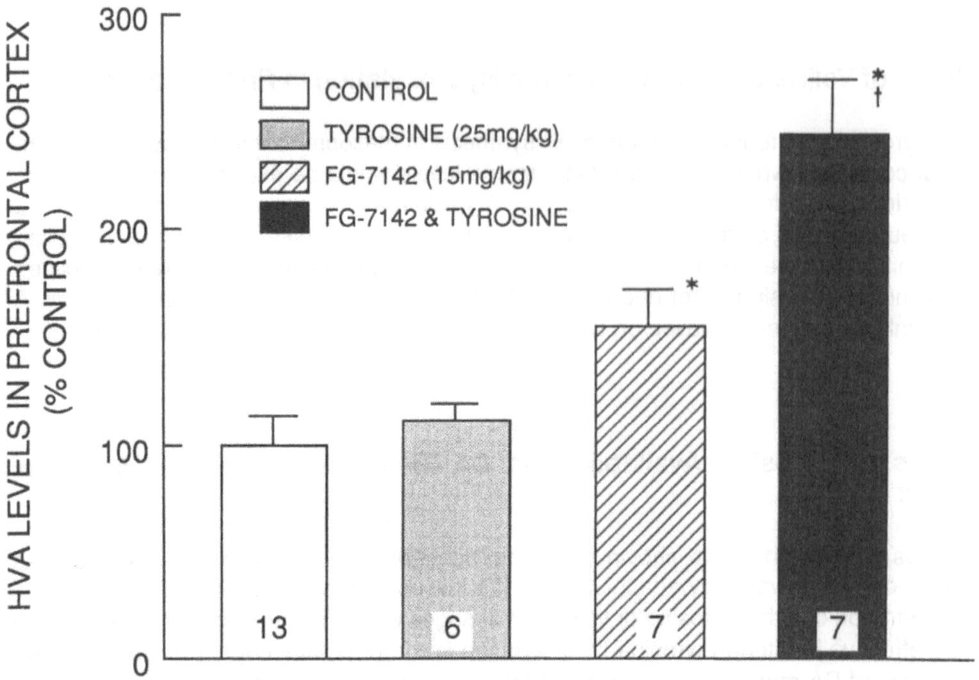

FIGURE 4

Tyrosine was administered, i.p., one hour before sacrifice. FG-7142 was given, i.p., 30 min after tyrosine administration. Data are expressed as the mean ±S.E.M.

* Differs significantly from saline controls ($p<0.05$).

*+ Differs significantly from FG-7142 alone ($p<0.05$).

mine also how responsive mesoprefrontal DA neurons were to alterations in the availability of precursor tyrosine under conditions in which these neurons were experimentally activated. The mesocortical DA neurons were selectively activated by administration of the anxiogenic beta carboline FG-7142, and the ability of small doses of systemically administered tyrosine to alter the drug-induced increase in DA metabolite levels examined as an index of the functional outflow of this system. As illustrated in Figure 4, tyrosine administration dramatically enhanced DA metabolites under conditions in which the mesoprefrontal DA system is pharmacologically activated. These findings suggest that enhanced activity of the DA projection to the prefrontal cortex renders these DA neurons much more dependent on tyrosine availability for maintenance of transmitter output.

INFLUENCE OF EXPERIMENTAL DIABETES ON THE FUNCTION OF MIDBRAIN DA NEURONS

The influence of experimental conditions which result in prolonged changes in the availability of brain tyrosine on transmitter dynamics in midbrain DA neurons was also investigated. Since diabetes in the rat has been shown to result in a significant decrease in the levels of brain amino acids, including tyrosine, we have employed this model (15); diabetes does not lead to a reduction in serum tyrosine.

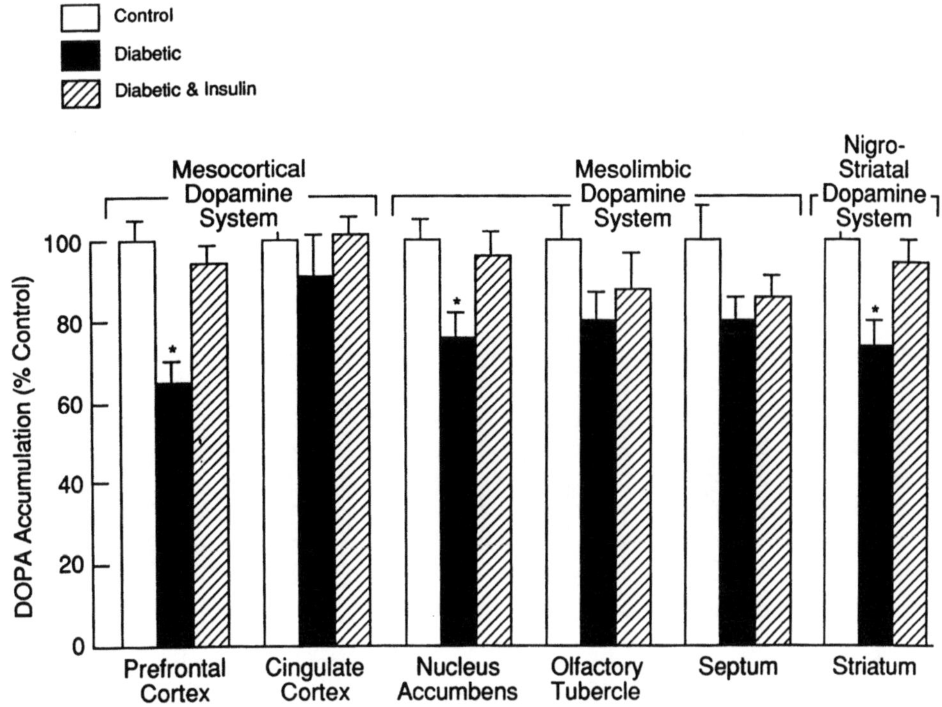

FIGURE 5

Animals were treated with STZ (65 mg/kg i.v.) 24 days prior to sacrifice. Where indicated, insulin treatment was begun 10 days prior to sacrifice. Values are expressed as percent of vehicle treated controls, mean ±S.E.M.

Diabetes was induced by administration of 65 mg/kg (i. v.) streptozotocin (STZ), and confirmed by blood glucose levels of > 400 mg% at the time of sacrifice. Tissue levels of L-DOPA 30 min after NSD-1015 treatment (100 mg/kg) were measured as an index of *in vivo* tyrosine hydroxylation in terminal projection fields of midbrain DA neurons, as well as in the DA cell body regions of ventral tegmental area (VTA) and substantia nigra (SN). Regional brain levels of tyrosine were also determined in the same animals. Dopamine metabolites were assessed in the same brain regions in a separate series of experiments. As illustrated in Figure 5, DOPA accumulation was reduced in several midbrain DA systems; the diabetes-related inhibitory effects observed on tyrosine hydroxylation were reversed by insulin treatment. The prefrontal cortical DA neurons appear to be the most susceptible subset of midbrain DA neurons to the effects of the diabetic state on tyrosine hydroxylation. DOPA accumulation in the VTA of diabetic rats was significantly decreased, while DOPA accumulation in the SN was unaffected.

Furthermore, there was a significant positive correlation between DOPA levels in the prefrontal cortex and the levels of cortical tyrosine. DA metabolite levels in certain midbrain DA projection fields were also reduced in diabetic rats. The changes observed in both *in vivo* tyrosine hydroxylation and DA metabolites appeared to be correlated with a

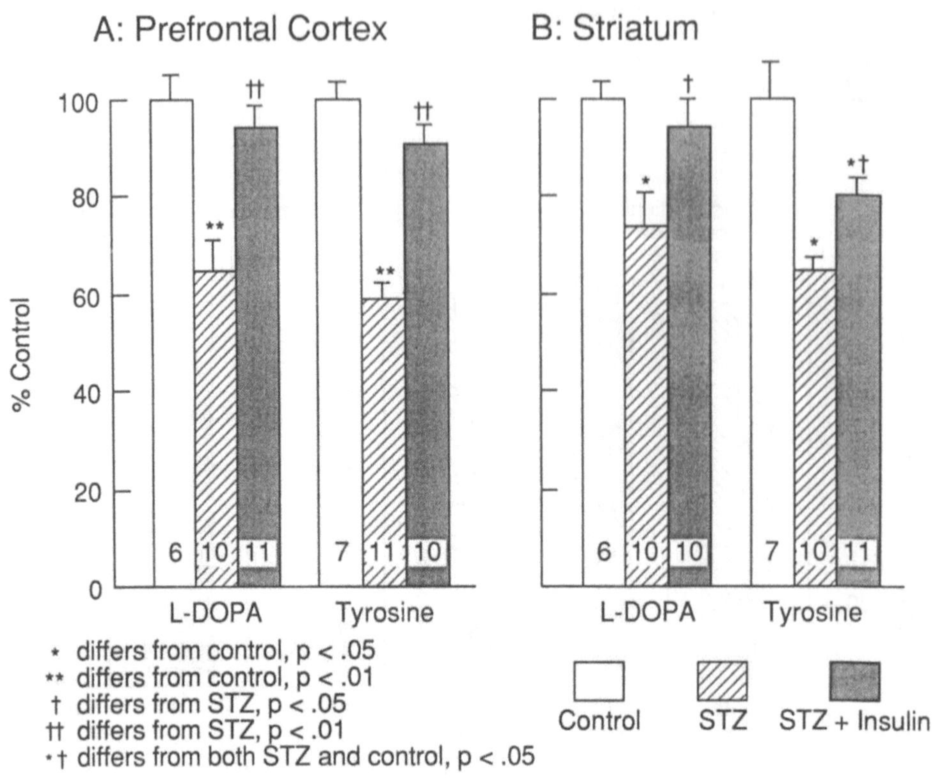

FIGURE 6

Results are expressed as the mean % of control ±S.E.M.
Numbers in columns refer to the number of individual animals.

reduction in brain levels of tyrosine. Insulin treatment of the diabetic rats normalized brain tyrosine levels, and restored tyrosine hydroxylation and DA metabolite levels to normal values (Fig. 6).

These observations suggest that in the diabetic state the diminished availability of brain tyrosine may contribute to decreased DA synthesis and metabolism in certain midbrain DA neurons. The observed decrease in transmitter availability could result in alterations in the function of neurons which have a small reserve capacity, or under conditions which enhance transmitter utilization. However, it is premature to assume that all of the changes in midbrain DA system observed in this experimental model can be attributed to alterations in the availability of brain tyrosine. Further studies will be required to determine the role played by other epiphenomena associated with the diabetic state on the function of midbrain DA neurons.

SUMMARY AND CONCLUSION

Subsets of midbrain DA neurons appear to possess different regulatory controls and to be differentially influenced by alterations in the brain levels of precursor tyrosine. Administration of physiologically relevant doses of tyrosine produces a significant increase in *in vivo* tyrosine hydroxylation in those neurons which have a high basal rate of activity (mesoprefrontal and mesocingulate), but is without effect on other midbrain DA systems. However, this effect of tyrosine on synthesis is relatively short-lived, probably due to the rapid elevation of endogenous levels of DA and the subsequent normalization of tyrosine hydroxylase activity. Under basal conditions tyrosine administered in dosages of 25—200 mg/kg has no significant effect on dopamine metabolite levels in midbrain DA neurons, including the DA projections to the prefrontal and cingulate cortex, suggesting that the basal functional output of these neurons is not normally influenced by precursor. However, under conditions which elicit an increase in the functional and metabolic activity of the mesoprefrontal DA neurons, precursor availability becomes an important factor in controlling transmitter output from these neurons.

In rats with experimental diabetes, brain levels of tyrosine are reduced and *in vivo* tyrosine hydroxylation and dopamine metabolite levels in certain subsets of midbrain DA neurons are decreased. The mesoprefrontal DA neurons appear to be the most susceptible to the diabetes-elicited effects exerted on *in vivo* tyrosine hydroxylation. There is a positive correlation between the brain levels of DOPA and the endogenous levels of cortical tyrosine. The inhibitory effects on DOPA accumulation in the prefrontal cortex of diabetic rats are reversed by treatment with insulin; this effect appears to be correlated with a restoration of cortical levels of tyrosine towards normal. The susceptibility of mesoprefrontal DA neurons to diabetes-induced alterations in cortical tyrosine is of special interest due to the possible role of this subset of DA neurons in cognitive function, and the knowledge that some cognitive deficits may occur in patients with uncontrolled or long-lasting diabetes.

ACKNOWLEDGEMENT

The research cited in this review was supported in part by grants from NIH, MH-14092 and the American Heart Association, Connecticut Affiliate.

REFERENCES

1) CHIODO, L. A., BANNON, M. J., GRACE, A. A., ROTH, R. H., BUNNEY, B. S. (1984). Evidence for the absence of impulse-regulating somatodendritic and synthesis-modulating nerve terminal autoreceptors on subpopulations of mesocortical dopamine neurons. *Neurosci.* **12:** 1—16.

2) ROTH, T. H., TAM, S.-Y., (1987). Regulatory control of midbrain dopamine neurons. In: *Aminoacids in Health and Disease: New Perspectives.* (Kaufman, S., ed.). Alan R. Liss. Inc., New York, pp. 159—178.

3) ROTH, T. H., TAM, S.-Y., (1987). Regulatory control of midbrain dopamine neurons. In: *Amino acids* by the anxiogenic beta-carboline FG 7142. *Biochem. Pharm.* **34:** 1595—1598.

4) ROTH, R. H., TAM, S.-Y., IDA, Y., DEUTCH, A. Y. (1988). Stress and the mesocorticolimbic dopamine systems. In: *Mesocorticolimbic Dopamine Systems.* Ann. N. Y. Acad. Sci. In Press.

5) BANNON, M. J., ROTH, R. H. (1983). Pharmacology of mesocortical dopamine neurons. *Pharmacol. Rev.* **35:** 53—68.

6) FERNSTROM, J. D. (1983). Role of precursor availability in control of monoamine biosynthesis in brain. *Physiol. Rev.* **63 (2):** 484—545.

7) MILNER, J. D., WURTMAN, R. J. (1986). Catecholamine synthesis: Physiological coupling to precursor supply. *Biochem. Pharmacol.* **35 (6):** 875—882.

8) IUVONE, P. M., DUNN, A. J. (1986). Tyrosine hydroxylase activation in mesocortical 3,4-dihydroxyphenylethylamine neurons following footshock. *J. Neurochem.* **47:** 837—842.

9) TAM, S.-Y., ROTH, R. H. (1984). Tyrosine preferentially increases dopamine synthesis in mesocortical dopamine neurons with high firing frequency. *Soc. Neurosci. Abstr.* **10:** 259.11.

10) TAM, S.-Y., ONO, N., ROTH, R. H. (1987). Precursor control and influence of aspartame on midbrain dopamine neurons. In: *Amino acids in Health and Disease: New Perspectives.* (Kaufman, S., ed.). Alan R. Liss. Inc., New York, pp. 421—435.

11) BADAWY, A. A.-B., WILLIAMS, D. L. (1982). Enhancement of rat brain catecholamine synthesis by administration of small doses of tyrosine and evidence for substrate inhibition of tyrosine hydroxylase activity by large doses of the amino acid. *Biochem. J.* **206:** 165—168.

12) THIERRY, A. M., TASSIN, J. P., BLANC, G., GLOWINSKI, J. (1976). Selective activation of the mesocortical DA system by stress. *Nature* **263:** 242—244.

13) DEUTCH, A. Y., TAM, S.-Y., ROTH, R. H. (1985). Footshock and conditioned stress increase 3,4-dihydroxyphenylacetic acid (DOPAC) in the ventral tegmental area but not substantia nigra. *Brain Res.* **333:** 143—146.

14) THIERRY, A. M., TASSIN, J. P., GLOWINSKI, J. (1984). Biochemical and electrophysiological studies of the mesocortical dopamine system. In: *Monoamine Innervation of Cerebral Cortex.* (Descarries, L., Reader, T. R., Jasper, H. H., eds.). Alan R. Liss., New York, pp. 233—261.

15) CRANDALL, E. A., FERNSTROM, J. D. (1983). Effect of experimental diabetes on the levels of aromatic and branched-chain amino acids in rat blood and brain. *Diabetes* **32:** 222—230.

MODIFICATION OF SYNTHESIS, RELEASE, AND FUNCTION OF CATECHOLAMINERGIC SYSTEMS BY PHENYLALANINE

Timothy J. Maher

Department of Pharmacology
Massachusetts College of Pharmacy
179 Longwood Avenue
Boston, MA., 02115 U.S.A.

The conversion of tyrosine to 3,4-dihydroxyphenylalanine (DOPA), via tyrosine hydroxylase (TH; EC.1.14.16.2), is the rate-limiting step in the synthesis of the catecholamines dopamine, norepinephrine, and epinephrine. Administration of supplemental tyrosine to normal animals under resting conditions usually fails to cause a sustained acceleration of catecholamine synthesis in the central nervous system, a notable exception being neurons within the prefrontal and cingulate cortices: these neurons, which normally fire very frequently, have been shown to be responsive to supplemental tyrosine (1).

In contrast, when selected groups of catecholamine-containing neurons are made to fire frequently, e. g., by increasing (2) or decreasing (3) blood pressure, by electrically shocking the rat's tail (4), by activating the retina with light (5) or by destroying 75 % of the nigrostriatal neurons , (so that the remaining neurons accelerate their firing in an attempt to compensate for the decreased neurotransmitter release (6)), tyrosine has been shown to increase catecholamine synthesis and/or augment catecholamine-associated behaviors. For example, while tyrosine administration decreases blood pressure in hypertensive rats (2), and increases blood pressure in hypotensive rats (3), tyrosine administration fails to alter blood pressure significantly in normotensive rats where homeostatic mechanisms that utilize catecholamines are not likely to be intensely activated. Studies which attempt to demonstrate an effect of tyrosine on catecholamine synthesis or function are destined to fail if the particular group of neurons being examined are not in an "activated" state (7).

Tyrosine is traditionally classified as a nonessential amino acid, since it can be formed from phenylalanine by the action of hepatic phenylalanine hydroxylase. Thus phenylalanine may support catecholamine synthesis by supplying the organism with tyrosine. Despite this potential role as a catecholamine precursor, phenylalanine in actuality is more likely to interfere with catecholamine synthesis by inhibiting TH directly (especially in species like humans, which hydroxylate phenylalanine very slowly). For instance, phenylalanine concentrations as low as 25 μM have been shown to inhibit TH activity by 64 % in homogenates of guinea pig heart (8), and 100 μM of the amino acid inhibited TH activity by 78 % in rat brain homogenates (9). Phenylalanine is also known to

NATO ASI Series, Vol. H20
Amino Acid Availability and Brain Function in Health and Disease. Edited by G. Huether

retard the entry of tyrosine into the brain by competing with it for access to the transport carrier for large neutral amino acids (LNAA) in the blood brain barrier (10). Thus phenylalanine might be expected to have dual dose-dependent effects on catecholamine synthesis; i. e., *supporting* synthesis at low doses (because of its conversion to tyrosine) while *inhibiting* it at slightly higher concentrations.

The ability of phenylalanine and tyrosine to modify dopamine release *in vitro* has recently been demonstrated in superfused rat brain striatal slices subjected to electrical stimulation (11, 12). Superfusion with low concentrations of tyrosine (20—50 μM) sustained the electrically-evoked (360 pulses; 12 Hz; 2 msec) release of dopamine in a dose-dependent manner, as reflected by maintenance of the S2/S1 ratio (the ratio of the release of dopamine following the second stimulus relative to that following the first). When tissues were superfused without tyrosine and similarly stimulated, tyrosine and dopamine levels 10 min. later, had decreased by 14 and 25 %, respectively. Superfusion with a tyrosine-containing solution (50 μM) completely prevented these decrements. While low concentrations of phenylalanine (50 μM) only partially sustained dopamine release in the absence of tyrosine, slightly higher concentrations of phenylalanine (200 μM; a value approximately 3 times that found in fasting plasma of rat) significantly *reduced* the S2/S1 ratio, even in the presence of tyrosine. This suppression by phenylalanine on dopamine release was not mimicked by another LNAA, leucine.

Using intracerebral microdialysis to monitor extracellular levels of dopamine and its major metabolites dihydroxyphenylacetic acid (DOPAC) and homovanillic acid (HVA) in the striatum of rats, During *et al.* (13) recently demonstrated that a low phenylalanine (200 mg/kg) dose which, in rats elevates the plasma tyrosine ratio more than the plasma phenylalanine ratio, increased dopamine release by 59 %. (The plasma ratio i. e., the ratio of a particular LNAA divided by the sum of the other competing LNAA, is a useful peripheral index of the transport of each LNAA from blood to brain. The plasma tyrosine ratio can be defined as the plasma tyrosine concentration divided by the summed concentrations of valine, leucine, isoleucine, tryptophan and phenylalanine).

An intermediate phenylalanine dose (500 mg/kg), which elevated both the plasma tyrosine and phenylalanine ratios, failed to alter dopamine release. However when a higher phenylalanine dose (1000 mg/kg), which increased the plasma phenylalanine ratio more than that of tyrosine was given, dopamine release was *decreased* by 26 %. No corresponding changes were observed in brain DOPAC or HVA concentrations with any of the treatments, suggesting that the altered dopamine synthesis may have largely occurred in a pool that was preferentially released into synapses. Thus, data exists which clearly demonstrate the dual dose-dependent effects of phenylalanine on catecholamine synthesis and release.

Aspartame (L-aspartyl-L-phenylalanine methyl ester), a recently approved artificial sweetener that is approximately 180 times as sweet as sucrose, contains about 56 % phenylalanine, and its ingestion rapidly raises blood levels of phenylalanine as well as its other constituents, methanol and aspartic acid. While it is true that the *absolute amount* of phenylalanine derived from the aspartame contained within a 12 oz can of diet soft drink is not dissimilar to the amount found in various commonly consumed protein sources (e. g., eggs), the changes that occur in the *plasma phenylalanine ratio,* and thus the brain concentration of phenylalanine, following aspartame ingestion are actually *opposite* to those following protein consumption, contrary to some published misconceptions (14, 15). This difference is due to the presence of other LNAA in proteins, and their absence from aspartame. An egg, and every other complete protein, contains most or all of the other 20 naturally-occurring amino acids, some of which, the LNAA, compete with phenylalanine and each other for passage across the blood-brain barrier. While *absolute* plasma phenylalanine levels can increase following protein consumption, the plasma phenylalanine ratio *decreases,* since protein contains much larger amounts of the other LNAA than of phenylalanine (16). Consumption of aspartame also increases plasma phenylalanine levels but, unlike the changes following consumption of proteins in food, it causes the plasma phenylalanine ratio *increases* (17). This effect is due to the lack of any other competing LNAA in aspartame. Hence, aspartame is the *only* dietary constituent available to man which selectively increases the entry of phenylalanine into brain. Additionally, if an aspartame-containing beverage is consumed along with a carbohydrate-rich, protein-poor food, its effects on brain phenylalanine are doubled, since carbohydrates elicit insulin release, which selectively lowers plasma levels of the branched-chain amino acids without having much of an effect on plasma phenylalanine (18). Should the hypothetical individual consuming the aspartame with carbohydrate also happen, unknowingly, to be heterozygous for the phenylketonuria (PKU) gene, so that his conversion of phenylalanine to tyrosine is greatly reduced, an additional doubling of aspartame's brain effects may occur (19). Additionally, impaired conversion of phenylalanine to tyrosine is known to occur in patients with iron-deficiency anemia (20).

Reports of adverse central nervous system effects (including seizures, headache, mood changes) associated with the consumption of aspartame have increased in parallel to the increased consumer availability of this artificial sweetener (21, 22). Catecholamines appear to play a role in the suppression of seizure activity in experimental animal models. Animals given drugs like reserpine or Ro 4-1284, which deplete their brains of particular monoamine neurotransmitters or which block the receptor-mediated effects of these transmitters, exhibit greater sensitivity to seizures in many animal models (23). In contrast, drugs like L-DOPA plus a monoamine oxidase (MAO) inhibitor, or 3,4-dihydroxyphenylserine (L-DOPS), thought to enhance monoaminergic neurotransmission, apparently protect rodents against the development of some types of experimentally-induced seizures. *Low* doses of phenylalanine or aspartame, — which, in rodents, raise plasma tyrosine levels more than those of phenylalanine — might be expected to have no effect on seizure thresholds, or even, conceivably, to protect animals against the epileptogenic effects of drugs like pentylenetetrazol (PTZ), possibly by enhancing catecholaminergic neurotransmission. In contrast, comparable doses given to humans (who hydroxylate phenylalanine slowly), or sufficiently-high aspartame doses (which transiently surpass the liver's capacity to hydroxylate the phenylalanine) given to rodents could enhance seizure susceptibility.

A proconvulsant effect of aspartame in mice treated with PTZ and fluorothyl has recently been demonstrated (24, 25). At 1000 and 2000 mg/kg aspartame, 78 and 100 % of the animals experienced PTZ-induced seizures, compared with 50 % in the water-pretreated control group. Other mice pretreated with a fixed dose (1000 mg/kg) of aspartame, or with water, and given various doses (50—75 mg/kg) of PTZ one hour later exhibited a significant leftward shift of the PTZ dose-response curve (CD_{50} lowered from a control value of 66 (64—68) to 59 (56—63) mg/kg). Enhanced suceptibility to PTZ-induced seizures was also observed among mice pretreated with phenylalanine (in doses equimolar to effective aspartame doses) but not among animals pretreated with aspartic acid or methanol. Co-administration of valine with aspartame protected mice from the seizure-promoting effects of the sweetener; in contrast, alanine, an amino acid which does not compete with phenylalanine for transport into the brain, failed to attenuate aspartame's effect on PTZ-induced seizures. Similar proconvulsant effects were observed when maximal electroshock, or when the convulsant fluorothyl were used. Other investigators have recently reported similar findings using different convulsants or species (26, 27), while some other investigators fail to see potentiation using other experimental systems (28, 29). Ultimately the effect of aspartame on seizure thresholds or brain functioning in humans will be of paramount importance. Spiers *et al.* (30) have recently examined the effects of Acceptable Daily Intake (ADI) doses (50 mg/kg) of aspartame on cognitive functioning and have found a significant impairment of behavior in *normal* volunteers. Unquestionably much future study is required in this field.

Lower doses of aspartame which, in the rodent, greatly *increase* brain tyrosine levels (31), have been shown to increase catecholamine synthesis in selected brain regions (32, 33), and to lower blood pressure in spontaneously hypertensive rats (33, 34). Results from studies using higher aspartame doses in the above systems have so far not been reported. Studies by Tam *et al.* (1) suggest that aspartame doses which can increase brain tyrosine may not invariably accelerate catecholamine synthesis in cingulate and prefrontal cortices. These results suggest that even *small* increases in brain phenylalanine, following these doses, were able to interfere with the otherwise-expected acceleration of catecholamine synthesis within these populations of neurons.

Evidence now exists which clearly demonstrates the ability of phenylalanine to affect catecholaminergic processes in several experimental paradigms. When phenylalanine is provided at low concentrations in the absence of tyrosine, or when aspartame is administered to rodents in doses that elevate brain tyrosine relative to phenylalanine, enhanced catecholaminergic indices have been observed. In contrast, when phenylalanine is present in higher concentrations (2—3 times normal), or when aspartame is administered in doses that elevate brain phenylalanine more than tyrosine (a situation which probably mimics the changes that would occur when humans consume moderate amounts of aspartame (35)), catecholaminergic indices have been reported to be impaired. With the increasing availability to consumers of aspartame-containing food and drug products, the effects of phenylalanine on catecholamine synthesis and in related behaviors now become extremely important issues. In order to assess effects of phenylalanine or its progenitors on catecholamine synthesis and on related neuronal functions, investigators must be aware of the dual dose-response curve for phenylalanine, and design experiments appropriately.

ACKNOWLEDGEMENTS

These studies were supported in part by a grant from the National Institutes of Neurological, and Communicative Diseases and Stroke (NS 21231) and the Center for Brain Sciences and Metabolism Charitable Trust.

REFERENCES

1) TAM, S. Y., ONO, N., ROTH, R. H. (1987). Precursor control and influence of aspartame on midbrain dopamine neurons. In: *Amino Acids in Health and Disease: New Perspectives* (Kaufman, S., ed.), **Vol. 55,** pp. 421—435.

2) SVED, A. F., FERNSTROM, J. D., WURTMAN, R. J. (1979). Tyrosine administration reduces blood pressure and enhances brain norepinephrine release in spontaneously hypertensive rats. *Proc. Natl. Acad. Sci. USA* **76:** 3511—3514.

3) CONLAY, L. A., MAHER, T. J., WURTMAN, R. J. (1981). Tyrosine increases blood pressure in hypotensive rats. *Science* **212:** 559—560.

4) REINSTEIN, D. K., LEHNERT, H., SCOTT, N. A., WURTMAN, R. J. (1984). Tyrosine prevents behavioral and neurochemical correlates of an acute stress in rats. *Life Sci.* **34:** 2225—2231.

5) FERNSTROM, M. H., VOLK, E. A., FERNSTROM, J. D. (1984). In vivo tyrosine hydroxylation in the diabetic rat retina: effect of tyrosine administration. *Brain Res.* **298:** 167—170.

6) MELAMED, E., HEFTI, F., WURTMAN, R. J. (1980). Tyrosine administration increases striatal dopamine release in rats with partial nigrostriatal lesions. *Proc Natl. Acad. Sci. USA* **77:** 4305—4309.

7) SARAW, G., BEHRENS, W. A., PEACE, R. W., MADERE, R., BOTTING, H. G. (1987). Lack of relationship between dietary tyrosine and sympathetic nervous system activity in rats fed normal protein diets. *Nutr. Rep. Interntl.* **35:** 471—478.

8) IKEDA, M., LEVITT, M., UDENFRIEND, S. (1967). Phenylalanine as a substrate and inhibitor of tyrosine hydroxylase. *Arch. Biochem. Biophys.* **120:** 420—427.

9) NAGATSU, T., LEVITT, M., UDENFRIEND, S. (1964). Tyrosine hydroxylase: the initial step in norepinephrine synthesis. *J. Biol. Chem.* **239:** 2910—2917.

10) PARDRIDGE, W. M. (1977). Regulation of amino acid availability to the brain. In: *Nutrition and the Brain* (Wurtman, R. J., Wurtman, J. J., eds), **Vol. 1,** pp. 141—204.

11) MILNER, J. D., IRIE, K., WURTMAN, R. J. (1986). Effects of phenylalanine on the release of endogenous dopamine from rat striatal slices. *J. Neurochem.* **47:** 1444—1447.

12) MILNER, J. D., WURTMAN, R. J. (1984). Release of endogenous dopamine from electrically stimulated slices of rat striatum. *Brain Res.* **301:** 139—142.

13) DURING, M. J., ACWORTH, I. N., WURTMAN, R. J. (1987). An in vivo study of dopamine release in striatum: the effects of phenylalanine. In: *Dietary Phenylalanine and Brain Function* (Wurtman, R. J., Ritter-Walker, E., eds.), in press.

14) FREEDMAN, M. (1987). Consumption of aspartame by heterozygotes for phenylketonuria. *J. Pediatr.* **110:** 662.

15) YELLOWLEES, H. (1983). Aspartame. *Brit. Med. J.* **287:** 162—163.

16) GLAESER, B. S., MAHER, T. J., WURTMAN, R. J. (1983). Changes in brain levels of acidic, basic, and neutral amino acids after consumption of single meals containing various proportions of protein. *J. Neurochem.* **41:** 1016—1021.

17) CABALLERO, B., MAHON, B., ROHR, F., LEVY, H., WURTMAN, R. J. (1986). Plasma amino acid levels after single dose aspartame consumption in phenylketonuria. *J. Pediatr.* **109:** 668—671.

18) WURTMAN, R. J. (1983). Neurochemical changes following high-dose aspartame with dietary carbohydrates. *N. Engl. J. Med.* **309:** 429—430.

19) STEGINK, L. D., KOCH, R., BLASKOVICS, M. E., FILER, L. J., BAKER, G. L., McDONNELL, J. E. (1981). Plasma phenylalanine levels in phenylketonuric heterozygous and normal adults administered aspartame at 34 mg/kg body weight. *Toxicology* **20:** 81—90.

20) LEHMANN, W. D., HEINRICH, H. C. (1986). Impaired phenylalanine- tyrosine conversion in patients with iron-deficiency anemia studied by a L-(2 H-5)-phenylalanine-loading test. *Amer. J. Clin. Nutr.* **44:** 468—474.

21) CENTER FOR DISEASE CONTROL (1984). Evaluation of consumer complaints related to aspartame use.

22) WURTMAN, R. J. (1985). Aspartame: possible effect on seizure susceptibility. *Lancet* **II:** 1060.

23) JOBE, P. C., DAILEY, J. W., REIGEL, C. E. (1986). Noradrenergic and serotoninergic determinants of seizure susceptibility and severity in genetically epilepsy-prone rats. *Life Sci.* **39:** 775—782.

24) PINTO, J. M. B., MAHER, T. J. (1986). High-dose aspartame lowers the seizure threshold to subcutaneous pentylenetetrazol in mice. *Pharmacologist* **28:** 201.

25) MAHER, T. J., PINTO, J. M. B. (1987). Aspartame administration potentiates fluorothyl-induced seizures in mice. *J. Neurochem.* **48:** S52.

26) KIM, K. C., KIM, S. H. (1986). Studies on the effects of aspartame and lidocaine interaction in central nervous system of mice. *Fed. Proc.* **46:** 705.

27) GARRATINI, S., CACCIA, S., ROMANO, M., DIOMEDE, L., GUISO, G., VEZZANI, A., SALMONA, M. (1987). Studies on the susceptibility to convulsions in animals receiving abuse doses of aspartame. In: *Dietary Phenylalanine and Brain Function* (Wurtman, R. J., Ritter-Walker, G. E., eds.), in press.

28) DAILEY, J., LASLEY, S. M., FRASCA, J., JOBE, P. C. (1987). Aspartame (ASM) is not pro-convulsant in the genetically epilepsy prone rat (GEPR). *Pharmacologist* **29:** 142.

29) NEVINS, M. E., ARNOLDE, S. M., HAIGLER, H. J. (1986). Aspartame: lack of effect on convulsant thresholds in mice. *Fed. Proc.* **45:** 1096.

30) SPIERS, P., SCHOMMER, D., SABOUNJIAN, L. (1987). Aspartame and human behavioral observations. In: *Dietary Phenylalanine and Brain Function* (Wurtman, R. J., Ritter-Walker, E., eds.), in press.

31) FERNSTROM, J. D., FERNSTROM, M. H., GILLIS, M. A. (1983). Acute effects of aspartame on large neutral amino acids and monoamines in rat brain. *Life Sci.* **32:** 1651—1658.

32) YOKOGOSHI, H., WURTMAN, R. J. (1986). Acute effects of oral or parenteral aspartame on catecholamine metabolism in various regions of rat brain. *J. Nutr.* **116:** 356—364.

33) THAKORE, K., CRANE, S. C. (1987). Blood pressure and regional brain tyrosine, norepinephrine and dopamine in spontaneously hypertensive rats fed aspartame or sucrose. *Fed. Proc.* **46:** 904.

34) KIRITSY, P. J., MAHER, T. J. (1986). Acute effects of aspartame on systolic blood pressure in spontaneously hypertensive rats. *J. Neural Transm.* **66:** 121—128.

35) WURTMAN, R. J., MAHER, T. J. (1987). Effects of oral aspartame on plasma phenylalanine in humans and experimental rodents. *J. Neural Transm.* (in press).

MONOAMINES AND SEIZURES IN GENETICALLY EPILEPSY-PRONE RATS: INFERENCES FROM PRECURSOR AVAILABILITY

Phillip C. Jobe, Ph.D., Stephen M. Lasley, Ph.D., John W. Dailey, Ph.D.

Department of Basic Sciences
University of Illinois College of Medicine at Peoria
P. O. Box 1649
Peoria, Illinois 61656
USA

INTRODUCTION

Norepinephrine, dopamine and serotonin have the potential to modulate seizure thresholds, seizure severity and other manifestations of seizure activity in non-epileptic animals (1—6). Noradrenergic influences appear to be anticonvulsant in several animal models including the genetically epilepsy-prone rat (7), electroshock seizures in non-epileptic rats (1), cobalt-induced seizures in rats (4), and the genetically epileptic baboon (8). Moreover, dopaminergic deficiencies or noradrenergic excesses may be determinants of seizure predisposition in the DBA/2J mouse (3, 9, 7) and tottering mouse (7, 10). In addition, at least some evidence supports the concept that abnormalities in noradrenergic and/or dopaminergic transmission may contribute to the epileptic state of humans as they do in some animal modes of epilepsy (11). The characteristics which make the genetically epilepsy-prone rat (GEPR) an excellent model for certain forms of human epilepsy have been reviewed elsewhere (12—14).

Evidence for Participation of Central Noradrenergic and Serotonergic Systems in Seizure Predisposition of GEPRs

The evidence that a noradrenergic deficit is a primary determinant of seizure severity in the GEPR has been obtained from a number of detailed pharmacological studies. The results of pharmacologic investigations reveal that seizure severity in the GEPR is inversely related to brain norepinephrine concentration (15). Accordingly, excessive formation of norepinephrine by precursor loading decreases the severity of seizures. This was demonstrated in a series of experiments in which L-dopa was administered to GEPRs characterized by moderate seizures. These GEPR-2s and GEPR-3s experienced an anticonvulsant effect when the L-dopa administration resulted in an increase in brain norepinephrine and dopamine. However, when the animals were pretreated with diethyldithiocarbamate, a dopamine-beta-hydroxylase inhibitor, the L-dopa induced increment in brain norepinephrine and the anticonvulsant effect were abolished. Similarly, intracere-

NATO ASI Series, Vol. H20
Amino Acid Availability and Brain Function in Health and Disease. Edited by G. Huether

broventricular (icv) administration of dopamine produced an anticonvulsant effect in moderate seizure GEPRs and this anticonvulsant effect was prevented by pretreatment with diethyldithiocarbamate (16). Activation of beta-receptors by icv administration of dobutamine or terbutaline and activation of alpha$_1$-receptors through icv administration of phenylephrine also attenuated seizure severity in the moderate seizure animals (17). In contrast, a series of experiments which were also carried out in moderate seizure GEPRs demonstrated that drug treatments which reduce norepinephrine concentrations also enhance seizure severity in the GEPR. For example, storage vesicle inactivation with reserpine or Ro 4-1284, a reserpine-like drug, resulted in a dramatic decrease in norepinephrine and serotonin concentration and an intensification of convulsions, an effect which was reversed when the monoamines were repleted (18, 19). In this same series of experiments, the repletion of the norepinephrine was delayed and the intensification of convulsions was prolonged when the animals were pretreated with drugs that inhibit either tyrosine hydroxylase (alpha methyl-p-tyrosine) or dopamine-beta-hydroxylase (diethyldithiocarbamate). Inhibition of tyrosine hydroxylase with alpha methyl-p-tyrosine did not result in intensification of convulsions unless the animals were also cold stressed or treated with other drugs so that a marked depletion of norepinephrine resulted (18, 19). Thus, inhibition of norepinephrine synthesis was, by itself, not sufficient to intensify seizures even though substantial depletion of norepinephrine occurred. Indeed, these experiments showed that norepinephrine stores in the brain must be severely depleted and the amount of norepinephrine available for synaptic release greatly diminished, before seizures were intensified. In another study, cytotoxic destruction of noradrenergic neurons with 6-hydroxydopamine resulted in intensification of seizures in the GEPRs (20).

Although dopaminergic neurons were affected by several of the treatments described above, the time course and the magnitude of the changes in this neurotransmitter system did not correspond to changes in seizure severity. Even large increases or decreases in dopamine concentrations failed to alter seizure patterns in the GEPR. Thus, dopaminergic systems appear not to be determinants of seizure intensity in the GEPR.

Serotonergic neurons have also been shown to exert an attenuating effect on sound-induced seizures in the GEPR (15, 21). Pharmacologic approaches similar to those described above showed that drug-induced incrememts in serotonin reduce seizure severity, whereas drug-induced depletion of serotonin intensifies the convulsive response to sound.

These pharmacologic data indicate that noradrenergic and/or serotonergic activity are inversely related to seizure *severity* but they say nothing about regulation of seizure *susceptibility* in the GEPR. The susceptibility issue was addressed by administering the monoamine depleting agent Ro 4-1284 to two types of rats: 1) non-susceptible progeny of neurologically normal parents; and 2) non-susceptible progeny of epileptic parents. Only in the progeny of epileptic parents was monoamine depletion associated with an increased susceptibility to seizures (22). Thus, monoaminergic decrements represent only a part of the total set of neurochemical defects which underlie the seizure-prone state in the GEPR. Non-susceptible progeny of normal parents do not become susceptible to seizures in response to decreased brain levels of norepinephrine because they do not carry the obligatory non-monoaminergic determinants. Conversely, non-susceptible progeny of GEPR parents do experience seizures when brain norepinephrine levels are decreased because these animals do possess such traits.

Pathophysiologic evidence also suggests that noradrenergic and/or serotonergic deficits within the brain are important determinants of the epileptic state of the GEPR. In studies performed in both seizure experienced and seizure naive GEPRs, widespread deficits in norepinephrine concentration and turnover rate were found to characterize the GEPR brains. In addition, deficits in tyrosine hydroxylase activity are present in selected brain areas (23, 24).

Three types of noradrenergic abnormalities characterize different parts of the adult GEPR brain (15). These abnormalities have been identified by comparing noradrenergic indices in discrete areas of the brains of non-epileptic control rats with the same areas of moderate seizure (GEPR-3) and severe seizure (GEPR-9) GEPRs. The types of abnormalities are: 1) equal noradrenergic deficits in GEPR-3s and GEPR-9s; 2) opposing abnormalities such that noradrenergic activity in the GEPR-3 is greater than control which, in turn, is greater than GEPR-9; 3) forward order graded deficits in which noradrenergic indices are highest in controls, intermediate in GEPR-3s and lowest in GEPR-9s.

Brain areas in which noradrenergic deficits appear to be of equal magnitude in GEPR-3s and GEPR-9s include the hippocampus, hypothalamus, inferior colliculus and the remainder of the midbrain. The area with opposing abnormalities in the two types of GEPRs is the cerebellum. Forward order graded deficits exist in the telencephalon minus the hippocampus and striatum as well as in the thalamus and the pons/medulla. Noradrenergic normality appears to characterize the striatum.

The different noradrenergic conditions in the various brain areas have led to a series of hypotheses regarding the roles of this neurotransmitter system in seizure regulation (23). Both GEPR-3s and GEPR-9s are susceptible to seizures precipitated by stimuli that do not trigger seizures in nonepileptic control rats and both share a common noradrenergic deficit in several brain areas. These commonalities suggested that noradrenergic deficits may be a primary determinant of seizure susceptibility.

The opposing noradrenergic abnormality in GEPR-3s and GEPR-9s occurs in the cerebellum. A role in regulating seizure severity is suggested by this result. Perhaps the noradrenergic increment in GEPR-3s prevents the moderate-seizure animals from experiencing the more severe tonic extensor convulsion characteristic of GEPR-9s.

Areas with forward order graded deficits may regulate seizure severity as well as seizure susceptibility. Deficits in these areas could determine susceptibility because both GEPR-3s and GEPR-9s share the deficit. These deficits could also determine seizure severity because they are greater in GEPR-9s than in GEPR-3s.

Innate serotonergic abnormalities in GEPR brains are in some instances similar and in other instances different from those identified for the noradrenergic systems. Based on our studies of serotonin only one area — the striatum — (15) appears to be a candidate for serotonergic regulation of seizure severity. Susceptibility but not severity may be determined by serotonergic abnormalities in the telencephalon (minus hippocampus), thalamus, hypothalamus, midbrain and the pons/medulla.

Effect of Precursor Availability on Seizures in Genetically Epilepsy-Prone Rats

The preceding overview summarizes the data that indicate that the GEPR is an effective model to study the effects on seizures of perturbations of central nervous system catecholaminergic and serotonergic neurotransmitter systems. GEPRs have widespread, inherent deficits in noradrenergic and serotonergic neurotransmission and drug-induced disturbances in these systems produce predictable changes in audiogenic seizure severity. Decreases in synaptic norepinephrine and/or serotonin produce increases in seizure severity whereas increases in synaptic norepinephrine decrease seizure severity in GEPRs.

Recently, concern has been expressed that the dipeptide sweetener, aspartame, might lower seizure thresholds in man and animals by interfering with monoamine synthesis in brain (25, 26). The hypothesized mechanism for this lowering of seizure threshold is based on the existence of a common transport system in brain for large neutral amino acids, and on the fact that aspartame is converted to aspartate and phenylalanine in the gastrointestinal tract so that these two amino acids are absorbed in large amounts. These facts led to the suggestion that plasma phenylalanine, which could be increased as a result of aspartame ingestion, might compete with tyrosine and/or tryptophan for entry into the central nervous system, thereby reducing the availability of precursors for norepinephrine and/or serotonin synthesis. In addition it was proposed that since tyrosine, and not phenylalanine, is the primary substrate for catecholamine synthesis in brain, an increase in the quantity of phenylalanine relative to tyrosine may result in competition for tyrosine hydroxylase and reduced synthesis of norepinephrine. Such reductions have been hypothesized to be responsible for the alleged lowering of seizure thresholds in individuals consuming large quantities of aspartame (25, 26).

TABLE 1
Effect of aspartame on audiogenic seizures in GEPRs[a]

Dose of Aspartame (Mg/kg PO)	Seizure Severity Score	
	GEPR-3s	GEPR-9s
0	3.2 (10) ±0.2	9.0 (10) ±0.0
50	3.0 (10) ±0.0	9.0 (10) ±0.0
100	3.0 (10) ±0.0	9.0 (12) ±0.0
250	3.0 (10) ±0.0	9.0 (12) ±0.0
500	3.0 (11) ±0.0	9.0 (11) ±0.0
1000	1.3 (10) ±0.0	9.0 (11) ±0.0
1330	3.2 (11) ±0.2	9.0 (11) ±0.0
2000	2.9 (10) ±0.1	9.0 (10) ±0.0

[a] Each animal was pre-tested three times at weekly intervals to insure consistency of response. Aspartame, in a methylcellulose vehicle, was given by gavage 1 hr before a fourth test. There were no significant changes in seizure score (Scheffe procedure). Number of animals in parenthesis.

In order to test this hypothesis, GEPRs were given a wide range of doses of aspartame by gavage and subjected to the standardized sound stimulus which is routinely used to elicit the phenotypic expression of their genetically determined seizure response (27, 19). The results of one such experiment are shown in Table 1.

The results shown in Table 1 suggest that even very large doses of aspartame do not perturb brain monoamines in a manner sufficient to alter seizure severity in the GEPR. This lack of effect may be the result of the fact that the monoaminergic neurotransmitter systems are steady state systems in which a large excess of neurotransmitter is stored within the nerve terminals. Because of the existence of this large excess of neurotransmitter, substantial inhibition of neurotransmitter synthesis must occur for a prolonged period before depletion of the transmitter reserves are sufficient to interfere with synaptic transmission. This observation is confirmed by the results of earlier experiments with GEPRs (18, 19). In these earlier experiments, the tyrosine hydroxylase inhibitor, alpha methyl-p-tyrosine, was administered to GEPRs in several different experimental paradigms. In no case was alpha methyl-p-tyrosine administration sufficient to enhance seizures unless some other drug or procedure such as cold stress was also employed in order to enhance depletion of the monoamines. In these experiments, marked norepinephrine depletion was required before enhancement of seizures took place. If, as hypothesized by Wurtman (1985), phenylalanine does act as a competitive inhibitor of tyrosine hydroxylase, its effects must be too weak to cause a reduction in norepinephrine of sufficient magnitude to enhance seizures in the GEPR.

In another experiment, a wide range of aspartame doses were administered to non-epileptic control rats and to GEPR-3s and GEPR-9s. One hour later, the animals were subjected to supramaximal electroshock. The results of this experiment are shown in Table 2.

TABLE 2

Effect of a single acute dose of aspartame on supramaximal electroshock in GEPRs and non-epileptic control rats: extension/flexion ratios for hindlimbs[a]

	Aspartame Dose (mg/kg)							
	2 ml/kg 0	50	100	250	500	1000	4 ml/kg 0	2000
	Mean Extension/Flexion Ratios ± SEM							
Control	1.53(5) ± 0.59	1.28(7) ± 0.23	2.29(7) ± 0.96	1.77(2) ± 1.04	1.48(7) ± 0.24	1.75(6) ± 0.43	1.58(9) ± 0.31	1.98(9) ± 0.39
GEPR-3	3.90(7) ± 0.35	3.47(9) ± 0.25	4.21(10) ± 0.28	3.83(9) ± 0.35	4.19(8) ± 0.21	4.20(10) ± 0.21	3.41(9) ± 0.27	4.28(9) ± 0.38
GEPR-9	32.98(9) ± 3.78	35.75(10) ± 4.23	34.62(10) ± 3.30	27.70(10) ± 2.87	30.50(10) ± 2.37±	29.00(10) ± 1.87	33.02(8) ± 2.20	29.20(10) ± 1.57

[a] Extension/flexion ratos were determined 1 hr after vehicle or aspartame gavage. Numbers in parentheses are numbers of animals that exhibited tonic extension. Doses from 50 to 1000 mg/kg were in 2 ml of vehicle. The 2000 mg/kg dose was in 4 ml of vehicle. Statistical comparisons(Scheffe's procedure for multiple doses, t-test for the single dose) were made with the appropriate vehicle control. There were no statistically significant differences between the vehicle and aspartame-treated groups.

The results of this experiment show that none of the aspartame doses altered the extension/flexion ratio within any of the groups. That is, aspartame did not alter the extension/flexion ratio in non-epileptic controls or in either of the types of epileptic animals. Neither did aspartame alter the fraction of animals that exhibited tonic extension in response to the electroshock procedure. However, the data do further indicate the nature and extent of the seizure-prone state of the GEPRs. As can be seen from the table, the extension/flexion ratio is, at each dose, greater in GEPR-3s than in controls and is substantially greater in GEPR-9s than in controls or GEPR-3s.

SUMMARY AND CONCLUSIONS

The Genetically Epilepsy-Prone Rat or GEPR is a useful model for studies of epilepsy and of the neurochemistry of the seizure-prone state. These animals are characterized by widespread deficits in central nervous system noradrenergic and serotonergic transmission, and drug-induced perturbations of these neurotransmitter systems produce predictable changes in seizure severity. Drugs or procedures which augment noradrenergic and/or serotonergic transmission are anticonvulsant whereas drugs or procedures which substantially diminish neurotransmission in these systems produce pro-convulsant effects.

Aspartame, given over a large range of oral doses (50—2000 mg/kg) produced no change in audiogenic seizure severity in either of two types of GEPRs. A similar dosage schedule failed to alter the extension/flexion ratio of electroshock seizures in non-epileptic controls or in either of the two types of GEPRs.

These initial investigations using the GEPR as an experimental model suggest that use of aspartame has little or no risk of facilitating convulsions in subjects with an innate predisposition to seizures coupled with innate noradrenergic and serotonergic deficits.

ACKNOWLEDGEMENTS

Neurochemical studies reviewed in this paper were partially supported by grants from the Veterans Administration (MRIS 4725) and the National Institute of Health (NS 16829). Recent work with aspartame has been supported by a grant from the NutraSweet Company.

REFERENCES

1) BROWNING, R. A. (1987). The role of neurotransmitters in electroshock seizure models. In: *Neurotransmitters and Epilepsy* (Jobe, P. C., Laird, H. E., eds., II.). Humana Press, Clifton, N. J., pp. 277—320.

2) BURLEY, E. S., FERRENDELLI, J. A. (1984). Regulatory effects of neurotransmitters on electroshock and pentylenetetrazol seizures. *Fed. Proc.* **43:** 2521—2524.

3) CHAPMAN, A. G., MELDRUM, B. S. (1987). Epilepsy prone mice: Genetically-determined sound-induced seizures. In: *Neurotransmitters and Epilepsy* (Jobe, P. C., Laird, H. E., eds., II.). Humana Press, Clifton, N. J., pp. 9—40.

4) CRAIG, C. R., COLASANTI, B. K. (1987). Experimental epilepsy induced by direct topical placement of chemical agents on the cerebral cortex. In: *Neurotransmitters and Epilepsy* (Jobe, P. C., Laire, H. E., eds., II.). Humana Press, Clifton, N. J., pp. 191—214.

5) JOBE, P. C. (1987). Spinal Seizures Induced by electrical stimulation. In: *Epilepsy and the reticular formation* (Burnham, W. M., Browning, R., Faingold, C., Fromm, G., eds.). New York, N. Y.: Alan R. Liss, pp. 81—91.

6) LAIRD, H. E., JOBE, P. C. (1987). The genetically epilepsy-prone rat. In: *Neurotransmitters and Epilepsy* (Jobe, P. C., Laired, H. E., eds., III.). Humana Press, Clifton, N. J., pp. 57—89.

7) JOBE, P. C., LAIRD, H. E. (1987). Neurotransmitter systems and the epilepsy models: Distinguishing features and unifying principles. In: *Neurotransmitters and Epilepsy* (Jobe, P. C., Laird, H. E., eds., II.). Humana Press, Clifton, N. J., pp. 339—366.

8) KILLAM, E. K., KILLAM, K. F. (1984). Evidence for neurotransmitter abnormalities related to seizure activity in the epileptic baboon. *Fed. Proc.* **43 (10):** 2510—2515.

9) DAILEY, J. W., JOBE, P. C. (1984). Effect of increments in the concentration of dopamine in the central nervous system on audiogenic seizures in DBA/2J mice. *Neuropharmacology* **23 (9):** 1019—1024.

10) NOBELS, J. L. (1984). A single gene error of noradrenergic axon growth synchronizes contral neurons. *Nature* **310:** 409—411.

11) KRESCH, M. J., SHAYWITZ, B. A., SHAYWITZ, S. B., ANDERSON, G. M., LECKMAN, J. L., COHEN, D. J. (1987). Neurotransmitters in human epilepsy. In: *Neurotransmitters and Epilepsy* (Jobe, P. C., Laird, H. E., eds., II.). Humana Press, Clifton, N. J., pp. 321—335.

12) JOBE, P. C. (1981). Pharmacology of audiogenic seizures. In: *Textbook, Pharmacology of Hearing Experimental and Clinical Bases* (Brown, R. D., Daigeneault, E. A., eds.). Wiley Interscience (A Division of John Wiley and Sons), pp. 271—304.

13) JOBE, P. C., LAIRD, H. E. (1981). Neurotransmitter abnormalities as determinants of seizure susceptibiliy and intensity in the genetic models of epilepsy. *Biochem. Pharmacol.* **30:** 3137—3144.

14) REIGEL, C. E., DAILEY, J. W., JOBE, P. C. (1986). The genetically epilepsy-prone rat: an overview of seizure-prone characteristics and responsiveness to anticonvulsant drugs. *Life Sci.* **39:** 763—774.

15) JOBE, P. C., DAILEY, J. W., REIGEL, C. E. (1986). Noradrenergic and serotonergic determinants of seizure susceptibility and severity in genetically epilepsy-prone rats. *Life Sci.* **39:** 775—782.

16) KO, K. H., DAILEY, J. W., JOBE, P. C. (1982). Effect of increments in norepinephrine concentrations on seizure intensity in the genetically epilepsy prone rat. *J. Pharm. Exp. Therap.* **222:** 662—669.

17) KO, K. H., DAILEY, J. W., JOBE, P. C. (1984). Evaluation of monoaminergic receptors in the genetically epilepsy prone rat. *Experentia* **40:** 70—73.

18) JOBE, P. C. (1970). Relationship of brain amine metabolism to audiogenic seizure in the rat. *Dissertation,* The University of Arizona.

19) JOBE, P. C., PICCHIONI, A. L., CHIN, L. (1973). Effect of the dopamine receptor stimulation and blockade on Ro 4-1284-induced enhancement of electroshock seizure. *J. Pharm. Pharmacol.* **25:** 830—831.

20) BOURN, W. M., CHIN, L., PICCHIONI, A. L. (1972). Enhancement of audiogenic seizure by 6-hydroxydopamine. *J. Pharm. Pharmacol.* **24:** 913—914.

21) LAIRD II., H. E. (1974). 5-Hydroxytryptamine as an inhibiting modulator of audiogenic seizures. *Doctoral dissertation.* The University of Arizona.

22) JOBE, P. C., BROWN, R. D., DAILEY, J. W. (1981). Effect of Ro 4-1284 on audiogenic seizure susceptibility and intensity in epilepsy-prone rats. *Life Sci.* **28:** 2031—2038.

23) DAILEY, J. W., JOBE, P. C. (1986). Indices of noradrenergic function in the central nervous system of seizure-naive genetically epilepsy-prone rats. *Epilepsia* **27 (6):** 665—670.

24) JOBE, P. C., KO, K. H., DAILEY, J. W. (1984). Abnormalities in norepinephrine turnover rate in the central nervous system of the genetically epilepsy-prone rat. *Brain Res.* **290:** 357—360.

25) MAHER, T. J. (1986). Neurotoxicology of food additives. *Neurotoxicol.* **7:** 183—196.

26) WURTMAN, R. J. (1985). Aspartame: Possible effect on seizure susceptibility (Letter to the editor). *Lancet* **II:** 1060.

27) DAILEY, J. W., LASLEY, S. M., FRASCA, J., JOBE, P. C. (1987). Aspartame (ASM) is not proconvulsant in the Genetically Epilepsy Prone Rat (GEPR). *The Pharmacologist*, p. 202.

REGULATION OF NEUROTRANSMITTER AMINO ACID FLUXES BY EXCITATORY AMINO ACID RECEPTORS IN DIFFERENT NEURAL CELL TYPES IN CULTURE

V. Gallo, C. Giovannini and G. Levi

Laboratory of Organ and System Pathophysiology
Istituto Superiore di Sanità
Viale Regina Elena 299
00161 Rome
Italy

INTRODUCTION

Glutamic acid is considered as the major excitatory neurotransmitter in the central nervous system (1). Its action is mediated by the activation of specific receptors which have been classified into three different subtypes, N-methyl-D-aspartate (NMDA), kainate (KA) and quisqualate (QA), according to the agonist which preferentially acts upon each of them (2). Due to the lack of specific antagonists for KA and QA, excitatory amino acid receptors are generally identified as NMDA or non-NMDA receptors.

In the cerebellar cortex, the granule cells represent the only excitatory interneurones and use glutamate as a neurotransmitter (3, 4). Granule cells themselves receive an excitatory input from the mossy fibre terminals, some of which have been recently suggested to be also glutamatergic (5). The definitive identification of an acidic amino acid as one of the neurotransmitters released from the mossy fibre terminals requires a study of the physiological effects produced by different excitatory amino acids and their structural analogues on granule cells and, possibly, the identification of those receptors whose activation can cause depolarization of the granule cells with subsequent release of their neurotransmitter.

Differentiation of Cerebellar Granule Cells in Culture and Expression of Excitatory Amino Acid Receptors

The cellular and structural complexity of the nervous system represents a drawback for studying the physiology of a given neuronal cell population. Therefore, we decided to use primary cultures greatly enriched in cerebellar granule cells to study whether the release of their transmitter, glutamate, is regulated by the activation of those receptors which, "in vivo", are presumably the sites of action of the neurotransmitter(s) released by the mossy fibre terminals.

NATO ASI Series, Vol. H20
Amino Acid Availability and Brain Function in Health and Disease. Edited by G. Huether

Granule cell cultures were prepared from 8-day postnatal rat cerebella as previously described (6) and were characterized immunocytochemically (7—9). The cultures comprised more than 95 % of granule cells, about 2 % of GABAergic interneurones and less than 2 % of astrocytes. Granule cells in culture differentiated morphologically by emitting long neurites, along which several varicosities filled with synaptic vescicles were present

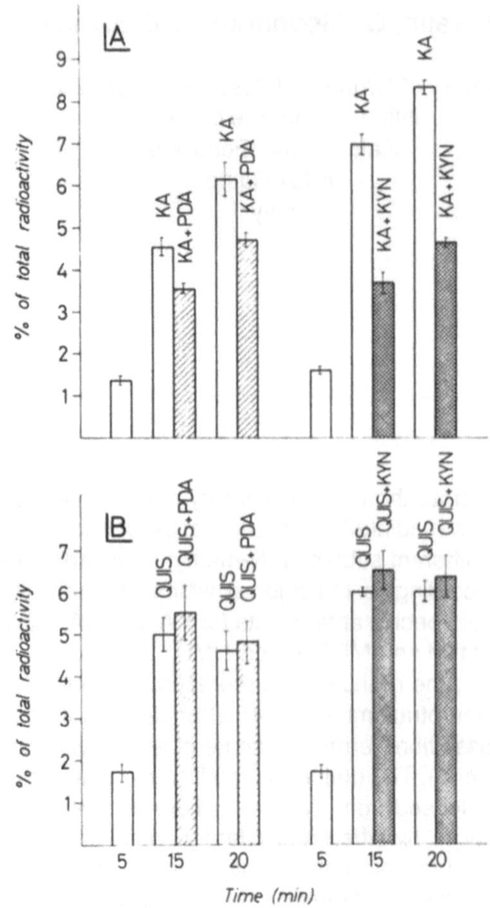

FIGURE 1

Effect of kainic acid and of quisqualic acid on ^{3}H-D-aspartate release from cerebellar granule cell cultures. Cells were cultured for 11 days, labelled with 0.5 μM ^{3}H-D-aspartic acid for 10 min and then subjected to 5 min washes with standard Krebs-Ringer medium to study ^{3}H-D-aspartate release. Kainic acid (50 μM), quisqualic acid (50 μM), PDA (100 μM) and kynurenic acid (100 μM) were present in the release medium from min 15 to min 25. The radioactivity recovered in each 5 min wash was expressed as a percent of that present in the cells at the beginning of the release experiment. Averages of 3 duplicate experiments are presented. Bars indicate S.E.M. Differences between kainate-induced ^{3}H-D-aspartate release in the absence or in the presence of kynurenic acid and PDA were statistically significant ($p < 0.001$ and $p < 0.05$ for kynurenic acid and PDA, respevtively) (13).

(7). "In vitro" granule cells progressively acquired a glutamatergic phenotype: after a week in culture they were able to synthesize glutamic acid from the direct metabolic precursor glutamine (3) and to release it in a Ca^{2+}-dependent way in response to depolarizing stimuli, such as high K^+ or veratridine (3). Furthermore, glutaminase activity was much higher in granule cell than in astrocyte cultures (10). Cultured granule cells also expressed a number of neurone-specific proteins, such as neurofilament proteins (Gallo and Levi, unpublished), the D_2 cell adhesion molecule (11) and synapsin I (9).

We have previously shown that cerebellar granule cells, after 5 days "in vitro" (DIV), were able to accumulate the radioactive, non metabolized glutamate analogue ^{3}H-D-aspartate, which utilizes the same membrane transport system as L-glutamate acid (12). When granule cell cultures maintained "in vitro" for 10 days were prelabelled with ^{3}H-D-aspartate and then exposed to micromolar concentrations of kainic acid, the excitatory amino acid receptor agonist caused a marked increase in the release of the labelled amino acid (Fig. 1A). The stimulatory effect of kainic acid was significantly reduced by the non-specific excitatory amino acid receptor antagonists cis-2,3-piperidine dicarboxylic acid (PDA) and kynurenic acid (Fig. 1A). Kynurenic acid was the more effective of the two antagonists and, at a concentration of 100 μM, inhibited by about 50 % the release of ^{3}H-K-aspartate induced by 50 μM kainate (Fig. 1A). Interestingly, the effect of kynurenic acid was concentration-dependent in the range of 50—200 μM, whereas that of PDA levelled off at 100 μM (13).

Of the other glutamate analogues (N-methyl-D-aspartate, quisqualic acid, dihidrokainic acid and quinolinic acid) tested at micromolar concentrations, only quisqualate produced a significant increase in ^{3}H-D-aspartate release from cerebellar granule cells (Fig. 1B). The effect of quisqualic acid was not counteracted by PDA or by kynurenic acid, nor by other excitatory amino acid receptor antagonists (glutamic acid diethylester, γ-D-glutamyglycine). Kainic acid and quisqualic acid significantly stimulated also the release of endogenous glutamate from granule cells and, also in this case, the effect of kainate was selectively blocked by kynurenic acid (13). Kainic acid and quisqualic acid showed a different dose-response curve on ^{3}H-D-aspartate release (Fig. 2). In fact, quisqualic acid caused a significant increase in ^{3}H-D-aspartate release at concentrations (5—10 μM) at which kainate was ineffective; maximal releasing activities were observed with 50—100 μM quisqualic acid and 100—200 μM kainic acid (Fig. 2) and, in the high dose range, kainic acid was much more effective than quisqualic acid (Fig. 2). Finally, the effects of kainic acid and quisqualic acid on ^{3}H-D-aspartate release exhibited different ionic requirements: in both cases the releasing effect was Na^+-independent, but, the effect of kainate was partially dependent on extracellular Ca^{2+}, whereas that of quisqualate was Ca^{2+}-independent (13). Interestingly, the endogenous agonist glutamate was also capable of enhancing ^{3}H-D-aspartate release. Part of this releasing effect could be accounted for by a heteroexchange process, and was abolished in the absence of Na^+. A substantial component, however, was receptor-mediated and was antagonized by kynurenic acid. The receptor involved seems, therefore, of the kainate type (14).

Excitatory Amino Acid Receptors in Cultured Astrocytes

Astrocytes have been shown to express receptors for a variety of neurotransmitters (15—17), including neuropeptides (18). We have used cerebellar primary cultures enriched in astrocytes to verify the possibility that these cells also expressed receptors for excitatory amino acids.

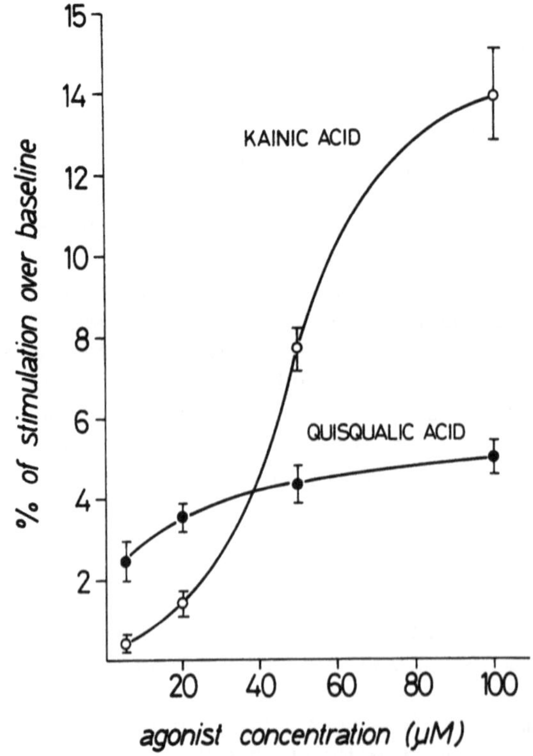

FIGURE 2
Dose-response curves for kainic acid and quisqualic acid-induced release of ^{3}H-D-aspartate from cerebellar granule cell cultures. Cells were cultured for 11 days and ^{3}H-D-aspartate release was studied as described in the legend for Fig. 1. The radioactivity recovered in each 5 min release fraction was expressed as a percent of that present in the cells at the beginning of the release experiment. Each point represents the percentage of stimulation of ^{3}H-D-aspartate release in the peak fraction (min 25) over the baseline release (min 10). Bars represent S.E.M. and the curves are obtained from 4 experiments run in duplicate.

At early stages "in vitro" (2—6 DIV) cerebellar astrocyte-enriched cultures comprised two morphologically distinct subpopulations of glial fibrillary acidic protein (GFAP)-positive cells, which could be distinguished also on the basis of their antigenic and functional features. In fact, the astrocytes characterized by a "stellate" morphology differed from the polygonal-epithelioid astrocytes for their capacity to bind the two monoclonal antibodies A2B5 and LB1 and for their ability to accumulate ^{3}H-GABA through a high-affinity transport system (19, 20). The structural specificity of this GABA uptake system was similar to that expressed by cultured neurones (19) and oligodendrocytes (21). At late stages in culture (8—10 DIV) only epithelioid GFAP-positive cells survived and actively proliferated. These cells showed a very low capacity of accumulating ^{3}H-GABA.

Micromolar concentrations of kainic acid and quisqualic acid stimulated ^{3}H-GABA release from cerebellar astrocyte cultures at 5 DIV (Fig. 3). The effect of kainic acid was greatly reduced by kynurenic acid, whereas PDA was ineffective (Fig. 3A), that of quisqualate was not antagonized by kynurenic acid, PDA, GDEE nor γ-DGG (Fig. 3B and not shown). The antagonistic effect of kynurenic acid on kainate-induced ^{3}H-GABA release was concentration-dependent: 50 μM kynurenic acid reduced by 50 % the effect of 50 μM kainic acid and a concentration of the antagonist of 200 μM blocked almost 80 % of the effect produced by 50 μM kainate (22, 23). Kainic acid was ineffective when tested on 12 DIV astrocyte cultures (22, 24), suggesting that stellate astrocytes were the main target for the glutamate analogue. NMDA and dihydrokainic acid did not affect ^{3}H-GABA release from 5 DIV astrocyte cultures, but glutamic acid had a strong releasing action which, similarly to that of kainate, was largely antagonized by kynurenic acid (23).

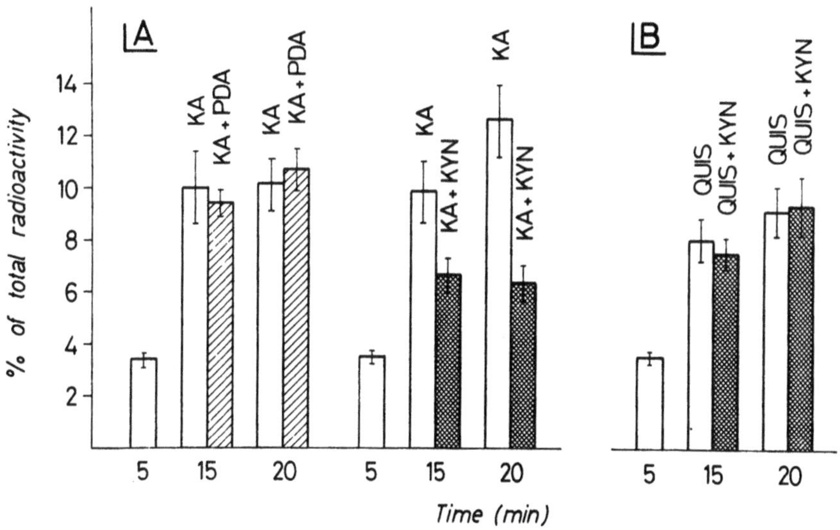

FIGURE 3

Effect of kainic acid and of quisqualic acid on ^{3}H-GABA release from cerebellar astrocyte cultures. Cells cultured for 5 days were incubated for 15 min in Krebs-Ringer containing 10 μM aminooxyacetic acid and 0.5 μM ^{3}H-GABA and were then used to study ^{3}H-GABA release. Kainic acid (50 μM), quisqualic acid (50 μM), kynurenic acid and PDA (both 100 μM) were applied for min 15 to min 25. Results are expressed as in Fig. 1. Averages of 3 duplicate experiments ± S.E.M. are presented. Kynurenic acid showed a significant inhibition ($p < 0.05$) of kainate-induced ^{3}H-GABA release (27).

The releasing effect of kainic acid was abolished when NaCl was totally replaced by sucrose in the incubation medium (Fig. 4A), and was largely maintained when NaCl was replaced by Na^+-isethionate (Fig. 4A). Furthermore, when the concentration of Ca^{2+}-ions was reduced to 0.2 mM and that of Mg^{2+} raised to 10 mM, the stimulation of ^{3}H-GABA release induced by kainic acid decreased by about 40 % (Fig. 4B).

Quisqualic acid (5—100 μM) also stimulated ^{3}H-GABA release from astrocyte cultures at 5 DIV (Fig. 5A). Its effect was concentration-dependent and, when tested in sister cultures, was always greater than that of kainate in the low concentration range (5—20 μM). At higher concentrations, the effect of quisqualate levelled off, whereas that of kainate continued to increase (Fig. 5A and Ref. 25). Similarly to kainate, the stimulatory effect of quisqualic acid on ^{3}H-GABA release was totally Na^+-dependent (Fig. 5B). However, the reduction of Ca^{2+} and the elevation of Mg^{2+} concentrations enhanced, rather than reduced, its ^{3}H-GABA-releasing effect (Fig. 5B).

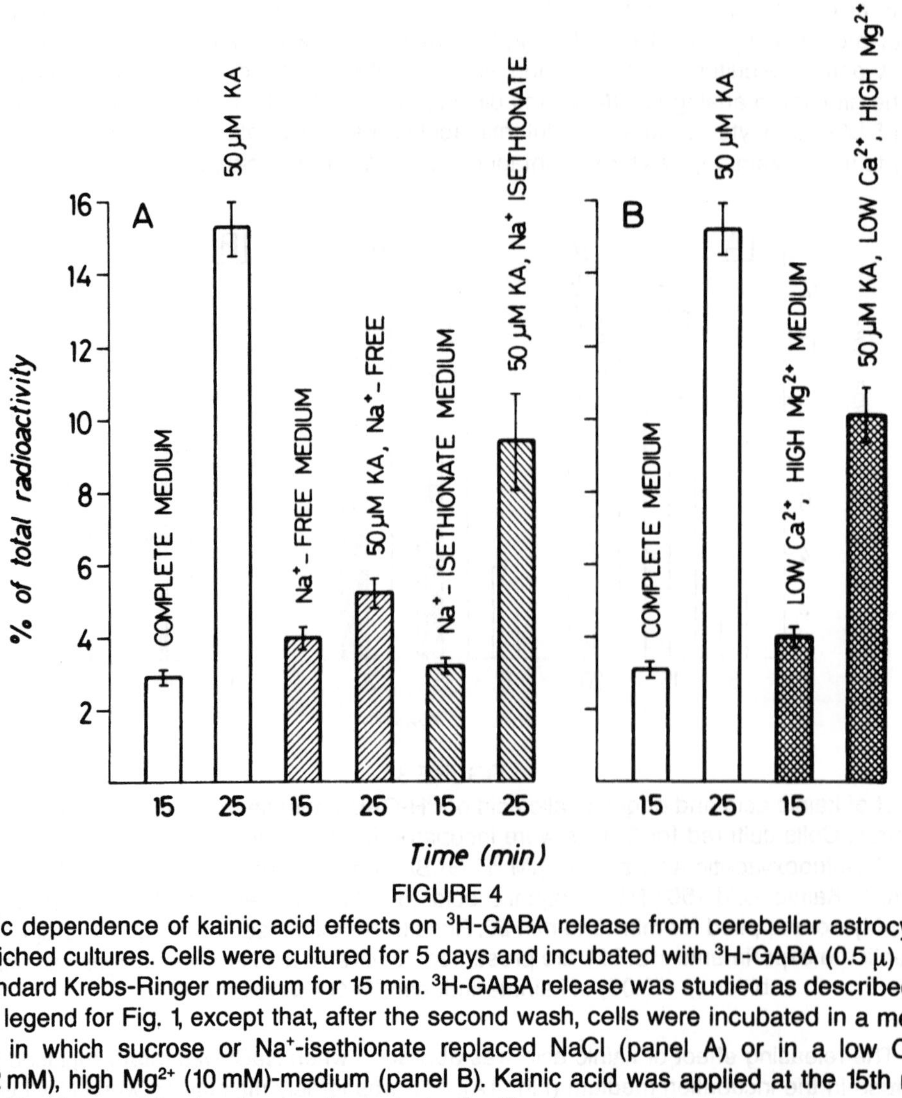

FIGURE 4

Ionic dependence of kainic acid effects on ^{3}H-GABA release from cerebellar astrocyte-enriched cultures. Cells were cultured for 5 days and incubated with ^{3}H-GABA (0.5 µ) in a standard Krebs-Ringer medium for 15 min. ^{3}H-GABA release was studied as described in the legend for Fig. 1, except that, after the second wash, cells were incubated in a medium in which sucrose or Na^+-isethionate replaced NaCl (panel A) or in a low Ca^{2+} (0.2 mM), high Mg^{2+} (10 mM)-medium (panel B). Kainic acid was applied at the 15th min for 10 minutes. The fractions collected at min 15 correspond to the baseline release, whereas those at min 25 correspond to the peak of the evoked release. Data are expressed as in Fig. 1. Averages of 4 duplicate experiments ± S.E.M. are presented.

The effect of kainate on ^{3}H-GABA accumulation into stellate astrocytes was not due to a direct interference of the glutamate analogue with the GABA transport system in these cells. In fact, neither kainate nor quisqualate inhibited the initial rate of uptake of ^{3}H-GABA into cultured astrocytes, but both agonists inhibited the accumulation of ^{3}H-GABA

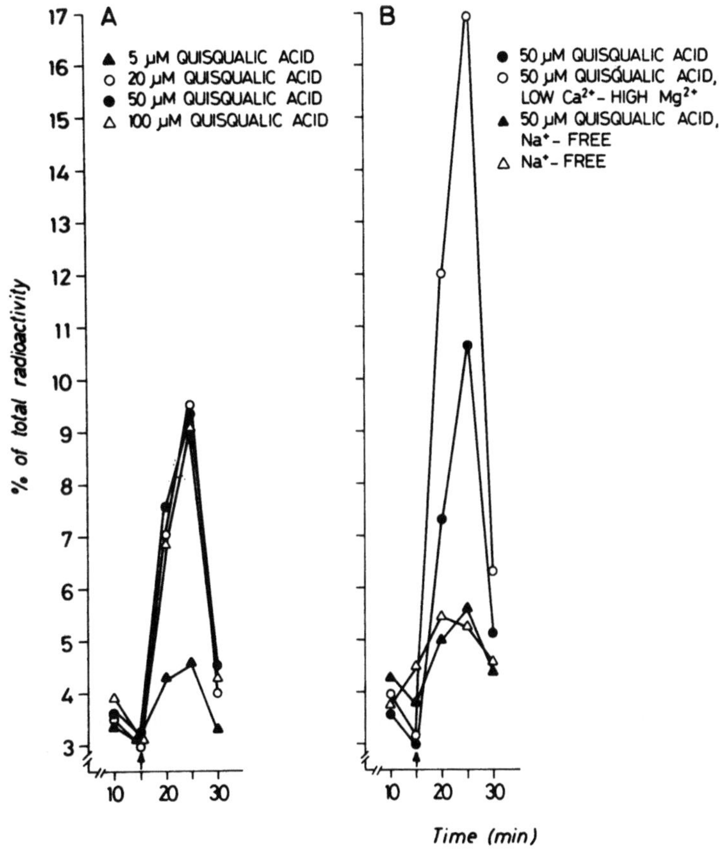

FIGURE 5

Effects of quisqualic acid on ^{3}H-GABA release from cerebellar astrocyte cultures. Cells were cultured for 5 days, incubated with 0.5 µM H-GABA for 15 min and then used to study ^{3}H-GABA release. During the release phase, after the second wash, the cells used for the experiments presented in panel B were incubated in a medium in which sucrose replaced NaCl (Na^+-free medium) or in a medium in which the concentration of Ca^{2+} was lowered to 0.2 mM and that of Mg raised to 10 mM (Low Ca^{2+}, high Mg^{2+} medium). Quisqualic acid and kynurenic acid were applied from the time indicated by the arrow for 10 min. Averages of 4 duplicate experiments, differing less than 20 % are presented.

in longer incubations (10 min) (23, 24). This inhibition can, therefore, be explained as an indirect effect due to the large ^{3}H-GABA releasing action of the two drugs. The ^{3}H-GABA depleting effect of kainic acid was confirmed by light microscopy autoradiography. When cultured astrocytes, at 5 days "in vitro", were incubated for 10 minutes with ^{3}H-GABA in the presence of 20—100 µM kainic acid, the radioactivity accumulated by the stellate astrocytes was drastically decreased (Fig. 6). Also in this case the effect of kainic acid appeared to be concentration-dependent (not shown).

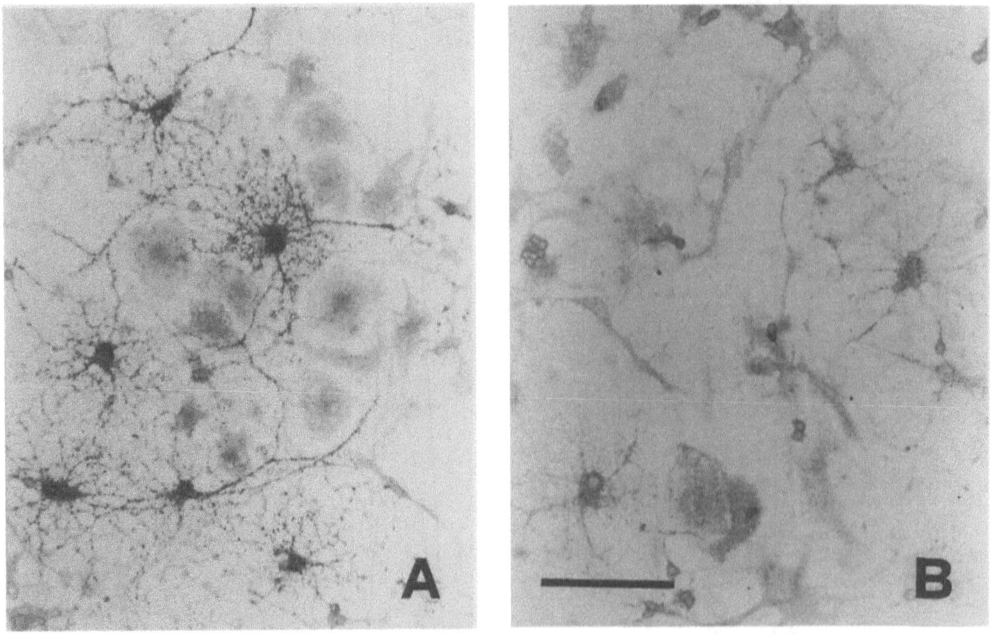

FIGURE 6
Effect of kainic acid on the autoradiographic accumulation of ^{3}H-GABA by cerebellar astrocyte enriched cultures. Astrocyte cultures at 5 DIV were incubated with 3 μCi/ml of ^{3}H-GABA in the absence (A) or in the presence (B) of 50 μM kainic acid, and then processed for autoradiography. The radioactivity accumulated by type-2 astrocytes decreased when kainic acid was present during the labelling period. Scale bar = 100 μm.

CONCLUSIONS

Excitatory amino acid receptor agonists have been described to increase Ca^{2+}-influx (25), cGMP levels (26) and phosphatidyl inositol turnover (27) in cultured cerebellar granule cells. Here we have shown that kainic acid and quisqualic acid activate receptors coupled to glutamate release from granule cells. The mechanism underlying the enhancement of glutamate release is, so far, unknown. The Na^{+}-independence of the releasing action of excitatory amino acid receptor agonists seems to exclude the involvement of Na^{+}-mediated depolarization. We are currently trying to determine what is the relationship (if any) between stimulation of second messenger systems and activation of neurotransmitter release. Our study does not provide any direct clue on the issue of the localization (pre- or postsynaptic) of kainate and quisqualate receptors in cerebellar granule cells. However, if the presence of excitatory amino acid receptors coupled to glutamate release in cerebellar granule cells were related to the excitatory innervation that these cells receive "in vivo" from the mossy fibres, the receptors described in the present study would appear to be postsynaptic.

Kainate and quisqualate receptors subtypes coupled to neurotransmitter amino acid release were present not only in neurones, but also in a subpopulation of astrocytes. Interestingly, the "glial" receptors differed from the "neuronal" receptors, both in terms of Na^+-dependence and sensitivity to the antagonist PDA. The difference in ionic dependence might be due to the activation of different receptor-gated channels in neurones or in astrocytes. Electrophysiological studies performed on cultured astrocytes have shown that kainic acid and glutamic acid cause a Na^+-dependent depolarization of the cells (15). Neurotransmitter release might be related to this depolarizing action. The presence of excitatory amino acid receptors on the membrane of astroglial cells may suggest that acidic amino acids exert a modulatory role in the central nervous system mediated by these cells.

REFERENCES

1) CURTIS, D. R., JOHNSTON, G. A. R. (1974). Amino acid transmitters in the mammalian central nervous system. *Ergebn. Physiol.* **69:** 97—188.

2) FAGG, G. E., FOSTER, A. C., GANONG, A. H. (1986). Excitatory amino acid synaptic mechanisms and neurological function. *Trends Pharmacol. Sci.* **9:** 357—363.

3) GALLO, V., CIOTTI, M. T., COLETTI, A., ALOISI, F., LEVI, G. (1982). Selective release of glutamate from cerebellar granule cells differentiating in culture. *Proc. Natl. Acad. Sci. (USA)* **79:** 7919—7923.

4) LEVI, G., ALOISI, F., CIOTTI, M. T., GALLO, V. (1984). Autoradiographic localization and depolarization-induced release of acidic amino acids in differentiating cerebellar granule cell cultures. *Brain Res.* **290:** 77—86.

5) SOMOGY, P., HALASY, K., SOMOGY, J., STORM-MATHIESEN, J., OTTERSEN, O. P. (1986). Quantification of immunogold labelling reveals enrichment of glutamate in mossy and parallel fibre terminals in cat cerebellum. *Neuroscience* **19:** 1045—1050.

6) THANGNIPON, W., KINGSBURY, A., WEBB, M., BALAZS, R. (1983). Observations on rat cerebellar cells in vitro: influence of substratum, potassium concentration and relationship between neurones and astrocytes. *Brain Res.* **11:** 177—189.

7) KINGSBURY, A. E., GALLO, V., WOODHAMS, P. L., BALAZS, R. (1985). Survival, morphology and adhesion properties of cerebellar interneurones cultured in chemically defined and serum-supplemented medium. *Dev. Brain Res.* **17:** 17—25.

8) ALOISI, F., CIOTTI, M. T., LEVI, G. (1985). Characterization of GABAergic neurones in cerebellar primary cultures and selective neurotoxic effects of a serum fraction. *J. Neurosci.* **5:** 2001—2008.

9) GALLO, V., CIOTTI, M. T., ALOISI, F., LEVI, G. (1986). Developmental features of rat cerebellar neural cells cultured in a chemically defined medium. *J. Neurosci. Res.* **15:** 289—301.

10) PATEL, A. J., HUNT, A., GORDON, R. D., BALAZS, R. (1982). The activities in different neural cell types of certain enzymes associated with the metabolic compartmentation of glutamate. *Dev. Brain Res.* **4:** 3—11.

11) MEIER, E., REGAN, C. M., BALAZS, R. (1984). Changes in the expression of a neuronal surface protein during development of cerebellar neurones "in vivo" and in culture. *J. Neurochem.* **43:** 1328—1335.

12) BALCAR, V. J., JOHNSTON, G. A. R. (1972). The structural specificity of the high affinity uptake of L-glutamate and L-aspartate by rat brain slices. *J. Neurochem.* **19:** 2657—2666.

13) GALLO, V., SUERGIU, R., GIOVANNINI, C., LEVI, G. (1987). Glutamate receptor subtypes in cultured cerebellar neurones: modulation of glutamate and GABA release. *J. Neurochem.*, in press.

14) LEVI, G., GALLO, V., GIOVANNINI, C., SUERGIU, R.. Modulation of glutamate and GABA release by excitatory amino acid receptor agonists in cultured cerebellar cells. In: *Modulation of Synaptic Transmission and Plasticity in Nervous System* (Spatz, H. C., et al., eds.). Springer Verlag, in press.

15) BOWMAN, S., KIMELBERG, H. K. (1984). Excitatory amino acids directly depolarize rat brain astrocytes in primary cultures. *Nature* **311:** 656—659.

16) BURGESS, S. K., McCARTHY, K. D. (1985). Autoradiographic quantitation of β-adrenergic receptors on neural cells in primary cultures. I. Pharmacological studies of ^{125}I pindol binding of individual atroglial cells. *Brain Res.* **335:** 1—9.

17) KETTENMANN, H., SCHACHNER, M. (1985). Pharmacological properties of GABA, glutamate and aspartate induced depolarizations in cultured astrocytes. *J. Neurosci.* **5:** 3295—3301.

18) TORRENS, Y., BEAUJOUAN, J. C., SAFFROY, M., DAGUET DE MONTETY, M. C., BERGSTRÖM, L., GLOWINSKI, J. (1986). Substance P receptors in primary cultures of cortical astrocytes from the mouse. *Proc. Natl. Acad. Sci. (USA)* **83:** 9216—9220.

19) JOHNSTONE, S. R., LEVI, G., WILKIN, G. P., SCHNEIDER, A., CIOTTI, M. T. (1986). Subpopulations of rat cerebellar astrocytes in primary culture: morphology, cell surface antigens and ^{3}H-GABA transport. *Dev. Brain. Res.* **24:** 63—75.

20) LEVI, G., GALLO, V., CIOTTI, M. T. (1986). Bipotential precursors of putative fibrous astrocytes and oligodendrocytes in rat cerebellar cultures express distinct features and "neuron-like" γ-aminobutyric acid transport. *Proc. Natl. Acad. Sci. (USA)* **83:** 1504—1508.

21) REYNOLDS, R., HERSCHKOWITZ, N. (1986). Selective uptake of neuroactive amino acids by both oligodendrocytes and astrocytes in primary dissociated culture: a possible role for oligodendrocytes in neurotransmitter metabolism. *Brain Res.* **371:** 253—266.

22) GALLO, V., SUERGIU, R., LEVI, G. (1987). Functional evaluation of glutamate receptor subtypes in cultured cerebellar neurones and astrocytes. *Eur. J. Pharmacol.* **138:** 293—297.

23) GALLO, V., SUERGIU, R., GIOVANNINI, C., LEVI, G. (1987). Expression of excitatory amino acid receptors by cerebellar cells of the type-2 astrocyte cell lineage. Submitted.

24) GALLO, V., SUERGIU, R., LEVI, G. (1986). Kainic acid stimulates GABA release from a subpopulation of cerebellar astrocytes. *Eur. J. Pharmacol.* **133:** 319—322.

25) WROBLEWSKI, J. T., NICOLETTI, F., COSTA, E. (1985). Different coupling of excitatory amino acid receptors with Ca^{2+} channels in primary cultures of cerebellar granule cells. *Neuropharmacology* **24:** 919—921.

26) NOVELLI, A., NICOLETTI, F., WROBLEWSKI, J. T., ALHO, H., COSTA, E., GUIDOTTI, A. (1987). Excitatory amino acid receptors coupled with guanylate cyclace in primary cultures of cerebellar granule cells. *J. Neurosci.* **7:** 40—47.

27) NICOLETTI, F., WROBLEWSKI, J. T., NOVELLI, A., ALHO, H., GUIDOTTI, A., COSTA, E. (1986). The activation of inositol phospholipid metabolism as a signal transducing system for excitatory amino acids in primary cultures of cerebellar granule cells. *J. Neurosci.* **6:** 1905—1911.

SUMMARY AND DISCUSSION-REPORT OF CHAPTER IV

John W. Commissiong

SUMMARY

The first four contributions in this session by Commissiong, Bradberry, Maher and Jobe all, in different ways, sought to determine possible relationships between precursor (PHE, TYR) supply, catecholamine synthesis/release/metabolism and neuronal function. I myself attempted to make a break with the past by presenting a scheme in which catecholamine metabolism is not closely linked, in any meaningful way, with the physiologic functioning of catecholaminergic neurons. The evidence is based primarily on the observation that under non-stressful conditions, electrical stimulation of the nigrostriatal tract did not cause any significant increase in dopamine metabolites (DOPAC, HVA) in the striatum; whereas under stressful conditions, the metabolites were increased. This observation, in turn, leads to the much broader question concerning the mechanism(s) by which catecholamines in the CNS are conserved (re-uptake and re-release versus release/catabolism/new synthesis) for neurotransmission. Charles Bradberry presented a stimulating paper from Dr. Roth's laboratory, in which the heterogenity of the mesotelencephalic dopaminergic neurons was stressed. The main thrust of this paper was that the basic electrophysiological properties (mean firing rate, bursting characteristics), synthesis control mechanisms (turnover rate, somatodendritic autoreceptor control mechanisms) and interaction with the environment (susceptibility to activation by stress) must all be factored into the equation, in any meaningful effort to arrive at a correct interpretation of control.

Timothy Maher and Phillip Jobe presented data on different seizure models. It emerged from Timothy Maher's results that in low to moderate doses, PHE causes an increase in the release of DA, whereas at higher doses, PHE causes a decrease in DA release. It also turns out, from Timothy Maher's results that at the highest dose range of PHE, the threshold for seizures in young mice is lowered. Phillip Jobe presented data on a very interesting strain of genetically epilepsy-prone rats (GEPRs), in which susceptibility to seizures appeared to correlate with several indices of reduced noradrenergic and serotonergic activity. L-DOPA and dopamine, both of which are precursors to norepinephrine, proved to be effective in suppressing seizure activity in this model. However, aspartame, which increases plasma concentrations of phenylalanine and tyrosine, does not appear to alter seizure responses in the GEPR. The last paper in this session by Vittorio Gallo brought a degree of precision to the session, dealing as it did, with the control of ^{3}H-D-aspartate release from cerebellar granule cells in culture, by kainic acid (KA), quisqualic acid (QA), NMDA, glutamate, and some of their analogues. The degree of precision described in this talk was clearly an envy to people working in *catecholamine release mechanisms.* Vittorio Gallo's presentation also made the indirect point that the mechanisms that control catecholamine release/synthesis/metabolism are likely to be much subtler than is presently imagined. Timothy Maher's interesting release data de-

NATO ASI Series, Vol. H20
Amino Acid Availability and Brain Function in Health and Disease. Edited by G. Huether

rived from using the in vivo dialysis technique, clearly indicates that measures of actual release in vivo will play a crucial role in the reconstruction of the correct sequence of events that occur in the homeostatic control of catecholamine availability to effect neurotransmission in the CNS.

DISCUSSION-REPORT

Can the relationship between electrophysiological activity of nigrostriatal dopamine neurons and neurotransmitter synthesis be explained by a unifying hypothesis?

Commissiong: By way of an answer, and also to get the discussion started, one ought to stress that monoaminergic neurotransmission is only one aspect of the much wider general process of chemical neurotransmission in the CNS. Information transfer in the CNS via chemical neurotransmission is a potential hazard, because of the possibility of transmitter depletion during periods of intense nerve activity. However, the experimental observation is that transmitter depletion rarely occurs, unless this is deliberately brought about by pharmacologic intervention. The nigrostriatal dopaminergic system has been studied very intensively, over a period of many years, and it is a measure of the difficulty of the problem that we still lack even an incomplete understanding of the homeostatic mechanisms that control the conservation of DA in the nerve terminal under conditions of varying degrees of transmitter release. The two extreme hypotheses are: 1) Released DA is taken up, repackaged and re-released. Little new synthesis or catabolism is envisaged in this scheme. 2) Released DA is primarily catabolized (DOPAC and HVA are increased), and synthesis is enhanced to replenish DA. There are published data that support both schemes. The problem is that the two extreme mechanisms might merge to varying degrees, under varying experimental conditions. In such a case, the conditions under which the experiment was done, become themselves important variables in determining the results obtained, and therefore in interpreting the data, and formulating a correct hypothesis about how the neuron functions "normally". The most important practical problem concerns changes in transmitter metabolism that are unrelated to release. It therefore seems to me that the study of the electrophysiological and related neurochemical activities stand the best chance of providing the results on which to base a correct hypothesis of the mechanisms that regulate synthesis and release of catecholamines.

Bradberry: John Commissiong claims to have offered a fundamentally new and unifying hypothesis of catecholamine neuronal function . In fact, he has examined the behaviour of one subset of midbrain dopamine (DA) neurons (the nigrostriatal system) and called his description of the behaviour of this one system a "unifying hypothesis". It is well documented that the nigrostriatal and mesolimbic DA systems are fundamentally different not only from other catecholamine systems, but also from other midbrain DA systems, because of a strong tonic modulatory control of synthesis and release exerted by nerve terminal autoreceptors. If he did have some general hypothesis for the regulatory control of neuronal function in the nigrostriatal system, he fails to address how it might apply to other, different, catecholamine systems. Even with respect to the nigrostriatal system, no new or insightful experimental information was presented concerning neuronal activity and transmitter disposition. The issue raised by John Commissiong "how directly is neuronal release related to measureable levels of metabolites?" is an important question. However, this is a question (rather than a hypothesis) which relates to a complex interaction between release, reuptake, synthesis, and catabolism of catecholamine in brain and transport of metabolites out of the brain. Many of these variables are in addition significantly influenced by the firing rate of the system under investigation. Thus, much more complete and involved experiments than those presented by John Commissiong will be necessary to address this important question and form a tenable hypothesis to explain the findings observed.

Commissiong: I really do not claim to have offered a fundamentally new an unifying hypothesis of catecholamine neuronal function. I merely suggest a model of the release, synthesis and metabolism processes that occur in dopaminergic (catecholaminergic) nerve terminals. The model suggested accommodates most of the published, conflicting data. The model specifies categories of activity e. g. physiologic and pharmacologic. The model further suggests that confusion may result from trying to interpret data across boundaries, e. g. trying to interpret purely pharmacologic data in a physiologic context. The model is posited on the assumption that the following querries are important.

1. The relationship between release and metabolism
2. The relationship between release and synthesis
3. The correct functional interpretation of changes in metabolite levels
4. The role of "stress" in changes in dopamine metabolism
5. The role of reuptake of DA into dopaminergic nerve terminals, e. g. transmitter conservation versus termination of transmitter action
6. The fact of intraneuronal synthesis and metabolism of transmitter unrelated to release

The model is based on the following data:

1. Under non-stressful conditions, there is not always an increase in DA metabolism associated with increased DA release, induced by electrical stimulations.
2. Under stressful conditions DA metabolism is usually increased when release is increased by electrical stimulation.
3. Under certain experimental conditions ongoing increased DA release and increased DA metabolism can be uncoupled.
4. Certain drugs cause marked increases in DA metabolism, apparently without a concomitant increase in DA release.
5. Metabolite levels may merely mirror levels of intraneuronal MAO activity under different experimental conditions. If this is the case, metabolite levels will bear little relationship to transmitter release.

The questions raised, transcend individual neuronal systems. There is a priori, no good reason for believing that dopaminergic neuronal systems, with different rates of firing, must regulate their synthesis, release/metabolism processes differently. The disposition of released ACh via hydrolysis, is independent of the rate of firing of cholinergic neurons.

There is an enormous amount of data already available on the dynamics of DA in different terminal regions. The mere accumulation of more data is unlikely to further understanding. Much of the confusion that currently exists may have resulted from doing experiments under stressful conditions, and using chloral hydrate, which itself activates tyrosine hydroxylase, as the anesthetic. Two factors may prove to be of crucial importance to the current discussion. One is the regulation of TH activity, and the other, the regulation of MAO activity.

Bradberry: Again, it is unclear just what new information John Commissiong's model entails, or what simplifying explanation it offers. It appears to be more of a restatement of the complexity of interaction between the many variable thought to affect measureable indices of dopamine metabolism. To say that the physiological function of neurons is likely to differ under pharmacological manipulation is to state the obvious. However, much insight into the metabolic regulatory features of neurons is possible by observing how they respond to pharmacological agents. The approach used by John Commissiong of employing a DA system well into a state of toxin-induced degeneration would not appear to be the most appropriate for the study of metabolic regulatory mechanisms.

It is quite true that there is no *a priori* reason for supposing there to be variations in the regulatory features of the different dopamine projection systems (e. g. the nigrostriatal vs. the mesoprefrontal) There is, however, compelling experimental evidence that such variations do exist between these systems. The paper from Dr. Roth's laboratory presented at this meeting indicated that tyrosine hydroxylase (TH) activity in the mesoprefrontal and mesocingulate DA systems is subject to precursor control in the basal unanaesthetized state, a finding not seen in any other DA system other than the retinal DA amacrine system.

Another important example of divergent regulatory features of the different mesotelencephalic DA projection systems is the striking difference in their response to impulse blockade by gamma-butyrolactone (GBL). A dramatic increase of *in vivo* TH activity is seen in the nigrostriatal system following GBL, while the mesoprefrontal DA system does not respond in this fashion (Bannon et al., Molecular Pharmacol, 19: 270—275). The increase in TH activity accompanying a cessation of impulse flow strongly argues for a tonically activated autoreceptor control over TH (for a review, see Roth, 1984, Ann. N.Y. Acad. Sci. 430: 27—53). In this scheme, the removal of ongoing DA release by GBL blockade of impulse flow prevents tonic activation of the synthesis modulating DA autoreceptor resulting in increased tyrosine hydroxylation. John Commissiong has stated here that such an increase in TH activity as a result of a cessation of impulse flow is illogical, because, as is well accepted, increases in dopaminergic neuronal activity also lead to increased TH activity. The problem with this interpretation is that John Commissiong has fallen into the trap which he has wisely warned others of. The use of GBL is a non-physiological manipulation of the dopaminergic system. It has presented the DA neuron with a situation which is unlikely in a physiological setting, namely, the complete cessation of neuronal firing rather than the graded increase or decrease in response to appropriate stimuli one would expect. Thus, it is not surprising that the result of this pharmacological manipulation appears on the surface to be paradoxical. This does not mean, as John Commissiong has stated it, that both increases and decreases in neuronal activity lead to increased TH activity. Rather, it would indicate that both increases, and a non-physiological complete cessation of impulse flow, can lead to increased TH activity. The concept of the synthesis modulating TH autoreceptor would appear at the present time to offer a mechanism by which these divergent data may be explained.

Are the neurochemical measurements adequate to monitor effects of altered transmitter synthesis on functional neurotransmission?

Lookingland: Neurochemical estimates of the activity of aminergic neurons are based upon the coupled relationship between neurotransmitter synthesis, release and metabolism. Accordingly, alterations in impulse flow and neurotransmitter release are accompanied by corresponding changes in neurotransmitter synthesis and metabolism, and this maintains a readily releasable pool of neurotransmitter within neurons. While it is generally accepted that alterations in neurotransmitter synthesis and metabolism reflect the activity of aminergic neurons, it should be noted that changes in the levels of neurotransmitters, and their precursors and metabolites, can occur independently of any change in neuronal activity (e. g. treatments that alter MAO activity, disruption of aminergic stores, etc.). Therefore, utilization of neurochemical measures to estimate that activity of aminergic neurons when the relationship between neurotransmitter synthesis, release and metabolism is uncoupled could lead to erroneous conclusions regarding functional neurotransmission.

Wurtman: Measuring neurotransmitter metabolites as an index of neurotransmitter release is neither more nor less solipsistic than any other experimental paradigm.

Biomedical scientists working in the twentieth century are empiricists, and choose to test their hypotheses by doing paradigmatic experiments. For example, to determine whether the release of dopamine from surviving nigrostriatal neurons is increased in experimental Parkinsonism, they might measure striatal levels of DOPAC or HVA, and then correct these levels for the apparent numbers of surviving neurons by dividing them by striatal tyrosine hydroxylase activity. This particular experiment succeeds (Melamed et al., 1980, Proc. Nat. Acad. Sci, 464: 4305—4309), and the scientist rejoices for two reasons: He has, for the time being, confirmed both his hypothesis and the utility of his experimental paradigm.

But if the experiment doesn't work, the scientist must decide whether his hypothesis or his paradigm is faulty. And if, on a sufficient number of occasions, the paradigm yields false negatives (or positives), such that its data would cause a hypothesis that later turned out to be "true" to be discarded, it is the paradigm and not the hypothesis that is thrown away. (Three decades ago scientists measured urinary norepinephrine as an index of brain catecholamine release. The paradigm was discarded when newer technologies showed that the turnover of brain norepinephrine bore no simple relationship to its levels in urine.) Of course "true" is simply a normative statement (Schrecker, P., 1948, Work and History, Princetown Univ. Press), meaning only that, at a certain moment in History, the then-favored paradigms yielded data that could be interpreted as confirming the "true" hypothesis. As Night follows the Day, virtually all such paradigms ultimately fail, and "truth" must then once again be tested using their replacements.

In my view, measuring neurotransmitter metabolites as an index of neurotransmitter release remains a pretty good experimental paradigm. Much of the time it works, the data it yields confirming to those obtained using other popular paradigms (for example, measurement of post-synaptic effects).

But, as shown in our paper (Acworth, et al., this volume) there are times when this paradigm seems to fail: For example, giving an otherwise-untreated rat supplemental tyrosine transiently enhances dopamine release (measured directly, via intracerebral dialysis) *without* concurrently elevating levels of dopamine's metabolites, DOPAC and

HVA, in the brain's ecf. So an investigator who measured *only* the levels of those metabolites might erroneously conclude that tyrosine has no effect on nigrostriatal dopamine release in such animals.

But good scientists never depend on a single paradigm. And it is jejune to promote as a revolutionary discovery the view that this particular paradigm isn't perfect. Of course not.

Jobe: Neurochemical measurements can provide estimates of functional monoaminergic transmission. For example, turnover rate can under defined circumstances serve as an index of the amount of norepinephrine utilized per unit time. However, it would be unwise in most investigations to rely on turnover rate as the only neurochemical estimate of transmission. Various factors can alter the rate of intraneuronal release of monoamines without altering the amount entering the synapse. Under such circumstances, a change in turnover rate would not reflect a change in neurotransmission. In my view neurochemical studies of functional monoaminergic neurotransmission should be multifaceted. Accordingly, they should be designed to address both presynaptic and postsynaptic questions. For example, if the effects of chronically administered precursors are being studied, the experiments should be designed to determine whether changes in the concentration of the neurotransmitter and its metabolites are accompanied by "homeostatic" changes in postsynaptic receptors. Presynaptic changes in levels accompanied by the anticipated postsynaptic receptor alterations would provide stronger evidence of functional neurotransmission than would changes in either type occurring singly.

Recent advances in neurochemical approaches are also proving valuable for studies of functional monoaminergic neurotransmission. For example, changes in norepinephrine released from noradrenergic terminals in the brain can now be detected through the use of intracerebral dialysis or push-pull perfusion. Alterations detected with either of these approaches upon chronic norepinephrine precursor administration would be valuable for interpretational purposes. However, an argument for a change in functional noradrenergic transmission would be strengthened if changes detected by one of these techniques were accompanied by studies showing homeostatic alterations in norepinephrine receptors. In summary, for studies of precursor-induced alterations in transmitter synthesis, no single neurochemical estimate of functional monoaminergic neurotransmission is adequate. Appropriately designed multifaceted estimates are preferable.

Commissiong: My frank answer to this question, is that at present, no meaningful functional significance can be assigned to monoamine metabolite measurements. At best, they probably reflect changes in synthesis. Synthesis may, in turn, be linked, or not linked to changes in release. Therefore, metabolite measurements should not be used as indices of release i. e. of functional neurotransmission. It is possible to take an entirely cynical view of this problem. It is that metabolite measurements have come to be equated with release/functional neurotransmission, not because scientists were persuaded of the validity of a causal relationship between changes in release and changes in metabolism, but simply because metabolites were (and still are) easy to measure. However, there is no justification of persisting with the practice.

Young: There is not necessarily any association between metabolite measurements and function of the parent amines. However, in some circumstances, such an association may exist, and may provide useful information. This is especially true for CSF amine metabolite measurements. E. g. a low level of HVA in the CSF of untreated Parkinsonian patients is certainly an indication of low dopaminergic function. A recent review indicates that 13 of 14 studies found an association between low CSF 5-HIAA, and suicidal behaviour or ideation (Asberg et al., 1987, in: Psychopharmacology, the Third Generation of Progress, ed. by H. Y. Melther, Raven Press, New York, pp 655—688). The idea that low 5-HT function will promote suicidal behaviour is consistent with what we know about the roles of 5-HT in the control of mood and aggression, therefore the CSF data suggest that drugs which potentiate 5-HT function may be a specific treatment of suicidal ideation. However, there are other situations in which CSF 5-HIAA does not reflect 5-HT function. E. g. the decline in CSF 5-HIAA when patients are treated with 5-HT uptake inhibitors does not indicate a decline in 5-HT function mediated by these drugs. Overall I would say that CSF amine metabolites can in some circumstances provide useful insights into CNS biogenic amine function in living humans, if the data are interpreted properly.

Curzon: In general, I agree with Simon Young that CSF metabolites can, in some circumstance, give information on transmitter function. This especially so if metabolite egress is blocked with probenecid, if accurate methods of determination are used (as in the work of Asberg et. al) or if CSF is available from cisternal or ventricular sites rather than from the lumbar region where relationships with brain metabolite values are less clear.

For example, we found that human ventricular 5-HIAA and HVA values correlated significantly with visual evoked potentials determined a day after CSF withdrawal from patients undergoing neurosurgery (Holder et al., 1980, Brain Res. 188: 582). This suggests that the metabolite values reflected release of the parent amines to receptors mediating characteristics of the evoked potentials rather than metabolism in a functionally irrelevant intraneuronal pool.

More recently, we have developed a cisternal CSF method for determining central transmitter amine turnover in individual rats without killing them (Hutson et al., 1984, J. Neurochem. 43: 151) and have used this to study the social behaviour of rats with known turnover values. As significant correlations where obtained between behaviour and turnover, the results suggest that turnover as measured using CSF provides an index of behaviour and hence of functional neuronal activity (Sahakian et al., 1986, Brain Res. 399: 162).

Does monoaminergic stimulation cause a depletion of precursor amino acids in the brain tissue?

Huether: Richard Wurtman has shown that stimulation of nigrostriatal brain slices causes a depletion of tyrosine in the slice. Similarily one would expect that stimulation of brain slices would also cause a depletion of tryptophan used for increased serotonin formation. Such a depletion, however, according to studies of Elks et al., 1979 (Brain Res. 172: 471) does not occur. The concentration of tryptophan in the slice, in the incubation medium as well as the uptake of tryptophan from the medium into the slice remained unchanged in spite of the fact that during stimulation more serotonin and hydroxyindole acetic acid were produced. Does anybody have a reasonable explanation for this phenomenon?

Wurtman: I think this study has to be redone. There is another possibility and that is increased breakdown of proteins which liberates aromatic amino acids that are then used as precursors for transmitter synthesis.

Curzon: I really think that in these slices we can get any kind of results we want to get simply by modulating the tryptophan concentrations. The really interesting situation however is, what happens in the whole animal and certainly in the whole animal one can do two things. One can make serotonin turnover go up by stimulating serotoninergic neurons or one can completely block serotonin turnover by destroying serotoninergic neurons. In both of these cases there is a very clear absence of changes in tryptophan concentrations. So we need tryptophan to make serotonin, but there is very little if any evidence that an increased or decreased activity of serotoninergic neurons will cause an increased or decreased concentration of tryptophan in the brain tissue. As Richard Wurtman has said himself: Tryptophan can influence serotonin synthesis, but this is an open loop not a closed one.

Fernstrom: One has to keep in mind in discussing such *in vivo* slice studies that the results are not very physiologic. Apart from being *ex vivo,* for example, the striatal slice studies of Milner and Wurtman (referred to by Gerald Huether) used 20 Hz to stimulate the slices, a frequency that is about twice what one would expect the nigrostriatal pathway to fire *in vivo,* even after haloperidol administration. Moreover, in the slice preparation, nomifensin is added to the incubation medium to block dopamine reuptake. Important chemical feedback mechanisms on tyrosine hydroxylase may thus have been disrupted as well. So the experimental paradigm is stacked toward pushing dopamine synthesis and release. Under such a situation, I guess it would not be so surprising to find a depletion of tyrosine concentrations in the slices. But this is not a paradigm that tells much about the *in vivo* situation, and is more generally not very physiologic.

Wurtman: We were surprised to find any change at all. Because if you by giving 6-hydroxydopamine completely destroy dopaminergic neurons in the striatum, you will find no measurable decrease in striatal tyrosine levels. There is no evidence that there is more tyrosine within a catecholaminergic terminal than there is in the other cells around it. And since catecholaminergic terminals are only 8 or 10 % of all cellular structures in these slices, in order to get a 50 % reduction there must have been a recruitment of tyrosine from other cells as well. In no sense can these data be interpreted as to apply to what happens in vivo. Their only interest is that they demonstrate the possibility of a tyrosine depletion under catecholaminergic stimulation. One other example might be relevant: the pineal organ is a structure consisting of about 99 % serotonin-producing cells, and if you use the pineal you can get some idea as to what percent of the tryptophan goes to make protein and what percent to make serotonin (which becomes melatonin). In fact 200 times as much goes to make the amine as goes to make the protein.

Lajtha: How good were the efforts in looking at the amino acids coming from proteins. Don't forget that it is more than 600 times more tryptophan which is protein bound compared to the tryptophan of the free amino acid pool. So a small amount of breakdown would release substantial amounts of tryptophan to be utilized for monoamines synthesis.

Harper: How can we extensively discuss quantitative relationships between decreasing precursor concentrations and increasing amine formations without knowing what the size of the available amino acid pool is and what the rate of the utilization of precursors for amine synthesis is. I wonder if anybody can comment on the relative size of the tryptophan pool and the pool of the other precursor amino acids in relation to the relative rates of their utilization for amine production under these various conditions.

Wurtman: It has been argued by a few authors that there will always be enough tryptophan molecules available in the brain for conversion to serotonin, and thus that changes in brain tryptophan levels aren't important in controlling the rate of serotonin synthesis. This would be true if enzyme kinetics weren't important, but they are.

There are also always enough cholesterol molecules in the adrenal cortex to make considerably more cortisol than the body needs, and enough thyroglobulin stored in the thyroid to liberate vast amounts of thyroxine, and so forth. But the presence of large amounts of cholesterol in the adrenal, or the thyroglobulin in the thyroid, doesn't help the investigator to guess how much cortisol or thyroxine are being released on a minute-to-minute basis, or whether such release occurs at a constant or variable rate.

What determines the release of these hormones, or the synthesis rate of brain serotonin for that matter, *is enzymes*. And the curious thing about serotonin and some other neurotransmitters (but, appearently, no hormones) is that it isn't *just* enzymes, but also *substrate* levels, i. e., the availability of their amino acid precursors. This is why physiologic changes in brain tryptophan levels can affect serotonin release.

The biochemical mechanism underlying this effect of tryptophan has to do with the kinetic properties of the enzyme, tryptophan hydroxylase, which initiates serotonin synthesis. This enzyme has a very high K_m for tryptophan, and functions optimally only when local tryptophan concentrations transcend that K_m . So even though, quantitatively, very few of the available tryptophan molecules are used to make serotonin, almost none will actually be converted to the neurotransmitter unless the tryptophan *concentration* is very high.

Curzon: It is not so much the amount of tryptophan that determines how much serotonin is released, certainly it influences the amount of serotonin that is produced and metabolized. But as we all know under in vivo conditions it has proved extremely hard to show that changes of brain tryptophan in the absence of any other drug do result in changes of extra neuronal serotonin. It may be that we are measuring this in the wrong situations. It may be that we have methodological problems. At the moment, all we really know is that physiological variations of tryptophan availability to the brain may influence behaviour and influence serotonin turnover, but we lack a clear effect on serotonin function.

Are newly synthesized transmitters preferentially released? Do neurons possess differentially releaseable transmitter compartments?

Lookingland: Basic to our understanding of the effects of precursors on functional neurotransmission is information regarding the fate of newly synthesized neurotransmitters within neurons. According to classical theory, newly synthesized neurotransmitters are stored in synaptic vesicles thereby replenishing neurotransmitter stores previously depleted through release and metabolism. The early studies of Javoy and Glowinski on nigrostriatal dopamine neurons (J. Neurochem. 18: 1305, 1971) have suggested, however, that there are two storage compartments for DA, and that newly synthesized amine may be preferentially released. We have reached the same conclusions in our studies with tuberoinfundibular dopamine neurons in the medial basal hypothalamus. Dopamine released from these neurons is transported in the hypophysial portal blood to the anterior pituitary gland, where it acts to inhibit prolactin secretion. Following blockade of dopamine synthesis (either at the tyrosine hydroxylase step or the DOPA decarboxylase step) there is a rapid increase in prolactin secretion, and this occurs prior to any change in dopamine concentrations in the median eminence. These results suggest that newly synthesized dopamine is rapidly released from tuberoinfundibular dopamine neurons and that neurotransmitter storage pools are conserved under these experimental conditions. This close coupling between de novo neurotransmitter synthesis and neurotransmitter release suggests that precursor-induced alterations in synthesis may produce similar changes in functional neurotransmission.

Youdim: Keith Lookingland is familiar with our studies where we showed the presence of two pools, one which is the cytoplasmic newly formed 5-HT storage site. This site can certainly release 5-HT into function via the action of parachloroamphetamine and is sensitive to inhibition by parachlorophenylalanine, the inhibitor of tryptophan hydroxylase.

Franklin: There is evidence from behavioural studies that dopamine release may be critically dependent on newly synthesized dopamine, possibly supplemented by recycled dopamine. It is well established that behavioural stimulation produced by catecholamine-releasing drugs depends primarily on dopamine release. However, these drugs are of two types. Amphetamine-like drugs block dopamine re-uptake and stimulate the release of newly synthesized dopamine, whereas drugs like methylphenidate displace dopamine from the "reserve" pool. If an animal is reserpinized, there is an initial failure of catecholamine transmission, but after vesicular uptake mechanism recovers, transmitter release becomes partially functional again. At this time amphetamine is behaviourally stimulant, but methylphenidate is inactive. In contrast, when synthesis is inhibited by α-methyl-p-tyrosine, the stimulant effect of amphetamine is blocked but methylphenidate remains a powerful stimulant. Now, when α-methyl-p-tyrosine is given to an animal, synthesis of catecholamine is inhibited very quickly though behavioural depression does not occur for some time. However, if the animal is given amphetamine soon after α-methyl-p-tyrosine one observes a few minutes of behavioural stimulation followed by sudden and severe behavioural depression. This behavioural depression can be immediately reversed by methylphenidate. These effects indicate that transfer of dopamine from the reserve pool to the releaseable pool is too slow to maintain dopamine release, so that when amphetamine speeds release and prevents re-uptake and recycling of transmitter, transmission rapidly fails as the newly synthesized pool is depleted. Methylphenidate mobilizes dopamine from the reserve pool and restores transmission. Thus the pool of dopamine which contains newly synthesized or recycled dopamine appears to be the functional pool, and is not readily replenished from the much larger "reserve" pool.

Youdim: If you reserpinize a rat and you deplete about 90 % of the serotonin from the brain, you can still induce serotoninergic behavioural responses by the 10 % that has remained using parachloroamphetamine. We have calculated that, out of the 10 % of serotonin that remains after reserpinization 30 % of that is responsible for the function. And I would think it is newly synthesized serotonin which is preferentially released into function and is primarily cytoplasmic. This is supported by the inhibition of functional activity by parachlorophenylalanine.

What is the relevance of an altered precursor availability for transmitter synthesis and functional neurotransmission under chronic conditions, e. g., in disease states

I. Smith: Every remark you make, ladies and gentlemen, is interesting to the disease that interests me, phenylketonuria. You talk about the consequences of a decreased or increased availability of precursors for functional neurotransmission. However, what you are producing in your experiments is trivial stuff compared to what is happening in the brain of a PKU-patient when its phenylalanine levels are high. As far as serotonin metabolism is concerned we know that there is a depletion of tryptophan and a severe depletion of serotonin in the brain. However these children are not demonstrating significant changes in appetite or in their sleep pattern. What they do demonstrate is a general hyperactivity, and therefore I would like to ask you: Is hyperactivity relevant to their serotonin disturbance?

Wurtman: I think I know what you would like, but what we cannot give you is things to look for. But we can tell you, over the short term, what the consequences of serotonin release are. Thereby one may guess what you would see when serotonin is depleted in the brain. The problem is that long term is very different from short-term for several reasons: First we know that there are many antidepressant drugs which have very different effects after 3 or 4 weeks than they have after one single administration. Secondly we know that the problem with neurotransmitter precursors is that they are metabolized very rapidly. That is why most of us do our experiments acutely. We are simulating what occurs normally by giving a treatment which lasts, let's say, 6 to 8 hours. You however are talking about pathology which is always a more or less long-term effect. Thirdly we know that in rats if we totally destroy the serotoninergic neurons by destroying the raphe neurons, and we wait a month or two and we simply look at the animals we cannot tell that we have deprived the animals of serotonin. Does that mean serotonin has no function? Of course, not!

Huether: Even though we do not clearly express it, we are certainly all aware of the fact that most effects of an altered precursor availability on transmitter synthesis, functional neurotransmission and finally, on mood and behaviour are seen after acute treatments. They are acute responses to acute modulations of the precursor supply. This situation is very different from what we see in a chronic PKU-patient. We need not to consider untreated PKU where, due to a severely impaired amino acid supply, brain function is affected as a result of irreversible deficits in the functional and structural organization of the brain. These are developmental deficits. But even a well controlled treatment cannot avoid a certain elevation of phenylalanine and a depletion of, among others, precursor amino acids for monoamine synthesis in the brain. But these are slow and long lasting shifts rather than acute changes and, therefore very different from all precursor loading studies. Under such circumstances, adaptations and counterregulation will occur. Long lasting changes of transmitter synthesis and transmitter release will cause subsequential changes in receptor density and modulations of pre- and postsynaptic mechanism to compensate for the altered transmitter release and neuronal activity. It may well be that the behavioural changes seen in PKU-patients when their phenylalanine levels are more or less elevated are in fact the consequences of such adaptations, and therfore not comparable to the consequences of the decreased supply of tryptophan or tyrosine to a normal brain.

Leathwood: One example from the sleep work. If you give para-chlorophenylalanine, you induce severe insomnia. After a while, however, in spite of the still deprived serotonin, the system begins to adapt and the sleep starts off again. This is one example were, even in spite of severe deficits of a transmitter in the brain, subsequent adaptation will occur to compensate for the deficit.

Chapter V.

INFLUENCE OF ALTERED PRECURSOR AMINO ACID SUPPLY ON PHYSIOLOGY AND BEHAVIOUR

ROLE OF AMINO ACIDS IN APPETITE CONTROL IN MAN

A. J. Hill and J. E. Blundell

BioPsychology Group
Psychology Department
Leeds University
Leeds LS2 9JT U. K.

INTRODUCTION

Amino acids constitute an essential element of man's food supply, and there are good biological reasons why mechanisms should have developed to ensure an adequate intake of these materials. It follows that there is likely to exist a mechanism (or mechanisms) which monitors the amount of protein (or amino acids) ingested. It is not necessary to argue that protein intake should be regulated at a fixed value but it does seem appropriate that protein ingestion should be controlled between upper and lower limits (1). If protein is being monitored and its intake adjusted between high and low boundaries, it is likely that in omnivores there is some link between the monitoring of protein being ingested and general appetite control. In this way, proteins or amino acids could play some role in the control of hunger and food consumption in man.

Over the years a number of mechanisms have been considered which could mediate the effect of proteins upon appetite. It has been suggested that amino acids in general are monitored by some type of amino-stat (2), or that qualitative adjustments in the amino acid supply may be reflected in particular profiles of amino acids in the blood and brain (3). Often associated with this latter view is the idea that certain critical amino acids are monitored and the amount of these limits the intake of amino acids (and therefore food) in general. Further, the quantitiy of protein ingested, or some amino acids in particular (e. g. phenylalanine), may directly influence short term post-ingestive mechanisms influencing food consumption. Of relevance here is the known action of phenylalanine (and other amino acids) to stimulate the release of the putative satiating hormone cholecystokinin (4). In addition, it is likely that certain privileged amino acids may be monitored via specific chemoreceptors situated along the gastro-intestinal tract (5). One further possibility is that particular amino acids such as tryptophan, tyrosine or phenylalanine may influence brain mechanisms through their roles as precursors of neurotransmitters — serotonin and the catecholamines (6). In turn, this action would depend upon the availability of these amino acids in the brain. This depends upon their ratios (one to another) in the plasma (7, 8).

Of course it should be recognised that these mechanisms are not mutually exclusive. Indeed, it is highly likely that they are all operative. This makes sense since it is useful for any appetite control system to receive information about the state and nature of ingested

NATO ASI Series, Vol. H20
Amino Acid Availability and Brain Function in Health and Disease. Edited by G. Huether

food at various stages during its passge through the body. Therefore, there may be expected a good relationship between the early and late stages of protein recognition by the system. Of particular importance will be links between sensory receptors at the site of ingestion (i. e. the mouth) and the post-ingestive and post-absorptive detection of protein by the body.

In order to investigate the role of proteins or amino acids in the control of appetite, it is necessary to have a conceptualisation of how an appetite control system could operate together with a sensitive and valid research methodology. In accordance with the proposals mentioned above, one approach to the understanding of appetite control is to evaluate the satiating capacity of food (in which the proportion of proteins or amino acids can be adjusted — between 0 and 100 %). Figure 1 illustrates diagramatically the processes involved in the satiety cascade. This model provides a rationale for developing research designs to manipulate the parameters of food (and pre-loads) and measuring

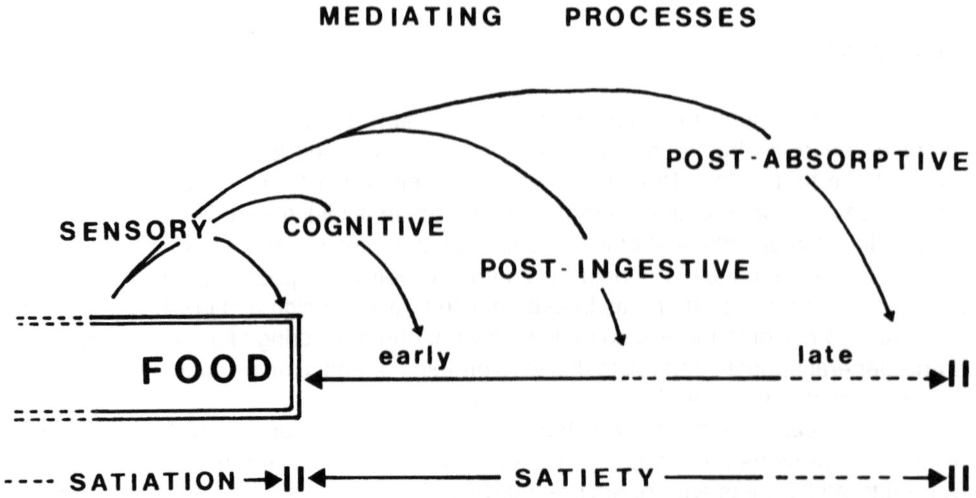

FIGURE 1
Conceptualisation of the contributions of sensory, cognitive, post-ingestive and post-absorptive stimuli to the time course of satiety.

devices to monitor temporal changes in eating and associated changes in the subjective motivation to eat (see 9). Techniques for measuring eating in the long and short term have been described previously (10, 11). This approach — conceptualisation plus methodology — has been used in our laboratory to measure the effect of protein on appetite control and to investigate specific effects generated by the amino acid tryptophan.

The Satiating Effect of Protein

It is a widely-held belief that calorie-for-calorie protein is more satiating than carbohydrate. Thus, "... many people find that the onset of hunger is delayed for a longer time when a meal containing ample proteins is taken, as compared to one rich in carbohydrate" (12, p. 296). Yet, until recently the empirical evidence for such an assertion has been weak.

Early studies examined the effects of breakfast nutritional composition and found that high protein breakfasts were followed by a consistent and prolonged 'sense of well-being' (13), and that daily energy intake computed from diaries was inversely related to the protein content of breakfast (14). Unfortunately though, the outcomes of both studies were compromised by methodological short-falls, such as not controlling for the energy content or bulk of the breakfast meals. Similar tantalizing but problematic findings were reported for the effect or protein on subjective hunger. Fryer *et al.* (15) noted that a high protein weight-loss diet was associated with the least reports of hunger, and Mellinkoff *et al.* (2) found an inverse correlational relationship between hunger and serum amino acid levels. However, the arbitrary and casual measurement of hunger was acknowledged by these authors.

More recent studies have enabled firmer conclusions to be drawn regarding the satiating power of protein. Booth *et al.* (16) evaluated the effects of equicaloric high and low protein composite meals on the intake of a nutritionally intermediate cornflower pudding three hours later. Although the palatability of the food offered must be questioned, the authors nevertheless described a 26 % reduction in voluntary intake after the high protein meal. Using a completely different methodology, Butler *et al.* (17) administered a small preload of an 8 g mixture of four amino acids or placebo (32 kcal) half an hour before offering a cooked midday meal. The amino acid preload led to a 10 % reduction in energy intake from this meal compared to placebo.

Other studies have described the effects of high protein loads or meals on carefully measured scales of subjective motivation to eat. Spring *et al.* (18) found that subjects reported feeling significantly more full after eating 227 g of turkey breast (high in protein) than after an equicaloric amount of high carbohydrate sherbert. Likewise, we found that eating a high protein lunchtime meal led subjects to rate themselves as having less desire to eat and as feeling more full than after an equicaloric and equivolume high carbohydrate meal (19).

There are two notable exceptions to the pattern of results described above. Geliebter (20) failed to find any differential effect on intake one hour after the ingestion of equicaloric single nutrient (egg albumin, corn starch or corn oil) liquid preloads. However, Geliebter's subjects were in the unusual state of being topically anaesthetised and wearing nose clips in order to disguise the (very unpalatable) preloads, a condition which may have influenced their later intake. Similarly, Sunkin and Garrow (21) failed to detect the influence of high or low protein cooked meals on later voluntary intake by obese subjects. However, three features of this study may account for this outcome. First, the low protein meal weighed over 20 % more than the high protein meal. Second, the test food was offered between $4\frac{1}{2}$ and $5\frac{1}{2}$ hours later. Third, the test food offered was exceptionally bland — milk, yoghurt and plain white bread.

One unforseen outcome from our study (19) was that the high protein meal brought about a qualitative change in food preference in addition to the quantitiative decrease in eating motivation. More specifically, the high protein lunch decreased preference for high protein foods. This change was not observed after the high carbohydrate meal, nor was it reciprocal, i. e. the high carbohydrate meal did not selectively reduce preference for high carbohydrate food. In order to test whether these measured subjective changes were present when real foods were offered, we have conducted a further series of studies. In the first, nine male and female subjects ate two fixed lunchtime meals (442 g, 580 kcal) on separate occasions. One was high carbohydrate (59 % of total energy) and low protein (13 %), the other was high protein (45 %) and low carbohydrate (28 %). Three hours later they returned and were offered buffet-style teatime meals from which they could choose to eat at least nine different (and generally liked) food items. Subjective motivation to eat and food preferences were measured throughout the experimental period and the amount of food eaten calculated to provide energy and macronutrient intakes. Table 1 shows that the high protein lunch significantly reduced both teatime energy intake (by 12 %) and protein intake (by 23 %), but left carbohydrate intake relatively unchanged.

TABLE 1
Summary of three studies evaluating the effects of high protein meals on energy and macronutrient intakes

Study	Subjects		Change in energy intake (kcal)	Change in protein intake (g)	Change in CHO intake (g)
1	Lean	N = 9	- 11.6 %*	- 22.8 %*	- 1.1 %
2	Lean	N = 9	- 21.7 %*	-31.7 %*	— 12.8 %*
3	Obese	N = 8	- 18.8 %*	- 15.0 %*	- 24.7 %*

*$p < .05$

The procedure for studies 2 and 3 was similar except for a small change in the composition of the lunchtime meals, the lunches being slightly smaller (422 g, 475 kcal) and different in composition (high protein lunch was 54 % of total energy, high carbohydrate lunch 63 %). In addition, the responses of lean and obese women were compared. Both groups of subjects significantly reduced their energy intakes following the high protein lunch (compared to the high carbohydrate lunch). However, they differed in the extent to which high protein and high carbohydrate foods were selected. The lean subjects showed a marked 32 % decrease in protein intake following the high protein meal, together with a significant 13 % decrease in carbohydrate intake. The obese subjects on the other hand, decreased their carbohydrate intake to a greater extent than protein intake, although both changes were statistically significant. One possible interpretation of the difference in behaviour of the lean and obese women lies in their stated preference for the foods providing these macronutrients. Whereas the lean subjects liked both the high protein and high carbohydrate foods, the obese women preferred the savoury protein and carbohydrate foods to the sweet carbohydrates. The obese women therefore, appear to have modified their choice of foods, in their state of decreased hunger, partly as a function of their initial liking of the foods and have sacrificed eating the sweet carbohydrates before the savoury items.

The majority of the studies therefore evaluating the short term effects of preloads or meals which are high in protein, clearly demonstrate the particular satiating power of protein. The superiority of protein over carbohydrate has been detected by subjective measures of eating motivation and in the voluntary intake of both lean and obese individuals. Moreover, there is some evidence that within the overall protein-induced anorexia, there is some specific avoidance of other high protein foods.

Tryptophan — the Action of a Single Amino Acid

Tryptophan occupies an almost unique place in studies of brain-behaviour relationships, and as such has received special scientific attention. As a precursor of a major neurotransmitter, the action of tryptophan has been investigated in many of the psychological domains thought to be incluenced by serotoninergic input e. g. feeding, sleep, pain and mood. Reviews of the involvement of serotonin in feeding behaviour have described a role for this neurotransmitter in overall energy regulation (22, 23). Other authors have placed more emphasis on the proposal that serotonin has some function in the control of nutrient intake (6, 24, 25). However, the evidence for the latter is tantalizing, but as yet, far from convincing (26).

Animal studies have indicated both properties of tryptophan, reducing overall energy intake (27, 28) and latering macronutrient selection (29), but these effects are not apparent simultaneously. In man, it appears much easier to highlight the anorectic potential of tryptophan, as long as the dose administered is 2 g or more. Rogers *et al.* (30) found 4 g tryptophan to reduce energy intake by 16 % compared to placebo when offered a test meal an hour after dosing. Hrboticky *et al.* (31) reported that doses of 1, 2 and 3 g tryptophan led to 5, 19 and 20 % decreases in energy intake 45 minutes after dosing. Similarly, Silverstone and Goodall (32) found no effect of 0.5 or 1 g, but 2 g tryptophan did significantly reduce energy intake by 10 % compared to placebo. However, in none of the above studies was the selective action of tryptophan on food choice clearly demonstrated, even though provision had been made in the experimental design for its detection. The only researchers to claim clear tryptophan-induced changes in macronutrient choice are the Wurtmans (33, 34). Unfortunately, subjects' responses are idiosyncratic and the paradigm chosen for study — carbohydrate craving — precludes a fair analysis as protein intake (by definition) is naturally suppressed compared to carbohydrate intake.

The physiology of tryptophan, once present in the body, suggests alternative methods of administration to that of acute loading in the manner of a drug. The nutritional context within which tryptophan may be given is a determinant of tha availability of tryptophan at the blood brain barrier. Tryptophan taken with a high carbohydrate drink markedly increases the plasma tryptophan/LNAA ratio for up to 3 hours compared to the high carbohydrate drink alone or a high protein drink (35). Mixing 2 g tryptophan with pure orange juice (94 % carbohydrate), we found tryptophan to potentiate the satiating effect of a meal, but when the meal was high protein rather than high carbohydrate (36).

To further investigate the action of tryptophan on satiation and satiety, we designed an experiment which included three additional methodological refinements. First, tryptophan was administered as part of a chocolate snack (54 % fat, 41 % carbohydrate) during mid-morning. Second, the procedure of temporal tracking (e. g. 37) was used to monitor subjective eating motivation before, during and after a lunchtime meal of fixed energy and nutritional content. This permitted the satiating effect of tryptophan to be described without overall differences in free food intake compromising the action of tryptophan. Third, any long-lasting effect of tryptophan was measured in diary records of food intake started once free eating was again allowed three hours after lunch. The fourteen female adults who took part in the study consumed 30 g chocolate (158 kcal) which contained either 1.5 g tryptophan or no addition (within subjects design), two hours before a small lunch (440 kcals, 16 % protein, 59 % carbohydrate) of salad sandwiches and a fruit yoghurt. Ratings of hunger, desire to eat, fullness and prospective consumption ('How much food do you think you could eat?'), mood and food preference measures were completed during the two hours before lunch, during lunch itself and for three hours after lunch.

The temporal profile of ratings of desire to eat and prospective consumption are shown in Figure 2. Consuming the chocolate snack brought about a significant decrease

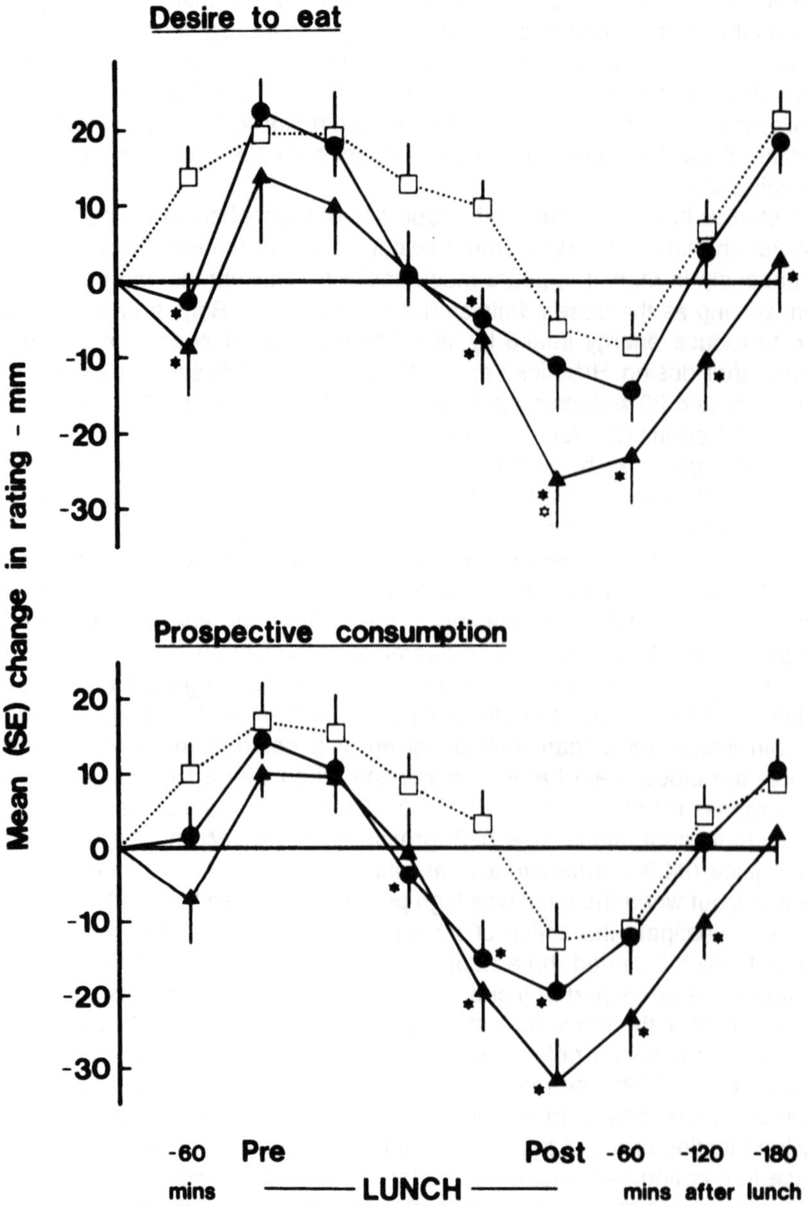

FIGURE 2

The effect of a tryptophan-supplemented chocolate snack (triangles), chocolate alone (circles) or no snack (squares) on the mean change in ratings of desire to eat and prospective consumption from pre-snack baseline. These subjective ratings were tracked across the course of a fixed lunchtime meal and for three hours after eating. Closed symbols indicate a significant change from pre-snack baseline ($p < .05$). Closed stars show a significant difference from no snack control, and open stars a significant difference from placebo chocolate snack ($p < .05$).

in desire to eat one hour later compared to the no preload profile, but eating motivation returned to control levels by the beginning of lunch. During the meal there was a steady decline in motivation to eat. There were significant with-in meal preload x time interactions for prospective consumption ($F(8,104) = 3.21$, $p < .01$) and desire to eat ($F(8,104) = 2.10$, $p < .05$). Post hoc comparisons revealed that both chocolate snacks significantly reduced motivational ratings compared to the no preload control about half way through the meal. It is clear from the profile of these ratings that the presence of tryptophan served to further decrease eating motivation during the latter part of the meal. This effect was persistent, and the significant main effect of preload on all ratings of motivation to eat (smallest $F(2, 26) = 5.31$, $p < .05$) during this three hour period is mostly due to the continued action of tryptophan.

The pattern of food preferences during the study followed that of the motivational ratings, but apart from the significant decrease in high carbohydrate foods selected from the checklist following chocolate consumption, no selective effect of the tryptophan supplement on macronutrient preference was observed. The diary records of food consumed once free-feeding was permitted revealed that during the next two hours, the prior ingestion of tryptophan reduced intake by approximately 100 kcals and carbohydrate intake by 17 g. However, the pattern of intakes had a high degree of associated variability and neither of these suppressions were statistically significant. Moreover, this 27 % drop in carbohydrate intake was accompanied by a 22 % decrease in protein intake. These effects of tryptophan were washed-out by intakes later in the evening.
tryptophan were washed-out by intakes later in the evening.

This study has clearly articulated the satiating effect of tryptophan and has provided a temporal profile of its action in given nutritional circumstances. Furthermore, it has shown tryptophan to amplify the satiating power of a small meal, an effect which may have practical applications. However, there was no evidence of a nutritionally selective influence. Is there a methodology more suited to the expression of changes in macronutrient preference? One possibility is to incorporate tryptophan into a meal and to compare the action of tryptophan embedded in a high carbohydrate meal with that of a high protein meal. Will tryptophan potentiate the satiating power of the high protein meal and will it, combined with carbohydrate, modify macronutrient preference? In fact we found that 1 g tryptophan (in chocolate, offered as part of the dessert) did add to the protein meal-induced suppression of hunger motivation and food intake in a self-selection test meal offered three hours after lunch (38). The addition of tryptophan brought about a further 9 % reduction in energy intake on top of the 12 % reduction brought about by the high protein lunch. Adding 1 g tryptophan to the high carbohydrate lunch however, resulted in a 2 % increase in energy intake, and no observable alteration in macronutrient preference of selection. Instead, it was the tryptophan plus high protein lunch which led to the only observed selective effect on macronutrient choice, significantly decreasing carbohydrate selection in the teatime meal. While unexpected, the particular appetite suppressant capacity of the tryptophan plus high protein meal condition is consisitent with previously mentioned findings (36). In addition, it is interesting to note that tryptophan did exert some selective action on macronutrient intake (a suppression of carbohydrate) at a time when there was a relative preference for carbohydrate (i. e. after a high protein meal).

To summarise, tryptophan has been shown to have mild anorectic properties. This amino acid, in amounts upward of 1.5 g, reduces voluntary food intake and hunger motivation. Careful analysis of the time course of its effects seems to reveal a likely mode of action — potentiating the process of satiation and prolonging the state of satiety. D-fenfluramine, an anorectic drug which enhances serotoninergic activity, has been shown to express its action in a similar manner (39, 40). It is likely therefore, that both compounds exert their effects through changing the availability and activity of serotonin. Selective effects of tryptophan (and other serotoninergic agents) on food intake appear to be detectable only when there is a high demand for carbohydrates.

CONCLUSIONS

The researches described in this chapter indicate that amino acids may influence appetite control in a number of ways. First, protein itself has been demonstrated to exert both general and specific effects on the pattern of food consumption. Second, research with a single amino acid — tryptophan — has revealed mild but definite effects on the processes of satiation and satiety. These effects require careful research designs and sensitive techniques in order to be disclosed. Under particular experimental circumstances which generate a high demand for carbohydrate, tryptophan may display a selective effect on macronutrient intake. The time course of this effect requires further analysis before the mechanism can be identified. Further researches may indicate that other single amino acids — in addition to tryptophan — will exert similar effects. At the present time it cannot be claimed with certainty that the effects reported above are specific for tryptophan. However, it is clear that a further understanding of the role of amino acids in appetite control should aim for a synthesis between the effects of protein in general and of individual amino acids in particular. The mechanisms underlying these various actions — both robust and subtle — remain to be identified and we suggest that special consideration be given to the balance between post-ingestive and post-absorptive processes.

ACKNOWLEDGEMENTS

The authors are grateful of Dr. Peter Leathwood for scientific collaboration with the tryptophan research and to Caroline Davey who carried out the tryptophan/chocolate snack study.

REFERENCES

1) BLUNDELL, J. E. (1983). Processes and problems underlying the control of food selection and nutrient intake. In: *Nutrition and the Brain* (Wurtman, R. J., Wurtman, J. J., eds), **Vol. 6,** Raven Press, New York, pp. 164—221.

2) MELLINKOFF, S. M., FRANKLAND, M., BOYLE, D., GREIPEL, M. (1956). Relationship between serum amino acid concentration and fluctuations in appetite. *J. Appl. Physiol.* **8:** 535—538.

3) HARPER, A. E. (1976). Protein and amino acids in the regulation of food intake. In: *Hunger: Basic Mechanisms and Clinical Implications* (Novin, D., Wyrwicka, W., Bray, G., eds.), Raven Press, New York.

4) GIBBS, J., FALASCO, J. D., McHUGH, P. R. (1976). Cholecystokinin-decreased food intake in rhesus monkeys. *Am. J. Physiol.* **230:** 15—18.

5) MEI, N. (1985). Intestinal chemosensitivity. *Physiol. Rev.* **65:** 211—237.

6) WURTMAN, R. F., HEFTI, F., MELAMED, E. (1981). Precursor control of neurotransmitter synthesis. *Pharmacol. Rev.* **32:** 315—335.

7) PARDRIDGE, W. M. (1983). Brain metabolism: a perspective from the blood-brain barrier. *Physiol. Rev.* **63:** 1481—1535.

8) FERNSTROM, J. D., WURTMAN, R. J. (1972). Brain serotonin content: physiological regulation by plasma neutral amino acids. *Science* **178:** 414—416.

9) BLUNDELL, J. E., ROGERS, P. J., HILL, A. J. (1988). Evaluating the satiating power of food: implications for acceptance and consumption. In: *Chemical Composition and Sensory Properties of Food and their Influence on Nutrition* (Solms, J., ed.). Academic Press London, in press.

10) BLUNDELL, J. E., HILL, A. J. (1986). Biopsychological interactions underlying the study and treatment of obesity. In: *The Psychosomatic Approach: Contemporary Practice of Whole-Person Care* (Christie, M. J., Mellet, P. G., eds.), Wiley, Chichester, pp. 113—138.

11) BLUNDELL, J. E., HILL, A. J. (1987). Descriptive and operational studies of eating in man. In: *Modern Concepts of the Eating Disorders: Diagnosis, Treatment, Research* (Blinder, B. J., Chaitin, B. F., Goldstein, R., eds.), Spectrum, New York, in press.

12) DAVIDSON, S., PASSMORE, R., BROCK, J. F., TRUSWELL, A. S. (1975). *Human Nutrition and Dietetics.* Churchill-Livingstone, Edinburgh.

13) ORENT-KEILES, E., HALLMAN, L. F. (1949). The breakfast meal in relation to blood-sugar values. *US Dept. Agriculture Circ.* **827:** 1—24.

14) OHLSON, M. A., HART, B. P. (1965). Influence of breakfast on total day's food intake. *J. Am. Dietet. Assoc.* **47:** 282—286.

15) FRYER, J. H., MOORE, N. S. WILLIAMS, H. H., YOUNG, C. M. (1955). A study of the interrelationships of the energy-yielding nutrients, blood glucose levels and subjective appetite in man. *J. Lab. Clin. Med.* **45:** 684—696.

16) BOOTH, D. A., CHASE, A., CAMPBELL, A. T. (1970). Relative effectiveness of protein in the late stages of appetite suppression in man. *Physiol. Behav.* **5:** 1299-1302.

17) BUTLER, R. N., DAVIES, M., GEHLING, N. J., GRANT, A. K. (1981). The effect of preloads of amino acid on short-term satiety. *Am. J. Clin. Nutr.* **34:** 2045—2047.

18) SPRING, B., MALLER, O., WURTMAN, J., DIGMAN, L. (1982/3). Effects of protein and carbohydrate meals on mood and performance. *J. Psychiat. Res.* **17:** 155—167.

19) HILL, A. J., BLUNDELL, J. E. (1986). Macronutrients and satiety: the effects of a high protein or high carbhydrate meal on subjective motivation to eat and food preferences. *Nutr. Behav.* **3:** 133—144.

20) GELIEBTER, A. A. (1979). Effects of equicaloric loads of protein, fat and carbohydrate on food intake in the rat and man. *Physiol. Behav.* **22:** 267—273.

21) SUNKIN, S., GARROW, J. S. (1982). The satiety value of protein. *Human Nutr.: Appl. Nutr.* **36A:** 197—201.

22) BLUNDELL, J. E. (1977). Is there a role for serotonin (5-hydroxytryptamine) in feeding? *Int. J. Obesity* **1:** 15—42.

23) BLUNDELL, J. E. (1984). Serotonin and appetite. *Neuropharmacol.* **23:** 1537—1551.

24) ANDERSON, G. H. (1979). Control of protein and energy intake: role of plasma amino acids and brain neurotransmitters. *Can. J. Physiol. Pharmacol.* **57:** 1043—1057.

25) ASHLEY, D. V. M. (1985). Factors affecting the selection of protein and carbohydrate from a dietary choice. *Nutr. Res.* **5:** 555—571.

26) BLUNDELL, J. E., HILL, A. J. (1987). Nutrition, serotonin and appetite: case study in the evolution of a scientific idea. *Appetite* **8:** 183—194.

27) LATHAM, C. J., BLUNDELL, J. E. (1979). Evidence for the effect of tryptophan on the pattern of food consumption in free feeding and food deprived rats. *Life Sci.* **24:** 1971—1978.

28) LI, E. T. S., MacMILLAN, M. L., ANDERSON, G. H. (1986). Food intake and selection after peripheral tryptophan. *Fed. Proc.* **45:** 590.

29) LI, E. T. S, ANDERSON, G. H. (1984). 5-Hydroxytryptamine: a modulator of food composition but not quantity? *Life Sci.* **34:** 2453—2460.

30) ROGERS, P. J., BINNES, D., McARTHUR, r. A., BLUNDELL, J. E. (1979). Effect of tryptophan and anorectic drugs on food intake, hunger, food selection and the micro-structure of eating in man. *Int. J. Obesity* **3:** 94.

31) HRBOTICKY, N., LEITER, L. A., ANDERSON, G. H. (1985). Effects of l-tryptophan on short term food intake in lean men. *Nutr. Res.* **5:** 595—607.

32) SILVERSTONE, T., GOODALL, E. (1984). The clinical pharmacology of appetite suppressant drugs. *Int. J. Obesity* **8 (Suppl. 1):** 23—33.

33) WURTMAN, J. J., WURTMAN, R. J. (1981). Suppression of carbohydrate consumption as snacks and at mealtimes by DL-fenfluramine or tryptophan. In: *Anorectic Agents — Mechanisms of Action and Tolerance* (Garattini, S., Samanin, R., eds.), Raven Press, New York, pp. 169—182.

34) WURTMAN, J. J., WURTMAN, R. J., GROWDON, J. H., HENRY, P., LIPSCOMB, A., ZEISEL, S. H. (1981). Carbohydrate craving in obese people: suppression by treatments affecting serotoninergic transmission. *Int. J. Eating Disorders* **1:** 2—15.

35) ASHLEY, D. V. M., LIARDON, R., LEATHWOOD, P. D. (1985). Breakfast meal composition influences plasma tryptophan to large amino acid ratios of healthy lean young men. *J. Neural Trans.* **62.**

36) BLUNDELL, J. E., MAVJEE, V., HILL, A. J. (1987). Complex interactions between tryptophan and macronutrients on appetite, mood and performance. In: *Progress in Tryptophan and Serotonin Research 1986* (Bender, A. E. et al., eds.), Walter de Gruyter, Berlin, pp. 95—98.

37) HILL, A. J., MAGSON, L. D., BLUNDELL, J. E. (1984). Hunger and palatability: tracking subjective experience before, during and after the consumption of preferred and less preferred food. *Appetite* **5:** 361—371.

38) BLUNDELL, J. E., HILL, a. J. (1987). Interaction between tryptophan and macronutrients on hunger motivation and dietary preferences. In: *Amino Acids in Health and Disease: New Perspectives* (Kaufman, S., ed.), Alan R. Liss, New York, pp. 403—419.

39) HILL, A. J., BLUNDELL, J. E. (1986). Model system for investigating the actions of anorectic drugs: effect of d-fenfluramine on food intake, nutrient selection, food preferences, meal patterns, hunger and satiety in healthy human subjects. In: *Advances in the Biosciences* **Vol. 60.** *Disorders of Eating Behaviour: a Psychoneurodendocrine Approach* (Ferrari, E., Brambilla, F., eds.), Pergamon, Oxford, pp. 377—389.

40) BLUNDELL, J. E., HILL, A. J. (1987). Serotoninergic modulation of the pattern of eating and the profile of hunger — satiety in humans. *Int. J. Obesity* **11 (Suppl. 3):** 141—155.

AMINO ACIDS IN THE REGULATION OF FOOD INTAKE AND SELECTION

G. Harvey Anderson, Robert J. Bialik, Edmund T. S. Li

Department of Nutritional Sciences
University of Toronto
Toronto, Ontario, Canada M5S 1A8

INTRODUCTION

Until approximately ten years ago the prevalent view of the role of amino acids in food intake regulation was that they acted primarily as inhibitors when consumed in excess or in imbalanced proportions (1). There is now, however, evidence to show that proteins and amino acids function in physiological amounts, as consumed in normal diets, to influence food intake and possibly food choice (2, 3).

DIETARY AMINO ACIDS AND FOOD INTAKE

A relationship between amino acid metabolism and the regulation of food intake was proposed in the 1940's and 50's (2, 4). These early studies provided a basis for the aminostatic theory of the control of feeding, which suggested that shifts in plasma amino acid patterns are monitored by the brain to mediate feeding behavior.

The aminostatic theory of food intake regulation received considerable attention during the 1960's. Most researchers used the experimental approach of feeding extremes in dietary protein content or composition to establish that the animal (usually the rat) responds to these extremes by decreasing food intake (1). Diets high (> 40 %) or low (< requirements) in protein, imbalanced with amino acid mixtures or excesses (e. g., 5 % of the diet) of single amino acids, bring about decreased food intake, or if a choice is available, the selection of a more balanced diet. Associated with these extremes in diet composition are distortions in plasma and brain amino acid patterns (1, 2, 5). Thus the data derived over the past thirty years have supported the original hypothesis that changes in plasma and brain amino acid concentrations and patterns influence feeding behavior.

These studies have not, however, identified a specific mechanism of action for amino acids in food intake regulation, nor do they provide support for an involvement of proteins or amino acids in the regulation of *normal* feeding behavior. For example, while the use of amino acid imbalanced diets for the purposes of elucidating food intake regulatory mechanisms is of interest as a model system, it should be recognized as a model of nutritional pharmacology. Even if studies of imbalanced diets lead to the description of a control mechanism, it does not necessarily mean that it would be the mechanism operating in determining the animal's feeding response to balanced diets. Similarly, it should be

NATO ASI Series, Vol. H20
Amino Acid Availability and Brain Function in Health and Disease. Edited by G. Huether

recognized that the mechanisms involved in controlling the animal's response to high- or low-protein diets may be different and that these remain to be elucidated.

In the mid 1970's we hypothesized that the animal's feeding behavior was influenced by protein content of diets (food) because of the essentiality of amino acids to its health (6). Under normal circumstances of foraging and food choice an animal must be able to detect nutrient content of foods in order to survive. Thus we felt that if protein intake regulation could be demonstrated, it would be possible to define those characteristics of protein that affect either food choice or food intake, or both (within the context of normal feeding patterns), and to define the control mechanism(s) operating under these conditions (2, 3).

PROTEIN INTAKE REGULATION

The concept of protein intake regulation is not new. In the early 1900's Osborne and Mendel (7) gave rats a choice of high- and low-protein diets and reported that the rats consumed a nutritionally adequate and constant intake of protein. Although studies on nutrient selection were conducted in the 1940's and 50's, the results of these were most likely influenced by a lack of knowledge of nutrient requirements of the rat and the consequent feeding of nutrient incomplete diets (2, 8). In the late 1960's Rozin (8) renewed interest in the subject when he reported that rats fed a choice of liquid forms of fat, carbohydrate and protein consumed a constant intake of protein. Even when the protein was fed in more dilute solutions, the rats increased the volume of intake to compensate.

Our investigations of protein intake regulation initially involved feeding a choice of nutrient defined diets to weanling rats for two to four weeks (6, 9, 10). The two isocaloric diets differed only in protein and carbohydrate content and contained 10 % fat and equal micronutrient concentrations. Over a wide range of dietary choices, rats regulated their absolute intake of protein as well as maintained their protein intake at a constant proportion of their caloric consumption. For example, when the dietary choices offered were 0 and 50, 5 and 45, 15 and 55, or 25 and 65 % casein, rats selected an average of 134 grams of protein over a period of four weeks. Protein intake constituted about 33 % of total energy consumption and was only slightly more variable within a group of rats than was the intake of total energy (6).

Many investigators have observed that protein intake is regulated at a relatively constant proportion of dietary energy if animals are given a choice of diets (2, 3, 10). However, the proportion of the diet selected as protein by rats varies from laboratory to laboratory, even when the same protein source is used. This may be due to the fact that protein intake regulation is easily influenced by many experimental conditions, including nutrient composition, texture and palatability of the diet, prior dietary experience and age of the rats. For example, rats consistently consume less protein and select a lower proportion of their energy as protein when given gel diets compared to powder or granular diets (11). The inclusion of fat in addition to carbohydrate and protein as a separate choice (12) also affects the pattern of food selection, usually reducing the proportion of energy consumed from carbohydrate and protein. Because high fat diets are known to be very palatable to rats and are comparatively energy dense, this result is perhaps not surprising. Although carbohydrate is the major choice displaced by fat, protein intake is also frequently reduced (12, 13).

Age and prior dietary experience of rats also affect their choice from high- and low-protein diets. For example, Leathwood and Ashley (14) found that five 260 g rats, maintained on rat chow and then introduced to a selection paradigm based on powdered diets, took several days to develop constant and adequate (but highly individual) intakes of protein. Similarly, if the amount of work required to obtain protein is increased, the rat readily economizes by reducing its intake to approximately requirement levels (15). However, rats will defend this lower level, perhaps in an effort to prevent malnutrition.

Further evidence for protein intake regulation arises from short-term (i. e. meal-to-meal) studies of macronutrient selection. We were the first to report that the composition of recently ingested food influences subsequent food intake and choice (16). Rats given a choice of high- (60 %) and low- (10 %) protein diets (also described as low- (23.5 %) and high- (73.5 %) carbohydrate diets, respectively) 30 minutes after the consumption of a 2 g high-protein meal decrease food intake, especially from the high-protein diet, resulting in a relative preference for carbohydrate in the next meal. These results are in keeping with associations that have been observed between protein and carbohydrate intake from meal to meal when rats are monitored under ad libitum conditions, and have access to low- and high-protein diets (17). After a voluntarily selected high-protein meal the rat takes a longer than average time to start another meal which is selected to contain relatively less protein and more carbohydrate calories than average.

Other laboratories have also provided evidence that rats can regulate their intake of protein and carbohydrate on a short-term meal-to-meal basis. In 1983, Wurtman *et. al.* (18) reported that rats fed a carbohydrate-free diet until ketosis developed and then given an opportunity to select their meal composition, ate significantly greater quantities of a high carbohydrate diet compared to controls. In the same study, the ability of rats to regulate carbohydrate intake was also tested by giving rats a choice of two isocaloric, isonitrogenous diets containing 25 or 75 % dextrin after they were prefed one hour earlier a meal of either dextrose or a diet containing protein. Rats eating the carbohydrate premeal subsequently ate as much total food as the mixed diet fed controls, but significantly less carbohydrate (30 % reduction) and more protein. In 1984, Arimanana and Leathwood observed an increase in preference for protein after rats were prefed with carbohydrate diets. They also suggested that protein and carbohydrate intakes are likely to be regulated on a short-term basis (19).

MECHANISMS OF ACTION OF DIETARY AMINO ACIDS

a) Chronic Intake and Choice

The observations that rats consume constant amounts of protein when given a choice of two diets for several days or weeks suggest that control mechanisms are at work. However, protein intake from dietary choices have been noted to range from requirement to near excess levels, suggesting many possible control mechanisms. It could be that when the choice is of high- (> 40 %) protein or low- (< 10 %) protein diets the rat selects a constant intake within a range of 15 to 40 % protein based on avoiding a pattern of feeding which leads to either excesses or limitations in amino acids. On the other hand, it could be that the constancy of intake observed over several days is achieved by a control mechanism that allows the animal to select for a specific intake level. At the present time the control mechanism(s) involved in protein intake regulation remain unresolved.

Based on the aminostatic hypothesis, we measured plasma amino acid levels in self-selecting rats, and found that patterns and levels were similar in rats selecting similar dietary protein concentrations, even though they were provided different dietary choices (9). It was only after making amino acid additions to the proteins, and thereby changing the protein intake selection level (20), that we found concurrent changes in the plasma amino acid patterns. The best marker of the protein intake selection level was the plasma tryptophan (TRP) to neutral (leucine, isoleucine, valine, phenylalanine, tyrosine) amino acid (NAA) ratio. The morning plasma TRP/NAA in rats allowed to eat overnight was inversely related to their habitual protein intake. For example, in groups of rats fed a choice of casein diets with graded increments of added methionine, the protein intake was reduced from 120 to 80 g per four weeks and the plasma TRP/NAA was increased from 0.07 to 0.11.

A major disadvantage of measuring plasma or brain amino acid patterns after the rat has been provided with dietary choices for several days and achieved a constancy of intake is that the amino acid patterns reflect adaptation in metabolic pathways. Thus it is very hard to determine if the plasma pattern is the result of the intake or if the intake selected is related in some meaningful way to the plasma or brain amino acid pattern. Nevertheless, we hypothesized (9) that the plasma TRP/NAA was not just the result of prior feeding, but was also involved in determining feeding behavior, because the plasma TRP/NAA influences brain TRP uptake and synthesis of brain serotonin, a neurotransmitter involved in the regulation of feeding behavior (3). This specific hypothesis linking a change in plasma amino acid pattern to a neurotransmitter system involved in control of feeding behavior has led to considerable investigation and extensive debate (21, 22).

b) Meal-to-Meal Regulation of Diet Composition

A role for amino acids in mechanisms which assist the animal in distinguishing protein from carbohydrate containing meals, and influence subsequent food intake, is appealing for several reasons. First, large protein and carbohydrate meals influence the concentration of many brain amino acids in opposite directions. A protein containing meal increases brain concentrations of branched-chain amino acids (BCAA). tyrosine and histidine, whereas it decreases brain phenylalanine and tryptophan (23, 24). The decrease in phenylalanine and tryptophan is explained by amino acid competition for uptake by a transport system at the blood-brain-barrier. Plasma BCAA increase much more than phenylalanine and tryptophan as a reflection of their greater concentration in protein combined with an absence of catabolism by liver (25). In contrast to protein, a carbohydrate meal brings about a decrease in BCAA and histidine but an increase in phenylalanine, tyrosine and tryptophan. Much of the increase of the aromatic amino acids in the brain after carbohydrate ingestion can again be explained on the basis of competition for transport, which is made more favourable due to a greater reduction in plasma BCAA. The reduction in the majority of amino acids and especially the BCAA is explained by the enhanced release of insulin after a carbohydrate meal, which stimulates the uptake of amino acids by peripheral tissues (25).

It must be noted, however, that these changes in plasma and brain amino acids have been documented after rats have been deprived of food overnight and then given access to food for two hours. Under these circumstances rats eat 6—10 g meals, which are much larger than the 2—3 g meals they consume in a normal eating pattern (17). Whether or not similar changes in pattern or magnitude would occur after normal meals consumed during ad libitum feeding remains to be determined.

A second reason for hypothesizing a role in food intake control mechanisms for some specific amino acids, namely tryptophan, histidine, tyrosine, and phenylalanine is because they exert some measure of precursor control over the brain neurotransmitters derived from them, serotonin, histamine and the catecholamines, respectively. These neurotransmitters are known to be involved in feeding mechanisms. Thus their sensitivity to diet induced variations in precursor availability would seem logical as part of a feed back loop involving feeding behavior (22). Of course, these and other amino acids could function directly through altered concentration in the brain and through some as yet undefined mechanisms.

Even though putative mechanisms can be formulated for the action of protein on food intake, or for the regulation of protein intake, it is not clear whether the important characteristic of a protein is its amino acid composition in total, or whether a group of amino acids or one or two single amino acids provide the key signals to the brain. For this reason we have undertaken studies of the effects of individual amino acids on food intake and selection.

Our observation that the composition of a meal influenced that of the next meal provided us with a feeding paradigm to use as a basis for investigation of the behavioral effects of the individual amino acids. In meal studies (16), rats are fasted during the light hours (lights on at 0700 h). At 1815 h, a meal of fixed composition is given for 15 minutes. Then at the onset of darkness (1900 h) rats are allowed to feed ad libitum from high-protein (low-carbohydrate) and high-carbohydrate (low-protein) diets. When amino acids are tested they are given via intraperitoneal (ip) injection usually 25—45 minutes prior to food presentation. In our studies each amino acid is tested at several dose levels, thereby avoiding the pitfall of arriving at an erroneous conclusion based on the results of a single dose (26). Furthermore, this is a necessary design because we wish to define the approximate lowest dose which brings about a feeding response under the circumstances of our experimental design.

AMINO ACIDS, FOOD SELECTION AND FOOD INTAKE

We were initially attracted to the notion of a central role for plasma and brain TRP in meal-to-meal influences on food intake and choice for the following reasons. First, twenty minutes after a self-selected meal the plasma TRP/NAA and brain TRP are inversely related to the protein content (ranging from 4—56 %) selected (27). Second, brain tryptophan concentration has a major influence over 5-HT synthesis (2). Third, serotonin is a neurotransmitter known to be involved in both food intake and choice control mechanisms (28).

Despite this rationale, the literature did not contain any consistency with respect to the effects of TRP on food intake. It had been reported that TRP injections both suppressed (29) and had no effect (26) on total food intake. Therefore, we conducted a food selection and intake study (30). In a dose-response (35, 55, 75, 95, and 115 mg/kg), repeated measures design we found that 75 mg/kg was a threshold dose suppressing food intake during the first hour of feeding. The reduction in carbohydrate intake (46 %) was statistically significantly greater than for protein intake (39 %) but the difference was small.

The effect on food selection of TRP injected ip at 75 mg/kg or higher doses was small and provides contrasting results to those obtained when a similar quantity is fed (31). TRP, when added to a 1 g carbohydrate meal enhances the rats preference for protein in the next hour of feeding, decreases its intake of carbohydrate, but leaves unchanged its total food intake. The amount of TRP required to decrease food intake when given with a meal has not been determined. Nevertheless, the food selection data indicate that the mechanism of action of amino acids may be different when they enter the gastrointestinal tract, with or without food, compared with direct entry into the peritoneal cavity.

HISTIDINE, PHENYLALANINE AND THE BRANCHED-CHAIN AMINO ACIDS

Brain uptake of histidine, phenylalanine, and the BCAA is influenced in opposite directions by protein and carbohydrate meals (23, 24). Thus these amino acids might be influential in determining food and macronutrient intake.

In dose response studies we have found that ip injections of either phenylalanine or tyrosine are highly potent in suppressing total food consumption during the first hour of feeding (32). Macronutrient selection is not affected by phenylalanine. The lowest effective doses of 60—90 mg/kg are equivalent to the amount of these amino acids in a 1.5—2 g meal of 20 % casein, thus raising the possibility of their role in appetite signals occurring from consumption of a single meal. At the present time we do not know if the observed effect of these amino acids is mediated via an enhanced synthesis of catecholamines, or by other mechanisms.

Histidine also decreases food intake and does not affect macronutrient selection. However, the lowest effective dose was between 250—375 mg/kg which is equivalent to four and a half to seven times the amount of histidine normally consumed by rats in a meal of a 20 % casein diet. This suggests that the histidine normally consumed by rats in a meal of a 20 % casein diet is unlikely to play a major role in the control of normal feeding responses to single meals.

Branched-chain amino acids influence both food intake and food selection in rats. Both plasma and brain BCAA levels increase after a protein containing meal, but decrease when the meal is of carbohydrate. Leucine, isoleucine and valine, when injected separately at doses of 96, 96 and 172 mg/kg, respectively, suppress total food intake during the first hour of feeding (34). When given in combination, even lower quantities of these amino acids suppress food intake (leucine and isoleucine: 48 mg/kg each, valine: 43 mg/kg). After the first hour, rats recover from the anorexia and exhibit a preference for carbohydrate over protein. These data suggest that the effects of protein on food intake and selection might be related to changes in plasma BCAA. If so, it will be important to determine their mechanism of action.

The quantities of amino acids which are effective, when injected ip, in reducing food intake in the first hour of feeding by rats given a choice of high- and low-protein diets are summarized in Figure 1. The values are expressed as a percentage of that contained in an average daily intake of food, based on the assumption that rats consume on the average 20 g of a 20 % casein diet. The potencies of leucine, tyrosine, phenylalanine and isoleucine are quite similar and are approximately 10 % of their content in a 20 g casein diet. The effective doses for valine, tryptophan, and histidine are 18, 60 and 100 % respectively, of the daily intake. If one assumes that the rat eats an average of even seven meals (17), then it can be calculated that approximately 50 % of the leucine, tyrosine of phenylalanine in a single meal is sufficient, when given by ip injection, to reduce food intake.

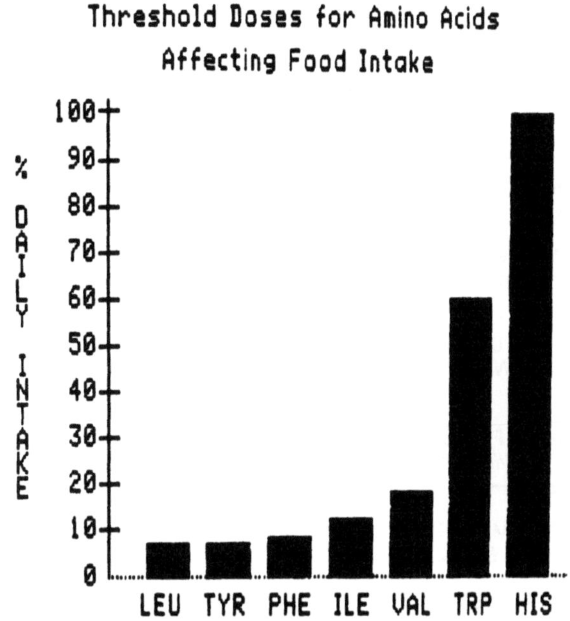

FIGURE 1
Threshold doses for amino acids affecting food intake.

Clearly, many amino acid affect the rat's feeding behavior, although their individual efficacy varies considerably. Whether or not combinations of amino acids in small amounts, consistent with the ingestion of single meals, can account for the food selection behavior of rats remains to be determined. Similarly, it remains to be determined if their mechanisms of action can be accounted for by changes in concentrations of free amino acid pools in plasma and brain and whether the same profile of sensitivity appears if the gastrointestinal tract is not bypassed, as it is by intraperitoneal injections. Nevertheless, the data available to date provide general support for the hypothesis that changes in plasma and brain amino acids during and after meals may be one of the signals influencing food intake and food choice of rats feeding from diets containing nutritionally balanced proteins.

REFERENCES

1) HARPER, A. E., BENEVENGA, N. J., WOHLHUETER, R. M. (1970). Effects of ingestion of disproportionate amounts of amino acids. *Physiol. Rev.* **50:** 428—558.

2) LI, E. T. S., ANDERSON, G. H. (1983). Amino acids in the regulation of food intake. *Nutr. Abstr. Rev. Clin. Nutr.* **53:** 169—181.

3) ANDERSON, G. H., LI, E. T. S., GLANVILLE, N. T. (1984). Brain mechanisms and the quantitative and qualitative aspects of food intake. *Brain Res. Bull.* **12:** 167—173.

4) MELLINKOFF, S. M., FRANKLAND, M., BOYLE, D., GREIPEL, M. (1956). Relationship between serum amino acid concentration and fluctuations in appetite. *J. Appl. Physiol.* **8:** 535—538.

5) ROGERS, Q. R., LEUNG, P. M. B. (1973). The influence of amino acids on the neuroregulation of food intake. *Fed. Proc.* **32:** 1709—1719.

6) MUSTEN, B., PEACE, D., ANDERSON, G. H. (1974). Food intake regulation in the weanling rat: self-selection of protein and energy. *J. Nutr.* **104:** 563—572.

7) OSBORNE, T. B., MENDEL, L. B. (1918). The choice between adequate and inadequate diets, as made by rats. *J. Biol. Chem.* **35:** 19—27.

8) ROZIN, P. (1968). Are carbohydrate and protein intakes separately regulated? *J. Comp. Physiol. Psychol.* **65:** 23—29.

9) ASHLEY, D. V. M., ANDERSON, G. H. (1975). Correlation between the plasma tryptophan to neutral amino acid ratio and protein intake in the self-selecting weanling rat. *J. Nutr.* **105:** 1412—1421.

10) LI, E. T. S., ANDERSON, G. H. (1987). Amino acids in food intake and selection. In: *Amino Acids in Health and Disease: New Perspectives* (Kaufman, S., ed.). New York, Alan R. Liss Inc., pp. 345—368.

11) BLUNDELL, J. E. (1983). Problems and processes underlying the control of food selection and nutrient intake. In: *Nutrition and the Brain* (Wurtman, R. J., Wurtman, J. J., eds.). New York, Raven Press, **Vol. 6,** pp. 163—221.

12) KANARAK, R. B., FELDMAN, P. G., HANES, C. (1981). Pattern of dietary self-selection in VMH-lesioned rats. *Physiol. Behav.* **27:** 337—343.

13) CASTONQUAY, T. W., ROWLAND, N. E., STERN, J. S. (1985). Nutritional influences on dietary selection patterns of obese and lean Zucker rats. *Brain Res. Bull.* **14:** 625—631.

14) LEATHWOOD, P. D., ASHLEY, D. V. M. (1983). Strategies of protein selection by weanling and adult rats. *Appetite* **4:** 97—112.

15) ASHLEY, D. V. M. (1985). Factors affecting the selection of protein and carbohydrate from a dietary choice. *Nutr. Res.* **5:** 555—571.

16) LI, E. T. S., ANDERSON, G. H. (1982). Meal composition influences subsequent food selection in the young rat. *Physiol Behav.* **29:** 779—783.

17) JOHNSON, D. J., LI, E. T. S., COSCINA, D. V., ANDERSON, G. H. (1979). Different diurnal rhythms of protein and non-protein energy intake by rats. *Physiol. Behav.* **22:** 777—780.

18) WURTMAN, J. J., MOSES, P. L., WURTMAN, R. J. (1983). Prior carbohydrate consumption affects the amount of carbohydrate that rats choose to eat. *J. Nutr.* **113:** 70—78.

19) ARIMANANA, L., LEATHWOOD, P. D. (1984). Effects of prior carbohydrate intake on protein/carbohydrate selection by the rat. *Intl. J. Vit. Nutr. Res.* **54:** 283.

20) ASHLEY, D. V. M., ANDERSON, G. H. (1975). Food intake regulation in the weanling rat: the effect of the most limiting essential amino acids of gluten, casein, and zein on the self-selection of protein and energy. *J. Nutr.* **105:** 1412—1421.

21) BLUNDELL, J. E., HILL, A. J. (1987). Nutrition, serotonin and appetite: Case study in the evolution of a scientific idea. *Appetite* **8:** 183—194.

22) ANDERSON, G. H., LI, E. T. S. (1987). Protein and amino acids in the regulation of quantitative and qualitative aspects of food intake. *Int. J. Obesity* (in press).

23) GLAESER, B. S., MAHER, T. J., WURTMAN, R. J. (1983). Changes in brain levels of acidic, basic and neutral amino acids after consumption of single meals containing various proportions of protein. *J. Neurochem.* **41:** 1016—1021.

24) FERNSTROM, J. D., FALLER, D. V. (1978). Neutral amino acids in the brain: Changes in response to food ingestion. *J. Neurochem.* **30:** 1531—1538.

25) MUNRO, H. N. (1983). Metabolism and functions of amino acids in man: overview and synthesis. In: *Amino Acids: Metabolism and Medical Applications* (Blackburn, G. L., Grant, J. P. Young, V. R., eds.). Boston. John Wright PSG Inc., pp. 1—12.

26) PETERS, J. C., BELLISSIMO, D. B., HARPER, A. E. (1984). L-tryptophan injection fails to alter nutrient selection by rats. *Physiol. Behav.* **32:** 253—259.

27) LI, E. T. S., ANDERSON, G. H. (1982). Self-selected meal composition, circadian rhythms and meal responses in plasma and brain tryptophan and 5-hydroxytryptamine in rats. *J. Nutr.* **112:** 2001—2010.

28) LEIBOWITZ, S. F., SHOR-POSNER, G. (1986). Brain serotonin and eating behavior. *Appetite* **7:** Supplement, 1—14.

29) LATHAM, C. J., BLUNDELL, J. E. (1979). Evidence for the effect of tryptophan on the pattern of food consumption in free feeding and food deprived rats. *Life Sci.* **24:** 1971—1978.

30) MORRIS, P., LI, E. T. S., MacMILLAN, M. L., ANDERSON, G. H. (1987). Food intake and selection after peripheral tryptophan. *Physiol. Behav.* **40:** 155—163.

31) LI, E. T. S., ANDERSON, G. H. (1984). 5-Hydroxytryptamine: a modulator of food composition but not quantity? *Life Sci.* **34:** 2453—2460.

32) GEOFFROY, P., ANDERSON, G. H. (1985). The effects of phenylalanine on the feeding behavior of rats. *Proc. Can. Fed. Biol. Soc.* **28:** PA 131.

33) SHEINER, J. B., MORRIS, P., ANDERSON, G. H. (1985). Food intake suppression by histidine. *Pharmacol. Biochem. Behav.* **23:** 721—726.

34) LI, E. T. S., MacMILLAN, M. L., ANDERSON, G. H. (1987). Food intake and macronutrient selection by rats receiving intraperitoneal injections of branched-chain amino acids. *Physiol. Behav.* (submitted).

CONTROL OF EATING BEHAVIOUR BY AMINO ACID SUPPLY

D. A. Booth and E. L. Gibson

Food and Nutrition Laboratory,
Department of Psychology,
University of Birmingham,
P. O. Box 363,
Birmingham B15 2TT.
U. K.

MECHANISMS OF ACTION OF AMINO ACID SUPPLY ON APPETITE

Sensor Specialisation versus Neurotransmission-wide Action

There are two distinct ways in which amino acid supply could influence behaviour. Amino acid supply might control eating behaviour specifically, or eating and other behaviour might be modulated by effects of amino acid supply on neurotransmitters in the pathways organising such behaviour. Given that behaviour is organised by multisynaptic networks, it is extremely unlikely that an effect of precursors on transmitters would have a specific consequence for eating behaviour. If we allow realistically for the nature of behaviour and for the way the brain works, we must expect any control of eating behaviour itself by amino acid supply to be via some specific receptor system. If monoamine precursor metabolism were critical to such a sensor, then these widespread metabolic processes would have to be set within a specialisation that makes the operation of the sensor specific to eating behaviour, preventing it from affecting all the other physiology and behaviour that would be modulated by those same neurotransmitter effects elswhere in the brain or indeed the rest of the body.

However, this is only the first of several distinctions between types of influence on eating behaviour which must be made if research is to advance our understanding of the mechanisms involved. The next simple but crucial distinction is between an effect on all behaviour towards foods and an effect which is selective for behaviour towards some foods and not towards others.

Food-general Effects of Amino Acid Supply

There is evidence for two sorts of effects of amino acid supply on general appetite, with no neurotransmitter precursor action involved in either.

NATO ASI Series, Vol. H20
Amino Acid Availability and Brain Function in Health and Disease. Edited by G. Huether

The first is the food-general anorexia that appears to be caused by excesses of amino acids and imbalances or deficiencies of the indispensable amino acids, whether large neutral or not (1, 2, 3). These may be toxic effects, not mediated by specialised receptors.

The other food-general action of amino acid supply is likely to arise from their oxidation for energy in the liver, which contributes to satiety (suppression of the tendency to eat any type of food), perhaps especially in the late stages of absorption of the previous meal (4). This contribution of protein-derived energy to normal satiety would in no sense be an "aminostatic" mechanism. There is no evidence that the energy from protein and the energy from carbohydrate have separable effects on general appetite (2, 5).

Food-specific Effects of Amino Acid Supply

Food-specific eating behaviour cannot occur unless the organism has mechanisms for recognising particular sensory charactersitics of the foods involved. Therefore, behaviour towards foods specifically for their nutrient content must be controlled by sensory characteristics which are reliable guides as to nutrient content that the organism has some mechanism for identifying. Failure to design experiments to distinguish such nutrient-specific sensory cues has vitiated much work on so-called nutrient selection (6).

There is as yet no evidence for a sensory characteristic which is in common amongst proteins, peptides and free amino acids, nor of any innate effect of amino acid metabolism on the selection of a sensory characteristic of foods, such as sweetness, saltiness, umami or some aroma. Therefore any food-specific behaviour controlled by amino acid supply must rely on an arbitrary connection between certain sensory characteristics of the amino-acid-supplying foods and some effect specific to the action of those amino acids in the body.

That is, the amino acids must have acted to reinforce or to condition a preference or an aversion for some distinctive sensory characteristic which is predictive of that amino acid supply signal to the internal sensor system.

There is indeed much evidence that amino acid imbalance or deficiency can associatively condition an aversion to any distinctive dietary cue (3, 7, 8) and that the repletion of a deficiency in an indispensable amino acid, or the provision of a balanced amino acid mixture (9, 10) or good quality protein (11, 12) to the mildly deprived animal, will condition a sensory preference. That is, after experiencing the amino acid-induced after-effect of exposure to the food cue, the animal will consistently avoid or accept food having such sensory qualities.

At its simplest, this training effect of amino acid supply could condition an aversion or preference that is once and for all. Nevertheless, a consistent change in the consequences of the sensory cue is liable to cause the learned response to extinguish.

By these mechanisms alone, an animal could avoid foods deficient in amino acids and positively select sources of good-quality protein. Toxicity from excess protein, or other deficiencies, could in turn cause the animal to forage elsewhere. In this way, the animal has a chance of maintaining the supply of amino acids in an adequate range over periods of several days.

Such simple aversion and preference conditioning is, however, incapable of making meal-to-meal dietary selection sensitive to current metabolic needs. To enable the current amino acid supply to influence the avoidance or selection of particular learned sensory characteristics, there must be a further, more complicated form of learning. This is the mechanism of an acquired protein-specific satiety or appetite. In the case of a truly protein-specific satiety or anorexia, the animal learns to avoid a dietary cue only while in a condition of amino acid excess, imbalance or deficiency. In the case of a genuinely protein-specific appetite, the animal learns to accept the preferred sensory cue only when amino acid supply is deficient (12, 13). Protein-specific dietary selection has been demonstrated only when the dietary protein-predictive sensory preference or aversion has been shown to depend on the current presence of a protein metabolism-specific deficit or excess.

MEASURING THE EFFECTS OF AMINO ACID SUPPLY ON EATING BEHAVIOUR

The above consideration of the nature of physiological influences on behaviour thus establishes the scientific requirements for determining whether amino acid supply has any control over eating behaviour as such. Amino acid supply must be manipulated without altering anything else in the body or in the sensory qualities of the diet. Eating behaviour must be measured, not as the intakes of diets merely of known nutrient composition, but as the effects of controlled sensory input on the ingestive acts. The usual design of varying the protein content of the diet in a way which also varies its sensory qualities and then weighing intake of the diet over short or long periods gives results which are uninterpretable (6). When an experimental design which measures the effect of amino acid supply on the sensory control of ingestion has been established, however, then evidence of the mediating mechanisms can be obtained either by measuring distinctive correlates of the observed behavioural process or by modulating the fully specified behaviour by interventions with drugs, surgery, etc. (14, 15).

Involvement of Serotoninergic Mechanisms in Dietary Intakes

Some of the dietary or pharmacological manipulations that affect plasma levels of the 5-HT precursor amino acid tryptophan can alter brain levels of tryptophan and 5-HT (16). Neither measurements of correlates nor causal interventions have provided satisfactory evidence, however, that central serotoninergic mechanisms are involved either in the normal variation of food-general appetite and satiety or in the learned food-specific mechanisms in eating behaviour involving genuine selection between protein and carbohydrate (6, 14).

Discussions of the effects of the serotoninergic drug fenfluramine on appetite and satiety have been confused by failures to distinguish between possible mechanisms affecting dietary intake — both those described above and other, more pathological effects. Food intakes, meal-pattern characteristics and dietary choices have been interpreted without adequate consideration of the controls operating in the eating behaviour.

The conventional test of short-term food intake following deprivation involves mainly the consumption of food in one very large meal. The measurement of food intake over several hours in free-feeding situations or after short periods of food deprivation involves a normal pattern of moderately sized meals, plus intermeal intervals which are partly controlled by the size of the preceding meal. Therefore there is no such thing as a single "control of meal size", let alone any justification for naming it "satiation". Also, intermeal interval is not a parameter which reflects a single control mechanism which can be called "satiety".

The suppression of food intake by fenfluramine in freely feeding rats is likely to arise mainly from a slowing of gastric emptying and a consequential increase, *not* in intermeal interval, but in the ratio of intermeal interval to meal size (14, 17—19). This is an action of the drug, not on a mysterious brain function called 'satiety' but on a peripheral physiological mechanism which normally contributes to the appetite-suppressant postingestional effects of food (5, 20). Previous failure to observe an interaction between the effects of fenfluramine on meal patterns in freely fed rats and the peripheral serotoninergic antagonist xylamidine (21) is attributable to use of a single dose and an inappropriate measure of gastric emptying, i. e., meal-to-meal duration instead of the postprandial ratio. With an effective dosing regimen and the correct measure, the peripheral mediation of the "anorexia" (prolonged postingestional satiety) caused by d-fenfluramine under free-feeding conditions has been confirmed (B. J. Baker and D. A. Booth, unpublished data).

The size of a meal after starvation in the usual test of an anorexigenic drug is probably affected by the high arousal of the food-deprived animal. Such behaviour is therefore highly susceptible to nonspecific effects that might be expected from serotoninergic involvement in brainstem mechanisms of activation and sleep (22) and of the sensorimotor control of mastication and swallowing (23) and hence of the response to food texture (24).

Signs of the sedative effects of brainstem serotoninergic activation have been seen in human subjects after administration of fenfluramine (25) and of tryptophan (26, 27), and also after the intake of foods which are liable to affect the transport of tryptophan into the brain (28). We ourselves have recently observed signs of a vigilance-augmenting effect of a food containing considerable amounts of large neutral amino acids with the exception of tryptophan (A. J. H. Gatherer and D. A. Booth, unpublished data).

The involvement of 5-HT in brainstem sensorimotor control of mastication and deglutition means that intake effects of fenfluramine are liable to depend on texture preferences. We have now observed such effects in rats eating chow that has been ground to different sizes of crumb: fenfluramine reduces the preference for larger crumbs (B. J. Baker and D. A. Booth, unpublished data). Furthermore, this effect of fenfluramine on texture preference disappears with the addition of dextrin to the chow crumbs at the level used in low-protein diets. Thus the various effects of fenfluramine on choice between high- and low-casein versions of casein/dextrin mixtures, observed by others, are likely to be confounded by such action of the drug on texture preferences and need not have anything to do with the diets' contents of carbohydrate or protein as such, as those experimenters have discussed.

Indeed, the critical tests for central or indeed peripheral involvement of 5-HT in the signalling of protein appetite, or in its modulation via the transmitter pathways between the protein need or the arbitrary dietary cue and the ingestive behaviour, have proved negative: the deprivation-dependent protein preference and carbohydrate preference in rats after conditioning are both unaffected by intraperitoneal injection of dl-fenfluramine 40 minutes beforehand (1.25 or 2.5 mg/kg) (14, 29, 30).

Mechanism of Action of Amino Acid Supply on Eating Behaviour

To illuminate the brain mechanisms of protein or indeed carbohydrate selection, variation in amino acid supply and its consequences must be localised in time and brain region to the process which controls behaviour. Chronic effects after brain lesions (31) and more direct but less precisely localised evidence from brain injections of conditioning amino acid mixtures (20) implicate the ventral forebrain of the rat, specifically the region of pyriform cortex, as the site of action of a plasma amino acid pattern which could control meal-to-meal dietary selection.

Furthermore, we find that the protein need state that triggers a learned sensory preference for protein is associated with increased ratios of tryptophan, phenylalanine and tyrosine to other large neutral amino acids in plasma and raised levels of those amino acids in ventral forebrain, relative to the replete state which suppresses the protein-induced food preference (32). Possibly relevant to our finding, decreases in noradrenaline content and increases in serotonin turnover have been observed in prepyriform cortex after brief exposure to certain imbalanced diets (33). It remains to be seen whether the cerebral precursor correlates of protein appetite have any causal role, either in signalling the protein need or in modulating it through specific or non-specific neurotransmitter pathway involvement. The limited pharmacological evidence suggests that these amino acid patterns are not signalling or modulating protein/carbohydrate selection through changes in neurotransmitter activity. Like the lack of effect of the serotoninergic drug fenfluramine already mentioned, the catecholaminergic drug d-amphetamine (0.5 and 1.0 mg/kg) has no effect on the learned appetites for either protein or carbohydrate when injected 40 minutes before a test of deficit-dependent sensory preference (29, 30).

Presumably these learned deficit-dependent food preferences (and perhaps learned excess-dependent food aversions) enable rats to control their protein intake at idiosyncratic levels over a very broad range when faced with choices between various mixtures of casein and starch (34). They lose this capacity after trigeminal nerve section and so it seems that this individual regulation of protein intake is achieved by learning to discriminate protein-repleting and non-repleting diets by their textures (35). We are therefore currently testing whether protein-conditioned protein-need dependent sensory preference can be obtained with independently controlled textures, as well as with the experimentally more convenient odours we have been using. The behavioural mechanism is exactly the same — the rat is learning to combine distinctive sensory characteristics of diets with jointly predictive bodily signals of acute nutritional status (36).

Distinguishing Nutrient-oriented from Other Control of Human Diet

The effects of amino acids and anorexigenic drugs on dietary selection in human subjects thus far described in the literature could be secondary to a general suppression of appetite. This is because people who are aware that they might be taking less food than usual could well be deliberately avoiding what they regard as "junk foods". Slimmers indeed would be wise to avoid the attractions of the usual foods eaten with drinks between mixed snacks or full meals (37). Both behavioural strategies are likely to produce decreases in intake of sugars, starches and fats, particularly between meals. Such effects therefore are not a suppressed 'carbohydrate craving' but are accounted for by an avoidance of foods categorised as tempting, non-nutritious or the like. Clearly marketers and health educators consider that nutrition information on food packs will influence some people's dietary selection. We have shown that mere guessing that a food is higher in calories suppresses appetite 20 minutes after it has been eaten (38). A genuine carbohydrate or fat craving can be distinguished from nutrient non-specific snacking by measuring the subjects' beliefs about and attitudes to the foods available and by assessing learned behaviour towards arbitrary sensory characteristics, which signal disguised variations in nutrient content of test foods. People learn to select starch-rich flavours when they are hungry and low-starch flavours when they are full (39, 40). We are currently testing whether this conditioning of appetite and satiety is specific for carbohydrate. Also, we are investigating whether protein appetite can be trained in people as we have shown it can be in rats (12, 13).

CONCLUSION

The existing evidence is that amino acid supply controls eating behaviour by a food-general suppression of appetite by the hepatic energy yield from the amino acids and by a food-specific preference and protein-deficit dependent appetite which has been learned from amino acid pattern-specific repletion. The evidence to date is that, although at least some forms of deficit in amino acid supply cause low brain levels of catecholamines and 5-HT and their precursors, these neurotransmitter effects do not signal or modulate the control of eating by the deficit cue. Neurotransmitter involvement in the appetite-training effect of amino-acid repletion remains to tested, although it may well be no more than removal of the deficit which serves as the internal cue in the learned protein appetite.

ACKNOWLEDGEMENT

This paper was prepared with support from AFRC grant FG 6/142 for research on mechanisms of macronutrient-specific food choices.

REFERENCES

1) HARPER, A. E., BENEVENGA, N. J., WOHLHEUTER, R. M. (1970). Effects of ingestion of disproportionate amounts of amino acids. *Physiol. Behav.* **50:** 428—558.

2) BOOTH, D. A. (1974). Food intake compensation for increase of decrease in the protein content of the diet. *Behav. Biol.* **12:** 31—40.

3) SIMSON, P. C., BOOTH, D. A. (1974). Dietary aversion established by a deficient load: specificity to the amino acid omitted from a balanced mixture. *Pharmacol. Biochem. Behav.* **2:** 481—485.

4) BOOTH, D. A., CHASE, A., CAMPBELL, A. T. (1970). Relative effectiveness of protein in the late stages of appetite suppression in Man. *Physiol. Behav.* **5:** 1299—1302.

5) BOOTH, D. A. (1972). Postabsorptively induced suppression of appetite and the energostatic control of feeding. *Physiol. Behav.* **9:** 199—202.

6) BOOTH, D. A. (1987). Central dietary "feedback onto nutrient selection": not even a scientific hypothesis. *Appetite* **8:** 195—201.

7) BOOTH, D. A., SIMSON, P. C. (1974). Taste aversion induced by an histidine-free amino acid load. *Physiol. Psychol.* **2:** 349—351.

8) SIMSON, P. C., BOOTH, D. A. (1974). The rejection of a diet which has been associated with a single administration of a histidine-free amino acid mixture. *Brit. J. Nutr.* **31:** 285—296.

9) BOOTH, D. A., SIMSON, P. C. (1971). Food preferences acquired by association with variations in amino acid nutrition. *Quart. J. Exp. Psychol.* **23:** 135—145.

10) SIMSON, P. C., BOOTH, D. A. (1973). Olfactory conditioning by association with histidine-free or balanced amino acid loads. *Quart. J. Exp. Psychol.* **25:** 354—359.

11) BOOTH, D. A. (1974). Acquired sensory preference for protein in diabetic and normal rats. *Physiol. Psychol.* **2:** 344—348.

12) BAKER, B. J., BOOTH, D. A., DUGGAN, J. P., GIBSON, E. L. (1987). Protein appetite demonstrated: learned specificity of protein-cue preference to protein need in adult rats. *Nutr. Res.* **7:** 481—487.

13) GIBSON, E. L., BOOTH, D. A. (1986). Acquired protein appetite in rats: Dependence on a protein-specific need state. *Experientia* **42:** 1003—1004.

14) BOOTH, D. A., GIBSON, E. L., BAKER, B. J. (1986). Gastromotor mechanism of fenfluramine anorexia. *Appetite* **7:** Suppl. 57—69.

15) BOOTH, D. A. (1987). How to measure learned control of food or water intake. In: *Methods and Techniques to Study Feeding and Drinking Behavior* (Toates, F. M., Rowland, N., eds.), pp. 111—149, in press.

16) FERNSTROM, J. D. (1987). Food-induced changes in brain serotonin synthesis: Is there a relationship to appetite for specific macronutrients? *Appetite* **8:** 163—182.

17) DAVIES, R. F., ROSSI, J., PANKSEPP, J., BEAN, N. J., ZOLOVICK, A. J. (1983). Fenfluramine Anorexia: A Peripheral Locus of Action. *Physiol. Behav.* **30:** 723—730.

18) ROWLAND, N., CARLTON, J. (1984). Inhibition of gastric emptying by peripheral and central fenfluramine in rats: correlation with anorexia. *Life Sci.* **34:** 2495—2499.

19) ROBINSON, P. H., MORAN, T. H., McHUGH, P. R. (1986). Inhibition of gastric emptying and feeding by fenfluramine. *Amer. J. Physiol.* **250:** R764—R769.

20) BOOTH, D. A., STRIBLING, D. (1978). Neurochemistry of appetite mechanisms. *Proc. Nutr. Soc. Lond.* **37:** 181—191.

21) FLETCHER, P. J., BURTON, M. J. (1986). Dissociation of the anorectic action of 5-HTP and fenfluramine. *Psychopharmacol.* **89:** 210—220.

22) JOUVET, M. (1968). Insomnia and decrease of cerebral 5-hydroxytryptamine after distruction of the raphe system in the cat. *Adv. Pharmacol.* **6:** 265—279.

23) KESSLER, J. P., JEAN, A. (1985). Inhibition of the swallowing reflex by local application of serotonergic agents into the nucleus of the solitary tract. *Eur. J. Pharmacol.* **118:** 77—85.

24) McARTHUR, R. A., BLUNDELL, J. E. (19869. Dietary self-selection and intake of protein and energy is altered by the form of the diets. *Physiol. Behav.* **38:** 315—319.

25) BLUNDELL, J. E., ROGERS, P. J. (1980). Effects of anorexic drugs on food intake, food selection and preference, hunger motivation and subjective experiences. *Appetite* **1:** 151—165.

26) HARTMANN, E., GRAVENS, J., LIST, S. (1974). Hypnotic effects of l-tryptophan. *Arch. Gen. Psychiat.* **31:** 394—397.

27) HRBOTICKY, N., LEITER, L. A., ANDERSON, G. H. (1985). Effects of l-tryptophan on short term food intake in lean men. *Nutr. Res.* **5:** 595—607.

28) LEATHWOOD, P. D., POLLET, P. (1983). Diet-induced mood changes in normal populations. *J. Psychiat. Res.* **17:** 147—154.

29) BOOTH, D. A., BAKER, B. J., GIBSON, E. L. (1986). Learned selection of protein-predictive dietary cues during protein-specific need, tested with 5-HT and CA agents. *Appetite* **7:** 244.

30) BOOTH, D. A., GIBSON, E. L., BAKER, B. J. (1986). Behavioral dissection of the intake and dietary selection effects of injection of fenfluramine, amphetamine or PVN norepinephrine. *Soc. Neurosci. Abst.* **12:** 593.

31) LEUNG, P. M. B., ROGERS, Q. R. (1971). Importance of prepyriform cortex in food-intake response of rats to amino acids. *Amer. J. Physiol.* **221:** 929—935.

32) GIBSON, E. L., BARBER, D. J., BOOTH, D. A. (1987). 5HT and CA precursor levels during protein selection. *Soc. Neurosci. Abstr.* **13:** 15.

33) GIETZEN, D. W., LEUNG, P. M. B., ROGERS, Q. R. (1986). Norepinephrine and amino acids in prepyriform cortex of rats fed imbalanced amino acids. *Physiol. Behav.* **36:** 1071—1080.

34) LEATHWOOD, P. D., ASHLEY, D. V. M. (1983). Strategies of protein selection by weaning and adult rats. *Appetite* **4:** 97—112.

35) MILLER, M. G. (1984). Oral somatosensory factors in dietary selfselection in rats. *Behav. Neurosci.* **98:** 416—423.

36) BOOTH, D. A. (1977). Appetite and satiety as metabolic expectancies. In: *Food intake and chemical senses* (Katsuki, Y., Sato, M., Takagi, S. F., Oomura, Y., eds.), pp. 317—330.

37) BOOTH, D. A., CONNER, M. T., MARIE, S. (1987). Sweetness and food selection: measurement of effects of sweetness on acceptance. In: *Sweetness* (Dobbing, J., ed.), pp. 143—158.

38) BOOTH, D. A., MATHER, P., FULLER, J. (1982). Starch content of ordinary foods associatively conditions human appetite and satiation, indexed by intake and eating pleasantness of starch-paired flavours. *Appetite* **3:** 163—164.

39) BOOTH, D. A., LEE, M., McALEAVEY, C. (1976). Acquired sensory control of satiation in Man. *J. Psychol.* **67:** 137—147.

40) BOOTH, D. A., TOASE, A. M. (1983). Conditioning of hunger/satiety signals as well as flavour in dieters. *Appetite* **4:** 235—236.

TRYPTOPHAN AVAILABILITY IN HUMANS: EFFECTS ON MOOD AND BEHAVIOR

Simon N. Young

Department of Psychiatry
McGill University
1033 Pine Avenue West
Montreal H3A 1A1
Canada

INTRODUCTION

Tryptophan has been tested for its action in a variety of disorders including depression, mania, insomnia, pain, aggression and eating disorders. The majority of the studies are concerned with the antidepressant and hypnotic actions of tryptophan, but although there are more than 30 published studies on each of these topics there is not complete consensus on the efficacy of tryptophan. Some of the controversy probably arises from methodological deficiencies in some of the studies. For example, a study which demonstrates that tryptophan is not significantly different in efficacy from imipramine has not demonstrated that tryptophan is significantly better than placebo, unless the sample size is large. However this is the conclusion that is often drawn from such studies. Another and possibly more important reason for the confusion surrounding the clinical actions of tryptophan is the fact that response to tryptophan may depend in part on the circumstances in which it is given and on the baseline state of the subjects. This, of course, applies to many drugs. Over twenty years ago it was shown that when oxazepam is given to normal subjects it reduces anxiety in high-anxious subjects and increases anxiety in low-anxious subjects (1), and similar effects have been seen with other drugs.

Tryptophan is given in situations where it is thought that potentiation of 5HT function may be therapeutic. However, complex aspects of brain function such as mood, sleep or aggression are the end results of the interaction of many neuronal systems. The end result of potentiating 5HT function will depend on the states of the other neuronal systems and the nature of the interactions between them. At the moment we are a long way from understanding how this can lead to different responses in different situations. However, because tryptophan is a dietary component control of 5-hydroxytryptamine function by tryptophan availability is likely to be influenced by many modulating factors. Thus, the behavioral effects of tryptophan should be influenced by the circumstances in which it is given at least to the same extent that occurs with a drug like oxazepam, or even more so.

The basic assumption behind the clinical use of tryptophan is that it increases 5-hydroxytryptamine (5HT) function. In humans tryptophan can increase the rate of 5HT synthesis up to two-fold (2), but an increase in function must be inferred from alterations in behavior that are believed to be mediated by 5HT. Function implies release of 5HT, and

NATO ASI Series, Vol. H20
Amino Acid Availability and Brain Function in Health and Disease. Edited by G. Huether

the extent to which tryptophan promotes release of 5HT is uncertain. Release is presumably dependent on the firing of 5HT neurons, and tryptophan administration to the rat causes a dose-related decline in raphe unit activity (3). We have approached the study of 5HT release by using CSF 5HT as an index of functionally active 5HT. Two antidepressants, the monoamine oxidase inhibitor pargyline and the uptake inhibitor amitriptyline, increased CSF 5HT about three-fold (4). These are compounds which are thought to potentiate 5HT function acutely, but which cause no gross behavioral changes in the rat. Three treatments which cause the 5HT behavioral syndrome, fenfluramine, pargyline plus tryptophan and 5-hydroxytryptophan plus a peripheral decarboxylase inhibitor, all caused a 16—20 fold increase in CSF 5HT. These results are consistent with the idea that CSF is an index of functionally active 5HT. Unlike the classical antidepressants, which increased CSF 5HT three fold, tryptophan increased CSF 5HT by only 15 % (K. L. Teff and S. N. Young, unpublished data). In these circumstances it is not surprising that the antidepressant effect of tryptophan is less clear cut than that of amitriptyline.

Tryptophan is likely to have a greater effect on 5HT release when 5HT neurons are firing at a faster rate. In the cat elevated levels of behavioral arousal are associated with higher firing rates of raphe neurons (5). Thus, the level of arousal of human subjects given tryptophan may be one of the factors that will determine whether there is a clinical effect.

The discussion below deals with the clinical effects of altered tryptophan levels on mood, aggression and pain. The effects of tryptophan on sleep and food intake in humans are discussed in the chapters by Leathwood and by Blundell and Hill.

Tryptophan as an Antidepressant

The first test of the use of tryptophan as an antidepressant was published by Coppen *et al.* in l963 (6). They showed that the addition of tryptophan to a monoamine oxidase inhibitor (MAOI) potentiated the therapeutic effect significantly. Three other studies produced similar results (7). However, although the ability of tryptophan to potentiate the therapeutic action of MAOI's is well established, it also potentiates the side effects of MAOI's and this combination is used only in treatment resistant patients. The combination of tryptophan with tricyclic antidepressants has been tested in several studies (7). Tryptophan produced little or no potentiation of the tricyclic antidepressants' therapeutic action. More than 30 studies have investigated the antidepressant effect of tryptophan when it is given by itself, but there is no consensus on its efficacy. My own interpretation of the literature is that tryptophan is not a particularly effective antidepressant in severely depressed patients, but that it may be useful in mildly depressed patients. A few small placebo controlled studies on severely depressed inpatients failed to demonstrate efficacy of tryptophan (7). However, the placebo-controlled study of tryptophan with the largest patient groups and the longest duration compared tryptophan with amitriptyline and placebo as given by general practitioners to mildly or moderately depressed outpatients (8) rather than to severely depressed inpatients. Tryptophan was significantly better than placebo and was comparable in efficacy to amitriptyline. Side effects with tryptophan were significantly less than with amitriptyline and similar to those which occurred with placebo.

The fact that tryptophan seems to be an effective antidepressant in mildly depressed patients but not in severely depressed patients shows similarities to its action as a hypnotic. It is effective in mild insomnia but not in severe insomnia (9). Thus, severity of the disease is a factor that helps to predict response in both depression and insomnia. The ability of tryptophan to potentiate the effect of monoamine oxidase inhibitors in severely depressed patients, while tryptophan is effective by itself only in mildly or moderately depressed patients, is in keeping with the biochemical studies mentioned above. Tryptophan potentiates release of 5HT more when it is given to rats which are receiving a monoamine oxidase inhibitor than when it is given by itself, so there may be a relationship between the efficacy of tryptophan and the extent to which it increases 5HT release.

Tryptophan in Mania

Several lines of thought have prompted trials of tryptophan in mania. In experimental animals elevated 5HT diminishes response to various stimuli, while exaggerated responses are a characteristic of manic patients. Tryptophan can act as a hypnotic and thus has sedative properties, while low 5HT has been suggested as one factor involved in the etiology of mania. An analysis of the four studies on the action of tryptophan in acute mania suggests that it has a definite but moderate therapeutic effect (7). If it is true, as suggested above, that a higher state of arousal will augment tryptophan-induced release of 5HT, then the clinical effects of tryptophan should diminish as the patients normalize. Thus, tryptophan may be acting in manic patients as a true mood stabilizer.

The data on the action of tryptophan in depression and mania suggest that it should be tested to determine if it can potentiate the action of lithium or if it can have a prophylactic action when given by itself to bipolar patients. One study suggests it can potentiate the action of lithium in manic and schizoaffective patients (l0), while two case reports indicate a prophylactic action of tryptophan alone in bipolar patients (11, 12). Further studies are needed to follow up on these important leads.

Possible similarities between the profiles of action of tryptophan and lithium are interesting. Both are active in acute mania, but it remains to be seen whether tryptophan has a prophylactic action like lithium. The action of lithium in depression, like that of tryptophan is controversial, while both may be useful in the treatment of aggression (l3, 14). If a similarity between the clinical profiles of tryptophan and lithium is established, it would have implications for our understanding of the mechanisms of action of both drugs.

Effect of Altered Tryptophan Levels on Mood in Normal Subjects

The rationale for the use of tryptophan in depressed patients derives from the indoleamine theory of depression. Low 5HT is obviously not a direct cause of depression, as subjects who have low CSF 5-hydroxyindoleacetic acid (5HIAA) when depressed still have low CSF 5HIAA when recovered. However, they have a higher frequency of depressive episodes than depressive patients without demonstrable 5HT disturbances (l5), suggesting that low 5HT predisposes patients to depression. If this is so, the mood elevating effect of tryptophan in depressed patients may be due to the removal of a factor (low 5HT) which predisposes the patients to depression rather than a direct effect of 5HT on mood. However, studies on altered tryptophan levels in normal subjects throw some light on whether 5HT can alter mood directly.

Of the six studies that looked at the acute effect of tryptophan administration on mood in normal subjects, two found that tryptophan produces euphoria in some subjects (16, 17) while four found no effect on mood (18—21). Thus, the effect on mood, if any, is small and inconsistent. In view of the fact that tryptophan has definite effects on social behavior in vervet monkeys (22), it would seem likely that tryptophan will have some effect on normal humans. Presumably mood scales are not the appropriate measure to detect such changes. In the monkeys tryptophan increased affiliative behaviors (22), so we decided to look at a measure that depends on interpersonal interactions (R. O. Pihl and S. N. Young, unpublished data). Normal males between the ages of 18 and 25 came into the laboratory and filled in various rating scales including the Social Avoidance and Distress (SAD) scale of Watson and Friend (23). This consists of 28 statements concerning response to social situations which the subject has to rate as true or false as it applies to himself. The subjects were given capsules containing 1 g of tryptophan or placebo. They were told to take two capsules at 6 pm and at 10 pm that evening and 2 the next morning with breakfast. They were given a further 2 capsules of the same type when they returned to the laboratory next morning to fill in the rating scales again. For the scores on the SAD scale there was a significant condition by time interaction with scores increasing for the subjects on placebo on the second day and decreasing with tryptophan. This is preliminary evidence suggesting that in humans as in monkeys tryptophan can promote a mental state which can enhance social interactions. It may be that tryptophan elevates mood in some normal subjects only as a secondary effect of its ability to promote affiliative tendencies. This suggestion is, of course, highly speculative and much more work will be needed to elucidate the exact effects of tryptophan in normal subjects and to determine which effects are primary and which secondary.

The studies on tryptophan ingestion in normal subjects leaves open the problem of whether 5HT can, in normal subjects, have a direct effect on mood. One obvious question is what effect a lowering of tryptophan might have on mood. We have studied this using a tryptophan deficient amino acid mixture which causes acute tryptophan depletion. Five hours after administration of the mixture plasma tryptophan was down to one quarter of its normal level (21). However, in control subjects, who ingested a nutritionally balanced mixture of amino acids containing the appropriate amount of tryptophan for a good protein source, plasma tryptophan did not decline. In our first study we found that five hours after ingestion of the mixtures the subjects who had the tryptophan-free mixture had a significant lowering of mood relative to those who ingested the control mixture (21). Lowering of mood was measured both by the depression scale of the Multiple Affect Adjective Checklist and by measuring how distracted the subjects were by a dysphoric passage read to them over headphones while they were performing a proof-reading task. Cognitive theories of depression predict greater distractability of depressed individuals by dysphoric themes.

In a second study using tryptophan-free amino acid mixtures we looked at the influence of environmental setting and cognition on the mood-lowering effect of tryptophan depletion (24). The influence of cognition was tested by either giving or not giving to the subjects information designed to account for any possible peripheral sensations that might be related to depressive affect. In an environmental manipulation the subjects were exposed either to a supportive and comfortable atmosphere (positive environment) or to an unrewarding environment (negative environment). We hypothesized that the mood lowering effect of tryptophan depletion might be diminished by the positive environment and by information about possible peripheral sensations induced by the amino acid mixtures. In fact the mood lowering effect was robust with respect to both cognitive and environmental manipulations. In both studies the lowering of mood was a definite effect but the state produced was not of the type seen in clinical depression. Indeed this would not be expected over the five-hour period of the experiment. Nonetheless our results suggest that low 5HT may contribute to the etiology of clinical depression in some patients, due to its mood-lowering action. If this is the case then tryptophan should be most effective as an antidepressant in patients with low 5HT. This has not yet been tested experimentally.

Tryptophan and Aggression

An extensive animal literature indicates that 5HT can have an inhibitory effect on aggression, making tryptophan an obvious condidate for testing in pathologically aggressive patients. In one small study, carried out at a hospital for mentally ill offenders, we looked at the effect of tryptophan and placebo in 12 male aggressive schizophrenics (14). When the patients were on tryptophan there were significantly less incidents on the ward, compared with when they were on placebo, as measured by a 17-item ward checklist which vervets were given tryptophan supplemented or depleted or nutritionally balanced amino acid mixtures, similar to those which we used with normal human subjects. Five hours after will be needed to determine conclusively whether tryptophan is an effective agent for the control of pathological aggression.

Because of the encouraging results in the clinical study on aggression we looked at the effect of tryptophan supplementation or depletion on aggression in normal subjects (25). Two measures were used. The first was the hostility scale of the Multiple Affect Adjective Checklist. The second was a modified Buss paradigm in which the subjects gave elective shocks to a (nonexistent) partner. The intensity and duration of shock administered were the measure of aggression. The tryptophan manipulations failed to influence response on these measures. This negative result may reflect the inadequacies of the measures used as a true measure of aggression. Alternatively the degree of alteration in 5HT function may have been too small to overcome the homeostatic effect of the other neuronal systems that influence aggression. If the latter explanation has some validity then producing greater alterations in 5HT function might influence aggression. In the introduction we suggested higher levels of arousal, which lead to greater rates of firing of 5HT neurons, might be expected to enhance changes in 5HT release in response to changes in tryptophan levels. This could explain why tryptophan manipulations failed to affect aggressive responding in the normal subjects, even though it seemed to be clinically useful in the pathologically aggressive subjects. The pathologically aggressive patients who showed a good response were impulsively aggressive and would thus be at a higher state of arousal when responding aggressively to stimuli. To test the validity of the hypothesis that level of behavioral arousal would influence the effect of altered tryptophan levels on aggression, we performed an experiment on vervet monkeys (26). The vervets were given tryptophan supplemented or depleted or nutritionally balanced acid mixtures, similar to those which we used with normal human subjects. Five hours after administration of the mixtures we measured spontaneous aggression and also competitive aggression. During the measurement of competitive aggression the animals were aroused by competition for food and the rate of aggression was elevated in the control animals, who received the nutritionally balanced mixtures. The results of the experiment were consistent with our hypothesis, as the tryptophan manipulations altered aggression more reliably during competitive aggression than during spontaneous aggression.

Tryptophan and Pain Perception

Animal studies indicate that spinal cord 5HT can have an inhibitory effect on pain perception, making tryptophan an obvious candidate for clinical studies of its analgesic action. In one study 30 patients with chronic maxillofacial pain were given tryptophan or placebo for four weeks (27). In the tryptophan group there was a greater reduction in reported clinical pain and a greater increase in pain tolerance than in the placebo group. This and several other small studies (reviewed in reference 7) suggest that tryptophan can have a useful analgesic effect in some circumstances. However, the type of pain may have an important influence on the effect of tryptophan. Animals studies indicate that in addition to the spinal cord 5HT, which can inhibit pain, there is an ascending 5HT projection which can antagonize morphine analgesia (27). In a recent clinical study we looked at pain perception and morphine requirements in the first few hours after abdominal surgery in patients who received intravenous infusions of either tryptophan or placebo (28). Tryptophan increased morphine requirements. Thus, in pain, as in other clinical conditions the state of the subject may have an important effect on the nature of the response to tryptophan.

CONCLUSIONS

Thirty years of work on the clinical psychopharmacology of tryptophan has produced a few clear answers and some pointers in the direction of the correct questions to ask. The important challenge for the future will be to define not only the clinical conditions which tryptophan can influence but the precise conditions in which they can be influenced and the ways in which the conditions influence the response. The long term goal will be to explain in neurochemical terms the variability in response caused by different baseline conditions.

ACKNOWLEDGEMENT

Work in the author's laboratory is funded by the Medical Research Council of Canada.

REFERENCES

1) DiMASCIO, A., BARRETT, J. (1965). Comparative effects of oxazepam in "high"and "low" anxious student volunteers. *Psychosomatics* **6:** 298—302.

2) YOUNG, S. N., GAUTHIER, S. (1981). Effect of tryptophan administration on tryptophan, 5-hydroxyindoleacetic acid and indoleacetic acid in human lumbar and cisternal cerebrospinal fluid. *J. Neurol. Neurosurg. Psychiatry* **44:** 323—328.

3) TRULSON, M. E., JACOBS, B. L. (1976). Dose-response relationships between systematically administered L-tryptophan or L-5-hydroxytryptophan and raphe unit activity in the rat. *Neuropharmacology* **15:** 339—344.

4) ANDERSON, G. M., TEFF, K. L., YOUNG, S. N. (1987).Serotonin in cisternal cerebrospinal fluid of the rat: a method for its measurement and its use as an index of functional serotonin. *Life Sci.* **40:** 2253—2260.

5) TRULSON, M. E., JACOBS, B. L. (1979). Raphe unit activity in freely moving cats: correlations with level of behavioral arousal. *Brain Res.* **169:** 135—150.

6) COPPEN, A., SHAW, D. M., FARRELL, J. P. (1963). Potentiation of the antidepressant effect of a monoamine-oxidase inhibitor by tryptophan. *Lancet* **i:** 79—81.

7) YOUNG, S. N. (1986). The clinical psychopharmacology of tryptophan. In: *Nutrition and the Brain* (Wurtman, R. J., Wurtman, J. J., eds.), **Vol. 7,** Raven Press, New York, pp. 49-88.

8) THOMSON, J., RANKIN, H., ASCROFT, G. W., YATES, C. M., McQUEEN, J. K., CUMMINGS, S. W. (1982). The treatment of depression in general practice: a comparison of L-tryptophan, amitriptyline, and a combination of L-tryptophan and amitriptyline with placebo. *Psychol. Med.* **12:** 741—751.

9) HARTMANN, E., GREENWALD, D. (1984). Tryptophan and human sleep: an analysis of 43 studies. In: *Progress in Tryptophan and Serotonin Research* (Schlossberger, H. G., Kochen, W., Linzen, B., Steinhart, H., eds.), Walter de Gruyter, Berlin, pp. 297—304.

10) BREWERTON, T. D., REUS, V. I. (1983). Lithium carbonate and L-tryptophan in the treatment of bipolar and schizoaffective disorders. *Am. J. Psychiatry* **140:** 757—760.

11) HERTZ, D., SULMAN, F. G. (1968). Preventing depression with tryptophan. *Lancet* **i:** 531—532.

12) BEITMAN, D. B., DUNNER, D. L. (1982). L-tryptophan in the maintenance treatment of bipolar II manic-depressive illness. *Am. J. Psychiatry* **139:** 1498—1499.

13) SHEARD, M. H. (1975). Lithium in the treatment of aggression. *J. Nerv. Ment. Dis.* **100:** 108—117.

14) MORAND, C., YOUNG, S. N., ERVIN, F. R. (1983). Clinical response of aggressive schizophrenics to oral tryptophan. *Biol. Psychiatry* **18:** 575—578.

15) VAN PRAAG, H. M., DE HAAN, S. (1979). Central serotonin metabolism and frequency of depression. *Psychiatr. Res.* **1:** 219—224.

16) SMITH, B., PROCKOP, D. J. (1962). Central-nervous-system effects of ingestion of L-tryptophan by normal subjects. *N. Engl. J. Med.* **267:** 1338—1341.

17) CHARNEY, D., HENINGER, G. R., REINHARD, J. F., STERNBERG, D. E., HAFSTEAD, K. M. (1982). The effect of IV L-tryptophan on prolactin, growth hormone, and moodin healthy sujects. *Psychopharmacology* **78:** 38—43.

18) GREENWOOD, M. H., LADER, M. H., KANTAMENENI, CURZON, G. (1975). The acute effects of oral (-)-tryptophan in human subjects. *Br. J. Clin. Pharmacol.* **2:** 165—172.

19) LEATHWOOD, P. D., POLLET, P. (1983). Diet-induced mood changes in normal populations. *J. Psychiatr. Res.* **17:** 147—154.

20) LIEBERMAN, H. R., CORKIN, S., SPRING, B. J., GROWDON, J. H., WURTMAN, R. J. (1983). Mood performance, and pain sensivity: changes induced by food constituents. *J. Psychiatr. Res.* **17:** 135—145.

21) YOUNG, S. N., SMITH, S. E., PIHL, R. D., ERVIN, F. R. (1985). Tryptophan depletion causes a rapid lowering of mood in normal males. *Psychopharmacology* **87:** 173—177.

22) RALEIGH, M. J., BRAMMER, G. L., YUWILER, A., FLANNERY, J. W., McGUIRE, M. J., GELLER, E. (1980). Serotonergic influences on the social behavior of vervet monkeys (Cercopithecus aethiops sabaeus). *Exp. Neurol.* **68:** 322—334.

23) WATSON, D., FRIEND, R. (1969). Measurement of social-evaluative anxiety. *J. Consult. Clin. Psychol.* **33:** 448—457.

24) SMITH, S. E., PIHL, R. O., YOUNG, S. N., ERVIN, F. R. (1987). A test of possible cognitive and environmental influences on the mood lowering effect of tryptophan depletion in normal males. *Psychopharmacology* **91:** 451—457.

25) SMITH, S. E., PIHL, R. D., YOUNG, S. N., ERVIN, F. R. (198). Elevation and reduction of plasma tryptophan and their effects on aggression and perceptual sensitivity in normal males. *Aggress. Behav.* **12:** 393—407.

26) CHAMBERLAIN, B., ERVIN, F. R., PIHL, R. D., YOUNG, S. N. (1987). The effect of raising or lowering tryptophan levels on aggression in vervet monkeys. *Pharmacol. Biochem. Behav.* **28:** in press.

27) ABBOTT, F. V., ENGLISH, M. J. M., FRANKLIN, K. B. J., JEANS, M. E., YOUNG, S. N. (1986).Effects of tryptophan loading on pain and morphine requirements after surgery. Paper presented at Fifth International Meeting of the International Study Group for Tryptophan Research. Cardiff, Wales, July l986.

28) ABBOTT, F. V., MELZACK, R. (1982). Brainstem lesions dissociate neural mechanisms of morphine analgesia in different kinds of pain. *Brain Res.* **251:** 149—155.

DIETARY MANIPULATION OF SEROTONIN AND BEHAVIOUR

P. Leathwood

Nestl Research Centre
Vers-Chez-Les-Blanc
CH-1000 Lausanne 26
Switzerland

INTRODUCTION

Tryptophan (TRP), an essential amino acid, is the dietary precursor of the neurotransmitter serotonin (5-Hydroxytryptamine; 5-HT). Brain neurotransmitter metabolism was long thought to be more or less independant of precursor supply, but in 1961 Hess and Doepfer (1) showed that peripheral administration of TRP could increase brain levels of TRP and 5-HT. A few years later, Fernstrom and Wurtman (2) found that in rats starved overnight and then fed a meal containing only carbohydrates, there was also an increase in brain TRP and 5-HT. If the meal was rich in protein, brain TRP and 5-HT were usually unchanged but occasionally fell slightly (2,3,4). They proposed that these dietary manipulations (including intake of TRP) influenced brain 5-HT via the following mechanism: the composition of food consumed influences plasma levels of the large neutral amino acids (LNAA). Since the LNAA all compete for transport across the blood-brain barrier, differential changes in their plasma levels influence the rate of TRP transport into the brain, brain tryptophan levels and the rate of 5-HT synthesis. It has been suggested that, via this mechanism, administration of tryptophan or even changes in diet might influence a range of behaviours linked to serotoninergic neurotransmission (5).

The aim of this review is to examine quantitatively the plausibility and the limits of these ideas, using as examples the arguments that, through the mechanism outlined above, (a) tryptophan acts as a mild sedative, and (b) carbohydrate or protein-rich meals can influence mood and behaviour in man. Current results seem to be coherent with the idea that tryptophan can influence serotoninergic function, although the composition of meals taken at about the same time as the TRP has a strong modulating influence, with carbohydrate potentiating, and protein counteracting tryptophan's effects. As for the suggestion that carbohydrate or protein meals alone can, via changes in plasma amino acids, lower brain TRP and 5-HT enough to have overt behavioural effects, the evidence available so far is not convincing.

NATO ASI Series, Vol. H20
Amino Acid Availability and Brain Function in Health and Disease. Edited by G. Huether

Serotonin and Sleep

It is perhaps appropriate to summarise current evidence as to the probable role of serotonin in the brain, especially with respect to sleep. Most central serotoninergic neurons are found in the Raphé nuclei of the brain stem. From there, fine, slowly conducting fibres project throughout the brain making multiple and often ill-defined connections with others neurons (4). This variability in anatomical specificity led to speculation that 5-HT might act both as a classical neurotransmitter (ie., functioning in a stimulus-response mode) and as a neuromodulator having broad tonic effects. This idea is to some extent supported by pharmacological, lesion and behavioural studies which show that 5-HT can influence a range of behaviours and brain functions including pain sensitivity, sexual behaviour, aggression, temperature regulation, and arousal, with a general tendency for an increase in 5-HT availability to be associated with a tonic "down regulation" of these functions (4,6,7,8).

One of the most consistent effects of treatments which might be expected to increase 5-HT availability or mimic serotoninergic neurotransmission, is an increase in sleepiness, or a facilitating of sleep onset (6,7). Injection of very small amounts of 5-HT into the fourth ventricle, the preoptic area, the medial hypothalamus or the solitary tract nucleus reduces sleep latency and enhances slow wave sleep in the the cat (7). Stimulation of the Raphé has a similar effect. Fenfluramine, a drug which increases serotonin release and prevents its reuptake, also has sedative effects, especially in the higher dose ranges.

Treatments which tend to decrease brain 5-HT consistently produce insomnia. Blocking 5-HT synthesis with parachlorophenylalanine (PCPA) leads to a dose-dependent decrease in 5-HT and, once serotonin supplies are exhausted, a decrease in sleep. Blocking central post-synaptic 5-HT receptors with metergoline or methysergide also reduces sleep (6). Lastly, in rats which have been made insomniac by pretreament with PCPA or with Raphé lesions, giving the serotonin pre cursor 5-hydroxytryptophan leads, after a delay of about 40 minutes, to resumption of sleep (6).

Thus there is abundant evidence that 5-HT is involved in sleep onset and slow wave sleep. It can also influence rapid eye movement sleep, apparently through an independent mechanism (6). Taken together, the results suggest that serotonin lowers arousal and facilitates sleep, but that any sleep effects depends on the total conditions. The mechanism of this sleep facilitation is still poorly understood although a plausible (but unproven) hypothesis has been put forward by Jouvet (6). Noting that 5-HT release and neuronal activity are high during arousal, decrease in slow wave sleep and are low during REM sleep, he proposed that 5-HT release during arousal, by inducing the synthesis and/or liberation of a stable hypnogenic factor, has a cumulative effect, gradually increasing the propensity to fall asleep (see reference 6 for a detailed discussion of this idea).

For the purpose of the present review we can conclude that, if tryptophan really does lead to an increase in 5-HT availability, there is a wealth of evidence that it could aid sleep onset. There is even a suggestion as to the change in 5-HT required. Wojcik et al (9) showed that ip injections of TRP enough to increase brain 5-HT + 5-HIAA by 10—20 % decreased sleep latency in rats without influencing any other sleep parameters. The experimental conditions likely to produce changes of this order are outlined below.

Tryptophan Availability and Serotonin Synthesis

The rate-limiting step in the synthesis of serotonin is the hydroxylation of TRP to 5-hydroxytryptophan by the enzyme tryptophan 5-monooxygenase. (EC 1.14.16.4). The rate of this reaction can be influenced by a variety of factors including TRP concentration, the firing frequency of the neuron, cofactor availability, and calcium-dependent phosphorylation of the enzyme (10). End-product inhibition can apparently occur when brain 5-HT levels are tripled or quadrupled by monoamine oxidase inhibitors, but not at physiological levels of 5-HT (10). Numerous experiments have demonstrated that changes in brain TRP levels can influence the rate of 5-HT synthesis and brain levels of 5-HT and 5-HIAA (4). In man it increases cerebrospinal fluid levels of 5-HIAA (11). Lowering brain TRP, both acutely (12) and chronically (13) can decrease brain concentrations of 5-HT and 5-HIAA.

Quantitatively, it seems that doubling brain TRP from say, 15 to 30 μM, produces a 20—30 % rise in 5-HT and 5-HIAA (3,4). The dose-response curve flattens off as brain TRP reaches about 90 μM (the level at which tryptophan monooxygenase is saturated). Small shifts in brain TRP do not produce reliable changes in brain 5-HT (14,15,16), probably because the minute effects that could be expected are swamped by other factors (listed above) which can also influence 5-HT synthesis.

It is appropriate to ask if there is any evidence that a small increase in neuronal 5-HT actually increases serotonin release (or neurotransmission). One mechanism by which release may be maintained constant in the face of changing intraneuronal 5-HT is that firing rates might adapt. Moderate to large (i. e., 50 mg/kg or more) doses of TRP do seem to slowfiring frequencies of Raphé neurons (10), but small increases in 5-HT (−26 to +12 %) do not (17) and so presumably can influence serotonin release. It is also possible that the "extra" 5-HT might be metabolised intracellularly without entering the functional pool. Experiments with synaptosomes (18), brain slices (19), or in vivo (20) suggest that this does in fact occur, and most excess 5-HT is deaminated to 5-HIAA without being released. If, however, the synaptosomes are depolarised (18), or the slices stimulated (Schaechter and Wurtman, unpublished observation) or, in whole animals, the Raphé is subjected to supramaximal stimulation (Lookingland, this volume), increasing TRP availability does accelerate 5-HT release. Thus current results, although still incomplete, suggest that when neuronal firing rates are high, increasing the rate of 5-HT synthesis may increase release. In addition, tryptophan's effects on a variety of brain functions provides further support for the notion that, under some circumstances, precursor-induced increases in 5-HT synthesis can also lead to increased release (10).

Transport of Tryptophan into the Brain

Tryptophan is carried across the blood-brain barrier by the LNAA transport system, and competes with leucine, isoleucine, valine, phenylalanine, tyrosine, and methionine for access to the carrier (21). The basic mechanisms of this system are described in detail in the introductory chapter of this book, and the relative importance of free and bound tryptophan have been dealt with in recent reviews (4,21), so only the quantitative aspects relative to the aim of manipulating brain TRP levels within the 15—30 μmolar range will be discussed here.

In 1972, Fernstrom and Wurtman plotted plasma TRP/LNAA ratio against brain TRP in the rat (3). The regression fitted closely to the linear equation: brain TRP = 16(TRP/

LNAA) + 2.5, where the two constants are expressed in μg TRP/g brain (4). Thus, a 4-fold rise in the ratio above baseline (i. e., from 0.1 to 0.4) will about double brain tryptophan levels. This relationship has been confirmed on several occasions (4,10), and even generalizes with surprising accuracy to monkeys (Leathwood and Fernstrom, in preparation). It is important to note that small changes in the ratio within the physiological range do not reliably influence brain TRP or 5-HT (4,10,15,16).

Effects of Tryptophan Ingestion on the Plasma TRP/LNAA Ratio: Interactions with Meals

The kinetic characteristics of blood-brain barrier transport of the LNAA appear to be very similar in rats and men (22), so the next step, assessing the effects of different TRP-food combinations on the plasma TRP/LNAA ratio will be examined directly in humans. 1—2 g TRP taken alone has little effect on plasma levels of the other LNAA, so in fasted subjects it produces consistent changes in the TRP/LNAA ratio (11). If, however, people eat a meal containing 20 g or more of protein, plasma levels of the LNAA increase and remain high for several hours. After a carbohydrate-rich, protein-free meal, plasma LNAA levels (including TRP) tend to fall, and remain low for at least 3 hours. Thus the composition of meals eaten at about the same time as, or several hours before, an oral dose of TRP could have a major influence on the plasma TRP/LNAA ratio (4). We examined this question by offering healthy volunteers protein-containing or carbohydrate-rich, protein-free supper, with or without the addition of a small (400 mg) amount of TRP (23). The meal was in the form of a soup with an energy content of 400 kcal (one contained 21 g protein, 50 g carbohydrate and 25 g fat; the other, 1 g protein, 70 g carbohydrate and 25 g fat). As expected the "protein" soup increased plasma levels of all the LNAA, while the protein-free soup lowered them. The TRP/LNAA ratio, however, did not change. With 400 mg TRP added to the carbohydrate-rich soup, the ratio doubled after 1 h and was almost trebled at 2 h (the values were: 0.11, 0.21, and 0.29). Calculated values for the ratio at 1 and 2 h after the protein-containing soup were 0.16 and 0.17 (23). In a more recent study we have shown that 1 g TRP, combined with a 30 g carbohydrate load, will quadruple the plasma TRP/LNAA ratio and produce a very consistant increase in drowsyness (Leathwood and Chauffard, in preparation).

Tryptophan and Sleep

There is a considerable number of publications showing that tryptophan seems to have mild sedative properties, but results have often been inconsistent. Hartman (24) showed that 4—5 g TRP given to hospitalized insomniacs significantly increased sleep time, reduced sleep latency and reduced the number of night awakenings, while in normal volunteers it decreased sleep latency and slighty increased total sleep time without altering qualitative aspects of sleep architecture. Several research groups have shown that large doses of tryptophan, given intravenously or orally, lead to feelings of being drowsy or "mellow" (25,26). In contrast, depletion of plasma tryptophan by feeding a large dose of mixed amino acids lacking tryptophan, produces a rapid lowering of mood in normal volunteers (27). Results with TRP loading have not been entirely consistent and some workers report that large doses of tryptophan increase slow wave (stages 3 and 4) sleep (24), or that rapid eye movement (REM) sleep is increased (28) or decreased (29), or unaffected (9,24). A partial explanation for these inconsistencies has been offered by Hartman and Greenwald (30) who reviewed over 40 studies of tryptophan and sleep. They noted that TRP was often ineffective in normal good sleepers and in patients with severe insomnia, but in people with mild insomnia and more particularly with a problem of getting to sleep (ie., with long sleep latencies), 1 g or more of TRP consistently reduced sleep latency. Good sleepers usually have such short sleep latencies that a small improvement would be extremely difficult to detect (31); patients with pathological insomnia may well be beyond the reach of the rather mild sleep facilitating effect it seems one can expect from small increases in serotonin (see 6,7 and above). Similarly, when researchers give TRP just before the subjects go to bed, and then report no effect on sleep latency (29), it should come as no surprise because, under these conditions, even subjects with a problem of sleep onset should be asleep before the pharmacological effects of TRP have come into play (4,23). Lastly, most studies on TRP and sleep do not report composition or timing of evening meals taken during the experiment. If the tryptophan really is acting via changes in its availability to the brain, composition of the accompanying meal could lead to major changes in plasma amino acids and hence diminish or potentiate its effects on sleep.

In summary, earlier experiments on the potential sedative effects of tryptophan led to conflicting results, and even to claims that TRP has no effect on sleep. On closer examination, it emerges that much of the work was carried out without any clear idea as to the type of effect (if any) one could expect from manipulating brain serotonin, the time-course and dose response of TRP effects, possible diet — TRP interactions, and suitability of the experimental subjects. Thus it should not be surprising that results have been somewhat inconsistent. If, as we have shown in two recent studies (32,33), the proposed chain of causality outlined above is taken into account, tryptophan combined with a small carbohydrate load does decrease subjective ratings of arousal, vigilance and sleep latency.

Effects of Dietary Carbohydrate and Protein on Mood and Behaviour

As noted above there have been suggestions that the composition of normal meals may, through effects on the plasma TRP/LNAA ratio, influence brain TRP levels, 5-HT synthesis and hence serotoninergic function. It has been argued that, via changes in the ratio: "A high carbohydrate, protein-poor meal elevates brain tryptophan, accelerating serotonin synthesis. In contrast, a high protein meal depresses serotonin synthesis." (5). These ideas are often repeated in reviews and have even led to specific dietary recommendations in popular books aimed at helping people control their own moods (34,35).

The implications of these ideas are extemely interesting, so it is worth examining the evidence on which they are based. The original publication by Fernstrom and Wurtman (2) showed that if rats are starved for 22 hours and then fed carbohydrate, brain TRP and 5-HT increase, peaking about 2 h after eating began. Later it was shown that the plasma TRP/LNAA ratio also rose (36). These observations have been confirmed on many occasions. If, however, the period of fasting is reduced to 3 hours (14) or the animals are free-feeding (Leathwood and Moennoz, unpublished observation), carbohydrate may produce a small change in the plasma ratio (and occasionally in brain TRP) but brain 5-HT and 5-HIAA do not change. Similarly, analysis of circadian rhythms show, if anything, a negative correlation between plasma TRP/LNAA, brain TRP and 5-HT (15,36). Thus animal studies suggest that although carbohydrate may influence serotonin metabolism through this mechanism after a period of starvation, the results cannot be extrapolated to any carbohydrate meal. Results from human studies show a similar pattern. Tryptophan or tryptophan/carbohydrate mixes can increase the plasma TRP/LNAA ratio, but carbohydrate meals either have no effect at all (23) or produce small (10—20 %) increases (37,38) which — extrapolating from animal studies — would at most increase brain 5-HT by 1—2 %. Since increases in the plasma TRP/LNAA ratio of at least 200—300 % are needed to generate detectable effects on mood or sleep (32,33), it is difficult to see how this mechanism could be involved in behaviour modulation by food.

The argument that protein lowers brain 5-HT via its effects on the plasma TRP/LNAA stands on even weaker ground. The original papers reported that protein did not lower brain TRP or 5-HT (2,3). These observations were confirmed in later studies (36). Similarly, in free feeding animals given a choice of high and low protein diets (15,16) or fixed levels of protein (16,36,37), no robust systematic relationships between dietary protein intake, plasma TRP/LNAA, brain TRP or 5-HT have emerged. Several studies have examined the effects of protein meals in man. With small amounts of protein (20 g) the ratio does not change at all (23). With larger amounts (80—100 g in one meal) the ratio can fall by 15—25 % (38). Again, in animal studies, changes of this magnitude do not seem to influence brain TRP or 5-HT metabolism (3,14,36,37). In the only comparable human behavioural study, where volunteers were given 100 g of amino acids minus tryptophan, very small mood changes occurred even though the ratio fell by 80—90 % (27).

Thus in man, carbohydrate and protein meals do, under some circumstances produce changes in the plasma TRP/LNAA ratio which might conceivably have subtle effects on serotoninergic function. Behavioural studies aimed at testing this idea have produced consistently negative results (32,38,39). In a double-blind experiment we compared the effects on mood, alertness and sleepyness of a carbohydrate breakfast with those of the same meal plus 500 mg tyrosine (ie enough competitor amino acid to cancel out the small rise in the TRP/LNAA ratio induced by the carbohydrate breakfast). For each rating, scores were indistinguishable. In contrast, the same breakfast combined with 500 mg TRP was perceived as being more sedating (32). In an open study, Spring et al (39) compared 75 g carbohydrate (as sherbert) with 75 g protein (as turkey meat). Meal composition did not significantly effect sleepiness although there was a significant sex x meal interaction (men were slightly less sleepy after sherbert as compared to turkey; women were slightly more sleepy). In contrast, men reported feeling less calm after sherbert and women more calm (again with no significant main effect). It is interesting to note that in subsequent reviews these findings have been interpreted as showing that eating carbohydrate made subjects "less alert" than after eating protein (40,41). In another study, Lieberman gave 40 volunteers 80 g protein (turkey) or 120 g carbohydrate (wheat starch bread) and administered mood and sleepiness questionniares and a large battery of performance tests. No significant changes were seen in mood or sleepiness (although there were differences in the baseline scores before the meals were consumed, with subjects feeling more sleepy before eating carbohydrate). Among the performance tests, auditory reaction time was a little slower 1.75 h after carbohydrate and the digit substitution test showed a small impairment at 3.5 hours (38). Other researchers have reported that carbo hydrate can improve performance. Thus Keul et al.(42), using a driving simulator (and a double-blind crossover design) reported a marked improvement in driving performance and concentration after carbohydrate (60 g glucose over 2 hours).

Taken together, these results do not provide any convincing support for the hypothesis that carbohydrate decreases alertness.

SUMMARY

When tryptophan is given, under limited and carefully controlled conditions, the results are coherent with the idea that serotoninergic function is being influenced through changes in precursor availability to the brain. However, as this review shows, the chain of proposed causality is very long, the measures are difficult to obtain, the extrapolations are often rather tenuous, and the current state of knowledge of the serotoninergic system so fragmentary, that it is impossible to be sure. With respect to claims that carbohydrate or protein meals influence serotoninergic function by the same mechanism, the idea is attractive but it is not supported at all by the currently available evidence.

REFERENCES

1.) HESS, S. M., DOEPFER, W. (1961). Behavioural effects and brain amine contents in rats. *Arch. Int. Pharmacodyn. Ther.* **134:** 89—99

2.) FERNSTROM, J. D., WURTMAN, R. J. (1971). Brain serotonin content: increase following ingestion of carbohydrate diet. *Science* **174:** 1023—1025

3.) FERNSTROM, J. D., WURTMAN, R. J. (1972). Brain serotonin content: physiological regulation by plasma amino acids. *Science* **178:** 414—416

4.) LEATHWOOD, P. D. (1987). Tryptophan availability and serotonin synthesis. *Proc. Nutr. Soc.* **46:** 143—156

5.) WURTMAN, R. J. (1986). Ways that foods can affect the brain. *Nutr. Rev.* **44 (suppl.):** 2—6

6.) JOUVET, M. (1984). Indolamines and sleep-inducing factors. *Exptl Br Res* **8 (suppl.):** 81—94

7.) KOELLA, W. P. (1984). Serotonin and sleep. In: *Sleep '84.* Koella, W. P., Rüther, E., Schulz (eds). Fischer, Stuttgart, pp. 6—10

8.) STEIN, L., WISE, C. D. (1974). Serotonin and behavioural inhibition. *Adv. Biochem. Psychopharmacol.* **11:** 281—291

9.) WOJCIK, W. J., FORNAL, C., RADOLOVACKI, M. (1980). Effect of tryptophan on sleep in the rat. *Neuropharmacol.* **19:** 163—167

10.) FERNSTROM, J. D. (1983). Role of precursor availability in the control of monoamine biosynthesis in the brain. *Physiol. Rev.* **63:** 484—5463

11.) GILLMAN, P. K., BARTLETT, J. R., BRIDGES, P. K., HUNT, A., PATEL, A. J., KANTERMANENI, B. D., CURZON, G. (1981). Indolic substances in plasma,cerebrospinal fluid, and frontal cortex of humans infused with saline or tryptophan. *J. Neurochem.* **37:** 410—417

12.) ARIMANANA, L., ASHLEY, D. V. M., FURNISS, D., LEATHWOOD, P. D. (1984). Protein/carbohydrate selection in rats following administration of tryptophan, glucose, or a mixture of amino acids. In: *Progress in Tryptophan and Serotonin Research* (Schlossberger, H. G., Kochen, W., Linzen, B., Steinhart, H. (eds.)). de Gruyter, Berlin, pp. 549—552

13.) LYTLE, L. D., MESSING, R. B., FISHER, L., PHEBUS, L. (1975). Effects of long-term corn consumption on brain serotonin and response to electric shock. *Science* **190:** 692—694

14.) ASHLEY, D. V. M., LEATHWOOD, P. D., MOENNOZ, D. (1984). Carbohydrate meal increases brain 5—hydroxytryptamine in the adult rat only after prolonged fasting. In: *Progress in Tryptophan and Serotonin Research* (Schlossberger, H. G., Kochen, W., Linzen, B., Steinhart, H. (eds.)). de Gruyter, Berlin, pp. 591—594

15.) BLATTER, F., LEATHWOOD, P. D., ASHLEY, D. V. M. (1986). Rats free to select protein and energy separately show decreased circadian rhythm of brain serotonin. *Ann.Rev. Chronopharmacol.* **3:** 5—8

16.) LEATHWOOD, P. D., Ashley D. V. M. (1983). Strategies of protein selection by weanling and adult rats. *Appetite* **4:** 97—112

17.) TRULSON, M. E. (1985). Dietary tryptophan does not alter the function of brain serotonin neurons. *Life Sci.* **37:** 1067—1072

18.) WOLF, W. A., KUHN, D. M. (1986). Uptake and release of tryptophan and serotonin: an HPLC method to study flux of endogenous 5-hydroxyindoles through synaptosomes. *J. Neurochem.* **46:** 61—67

19.) ELKS, M. L., YOUNGBLOOD, W. W., KISER, J. S. (1979). Serotonin synthesis and release in brain slices: independence of tryptophan. *Brain Res.* **172:** 471—486

20.) LOOKINGLAND, K. J., SHANNON, N. J., CHAPIN, D. S., MOORE, K. E. (1986). Exogenous tryptophan increases synthesis, storage and intraneuronal metabolism of 5—hydroxytryptamine in the rat hypothalamus. *J. Neurochem.* **47:** 205—212

21.) PARDRIDGE, W. M. (1983). Brain metabolism: a perspective from the blood—brain barrier. *Physiol. Rev.* **63:** 1481—1535

22.) CHOI et al (1986). Phenylalanine transport at human blood-brain barrier. *J. Biol. Chem.* **261:** 6536

23.) ASHLEY D. V. M., BARCLEY, D. V., CHAUFFARD, F., MOENNOZ, D., LEATHWOOD, P. D. (1982). Plasma amino acid responses to evening meals of differing nutritional composition. *Am. J. Clin. Nutr.* **36:** 143—153

24.) HARTMANN, E. (1971). Tryptophan and sleep. *Psychopharmacol. (Berl.)* **19:** 114—127

25.) CHARNEY, D. S., HENIGER, G. R., REINHARD, J. F., STERNBERG, D. E., HAFSTED, K. M. (1982). The effect of l-tryptophan on prolactin, growth hormone and mood in healthy subjects. *Psychopharmacol.* **78:** 38—43

26.) LEATHWOOD, P. D., CHAUFFARD, F. (1982). Quantifying the effects of mild sedatives. *J. Psychiatr. Res.* **17:** 115—122

27.) YOUNG, S. N., SMITH, S. E., PIHL, R. O., ERVIN, F. R. (1985). Tryptophan depletion causes a rapid lowering of mood in normal males. *Psychopharmacol.* **87:** 173—177

28.) NICHOLSON, A. N., STONE, B. M. (1979). Tryptophan and sleep in healthy man. *Electroenceph. Clin. Neurophysiol.* **47:** 539—545

29.) BROWN, C. C., HORROM, N. J., WAGMAN, A. M. I. (1979). Effects of tryptophan on sleep onset insomniacs. *Waking Sleeping* **3:** 101—108

30.) HARTMANN, E., GREENWALD, D. (1984). Tryptophan and human sleep: an analysis of 43 studies. In: *Progress in Tryptophan and Serotonin Research* (Schlossberger, H. G., Kochen W, Linzen B, Steinhart H (eds.)). de Gruyter, Berlin, pp. 297—304

31.) LEATHWOOD, P. D., CHAUFFARD, F., MUNOZ-BOX, R. (1983). Effect of Valeriana officinalis L. on subjective and objective sleep parameters. In: *Sleep 1982* (Koella, W. P. (ed.)). Karger, Basle pp. 402—405

32.) LEATHWOOD, P. D., POLLET, P. (1983). Diet-induced mood changes in normal populations. *J. Psychiatr. Res.* **17:** 147—154

33.) LEATHWOOD, P. D., POLLET, P. (1984). Tryptophan (500mg) decreases subjectively perceived sleep latency and increases sleep depth in man. In: *Progress in Tryptophan and Serotonin Research* (Schlossberger, H. G., Kochen W, Linzen B, Steinhart H (eds.)). de Gruyter, Berlin, pp. 311—314

34.) EHRET, C. F., SCANLON, L. W. (1986). Overcoming jet lag. Berkley, New York

35.) WURTMAN, J. J. (1986). Managing your mind and mood through food. Rawson, New York

36.) FERNSTROM, J. D. (1986). Acute and chronic effects of protein and carbohydrate ingestion on brain tryptophan levels and serotonin synthesis. *Nutr. Rev.* **44 (suppl.):** 25—36

37.) PETERS, J. C., HARPER, A. E. (1981). Protein and energy consumption, plasma amino acid ratios, and brain neurotransmitter concentrations. *Physiol. Behav.* **27:** 287—298

38.) EIBERMANN, H. R., SPRING, B. J., GARFILED, G. S. (1986). The behavioural effects of food constituents: strategies used in studies of amino acids, protein, carbohydrate and caffeine. *Nutr. Rev.* **44 (suppl.).** : 61—70

39.) SPRING, B. J., MALLER, O., WURTMAN, J. J., DIGMAN, L., COZOLINO, L. (1983). Effects of protein and carbohydrate meals on mood and behaviour. *J. Psychiatr. Res.* **17:** 155—167

40.) SPRING, B. J. (1986). Effects of foods and nutrients on the behaviour of normal individuals. In: *Nutrition and the Brain Vol 7:* (Wurtman, R. J. and Wurtman, J. J. (eds.)). Raven, New York pp. 1—48

41.) SPRING, B. J. (1986). Effects of carbohydrates on mood and behaviour. *Nutr. Rev.* **44 (suppl.).:** 51—60

42.) KEUL, J., HUBER, G., LEHMAN, M., BERG, A., JACOB, E.-F. (1982). Einfluss von Dextrose auf Fahrleistung, Konzentrationsfähigkeit, Kreislauf und Stoffwechsel im Kraftfahrzeug-Simulator (Doppelblind-Studie im cross-over design). *Akt. Ernähr. Med.* **7:** 7—14

NEUROENDOCRINE RESPONSES TO TRYPTOPHAN AS AN INDEX OF BRAIN SEROTONIN FUNCTION

P. J. Cowen

MRC Unit of Clinical Pharmacology and
University Department of Psychiatry
Littlemore Hospital
Littlemore
Oxford OX4 4XN

INTRODUCTION

Drugs that increase brain serotonin (5-hydroxytryptamine, 5-HT) function elevate plasma levels of certain anterior pituitary hormones, notably prolactin (PRL), ACTH and growth hormone (GH) (1). Administration to humans of the serotonin precursor L-tryptophan (LTP), reliably elevates plasma PRL concentration (1). The present chapter reviews the evidence that this neuroendocrine response is mediated by increases in brain 5-HT function and also examines whether abnormal PRL responses to LTP are seen in depressed patients.

Neuroendocrine Challenge Tests

It has been apparent for some years that the secretion of anterior pituitary hormones is partly controlled by monoamine neurones and therefore alterations in hormone secretion could provide evidence of abnormalities in monoamine transmission (2). Physiological hormone secretion is closely controlled by several overlapping feedback systems and perhaps for this reason, measurement of basal hormone output seldom gives illuminating insights into neurotransmitter function.

Recently it has been demonstrated that more consistent effects can be obtained from dynamic tests of hormone release in which a specific monoamine pathway is stimulated by a selective drug (2). The size of the hormonal response to a given stimulus provides an index of the functional activity of the neurotransmitter pathway with which the drug interacts. Such investigations require careful control and even under well-standardised conditions, variation in hormonal response between subjects is substantial. The usefulness of neuroendocrine tests is increased if the drug challenge has a specific action on the monoamine pathway it stimulates and is well-tolerated with an absence of stressful side-effects (since these themselves can elevate plasma hormone levels). How well does LTP meet these requirements?

NATO ASI Series, Vol. H20
Amino Acid Availability and Brain Function in
Health and Disease. Edited by G. Huether

L-Tryptophan as a Neuroendocrine Probe

Oral LTP, even in doses up to 10 g, does not reliably alter plasma hormone concentrations. In contrast, intravenous LTP in doses of 5 g or greater increases plasma PRL and GH. Plasma cortisol is usually unchanged (1). The increase in PRL and GH which follow intravenous LTP is dose-related (3) and shows acceptable intrasubject reliability (4). In addition LTP infusion is well-tolerated and appears safe. The crucial question remaining is that of specifity: does LTP produce increases in GH and PRL solely through enhancement of brain 5-HT function?

This question has yet to be definitely resolved. It must be acknowledged that there is little direct evidence that LTP increases brain 5-HT function rather than intraneuronal 5-HT metabolism (5). It is possible that LTP could be converted to an active metabolite other than 5-HT, for example tryptamine, which could itself alter hormone levels (5). In an attempt to resolve this issue we have carried out a number of drug interaction studies.

The PRL response to LTP was markedly enhanced by a single pre-infusion dose of the selective 5-HT uptake blocker, clomipramine (6), which is consistent with this hormonal response being mediated through brain 5-HT pathways. Further, the 5-HT receptor antagonist, metergoline, blocked the PRL response to LTP (6), also suggesting that 5-HT receptors are involved. However, caution is needed in assessing hormonal responses when using metergoline because at higher doses this drug can lower plasma PRL by a direct action on pituitary dopamine receptors.

Metergoline blocks both 5-HT_1 and 5-HT_2 receptors. Administration of the selective 5-HT_2 receptor antagonist, ritanserin, actually enhanced the PRL response to LTP (6). This suggests that the post-synaptic receptor mediating the PRL response to LTP is of the 5-HT_1 subtype, and in addition that there may be functional interactions between 5-HT_1 and 5-HT_2 receptors; interestingly there is some evidence for this from electrophysiological studies in the rat (6).

While the GH response to LTP is enhanced by clomipramine it is not attenuated by metergoline or ritanserin (7). The role of 5-HT in this hormonal response to LTP, therefore, requires further study. The remainder of this chapter will be restricted to discussion of the PRL response to LTP. At present the balance of evidence suggests that this effect of LTP is caused by increased brain 5-HT function, perhaps mediated by post-synaptic 5-HT_1 receptors.

Effect of Weight Loss on Prolactin Response to L-Tryptophan

Neuroendocrine tests in depressed patients can give misleading results unless appropriate control is made for variables such as age, sex and weight (2). A particular difficulty is that the syndrome of depressive illness is associated with certain physiological changes such as sleep disturbance and weight loss which could themselves alter brain monoamine function. 5-HT pathways, for example, are known to influence feeding behaviour and also reduced dietary intake, which is a common feature of depressive illness, may alter 5-HT metabolism (8). In order to assess whether moderate weight loss might alter 5-HT-mediated neuroendocrine responses, we studied the effect of a 3 week carbohydrate-reducing diet (1000 Kcal daily) on the PRL responses to LTP in volunteers of normal body weight.

All subjects lost weight, usually between 3—4 kg. In the men dieting did not change the PRL response to LTP (8). In contrast, in female subjects there was a marked increase in PRL responses. In neither males nor females was the PRL response to Thyrotropin Releasing Hormone (TRH) altered (8). Since TRH is believed to increase plasma PRL by a direct action on pituitary lactotrophs, it appears that the ability of dieting to enhance PRL responses to LTP in females is not due to a simple increase in the pituitary reserve of PRL. The mechanism of this interesting effect requires further study. However, our findings suggest that in women a modest period of dieting can alter 5-HT-mediated neuroendocrine responses. Clearly this must be taken into account when assessing findings in depressed patients.

Effect of Depression on Prolactin Response to L-Tryptophan

It has long been proposed that the function of brain 5-HT pathways is impaired in depressive disorders, but results from different kinds of investigations are conflicting (9). If the increase in plasma PRL following LTP does indeed reflect transmission through 5-HT_1 synapses then PRL responses to LTP in depressed patients are of considerable interest.

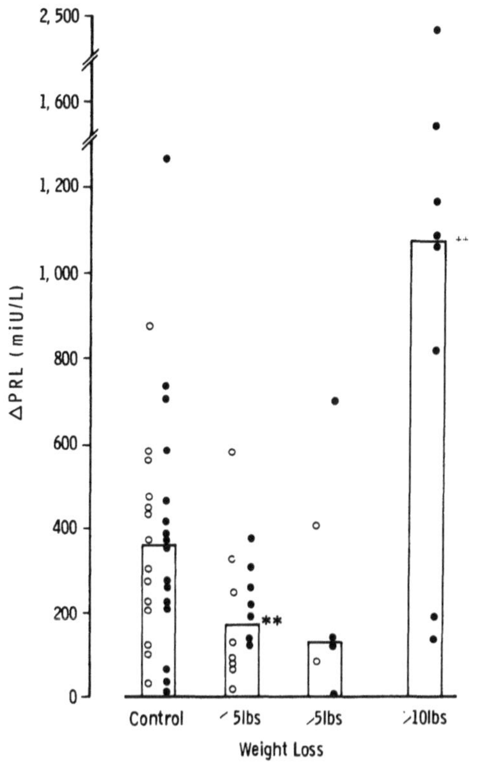

FIGURE 1

Peak (Δ) PRL responses to LTP in control subjects and depressed patients categorised by weight loss on the Beck Scale (open circles men, closed circles women). ΔPRL of control subjects is significantly different from patients with < 5 lbs weight loss ($p < 0.01$) and patients with > 10 lbs weight loss ($p < 0.02$). Reproduced from (9).

We studied 30 patients who met DSM III criteria for major depression. The subjects had been free of antidepressant treatment for at least 3 weeks (and most for considerably longer) but half were receiving night sedation with benzodiazepines. After an overnight fast subjects received intravenous tryptophan (100 mg/kg) and plasma PRL was measured over the next 120 min. The PRL response to LTP was expressed as peak increase over baseline (Δ) which correlated highly ($r > 0.95$) with area under the curve of PRL secretion. Because of our findings in normal dieters we split the patients into two groups, those without recent weight loss, and those with severe recent weight reduction (> 10 lbs on the Beck rating scale) (9). We found that patients without severe weight loss had blunted PRL responses compared to age-matched controls. In contrast, patients with weight loss of greater than 10 lbs had PRL responses which were significantly greater than either controls or the patients without weight loss (Fig. 1) (9).

Interestingly the area under the curve of tryptophan following LTP infusion did not differ between the two groups of depressed patients or the normal controls (Fig. 2). This suggests that changes in total tryptophan disposition are not responsible for the differences in neuroendocrine responses we have identified. Finally benzodiazepine treatment did not apparently produce independent effects on PRL responses (9).

Our findings thus suggest that depressed patients have different kinds of abnormal 5-HT-mediated neuroendocrine responses depending on whether or not they have severe

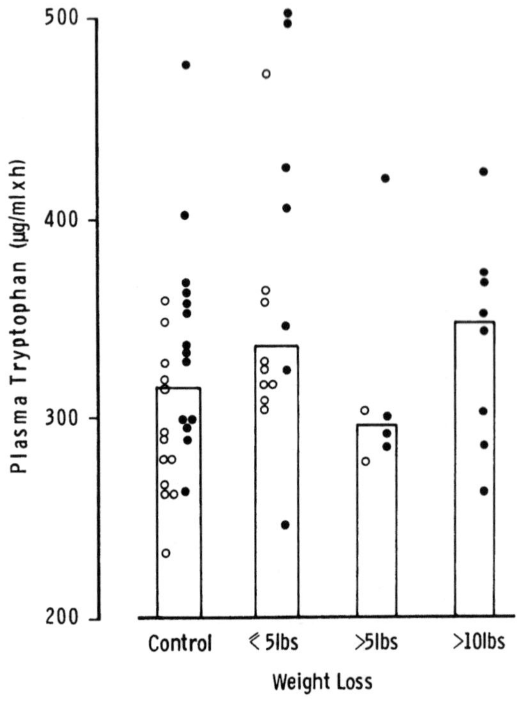

FIGURE 2

Area under the curve of plasma tryptophan following LTP infusion in control subjects and depressed patients categorised by weight loss on the Beck Scale (open circles men, closed circles women). There is no significant difference between the groups. Reproduced from (9).

recent weight loss. The fact that normal women who diet also develop exaggerated PRL responses to LTP suggests the increased PRL responses in depressed patients with weight loss are caused by weight loss itself rather than the depressive disorder. All of our depressed patients with severe weight loss were female; it is not yet clear whether male depressives who lose weight would also have increases PRL responses to LTP. If their pattern follows that of the male dieters then the responses of these patients should not be abnormally increased.

How do our findings compare with other neuroendocrine tests of 5-HT function in depression (Table 1)? Two other groups have studied PRL responses to LTP. Heninger *et al.* (10) found that depressed patients as a group had decreased PRL responses to LTP. In this study subjects with symptoms of melancholia were in a minority; relatively few of them may have experienced recent weight reduction. A study by Pennell and Deakin (personal communication) also found blunted PRL responses to LTP in depressed patients but as in our investigation only when subjects with acute recent weight loss were excluded. In both our study (9) and that of Pennell and Deakin (personal communication) GH responses to LTP were blunted in depressed patients as a group, though as mentioned above it is less certain that this hormonal response is mediated by 5-HT pathways.

TABLE 1
Endocrine Responses to 5-HT Neuroendocrine Probes in Depressive Illness

Author	Drug	Response
Takahasi et al. (12)	5-HTP	↓ GH
Siever et al. (11)	Fenfluramine	↓ PRL
Heninger et al. (10)	LTP	↓ PRL
Cowen & Charig (9)	LTP	↓ PRL (patients without weight loss) ↓ GH
Deakin & Pennell (personal communication)	LTP	↓ PRL (patients without weight loss) ↓ GH
Meltzer et al. (13, 14)	5-HTP	↑ Cortisol

A study using the 5-HT releasing agent, fenfluramine, also found reduced PRL responses in depressed patients (11). This is consistent with investigations using LTP, and if the result with fenfluramine is accepted it would imply that the abnormality responsible for the blunted PRL response to LTP occurs at least at the level of the presynaptic 5-HT neurone.

The 5-HT precursor 5-hydroxytryptophan (5-HTP) has also been employed as a neuroendocrine challenge test in depressed patients. Reservations have been expressed about the ability of 5-HTP to increase plasma hormone concentrations reliably and specifically (5). In an early study Takahashi *et al.* (12) found that the GH response to 5-HTP was reduced in depressed patients compared to a separately studied control group. Rather different findings were obtained by Meltzer *et al.* (13) who reported that cortisol responses to 5-HTP were abnormally increased in depressed patients. The authors argued that although 5-HT-mediated neuroendocrine responses were enhanced in depressed patients the underlying deficit was a reduction in 5-HT transmission which resulted in postsynaptic receptor supersensitivity. While this explanation could be correct it shows that somewhat flexible explanations can be offered for neuroendocrine abnormalities found in depressed patients. Interestingly, in Meltzer's study the cortisol response to 5-HTP correlated with weight loss (14) suggesting a parallel with our own findings.

In conclusion, studies of the endocrine responses to LTP and 5-HTP in depression suggest that some patients do have abnormal 5-HT-mediated neuroendocrine responses. At least some of these abnormalities might be attributable to epiphenomena of depression such as weight loss. It remains to be seen whether any of the changes detected are an integral part of the depressive syndrome, or have implications for treatment with different antidepressant drugs.

REFERENCES

1) COWEN, P. J., ANDERSON, I. M. (1986). 5-HT Neuroendocrinology. In: *The Biology of Depression.* (Deakin, J. F. W., ed.). Royal College of Psychiatrists special publication, pp. 71—89.

2) CHECKLEY, S. A. (1980). Neuroendocrine tests of monoamine function in man: a review of basic theory and its application to the study of depressive illness. *Psychol. Med.* **10:** 35—53.

3) COWEN, P. J., GADHVI, H., GOSDEN, B., KOLAKOWSKA, T. (1985). Responses of prolactin and growth hormone to L-tryptophan infusion: effects in normal subjects and schizophrenic patients receiving neuroleptics. *Psychopharmacology* **86:** 164—169.

4) NUTT, D. J., COWEN, P. J. (1987). Diazepam alters brain 5-HT function in man: Implications for the acute and chronic effects of benzodiazepines. *Psychological Medicine* (in press).

5) VAN PRAAG, H. M., LEMUS, C., KAHN, R. (1986). The pitfalls of serotonin precursors as challengers in hormonal probes of central serotonin activity. *Psychopharmacology Bulletin* **22:** 565—570.

6) COWEN, P. J. (1987). Psychotropic drugs and human 5-HT neuroendocrinology. *Trends Pharmac. Sci.* **8:** 105—108.

7) McCANCE, S. L., COWEN, P. J., WALLER, H., GRAHAME-SMITH, D. G. (1987). The effect of metergoline on endocrine responses to L-tryptophan. *Journal of Psychopharmacology* (in press).

8) GOODWIN, G. M., FAIRBURN, C. G., COWEN, P. J. Dietary changes serotonergic function in women not men: implications for the aetiology of anorexia nervosa. *Psychological Medicine* (in press).

9) COWEN, P. J., CHARIG, E. M. (1987). Neuroendocrine responses to intravenous tryptophan in major depression. *Archives General Psychiat.* (in press).

10) HENINGER, G. R., CHARNEY, D. S., STERNBERG, D. E. (1984). Serotonergic function in depression: prolactin response to intravenous tryptophan in depressed patients and healthy subjects. *Arch. Gen. Psychiat.* **41:** 398—402.

11) SIEVER, L. J., MURPHY, D. L., SLATER, S., DE LA VEGA, E., LIPPEN, S. (1984). Plasma prolactin following fenfluramine in depressed patients compared to controls: an evaluation of central serotonergic responsivity in depression. *Life Sci.* **34:** 1029—1039.

12) TAKAHASHI, S., KONDO, H., YOSHIMURA, M., OCHI, Y., YOSHIMI, T. (1973). Growth hormone responses to administration of L-5-hydroxytryptophan (L-5-HTP) in manic depressive psychoses. *Folia Psychiatrica Neurologica Japonica* **27:** 197—206.

13) MELTZER, H. Y., PERLINE, R., TRICOU, P. J., LOWY, M., ROBERTSON, A. (1984). Effect of 5-hydroxytryptophan on serum cortisol levels in major affective disorders: 1. Enhanced response in depression and mania. *Arch. Gen. Psychiat.* **41:** 366—374.

14) MELTZER, H. Y., PERLINE, R., TRICOU, B. J., LOWY, M., ROBERTSON, A. (1984). Effect of 5-hydroxytryptophan on serum cortisol levels in major affective disorders: 2. Relation to suicide, psychosis and depressive symptoms. *Arch. Gen. Psychiat.* **41:** 379—387.

STRESS, AMINO ACIDS AND THE BEHAVIOURAL EFFECTS OF DRUGS

K. B. J.Franklin

Department of Psychology
McGill University
Montreal, Quebec Canada

INTRODUCTION

Most of the time the functional pool of monoamine neurotransmitters is tightly regulated by a number of control mechanisms such as end product inhibition of synthesis, the existence of a large "reserve" pool, restricted transport of precursors, feedback inhibition of cell firing and transmitter release, and intraneuronal degradation of transmitter. Thus moderate fluctuations in the availability of amino-acid precursors may have little effect on behaviour. In this paper I will review some recent evidence that monoamine releasing drugs and stress are treatments that may make transmitter release dependent on the supply of amino acids so that changes in amino acid availability are rapidly reflected in behaviour.

Tryptophan and Monoamine Oxidase Inhibitors

The results of several biochemical studies suggest that while tryptophan loads are effective in stimulating serotonin synthesis, the additional serotonin thus formed may be degraded intraneuronally (see papers by Commissiong and Trulson in this volume). The importance of intraneuronal degradation in damping the consequences of fluctuations in tryptophan availability is clearly demonstrated by the dramatic effect of tryptophan in animals pretreated with a monoamine oxidase inhibitor (1). Twenty to 30 min after tryptophan administration rats begin to move about compulsively with an abnormal posture and gait. The hind legs become ataxic and the animals make padding movements of the forepaws accompanied by head weaving. Rats also become hyperthermic. Doses of tryptophan as high as 1 g/kg do not produce this syndrome, though in the presence of MAOI as little as 2.5 mg/kg tryptophan can produce hyperactivity and hyperpyrexia (1). Moderate doses of tryptophan plus MAOI, which produce mild hyperactivity, suppress operant behaviour reinforced by rewarding brain stimulation (2). Thus the tryptophan/MAOI combination is not a general behavioural stimulant, rather it induces a specific behavioural syndrome while other voluntary behaviours are suppressed.

The absence of detectable behavioural effects of large doses of tryptophan in animals not otherwise stimulated is consistent with the neuropharmacological studies indicating that excess serotonin is degraded intraneuronally. However there is evidence that increased amino acid availability can modify behaviour when drugs, or the behavioural

NATO ASI Series, Vol. H20
Amino Acid Availability and Brain Function in Health and Disease. Edited by G. Huether

situation, would be expected to stimulate the release of the monoamine neurotransmitters. Thus tryptophan, and in restricted circumstances, tyrosine, can modify the effects of the psychomotor stimulant amphetamine, morphine, and the behavioural and biochemical effects of stress. It is interesting to note that these treatments also increase brain tryptophan concentrations (1, 3—5).

Amino Acids and the Effects of Amphetamine

Amphetamine stimulates locomotor and exploratory behaviour, facilitates the performance of operant behaviour such as self-stimulation, and is avidly self-administered. These behavioural stimulant effects are believed to be mediated primarily by the release of dopamine in the brain (6, 7). Amphetamine also stimulates the release of serotonin (8), and in large doses (> 15 mg/kg) can induce a syndrome which resembles that produced by l-tryptophan plus an MAOI (9).

Administration of tyrosine does not alter the general stimulant effect of amphetamine (10) probably because tyrosine hydroxylase is normally saturated with the substrate. However, chronic treatment with amphetamine leads to depletion of brain catecholamines which can be reversed by administration of tyrosine (11). Recently Geis *et al.* (12) studied the effect of l-tyrosine loads on amphetamine self-administration. L-tyrosine (100 mg/kg) had no effect on amphetamine self-administration during the first 35 days of amphetamine exposure though it antagonized the depletion of brain noradrenaline pro-

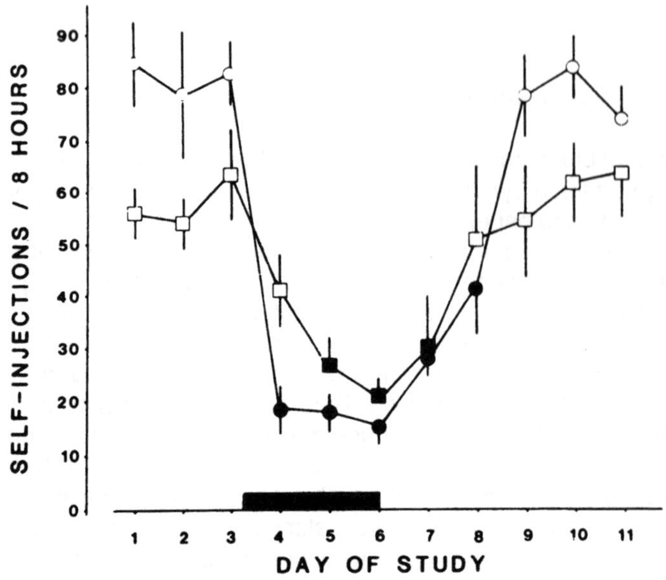

FIGURE 1
Daily rate of amphetamine self-administration before, during and after rats were placed on diets enriched with 2.0 % tryptophan (squares) or 4 % tryptophan. Darkened symbols show significant differences from the control diet period. Vertical lines are standard errors. (Reproduced from Ref. 11 with permission).

duced by amphetamine. However, after four months of exposure l-tyrosine produced marked depression of self-administration. Unfortunately there were too few subjects remaining to perform neurochemical analyses at this time so the depressant effect of tyrosine is difficult to interpret.

In contrast to the weak effects of tyrosine manipulation, tryptophan administration can readily modify amphetamine effects. In general, serotonin appears to act antagonistically to the catecholamines in the control of motor activity, and serotonin depletion potentiates the locomotor stimulant effect of amphetamine (13). More recently Lyness and his colleagues have reported that tryptophan loading reduces the rate of amphetamine self-administration in rats with well established self-administration patterns (see Fig. 1). This effect can be obtained with both intraperitoneal injections of tryptophan (14) and with diets enriched with 2—4 % l-tryptophan (11). The depression of responding is not a general depression because tryptophan loading does not reduce the responding that occurs when saline is substituted for amphetamine. It is suggested that increased serotonin release by amphetamine potentiates the aversive effects of amphetamine so that the animals come to avoid self-administration.

Tryptophan, Stress and Morphine Analgesia

There is a very large body of evidence showing that serotonin systems in the brain and spinal cord are involved in modulating afferent input from nociceptors, and the release of serotonin is believed to play an important role in analgesia. A variety of treatments which are analgesic in man or animals, increase tryptophan uptake by the brain and increase the turnover of serotonin. These include morphine (1), amphetamine (5, 15), salicylates (16), stress (3, 17). The evidence directly linking tryptophan and serotonin to analgesia is less consistent. Inhibition of serotonin synthesis or blockade of serotonin receptors has frequently been reported to attenuate analgesia (18—20) but there are also many reports of no effect, or even potentiation of analgesia (21—23). Tryptophan loading has been reported to increase pain tolerance in humans (24, 25) but in animal experiments it has been ineffective (26). In a clinical trial (27 and see paper by S. N.Young in this volume) we have found that tryptophan does not potentiate morphine analgesia in post-operative pain in humans. Nevertheless, in spite of the contradictions in the literature,there is general agreement that serotonin is involved in the neurological mechanisms that modulate nociceptive input.

An important biological stimulus for brain tryptophan uptake and for serotonergic systems in the brain is stress. It is well known that stressors increase the uptake of tryptophan into the brain and the turnover of serotonin (see paper by G. Curzon in this volume). The increased availability of serotonin appears to be one of the factors modulating output of adrenal corticosteroids (28). Behaviourally, a major role of serotonin systems seems to be to inhibit spontaneous and operant behaviours in the presence of fear-producing stimuli, and other aversive stimuli (29). Severe stress also inhibits, or raises the threshold for, withdrawal reflexes that are the behavioural indices of acute pain in animals (30). This stress-induced analgesia involves an interaction between endogenous opioids and serotonin in which the stress-induced increase in tryptophan availability plays an important role.

When stress is brief or mild, the interaction between opioids and serotonin is revealed by treatment with exogenous opioids such as morphine. If rats are placed in wire mesh tube restrainers during testing for morphine analgesia with the tail flick test they show more analgesia than rats that are briefly handheld for each test (31—33). This stress-potentiation of morphine analgesia is not due to increased diffusion of morphine into the brain (31). The effect is, however, blocked by the serotonin antagonist, methysergide (33). More interesting in the present context, the stress potentiation of morphine analgesia is blocked if animals are pretreated with 200 mg/kg l-valine (33, 34) to prevent the rise in brain tryptophan concentration produced by restraint stress (35, 36). The critical role of tryptophan in this effect is shown by the fact that when the preload contains tryptophan to balance the excess valine stress potentiation of morphine analgesia is restored. In contrast, as can be seen in Figure 2, adding l-tyrosine to the valine preload does not restore the analgesia. These treatments have no effect on analgesia in rats that are not subjected to situational stress (33, 34) though morphine analgesia in the tail flick test is very sensitive to stress. If control subjects are very well adapted to handling, even

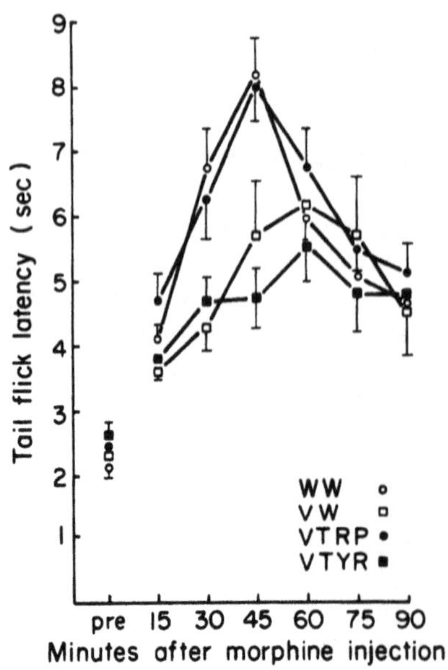

FIGURE 2

Morphine-induced analgesia exhibited by rats that were habituated to the laboratory and tested while restrained. Rats were injected with morphine (5 mg/kg s. c.) 20 min after injections of water (WW), valine (VW), valine and tryptophan (VTRP) or valine and tyrosine. Vertical bars are standard errors. (Reproduced from Ref. 28 with permission).

bringing rats into a strange room will produce a stress potentiation of analgesia that can be blocked by valine pretreatment (34). In the experiments described above we were not able to measure brain tryptophan concentration directly but in later experiments we found that the restraint stress increased brain tryptophan concentration by about 20 % while valine pretreatment reduced tryptophan content by 25 %. The stress-induced rise in brain tryptophan is probably mediated by sympathetic catecholamine release because if rats are treated for 5 days with the sympatholytic drug guanethidine, the effect of stress on both brain tryptophan and morphine analgesia are abolished (37).

More severe stress itself induces analgesia which has been shown to depend on endogenous opioids (30). We found that three hours of restraint produced a significant increase in tail flick latency which is abolished by pretreatment with l-valine or l-valine plus l-tyrosine but not by l-valine plus l-tryptophan. This stress analgesia is opioid in nature by the criterion that it is also blocked by the opioid antagonist naltrexone (34, 38). Naltrexone does not antagonize the stress-induced increase in brain tryptophan level (38) so that stress appears to stimulate tryptophan uptake, brain serotonin activity and brain opioid activity by different mechanisms whose influence combines to produce analgesia.

It was noted above that the literature on serotonin and analgesia shows a considerable degree of inconsistency. The stress-induced interactions between brain serotonin and opioids provides one explanation for these inconsistencies. It might be expected that experiments which were run under conditions which were stressful to rats would show a strong role for serotonin in analgesia while other, less stressful experiments, might fail to find it. However the effects of situational stress are not determined simply by the experimental treatment, but by the response of the animal to that treatment.Thus individual differences in reactivity may modify the effect of stressors. In laboratory animals the environment and caretaking practices in breeding colonies may be an important source of variability in the stress reactivity of subjects. In wild rodents the temperature of the nest is maintained at 30 to 33 °C and the nest is isolated from the outside world. In contrast laboratory animals are usually bred in noisy rooms, maintained at temperatures below 22 °C and are often handled by human caretakers. Both early handling and rearing temperature are known to affect the responses of adult animals to laboratory environments (39, 40).

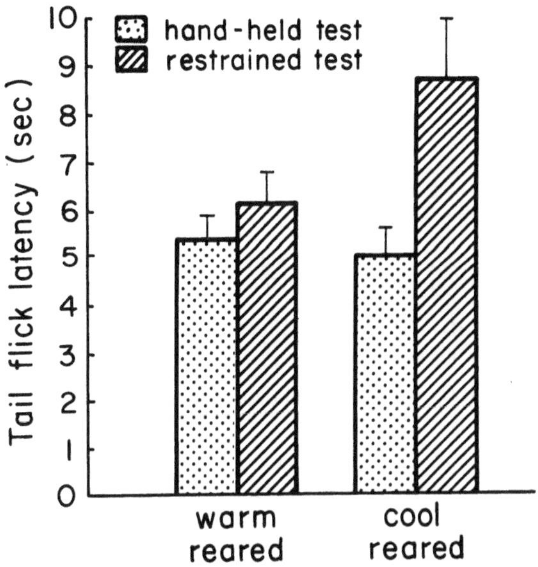

FIGURE 3

The influence of restraint stress on morphine analgesia in rats that had been reared in a warm (33 °C) or cool (22 °C) environments during the first 21 days of life. All rats had been handled during infancy. Scores represent the mean of the tail flick latencies taken 40, 50 and 60 min after injection of morphine (4 mg/kg s. c.).

Drs J. Jans and B. Woodside (Dept. of Psychology, Concordia University, Montreal) maintain small colonies of rats that are reared in temperatures approximating those in the nest area of the wild rodent and others that are more typical of laboratory conditions. In sub-colonies infant rats are either left undisturbed or subjected to daily handling. With their collaboration we have tested rats from some of these colonies for their sensitivity to stress-potentiation of morphine analgesia. As can be seen in Figure 3, with rats reared in warm (33 °C) nest boxes and subjected to early handling, restraint did not potentiate morphine analgesia in the tail flick test. In contrast rats reared in a cool (22 °C) environment showed clear potentiation. Assay of the brains of rats killed at the end of the experiment (60 min after morphine) showed that restraint increased the tryptophan level in cool-reared rats but not in warm-reared rats. Thus the biochemical indicators of the response to stress agree with the behavioural results in showing that animals reared under conditions known to develop stress resistance are indeed less sensitive to stress-induced analgesia.

SUMMARY AND CONCLUSIONS

The behavioural effects of a number of drugs that act through stimulation of monoamine systems can be modified by changes in the availability of the amino acid precursors of monoamines, particularly of tryptophan. However, behavioural effects are not an invariable outcome of fluctuations in amino acid availability but seem to depend on the state of the monoamine system involved. Serotonergic mechanisms appear to be sensitive to changes in tryptophan availability when animals are subjected to stress and, in turn, the response to stress is determined both by the experimental situation and by the characteristics of the animal subject.

REFERENCES

1) GRAHAME-SMITH, D. G. (1971). Studies in vivo on the relationship between brain tryptophan, brain 5-HT synthesis and hyperactivity in rats treated with a monoaminoxidase inhibitor and l-tryptophan. *J. Neurochem.* **18:** 1053—1066.

2) HERBERG, L. J., FRANKLIN, K. B. J. (1976). The "stimulant" action of tryptophan-monoamine oxidase combinations: suppression of self-stimulation. *Neuropharmacol.* **15:** 349—351.

3) CURZON, G., JOSEPH, M. H., KNOTT, P. J. (1972). Effects of immobilization and food deprivation on rat brain tryptophan metabolism. *J. Neurochem.* **19:** 1967—1974.

4) SCHUBERT, J., SEDVALL, G. (1972). Effect of amphetamines on tryptophan concentrations in mice and rats. *J. Pharm. Pharmacol.* **24:** 53—62.

5) VALZELLI, L., BERNASCONI, S., COEN, E., PETKOV, V. V. (1980). Effect of different psychoactive drugs on serum and brain tryptophan levels. *Neuropsychobiol.* **6:** 224—229.

6) CREESE, I., IVERSEN, S. D. (1972). Amphetamine response after dopamine neurone destruction. *Nature NB* **238:** 247—248.

7) YOKEL, R. A., WISE, R. A. (1975). Increased lever pressing for amphetamine in rats: Implications for a dopamine theory of reward. *Science* **187:** 547—549.

8) HOLMES, J. C., RUTLEDGE, C. D. (1976). Effects of the d-and l-isomers of amphetamine on uptake, release and catabolism of norepinephrine, dopamine, and 5-hydroxytryptamine in several regions of the rat brain. *Biochem. Pharmacol.* **25:** 447—451.

9) SLOVITER, R. S., DRUST, E. G., CONNOR, J. D. (1978). Evidence that serotonin mediates some behavioural effects of amphetamine. *J. Pharmacol. Exp. Ther.* **206:** 348—352.

10) FERNANDO, J. C. R., CURZON, G. (1981). Behavioural responses to drugs releasing 5-hydroxytryptamine and catecholamines: Effects of treatments altering precursor concentrations in brain. *Neuropharmacol.* **20:** 116—122.

11) SMITH, F. L., YU, D. S. L., SMITH, D. G., LECCESE, A. P., LYNESS, W. H. (1986).Dietary tryptophan supplements attenuate amphetamine self-administration in the rat. *Pharmacol. Biochem. Behav.* **25:** 849—855.

12) GEIS, L. S., SMITH, D. G., SMITH, F. L., YU, D. S. L., LYNESS, W. H. (1986). Tyrosine influence on amphetamine self-administration and brain catecholamines in the rat. *Pharmacol. Biochem. Behav.* **25:** 1027—1033.

13) BREESE, G. R., COOPER, B. R., MUELLER, R. A. (1974). Evidence for involvement of 5-hydroxytryptamine in the actions of amphetamine. *Br. J. Pharmacol.* **52:** 307—314.

14) LYNESS, W. H. (1983). Effect of l-tryptophan pretreatment on d-amphetamine self administration. *Subst Alcohol Actions Misuse* **4:** 305—312.

15) ROBERTSON, J., WESTON, R., LEWIS, M. J., BARASI, S. (1981). Evidence for the potentiation of the antinociceptive action of morphine by bromocriptine. *Neuropharmacol.* **20:** 1029—1032.

16) TAGLIAMONTE, A., BIGGIO, G., VARGIN, L., GESSA, G. C. (1973). Increase of brain tryptophan and stimulation of serotonin synthesis by salicylate. *J. Neurochem.* **20:** 909—912.

17) JOSEPH, M. H., KENNETT, G. A. (1983). Stress-induced release of 5HT in the hippocampus and its dependence on increased tryptophan availability: an in vivo electrochemical study. *Brain Res.* **270:** 251—257.

18) GORLITZ, B.-D., FREY, H.-H. (1972). Central monoamines and antinociceptive drug action. *Eur. J. Pharmacol.* **20:** 171—180.

19) TABER, R. I., LATRANYI, M. B. (1981). Antagonism of the analgesic effect of opioid and non-opioid agents by p-chlorophenylalanine (PCPA). *Eur. J. Pharmacol.* **75:** 215—222.

20) TENEN, S. S. (1968). Antagonism of the analgesic effect of morphine and other drugs by p-chlorophenylalanine, a serotonin depletor. *Psychopharmacologia (Berl.)* **12:** 278—285.

21) SUGRUE, M. F. (1979). Effect of depletion of rat brain 5-hydroxytryptamine on morphine-induced antinociception. *J. Pharm. Pharmacol.* **31:** 253—255.

22) REINHOLD, K., BLASIG, J., HERZ, A. (1973). Changes in brain concentrations of biogenic amines and the antinociceptive effect of morphine in rats. *Naunyn-Schmied. Arch. Pharmacol.* **278:** 69—80.

23) LONG, J. B., KALIVAS, P. W., YOUNGBLOOD, W. W., PRANGE, A. J. Jr., KIZER, J. S. (1984). Possible involvement of serotonergic neurotransmission in neurotensin but not morphine analgesia. *Brain Res.* **310:** 35—43.

24) SHPEEN, S. E., MORSE, D. R., FURST, M. L. (1984). The effect of tryptophan on postoperative endodontic pain. *Oral Surg.* **58:** 446—449.

25) SELTZER, S., DEWART, D., POLLACK, R. L., JACKSON, E. J. (1983). The effects of dietary tryptophan on chronic maxillofacial pain and experimental pain tolerance. *J. Psychiat. Res.* **17:** 181—185.

26) HOLE, K., MARSDEN, C. A. (1975). Unchanged sensitivity to electric shock in L-tryptophan treated rats. *Pharmacol. Biochem. Behav.* **3:** 307—309.

27) ABBOTT, F. V., ENGLISH, M. J. M., FRANKLIN, K. B. J., JEANS, M. E., YOUNG, S. N. (1987). Effects of tryptophan loading on pain and morphine requirements after surgery. *Abs. Canadian Coll. Neuropsychopharmacol. 10th Ann. Meeting.*

28) JOSEPH, M. H., KENNETT, G. A. (1983). Corticosteroid response to stress depends upon increased tryptophan availability. *Psychopharmacol.* **79:** 79—81.

29) IVERSEN, S. D. (1984). 5-HT and anxiety. *Neuropharmacol.* **23:** 1553—1560.

30) AKIL, H., MADDEN, J., PATRICK, R. L., BARCHAS, J. D. (1976). Stress-induced increase in endogenous opiate peptides:concurrent analgesia and its partial reversal by naloxone. In: *Opiates and Endogenous Opioid Peptides* (Kosterlitz, H. W., ed.) Elsevier, Amsterdam, pp. 63—70.

31) APPELBAUM, B. D., HOLTZMAN, S. G. (1984). Characterization of stress-induced potentiation of opioid effects in the rat. *J. Pharmacol. exp. Ther.* **231:** 555—565.

32) SCHLEN, H., BENTLEY, G. A. (1980). The possibility that a component of morphine analgesia is contributed directly via the release of endogenous opioids. *Pain* **9:** 73—84.

33) KELLY, S. J., FRANKLIN, K. B. J. (1984). Evidence that stress augments morphine analgesia by increasing brain tryptophan. *Neurosci. Lett.* **44:** 305—310.

34) KELLY, S. J., FRANKLIN, K. B. J. (1985). An increase in tryptophan may be a general mechanism for the effect of stress on sensitivity to pain. *Neuropharmacol.* **24:** 1019—1025.

35) KENNETT, G. A., JOSEPH, M. H. (1981). The functional importance of increased brain tryptophan in the serotonergic response to restraint stress. *Neuropharmacol.* **20:** 39—43.

36) MESSING, R. B., FISHER, L. A., PHEBUS, L., LYTLE, L. D. (1976). Interaction of diet and drugs in the regulation of brain 5-hydroxyindoles and the response to painful electric shock. *Life Sci.* **18:** 707—714.

37) FRANKLIN, K. B. J., KELLY, S. J. (1986). Sympathetic control of tryptophan uptake and morphine analgesia in stressed rats. *Eur. J. Pharmacol.* **126:** 145—150.

38) KELLY, S. J., FRANKLIN, K. B. J. (1987). Role of peripheral and central opioid activity in analgesia induced by restraint stress. *Life Sci.* **41:** 789—794.

39) LEVINE, S. (1957). Infantile experience and resistance to physiological stress. *Science* **126:** 403.

40) DALY, M. (1973). Early stimulation of rodents: a critical review of present interpretations. *Br. J. Psychol.* **64:** 435—460.

41) JANS, J. E., ABBOTT, F. V., FRANKLIN, K. B. J., WOODSIDE, B. C. (1987). The temperature of the rearing environment determines sensitivity to stress potentiation of morphine analgesia and brain tryptophan metabolism. *Soc. Neurosci. Abs.* **13:** 408.

BEHAVIORAL AND COGNITIVE EFFECTS OF ALTERED TRYPTOPHAN AND TYROSINE SUPPLY

Michael J. Raleigh, Michael T. McGuire, and Gary L. Brammer

Department of Psychiatry and Biobehavioral Sciences
UCLA School of Medicine
760 Westwood Plaza
Los Angeles, CA 90024-1759, USA

INTRODUCTION

Biochemical, behavioral, and pharmacological data have enabled us to document many complex behavior-physiology interactions in vervet monkeys *(Cercopithecus aethiops sabaeus)*. Our research has been guided by the view that there is a bidirectional relationship between behavior and physiology: behavior, particularly social behavior, is both a cause and a consequence of physiological changes (1). Thus, adult animals living in social groups have distinct and replicable behavioral profiles which strongly influence the metabolism and behavioral effects of amino acids. Alterations in group composition can predictably change individuals behavioral profiles which result in corresponding physiological changes. Conversely, administration of tryptophan, tyrosine, and other amino acids can profoundly affect social behavior and cognitive performance. While the bulk of our research has focused on vervet monkeys, preliminary observations indicate that many of these findings generalize to other nonhuman primate species and to humans (2, 3).

In the present chapter, we will enumerate several features of vervet monkeys that make them well suited to our research. Subsequently, we will report on the behavioral effects of exogenous tryptophan and tyrosine. These data strongly support the view that the effects of these amino acids arise from their influences on central monoaminergic neurotransmitter systems. The final section of this chapter will describe how social factors and housing conditions can dramatically constrain the metabolism of amino acids.

Vervet Monkeys

At least three characteristics of vervet monkeys make them ideal research subjects. Vervets are Old World monkeys and hence are phylogenetically much closer to humans than are nonprimates, prosimians, or New World monkeys. Vervets exhibit a rich behavioral repertoire that parallels many aspects of human nonverbal behavior (4). Found throughout sub-Saharan Africa and in the Caribbean islands of Barbados, St. Kitts, and Nevis, vervets are among the most widely distributed and numerous of the nonhuman pri-

NATO ASI Series, Vol. H20
Amino Acid Availability and Brain Function in Health and Disease. Edited by G. Huether

mates. Unlike many nonhuman primates, vervets are adaptable rather than adapted to a particular circumscribed habitat. Appropriately, they have been called opportunistic omnivores (5). Because of their morphological and behavior generality and because of their evolutionary proximity to humans, conclusions about behavior-physiology interactions in vervets may be more likely to generalize to humans than are those derived from rodents or felines.

A second feature is that vervets' life histories make it easy to investigate state-trait relationships. In free-ranging settings vervets live in multimale, multifemale groups ranging from 8 to 64 individuals (6). At puberty, males migrate out of their natal groups and attempt to join another group. As adults, males will migrate to other groups at one to three year intervals; thus, it is not unlikely for a male to have been a member of four or more different groups (7). Paralleling the free ranging situation, in captivity, males can be moved from group to group. We have utilized this strategy to determine the extent to which behavioral and physiological features are state dependent. Thus, for example, interindividual differences in whole blood serotonin concentration, in the effects of tryptophan on affiliative behavior, and in behavioral responses to standardized stimuli are largely state, rather than trait, dependent (8, 9, 10).

The third feature is that in both captive and free-ranging settings dominance relationships are an important axis of social organization. Dominance relationships occur when the animals exhibit consistent, asymmetrical interactions. In vervets, dominance relationships are often recognized on the basis of consistent approach-avoidance patterns or success in agonistic encounters. It is important to underscore that in vervet monkeys as well as in other species that dominance is not synonymous with aggressivity. In fact, the dominant male is often a focal point for positive social actions such as juvenile play, huddling, or grooming. Table 1 shows the rates of several types of aggressive behavior as well as the likelihood of females being in proximity to dominant and subordinate males. Once established, dominance relationships persist for long periods of time: in our undisturbed groups, the average time a male remains dominant is 1.6 ± 0.3 years ($\bar{X} \pm$ SEM, $n = 30$). However, when a dominant male is removed from his troup the remaining subordinate males compete to become dominant. The initial phase of this competition involves the formation of alliances with females and is sensitive to tryptophan and other serotonergic agonists. Thus, treatment of one of the remaining subordinate males with tryptophan is uniformly associated with that animal becoming dominant. Conversely, reduction of serotonergic function is uniformly associated with the treated subject remaining subordinate (11).

TABLE 1

Status Differences in Aggressive and Affiliative Behavior

	Dom	Sub
Threat	8 ± 3	9 ±5
Contact	1 ± 1	4 ±1*
Chase	2 ± 1	4 ± 3
Proximity	51 ± 10	21 ± 8*

Data represent the rate/hour of three of three types of aggression: threatening, contact aggressing (hitting, biting), and chasing as well as the likelihood of being in proximity to a female. The proximity score represents the percentage of time a female is within a meter of a male. Data are from 13 dominant and 28 subordinate males and are in $\bar{X}$±SEM. An asterisk indicates that the difference between the two groups is significant ($p<.01$).

Behavioral Effects of Tryptophan and Tyrosine

Reflecting our historical interests in serotonergic systems, the bulk of our data comes from studies of tryptophan administration. However, similar strategies and procedures are presently being utilized to examine the effects of tyrosine. This section describes the rationale and results of these studies.

We have assumed that vervet monkeys resemble other species in that their serotonin biosynthesis is substrate-sensitive as well as enzyme-limited and that tryptophan administration increases brain serotonin content and turnover (12, 13, 14). The elevated serotonin produced by tryptophan administration has the same anatomical distribution as endogenous serotonin (15). Thus, the behavioral effects of tryptophan are most likely due to enhancement of central serotonergic function. However, tryptophan has many other metabolic effects: it may even alter catecholamine biosynthesis by competing with tyrosine for access to the blood-brain barrier's neutral amino acid carrier (16, 17). Consequently we have tested the assumption that tryptophan's behavioral effects are due to increased serotonergic function by comparing the effects of tryptophan to those of other drugs that primarily affect serotonergic systems. For example, we have utilized other drugs — including quipazine and fluoxetine — that increase serotonergic function but by different mechanisms than does tryptophan (18). The common behavioral effects produced by these drugs are most readily attributed to their common impact on central serotonergic transmission. Furthermore, we have examined drugs that decrease serotonergic function including PCPA, cyproheptadine, and chronic fenfluramine (19, 20). Because these drugs have the behavioral effects opposite to those of tryptophan, the assumption that tryptophan's effects are due to enhanced serotonergic function is supported.

Another assumption we have made is that behavior must be documented in great detail. This approach has been used successfully by others in nonprimates but somewhat surprisingly has only recently been applied to primates (21, 22, 23, 24, 25). Thus, for example, when our studies began, global, undifferentiated categories such as initiating social behavior were commonly used. However, tryptophan differentially affects components of such broad behavioral categories. Some more discrete behaviors such as social grooming are increased while others such as avoiding are decreased. Consequently, lumping these and other behaviors into a broad category may mask the behavioral effects of tryptophan (26, 27, 28, 29).

In our studies of the social behavioral effects of tryptophan, we utilized stable groups containing at least three adult males, three adult females, and their immature offspring. Tryptophan (or tyrosine) was administered at 0600 and behavioral observations conducted between 0645 and 0845. As shown in Table 2, repeated acute tryptophan (20 mg/kg/day) administration induced changes in many behaviors. Social grooming, approaching, resting, and eating increased while locomoting, avoiding, being solitary, and being vigilant decreased. Huddling, aggressing, being aggressed, and sexual behavior were not significantly altered. Tryptophan's failure to affect sexual and aggressive behavior may be due to the low rates at which these behaviors occur. Both behaviors were reduced by about 20 %. Subsequent studies have shown that fluoxetine and quipazine produce similar behavioral effects while cyproheptadine, PCPA, and fenfluramine produce essentially the opposite results (30, 31, 32). These observations support the contention that tryptophan exerts its behavioral effects via serotonergic mechanisms.

In our initial studies we were struck by the interindividual variability in the behavioral effects of tryptophan. One factor that often contributes to the behavioral effects of drugs

TABLE 2
Effects of Tryptophan and Other Treatments on Social Behavior

BEHAVIOR	TR	FL	QP	PCPA	CYP	FEN
Grooming	154	166	168	22	25	32
Approaching	146	143	166	62	51	48
Resting	139	155	153	51	39	46
Eating	138	144	129	45	48	29
Locomoting	68	63	70	153	162	159
Avoiding	72	77	68	149	231	258
Being Vigilant	49	65	67	148	201	149
Being Solitary	69	58	50	136	183	149
Huddling	125	101	93	58	81	101
Aggressing	87	83	81	224	308	521
Being Agressed	77	88	79	158	296	191
Sexual	75	63	102	104	91	112

Values are the rate of behavior during treatment divided by its rate during baseline. A number >1.00 indicates that the behavior increased during treatment. TR, FL, QP, PCPA, CYP, and FEN refer to tryptophan, fluoxetine, quipazine, para-chlorophenylanine, cyproheptadine, and fenfluramine. Dose were TR (20 mg/kg), FL (1.0 mg/kg), QP (0.5 mg/kg), PCPA (80 mg/kg), CYP (60 g/kg), and FEN (2 mg/kg).

is an animal's social status (33, 34, 35). In a series of studies we showed that there were status-linked differences in the behavioral consequences of tryptophan and that these result from status-characteristic differences in central serotonergic systems. In these studies we examined the effects of tryptophan (10, 20, 40 mg/kg), fluoxetine (0.5, 1.0, and 2.0 mg/kg), and quipazine (0.25, 0.50, 1.0 mg/kg). Relative to subordinate males, domi-

TABLE 3
Qualitative Effects of Tryptophan, Fluoxetine, and Quipazine

	Direction of Change			Status Interaction		
BEHAVIOR	TR	FL	QP	TR	FL	QP
Groom	+	+	+	Yes	Yes	Yes
Approach	+	+	+	Yes	Yes	Yes
Rest	+	+	+	No	No	No
Eat	+	+	+	No	No	No
Locomote	-	-	-	Yes	Yes	Yes
Avoid	-	-	-	Yes	Yes	Yes
Be Vigilant	-	-	-	Yes	Yes	Yes
Be Solitary	-	-	-	Yes	Yes	Yes
Huddle	0	+	0	No	No	No
Aggress	0	0	0	No	No	No
Be Aggressed	0	0	0	No	No	No
Sex	0	-	0	No	Yes	No

In the Direction of Change column, a plus (+), a minus (−), or a zero (0) indicates that the treatment increased, decreased, of had no effect on the behavior respectively. The Status Interaction column indicates that there was a significant status x dose interaction. In this case, the dominant males exhibited a larger change from vehicle levels than did subordinate males.

nant males are more responsive to all three drugs and this heightened responsivity is largely confined to behaviors known from previous studies to be serotonergically influenced (30). Table 3 qualitatively summarizes these studies. Serotonergic neurons have a restricted origin in or near the raphe and project throughout the telencephalon and diencephalon. Consequently, it is unlikely that the behavioral effects of these treatments arise from direct, specific serotonergic influences on the final motor pathways that mediate the affected behaviors. It is more likely that enhanced serotonergic function contributes to an affective state that promotes tranquil, quiescent behavior. Thus, these behavioral changes are of the type that might be anticipated in calm, unanxious animals (36). This view is consistent with reports of the effects of tryptophan on mood in humans (37, 38).

In a more recent set of studies we examined the effects of tryptophan on individual cognitive behavior tests. In these tests animals are briefly removed from their home cage and exposed to such standard stimuli as a maze, novel objects, an unfamiliar juvenile conspecific, or a potential predator. As shown in Table 4, there are status-related differences in these measures. Thus, in basal conditions, dominant males traveled less than subordinate males in the maze. Compared to subordinate males, dominant males appeared relaxed and systematic in solving the maze. There were also status-linked differences in the exploration of novel objects. Dominant males responded more aggressively and located themselves closer to a dog than did subordinate males. The table also shows that tryptophan and cyproheptadine produced opposite effects on three of the four sets of measurements. Thus, tryptophan improved maze performance, increased affiliative behavior, and resulted in animals being in closer proximity to the dog. The magnitude of these alterations was linked to the subjects' social status. Finally, it is of interest that when dominant individuals received cyproheptadine, their behavior was indistinguishable from that of tryptophan-treated subordinate males. These observations suggest that differences in serotonergic systems may underlie some of the status-linked differences in individual test performance.

TABLE 4
Individual Behavior: Status and Drug Effects

	VEHICLE		TRP		CYP	
Test	Dom	Sub	Dom	Sub	Dom	Sub
Maze-travel	84	221	79	110	121	235
Object-touch	9	22	6	13	15	23
Object-distance	1.5	4.5	1.0	2.5	2.7	4.8
Juvenile-threat	3	11	2	6	7	18
Juvenile-affiliate	13	3	18	5	6	2
Dog-threat	18	6	19	14	14	5
Dog-distance	1.0	4.0	1.0	2.0	1.9	3.9

Tests are described in the text. Units are the average number of events/test except for the object-distance and dog-distance which are in meters. Data are from 7 dominant and 12 subordinate males. These were observed after receiving vehicle, tryptophan (40mg/kg), and cyproheptadine (60g/kg).

In brief, then, our data indicate that tryptophan administration can exert a wide range of social, behavioral, and cognitive effects, that the magnitude and time course of these effects are linked to an animal's social status, and that tryptophan's behavioral effects are mediated by central serotonergic systems.

TABLE 5
Effects of Tyrosine on Social Behavior

BEHAVIOR	PERCENT	STATUS
Groom*	63	Yes-Sub
Approach*	181	No
Rest*	63	No
Eat	103	No
Locomote*	181	Yes-Sub
Avoid	103	No
Be Vigilant*	152	Yes-Sub
Be Solitary*	74	Yes-Dom
Huddle	92	No
Aggress*	403	Yes-Sub
Be Aggressed*	269	Yes-Sub
Submit*	71	Yes-Sub
Sex	113	No

The percent column indicates the rate of a behavior during treatment relative to its rate during baseline. Conventions are as in Table 2. The Status column indicates whether social status contrained the behavioral response to tryposine. A Sub ar a Dom indicates that subordinate or dominant males were more affected. An asterisk after the behavior indicates that the behavior was significantly altered by tyrosine.

At present, our studies of the behavioral effects of tyrosine are in a nascent stage. As Table 5 shows, tyrosine affected many social behaviors. Thus, it increased approaching, locomoting, being vigilant, aggressing and being aggressed, and decreased grooming, resting, and submitting. As with tryptophan, the behavioral effects were constrained by an individual's social rank.

Social Factors and Amino Acid Metabolism

Because tryptophan produced greater behavioral effects in dominant than in subordinate males, we examined some of the physiologic consequences of tryptophan administration. In addition to its effects on central serotonergic mechanisms, tryptophan also augments serotonin concentrations in peripheral tissues (39). A readily measured portion of peripheral serotonin is sequestered in the platelets and in these studies we measured serotinin in whole blood (40).

TABLE 6
Whole-Blood Serotonin Response to Tryptophan

	Dominant (n=7)	Subordinate (n=9)
Basal level (ng/ml)	1,136 ± 40	668 ± 26
Post tryptophan	2,428 ± 88	1,029 ± 48
Absolute change	1,292 ± 70	361 ± 36
Relative change	2.14 ± 0.07	1.54 ± 0.06

Data are in X ± SEM

Table 6 illustrates the effect on an acute dose of 20 mg/kg of tryptophan on whole blood serotonin concentrations. Seven dominant and nine subordinate males were investigated (41). A basal blood sample was obtained immediately before administering 20 mg/kg of tryptophan. Sixty minutes later, a time at which status-linked behavioral differences are apparent, another blood sample was collected. Dominant and subordinate males differ substantially in their whole blood serotonin response to tryptophan. The measure that most clearly illustrates this difference is the ratio of post-tryptophan to basal whole blood serotonin. Despite having basal whole blood serotonin levels approximately twice those of subordinate males, dominant males exhibited a larger relative rise. Should such a status-related difference in tryptophan metabolism occur in the central nervous system, it might account for the greater behavioral change shown by dominant males treated with tryptophan.

Recently we extended these findings by showing that normally raised adult males who are isolated socially for 10 weeks become nonresponsive to tryptophan. These 8 individually housed animals show essentially no increase in whole blood serotonin levels following either 20 mg/kg or 40 mg/kg. In these studies, blood samples are obtained 30, 60, and 90 minutes after tryptophan administration. These nonresponsive animals show less than a 10 % rise in whole blood serotonin at any of these time points.

Prior to isolation these animals had responded to tryptophan and switching from a tryptophan responder to a nonresponder is not due to illness, exercise, or dietary alterations. Isolated animals have been tested repeatedly (all more than 5 times) and they are nonresponsive on all tests.

The basal whole blood serotonin levels in nonresponders range from 500 to 1,300 ng/ml, which spans the low to high/middle range seen in socially-living responders. Both responders and nonresponders show similar (5 to 10-fold) increases in serum tryptophan following tryptophan administration.

Both sets of animals exhibit elevated plasma 5-HIAA concentrations following concurrent tryptophan and probenecid (100 mg/kg) treatment, but the nonresponders' levels exceed those of the responders ($t = 3.01$, $df = 12$, $p < .05$). Thus, nonresponsivity to tryptophan does not result from the failure to convert tryptophan to serotonin alone. Concurrent tryptophan and pargyline (at a dose that reduces MAO activity by 50 %) does not result in nonresponders becoming responders. Thus, differential MAO activity alone also does not appear to explain nonresponsivity.

The mechanisms underlying the failure to show a whole blood serotonin response to tryptophan are unknown. At present we are examining three sets of hypotheses. One concerns platelets. It is possible that in nonresponders platelets have reduced 5-HT uptake and/or a large portion of 5-HT "leaks" back into the plasma (42). In either case, the nonsequestered 5-HT would rapidly be degraded and converted to 5-HIAA and tryptophan might result in essentially no rise in whole blood serotonin concentration. A second set of possible explanations deals with the metabolism of tryptophan to serotonin. If less tryptophan is converted to 5-HTP and/or less 5-HTP is converted to 5-HT in nonresponders, then they may not show an elevation in blood serotonin. The third hypothesis deals with differential MAO activity. Preliminary data indicate that MAO activity may be increased in nonresponders.

In brief, then, social status and housing conditions substantially influence tryptophan metabolism. In humans, age, gender, and genetic background have been identified as factors that may influence amino acid metabolism. The present observations suggest that social status and social interactions may also be potent predictors of amino acid metabolism in humans.

CONCLUSIONS

To conclude we want to reiterate the importance of viewing behavior-amino acid metabolism interactions as bidirectional. Behavior is a particularly sensitive means of evaluating the significance of alterations in amino acid supply. Conversely, behavioral alterations can strikingly affect amino acid metabolism. As the physiological mechanisms which underlie these behaviorally-induced metabolic changes become known, interest in behavior-metabolism interactions will mushroom.

ACKNOWLEDGEMENTS

Our research was supported by grants from the United States government's National Institute of Health, from the H. F. Guggenheim Foundation, and from the Giles and Elise Mead Foundation. David Torigoe prepared and nurtured the manuscript. Nuria Kimble and Art Yuwiler provided helpful suggestions.

REFERENCES

1) McGUIRE, M. T., RALEIGH, M. J. (1985). Serotonin-behavior interactions in vervet monkeys. *Psychopharmacol. Bull.* **21:** 458—463.

2) STEKLIS, H. D., BRAMMER, G. L., RALEIGH, M. J., and McGUIRE, M. T. (1985). Serum testosterone, male dominance, and aggression in captive groups of male vervet monkeys *(Cercopithecus aethiops sabaeus). Hormones and Behavior,* **19:** 154—163.

3) McGUIRE, M. T., RALEIGH, M. J., and BRAMMER, G. L. (1982). Sociopharmacology. *Ann. Rev. Pharmacol. Toxicol.* **22:** 643—661.

4) CHENEY, D. L., SEYFARTH, R., SMUTS, B. (1986). Social relationships and social cognition in nonhuman primates. *Science* **234:** 1361—1366.

5) STRUSHAKER, T. T. (1967). Behavior of vervet monkeys. *Univ. Calif. Publ. Zool.* **82:** 1—64.

6) McGUIRE, M. T., RALEIGH, M. J., and JOHNSON, C. (1983). Social dominance in adult male vervet monkeys: General considerations. *Social Science Information* **22:** 89—129.

7) McGUIRE, M. T., and RALEIGH, M J. (1987). Serotonin, social behavior, and aggression in vervet monkeys. In: *Ethopharmacology of Agonistic Behaviour.* (Mos. J., Olivier, B., Poshivalov, V., eds.). New York, Alan R. Liss. In press.

8) RALEIGH, M. J., McGUIRE, M. T., BRAMMER, G. L., and YUWILER, A. (1984). Social and environmental influences on blood serotonin concentrations in monkeys. *Arch. Gen. Psychiat.* **41:** 405—410.

9) McGUIRE, M. T., RALEIGH, M. J., and JOHNSON, C. (1983). Social dominance in adult male vervet monkeys: Behavior-biochemical relationships. *Social Science Information* **22:** 311—328.

10) McGUIRE, M. T., RALEIGH, M. J., and BRAMMER, G. L. (1984). Adaptation, selection, and benefit-cost analysis: implications from behavior-physiology studies of social dominance in vervet monkeys. *Ethol. Sociobiol.* **5:** 269—277.

11) RALEIGH, M. J., and McGUIRE, M. T. (1987). 5-HT metabolism and social behaviour in the monkey. In: *Proceedings of the 5th International Meeting on Tryptophan Research.* (Bender, D. A., ed.). Walter de Gruyter, Berlin. In press.

12) FERNSTROM, J. D., and WURTMAN, R. J. (1974). Control of brain serotonin levels by the diet. *Adv. Biochem. Psychopharmacol.* **11:** 134—142.

13) MURPHY, D. L. (1986). Serotonin neurochemistry: a commentary on some of its quandaries. *Neurochem. Int.* **8:** 161—163.

14) WURTMAN, R. J., HEFTI, F., MELAMED, E. (1981). Precursor control of neurotransmitter synthesis. *Pharmacol. Rev.* **32:** 315—335.

15) MOIR, A. T. B., ECCLESTON, D. (1968). The effects of precursor loading in the cerebral metabolism of 5-hydroxyindoles. *J. Neurochem.* **15:** 1093—1108.

16) YOUNG, S. (1986). The clinical psychopharmacology of tryptophan. In: *Nutrition and the Brain.* (Wurtman, R. J., Wurtman, J. J., eds.). **Vol 7.** Raven, New York, pp. 49—88.

17) YUWILER, A. (1973). Conversion of d- and l-tryptophan to brain serotonin and 5-hydroxy-idoleacetic acid to blood serotonin. *J. Neurochem.* **20:** 1099—1109.

18) FULLER, R. W., CLEMENS, J. A., SLATER, I. H., and RATHBUN, R. C. (1978). Neuroendocrine and behavioral studies with fluoxetine, an inhibitor of serotonin uptake in brain. In: *Neuropharmacology.* (Deniker, P., Radouco-Thomas, C., and Villeneuve, A., eds.). Pergamon, Oxford, pp. 641—646.

19) RALEIGH, M. J., BRAMMER, G. L., YUWILER, A., FLANNERY, J. W., McGUIRE, M. T., and GELLER, E. (1980). Serotonergic influences on the social behavior of vervet monkeys *(Cercopithecus aethiops sabaeus). Exp. Neurol.* **68:** 322—334.

20) RALEIGH, M. J., BRAMMER, G. L., RITVO, E. R., GELLER, E., McGUIRE, M. T., and YUWILER, A. (1986). Effects of chronic fenfluramine on blood serotonin, cerebrospinal fluid metabolites, and behavior in monkeys. *Psychopharm.* **90:** 503—508.

21) DICKINSON, S. L., CURZON, G. (1983). Roles of dopamine and 5-hydroxytryptamine in stereotyped and non-stereotyped behaviour. *Neuropharmacology* **22:** 805—812.

22) GOODWIN, G. M., and GREEN, A. R. (1984). A behaviour and biochemical study in mice and rats of putative selective agonists and antagonists for 5-HT_1 and 5-HT_2 receptors. *Br. J. Pharmacol.* **84:** 743—753.

23) GREEN, A. R. (1984). 5-HT-mediated behaviour: animal studies. *Neuropharmacology* **23:** 1521—1528.

24) JACOBS, B. L., and COHEN, A. (1976). Differential behavioral effects of lesions of the median dorsal raphe nuclei in rats: open field and pain-elicited aggression. *J. Comp. Physiol. Psychol.* **46:** 102—112.

25) REITE, M., and SHORT, R. (1986). Behavior and physiology in young bonnet monkeys. *Devel. Psychobiol.* **19:** 567—579.

26) BOELKINS, R. C. (1973). Effects of parachlorophenylalanine on the behavior in monkeys. In: *Serotonin and Behavior.* (Barchas, J. D., and Udsin, E., eds.). Academic Press, New York, pp. 357—364.

27) MAAS, J. W., REDMOND, D. E., and GREEN, R. (1973). Effects of serotonin depletion on behavior in monkeys. In: *Serotonin and Behavior.* (Barchas, J. D., Udsin, E., eds.). Academic Press, New York, pp. 351—356.

28) RALEIGH, M. J., YUWILER, A., BRAMMER, G. L., McGUIRE, M. T., GELLER, E., and FLANNERY, J. W. (1981). Peripheral correlates of serotonergically-influenced behaviors in vervet monkeys *(Cercopithecus aethiops sabaeus). Psychopharmacology* **72:** 241—246.

29) RALEIGH, M. J. (1987). Differential behavioral effects of tryptophan and 5-hydroxytryptophan in vervet monkeys: influence of catecholaminergic systems. *Psychopharmacology* **93:** 44—50.

30) RALEIGH, M. J., BRAMMER, G. L., McGUIRE, M. T., and YUWILER, A. (1985). Dominant social status facilitates the behavioral effects of serotonergic agonists. *Brain Res.* **348:** 274—282.

31) REDMOND, D. E., MAAS, J., KLING, A., GRAHAM, C., and DERKIRMENJIAN, H. (1971). Social behavior of primates selectively depleted of monoamines. *Science* **174:** 428—430.

32) SMITH, E., and BYRD, L. D. (1983). Studying the behavioral effects of drugs in group-living nonhuman primates. In: *Ethopharmacology: Primate Models of Neuropsychiatric Disorders.* (Miczek, K. A., ed.). New York: A. R. Liss., pp. 1—31.

33) HABER, S., BARCHAS, P. R., and BARCHAS, J. D. (1977). Effects of amphetamines on social behavior of rhesus macaques: an animal model of paranoia. In: *Animal Models in Psychiatry and Neurology.* (Hamin, I., and Udsin, E., eds.). Pergamon, Oxford, pp. 107—114.

34) KRAEMER, G. W. (1985). The primate social environment, brain neurochemical changes and psychopathology. *Trends Neuroscience* **8:** 339—340.

35) MICZEK, K. A., GOLD, L. (1983). D-Amphetamine in squirrel monkeys of different social status: effects of social and agonistic behavior, locomoting, and stereotypies. *Psychopharmacology* **81:** 183—190.

36) RALEIGH, M. J., and McGUIRE, M. T. (1986). Animal analogues of ostracism: biological mechanisms and social consequences. *Ethol. Sociobiol.* **7:** 201—214.

37) VAN PRAAG, H. M. (1983). CSF 5-HIAA and suicide in non-depressed schizophrenics. *Lancet* **2:** 977—978.

38) YOUNG, S. N., and SOURKES, T. L. (1977). Tryptophan in the central nervous system: regulation and significance. *Adv. Neurochem.* **2:** 133—191.

39) YUWILER, A., BRAMMER, G. L., MORLEY, J. E., et al. (1981). Shortterm and repetitive administration of oral tryptophan in normal men. *Arch. Gen. Psychiat.* **38:** 619—626.

40) BRAMMER, G. L., McGUIRE, M. T., and RALEIGH, M. J. (1987). Whole blood serotonin level is determined by platelet uptake sites. *Life Sciences* **41:** 1539—1546.

41) BRAMMER, G. L., RALEIGH, M. J., and McGUIRE, M. T. (1982). Blood platelet properties, response to tryptophan load, and CSF 5-HIAA in relation to dominance in vervet monkeys *(Cercopithecus aethiops sabaeus). Int. J. Primatol.* **3:** 265.

42) STAHL, S., MELTZER, H. Y. (1978). A kinetic and pharmacologic analysis of 5-hydroxytryptamine transport by human platelets and platelet storage granules: comparison with central serotonergic neurons. *J. Pharmacol. Exp. Ther.* **205:** 118—132.

SUMMARY AND DISCUSSION-REPORT OF CHAPTER V

John D. Fernstrom and Peter D. Leathwood

SUMMARY

Leathwood: The common conclusion of all the presentations dealing with behavioral effects of tryptophan was that tryptophan does seem to influence behaviour and that the changes are compatible with involvement of serotoninergic mechanisms. Most speakers found little experimental support for the suggestion that the proportions of carbohydrate and protein in a normal meal can (via changes in the plasma TRP/LNAA ratio) also influence serotoninergic function.

Fernstrom: Several of the presentations considered, more specifically, the notion that the ingestion of particular macronutrients, by indirectly modifying brain neurotransmitter synthesis and release, alters subsequent selection of dietary macronutrients. As was discussed, this concept is most highly developed as relates to serotonin synthesis and the intake of carbohydrates and protein. But the validity of this latter hypothesis depends on an interpretation of a set of pharmacologic data, not very often discussed, that in actuality is incorrect. That is, the hypothesis considers that carbohydrate appetite is regulated by a negative feedback loop involving brain serotonin. When a subject consumes carbohydrates, the notion is that the plasma neutral amino acid pattern changes so as to favor tryptophan uptake into brain, and thus to stimulate serotonin synthesis. As a consequence, serotonin release from nerve terminals in increased, which ultimately causes the subject to reduce his intake of carbohydrates (a negative feedback effect). By doing this, the idea is that the subject then consumes proportionally greater amounts of protein, which has the effect of changing the plasma neutral amino acid pattern again so as to reduce tryptophan uptake into brain, and thus also to reduce serotonin synthesis. As a consequence, less serotonin is released, and the negative feedback effect to reduce carbohydrate intake is diminished. The subject then begins again to increase his intake of carbohydrates, and the feedback loop begins to work again. The nutritional-biochemical-neurochemical portions of this loop are certainly sustained by published data, at least in single meal paradigms in fasting rats (see Fernstrom, 1983, Physiol Rev 63: 484—546). I would like to focus a few comments on the neurochemical-behavioral portion of the loop, which is founded on a set of pharmacologic results. These results have been obtained largely with fenfluramine, a drug that induces serotonin release and blocks presynaptic reuptake. The claim has been that fenfluramine administration to rats *selectively* suppresses the ingestion of carbohydrates. With such data in hand, the claim has been that serotonin neurons, when active, inhibit carbohydrate ingestion, thus making the negative feedback component of the regulatory loop described above. But it is important to note that if fenfluramine is not selective in its effects on macronutrient intake, then the feedback model cannot work. And in fact, ample data exist showing fenfluramine not to be the selective anorectic it is touted to be (see Fernstrom, 1987, Appetite 8: 163—182). Its administration can lead to a suppression of the intake of three macronutrients, not just

NATO ASI Series, Vol. H20
Amino Acid Availability and Brain Function in
Health and Disease. Edited by G. Huether

carbohydrates. And there is great variability in the specific anorectic actions of fenfluramine, depending on the environmental and dietary conditions of the study (see Fernstrom, 1987, Appetite 8: 163—182). Hence, based on such pharmacologic data, the carbohydrate intake regulatory loop hypothesis is seriously flawed.

Other, phenomenologic data are often used as additional support for the carbohydrate regulation model. In particular, it is noted that the ingestion of a meal of carbohydrate will inhibit the intake of carbohydrate at a subsequent meal, and thus increase protein intake (Wurtman et al., 1983, J. Nutr 113: 70—78, Li and Anderson, 1982, Physiol Behav 29: 779—783). It is said that such data tangentially support the above serotonin feedback loop. But such data need not have *anything* to do with the serotonin loop noted above. Rats normally select for novelty, and thus will certainly avoid carbohydrates if recently consumed (i. e., select a protein diet). In an experimental context in which the choice is limited to protein and carbohydrates, the animal thus appears to go from one macronutrient to the other; i. e., to regulate carbohydrate (and/or protein) intake. Instead, he may simply be demonstrating a desire for novelty.

As John Blundell and others note, serotonin neurons must be imbedded in brain circuits that regulate food intake and appetite, because pharmacologic agents that influence these neurons do inhibit (and sometimes stimulate) food intake. Current evidence thus favors a participation by serotonin neurons in food intake regulation, but does not generally support the above notion that these neurons participate in a negative feedback loop regulating carbohydrate intake.

Leathwood: I would like to comment on the same point. The results presented by the speakers in this session lend support to the idea that serotonin is (directly or indirectly) one of the factors involved in control of food intake, with increases in serotoninergic tone tending to induce satiety or decreases in food intake. They (along with the presentations of Harper and Gietzen), also underline the point that availability of other amino acids can influence food intake quite independently of precursor effects on serotoninergic function.

Overall, the different presentations cast serious doubts on the model discussed above by John Fernstrom. As several speakers pointed out, the hypothesis is that protein meals are supposed to lower and carbohydrate to increase the plasma TRP/LNAA ratio, influencing brain TRP and 5-HT. 5-HT is then presumed to act as a "variable ratio sensor-regulator" such that a fall in 5-HT increases preference for carbohydrate, which in turn, increases brain 5-HT and so diminishes preference for carbohydrate, while an increase in brain 5-HT is supposed to increase preference for protein. (The mechanisms by which 5-HT is presumed to influence choice are rarely discussed). It emerged that the evidence supporting this idea is extremely fragile.

First, a CHO meal fed to overnight fasted rates usually increases the plasma TRP/LNAA ratio, brain TRP and 5-HT (Fernstrom and Wurtman, 1971, Science 174: 1023—1025). If, however, the rat is fasted only for a short time or is free feeding, most studies show negliable or small changes in the ratio and no effect on 5-HT (Ashley, et al., 1984, In: Progress in Tryptophan and Serotonin Research, de Gruyter, Berlin, 591—594; Leathwood and Ashley, 1983, Appetite 4:97—112). In free-feeding animals, selecting from a choice of diets or with imposed levels of protein, there is no consistent relationship between changes in protein/carbohydrate selection, plasma amino acids, brain TRP and 5-HT (Leathwood and Ashley, 1983, Appetite 4: 97—112). With protein, the results fit even less well with the hypothesis. The original findings on overnight starved rats fed a protein-containing meal showed no decrease in the plasma TRP/LNAA ratio, brain TRP or

5-HT and most subsequent observations have confirmed this (Fernstrom and Wurtman, 1971, Science 174: 1023—1025; Fernstrom and Faller, 1978, J. Neurochem 30: 1531—1538).

Secondly, since this cycle is supposed to be regulating protein/carbohydrate intake it is worth looking at the factors which influence selection. Appetite for protein seems to be learned and, although individual animals offered a choice of 2 diets often come to select constant proportions of each, the range of spontaneous selection is wide (from 10—55 % or more) and is influenced profoundly by texture, taste, type of protein, work required to obtain the foods and quantities of protein in the diets offered (Leathwood and Ashley, 1983, Appetite 4: 97—112; Fernstrom and Faller, 1978, J. Neurochem 30: 1531—1538). In addition, changing the diets offered leads to marked (and unforced) changes in proportions selected (Leathwood and Ashley, 1983, Appetite 4: 97—112; Ashley, 1985, Nutr. Res. 5: 555—571). This suggests that while rats "regulate" protein intake in the sense that they tend to avoid extremes, there is certainly no tight control of intake. (It should not be forgotten that metabolic adaptation can take care of a wide range of protein intakes).

Thirdly, the observation that animals prefed protein or carbohydrate tend to select less of that macronutrient in the next meal does not necessarily mean that they are selecting specific macronutrients. Rats and other omnivores show a powerful drive for variety among familiar foods, so when they chose carbohydrate after being prefed protein they might well be choosing for variety rather than *against* protein. Even when the rat's choice does seem to be influenced by macronutrient content, it is not necessary to invoke changes in precursor availability as an explanation because there are amino acid receptors in the gut which relay information to the brain via the vagus.

Lastly, if the hypothesis outlined above really does not stand up to critical evaluation, what should one make of the small but consistent changes in selection which seem to follow certain manipulations of 5-HT metabolism? (Arimanana et al., 1984, In: Progress in Tryptophan and Serotonin Research, de Gruyter, 4549—5525). Two possibilities are: (a) that serotoninergic relays are involved in the brain circuits influencing food choice, texture and taste preferences, and that pharmacological manipulations of 5-HT "key in" to these circuits. The resulting changes in food choice are then interpreted by the experimenter as changes in protein/carbohydrate preference. (b) it is equally possible that changes in "protein" selection are in fact side effects (due dryness of the mouth, problems in swallowing...) consequent upon changes in brain stem serotoninergic function which are again fortuitously interpreted in terms of varying "protein" appetite.

In summary, although the scheme outlined above seems at first sight seductively simple and plausible, it is not supported by the evidence and is unlikely to play more than a marginal role in food selection.

Comment by Gibson: The prefeeding results do not even fit the model, because the observed switch between diets does not depend on carbohydrate and protein content. You can get it with two carbohydrate diets (or two protein diets) of contrasting sensory characteristics. Any omnivore gets temporarily bored with the diet it last ate (Booth, 1976, In: T. Silverstone (Ed.) Appetite and Food Intake, Dahlem Konferenzen, pp 415—478; Booth, 1985, Ann. N. Y. Acad. Sci. 443: 22—37). Le Magnen (In: C. F. Code (Ed.) Handbook of Physiology, Section 6, Vol. 1, 11—33) called this "sensory-specific satiety" and it is a robust effect in people and in rats. Taking a main-course (which is high in protein in our culture) will temporarily inhibit the desire for first-course menus and increase the desire for desserts (which in our culture are often lower in protein). It has nothing to do with selection of diets for their protein or carbohydrate content.

DISCUSSION-REPORT

What are the current problems in studying the involvement of tryptophan and serotonin in food intake regulation?

Anderson: I don't think anyone should go away thinking that one could completely sequence how macro-nutrient intake is regulated. You can completely weep out the serotonin system and the organism will still survive and competently select its food according to its requirements. If we concentrate our search on one individual neurotransmitter system and its role in the regulation of food intake we may easily overestimate its importance; it may later turn out not to have that kind of priority. What we should try to do, is to ask the question: Does this system participate in the whole process? Which is very different from saying it is the only one regulator. It is clear that serotonin is involved in appetite regulation, but I do not know whether the signal is a metabolic shift in the tryptophan versus large neutral amino acid ratio or it is an increased release of serotonin independent of precursor availability.

Wurtman: I would like to draw particular emphasis to the fact that although originally there were disagreements as to whether the brain regulated carbohydrate or protein appetite via serotonin it has long since been apparent that it regulates both, and both via serotonin.

If one is setting up an experiment to examine the possible effect of serotonin release on nutrient selection, the experiment is much likely to work if the animal's (or human's) situation is such that it would then be electing to consume carbohydrates. (It took us a long while to realize that one cannot suppress carbohydrate intake when the animal is not taking in carbohydrates.) In that context, John Blundell's seemingly-paradoxical finding becomes understandable: He showed that adding tryptophan to a protein-rich meal suppressed subsequent carbohydrate intake (i. e., "dessert"). The protein meal, in itself, would have generated a desire for carbohydrates, and the tryptophan — by decreasing the protein-induced fall in the plasma tryptophan ratio — blocked that response. This generalization also explains, I think, why d-fenfluramine suppresses snack carbohydrate intake so much more than mealtime carbohydrate intake, and why the rat nutrient-choice studies work so much better using 5 % vs 45 % protein meals than 25 % vs 70 % carbohydrate meals which also, by the way, contained about 20 % protein.

Another point which is important in the discussion of the problems to study the involvement of tryptophan and serotonin in food intake regulation: Michael Raleigh found that tryptophan actually increased feeding in his dominant vervets. That suggests that the role of serotonin in feeding in that species differs from its role in man. One is reminded of the situation vis-a-vis my first love, melatonin: In all species it is secreted at night, and more (per 24 hrs) appears to be secreted in the Fall than the Spring. Moreover, in all species studied its rhythm apparently is the photoperiodic signal linking the lighting environment to gonadal function. But in most animals, the decrease in melatonin (springtime) "turns on" gonadal recrudescence, while in some others (e. g., sheep), melatonin is pro-gonadal, and pregnancy occurs in the fall. So it may well be that tryptophan always enhances the formation of serotonin, but that the actual functions of serotonin may differ (and even be opposite) in different species.

van Gelder: In the sixties very interesting experiments were performed studying the regulation of food intake by various regions of the brain where serotoninergic pathways have selectively been destroyed. Are the conclusions reached by these experiments consistent with the ideas discussed here or are they contradictory?

Fernstrom: Such brain lesion studies, it seems to me, are about as physiologic as lesioning the tongue when studying the fine points of food intake behavior. It is important to be as physiologic as possible with our concepts and treatment strategies.

van Gelder: There is one subject that has not been touched in this discussion until now and that is, what is taste and what is the role of taste in the regulation of food intake. We know that mice and rats and rodents in general can taste amino acids much better than we do and human beings who have had a high carbohydrate diet have often an subsequent aversion to something that tastes sweet.

Harper: We certainly must consider taste as an important component of food selection and we have certainly to consider another component and that is food selection based on experience and learning. The question is to what extent is a learned preference process involved in food selection, either in our animal studies or in patients.

Leathwood: Certainly learning is involved in food selection. There are also rather simple straight forward innate behaviors involved. For example the preference for variety among familiar foods. If one offers a rat a choice of diets it will tend to select from both. If one diet is rich, and the other poor, in protein, the rat will appear to "select" a given level of protein.

Gibson: We would like to emphasize again that there is no evidence that genuine nutrient selection is controlled by innate reactions to taste. Proteins do not taste of free amino acids, and so an ability to taste individual amino acid in a diet cannot explain how rats self-select consistent and adaptive levels of protein. Likewise, starches and sugars have no characteristic taste, and so selection of carbohydrate cannot simply depend on an innate reaction to taste. Rather, if protein (or carbohydrate) intake is controlled or "regulated", not just fortuitous, the animal must learn to accept or reject diets on the basis of their sensory qualities that have become associated, through experience, with postingestional effects specific to protein (or carbohydrate). In this respect, taste is probably less relevant than other sensory characteristics such as smell or texture. We have shown that rats can learn to prefer odors that predict removal of a need for protein (Baker et al., 1987, Nutr. Res, 7: 481—487, Booth, 1974, Physiol. Psychol, 2: 344—348; Gibson & Booth, 1986, Experientia, 42: 1003—1004). Also, with solid casein/dextrin diets, texture is crucial: partial deafferentiation of the trigeminal nerve, which impairs oral somatosensory input, disrupted (Miller, 1984, Behav. Neurosci., 98: 424—434) or prevented (Miller & Teates, 1985, Physiol. Behav., 34: 401—408) acquisition of the stable dietary selection seen in unoperated rats. Serotonin affects the reflexes of eating (Hashim & Bieger, 1987, Brain Res. Bull., 18: 355—363) and so textural differences will affect dietary selection, as we and others have observed (Booth & Gibson, this issue; Mc Arthur & Blundell, 1986, Physiol. Behav., 38: 315—319).

Rogers: I would like to make three comments related to food intake in dietary choice. Firstly, I can only agree that animals will choose protein within a certain dietary range. I would like to reemphasize something Harvey Anderson mentioned without much emphasis, that is that the percentage of energy chosen as protein by weanling rats is dependent upon the nutritional quality of the protein. He found rats chose about 33 % of their energy-intake as protein when this protein was wheat gluten or unsupplemented casein. The percentage of energy selected as protein is found to be near 15 % when the casein is supplemented with methionine or when a near ideal protein or amino acid mixture is fed. Secondly, I think one has to be very careful in determining how far the choice is higher than random choice. Harvey Anderson, for instance, gave a choice of protein content in the diet between 15 % and 55 % and the rats chose 33 %. The random choice would be 35 %. A last point with regard to taste: We have found that the vitamin-free casein we have been using has a general negative palatability. If you increase casein above the requirement the animal will start avoiding it, so one should clearly dissociate taste from the metabolic effect of a diet.

Anderson: I think the issues of taste, learning, food choice, and whatever is involved in the regulation in this process are so complex that we all can have several hypotheses. Mark Fredman for example suggested that liver metabolism would be involved in the regulation of food intake. And this is all very interesting. There are lots of regulatory mechanisms which all have to be considered to gain a full picture of the regulation of food intake under physiological conditions.

Leathwood: I think this is a terribly important point. So often we think in terms of a mechanical engineer who constructed an automatic feedback system into his machine. It is perfectly obvious, as you say, that getting adequate food is the first priority any organism relies upon. So it is clear that each animal must be bound to a host of behavioural and metabolic adaptations which have been developed throughout evolution and if you wipe out one, there are several others to take over. In this context, let me mention that even the most sophisticated methodology to measure the mechanisms of food selection in animal models may still be inadequate for a living system. Delivering food by gastric cannula, for instance, is thought to be a highly objective approach. But as you may know delivering food by gastric cannula can itself be aversive and it is quite possible that different macronutrients have differentially aversive effects.

Gibson: Of course the system is complex, but we should not confuse ourselves into missing the simple crucial point which is specifically about one small part of the system. Rats and perhaps people can learn to select protein and perhaps carbohydrate, whether taken voluntarily by mouth or in non-aversive amounts by gastric cannula. Yet when the doses of fenfluramine alleged to affect nutrient selection are tested on this learned actual selection of the nutrients as such, absolutely no effect is seen on the selective behavior (Gibson & Booth, 1988, Behav. Brain Res., in press). It matters not a whit how complex appetite is or what tryptophan does to brain serotonin. The evidence is conclusive that serotonin is not specifically involved in protein intake regulation (and carbohydrate intake regulation may not even exist).

Are there specific circumstances under which an altered precursor supply can affect behaviour whereas under other circumstances it does not?

Bruinvels: In our discussions on the effects of tryptophan on behaviour and mental functions we always think in terms of homogeneous groups. But such homogeneous groups do not exist in reality. For instance, sleep disorders may have various causes. There might be some subgroups which respond to tryptophan whereas others will not. The same is true for the treatment of depression. First of all, we have always to start with a detailed diagnosis to identify the subpopulation that would respond to our treatment and then, in these subgroups, we might see effects to tryptophan which may eventually not be seen in the whole population.

Wurtman: Even more, also within the group of depressed persons, subgroups seem to exist which respond to changes in tryptophan supply, whereas others do not. Philip Cowen (this volume) showed that prolactin response to tryptophan among depressives depended on whether or not they had lost weight. The weight loss might be a marker for two different diseases (i. e., two types of depression), of which one, the atypical variety (which includes but is not limited to SADS), without weight loss, may be the more serotonin-responsive of the two.

Commissiong: It seems to me that one ought to put the use of tryptophan to treat depression in a subset of depressed patients, into a correct biological perspective. One general hypothesis could be that tryptophan (acting via 5-HT), affects mood in humans. If the mood of some depressed patients is improved by tryptophan, then one ought to expect, perhaps that some normal people would be made euphoric by tryptophan. If all that can be said, is that tryptophan is effective in treating depression in a subset of depressed patients, then we are into the realm of observed clinical effectiveness, without defining a tenable mechanism, or a working hypothesis.

Fernstrom: One can imagine a situation at the neuronal level which might be in accord with Jaques Bruinvels suggestion, and would lend itself to testing; that is, a case in which tryptophan would have a functional effect at some times, but perhaps not at others. It is known that raphe unit firing is not uniform over the 24-hour period, but is high at certain times, and almost absent at others (McGinty and Harper, 1976, Brain Res, 101: 569—575; Trulson and Jacobs, 1979, Brain Res 163: 135—150). It is also known that the raphe nuclei project to the facial motor nucleus in the rat, and synapse directly on motor neurons. The function of this serotonin connection appears to be to modulate (enhance) impulse flow from upper motor neurons onto the facial nucleus motor neurons (McCall and Aghajanian, 1979, Brain Res 169: 11—27). Given these two facts, one could hypothesize (and test) that at the time of day raphe neurons are off (or firing at a very low rate), a tryptophan-induced stimulation of serotonin production would have little impact on the facial motor neurons, since even though more serotonin would be produced, very little would actually be released. In contrast, at a time of day when the raphe neurons are very active, it seems likely that driving serotonin synthesis by administering tryptophan would lead to increased serotonin release, and thus enhanced effects on the facial motor neurons.

Is precursor availability an important factor for blood-pressure-regulation?

Lehnert: A role for serotonin-containing neurons and their projections within the central nervous system in cardiovascular regulation is suggested by their paralleling central pathways relevant for cardiovascular function. The precise role of serotonin in the regulation of blood pressure has remained inconclusive. Specifically, we have to deal with different cardiovascular effects of serotonin neurons depending on what set of neurons has been stimulated and what their relative contribution to sympathetic nervous system activity is. For example, injection of serotonin into the anterior hypothalamus or preoptic area as well as iontophoretic application to the spinal cord appears to exert pressor effects, while on the other hand a large bulk of evidence (pharmacological studies, stimulation of raphe nuclei) clearly suggests depressor effects. There appears to be agreement that modulation of brain serotonin activity affects the cardiovascular system directly through changes in preganglionic sympathetic activity. We have shown (J. Cardiovasc. Pharmacol., 1987, 10: 389) that administration of 5-HTP in conjunction with carbidopa increased cerebrospinal fluid concentrations of serotonin and 5-HIAA significantly (without alterations in NE, MHPG or DOPAC) and concomitantly decreased arterial blood pressure, heart rate and efferent sympathetic neural activity in the cat. In addition, it attenuated the surge in sympathetic activity associated with acute myocardial ischemia and elevated the ventricular fibrillation threshold. Thus, neurochemical changes indicative of enhanced brain serotonin turnover were associated with a decrease in sympathetic nervous tone and cardiovascular depressor effects at least in the animal model.

As to the role of direct (and peripheral) precursor administration, Fernstrom's group has performed various pharmacological maneuvers demonstrating a centrally mediated antihypertensive effect of l-tryptophan (JPET, 1982, 221:329). A peripheral mechanism for tryptophan to lower blood pressure has been suggested by Wolf and Kuhn (Brain Res., 1984, 295:356) but still has to be validated. That indeed precursor administration is capable of lowering blood pressure is further substantiated by a recent finding of our group, where we demonstrated a significant antihypertensive effect of l-tryptophan in conjunction with a high-carbohydrate diet administered to essential hypertensive patients. At the end of four week trial, blood pressure levels were significantly lower in the tryptophan than in the control group. The group receiving tryptophan and a high-carbohydrate diet (as compared to a high-protein diet) had even lower blood pressure levels and thus predominantly contributed to the observed decrease. A strong inverse relationship was observed between the amino acid ratio and blood pressure levels. In addition, a negative correlation was found between pretreatment renin levels and posttreatment blood pressure data, suggesting that the treatment might be effective in "high-renin hypertension". Also, negative correlations were observed between the amino acid ratio and plasma renin activity and aldosterone concentrations, with renin activity significantly lower in those receiving tryptophan and carbohydrates versus tryptophan and proteins.

In conclusion, profound cardiovascular (and endocrine) effects can be observed following strategies that enhance brain serotonin biosynthesis; the majority of available data suggests a centrally mediated antihypertensive mode of action of the precursor amino acid l-tryptophan while nevertheless different sets of brain serotonin neurons might exert opposite cardiovascular effects.

Fernstrom: I suspect that precursor availability is not an important physiologic factor in the moment-to-moment regulation of blood pressure in animals or humans with normal blood pressure. Precursor (tryptophan, tyrosine) administration at pharmacologic doses does lower blood pressure in hypertensive rats (Bresnahan et al., 1980, Am. J. Physiol, 239: H206—H211; Sved et al., 1979, Proc. Natl. Acad. Sci. US 76: 3511—3514; 1982, J. Pharmacol. Exp. Ther. 221: 329—333; Yamori et al., 1980, Eur. J. Pharmacol. 68: 201—204) as well as apparently to raise blood pressure in rats with dangerously low blood pressure (Conley et al., 1981, Science 212: 559—560). Effects of tyrosine have not been observed in humans with borderline hypertension, however, and it is thus not certain if precursor effects in rats also occur in man (Sole et al., 1985, Hypertension 7: 593—596) (there are simply too few published studies at present to allow a proper evaluation of this issue). Regardless, it is clear that whatever precursor-related effects are present on blood pressure regulation, they are confined to a pharmacologic context. And, even if present, it appears that tryptophan and tyrosine are relatively "weak" as antihypertensive drugs, and thus unlikely to find much acceptance as therapeutically useful agents in man.

Is it possible that, just because nutrients are used as pharmacological agents, a confusion has arisen between pharmacological and nutritional, physiological effects in nutrients?
(Question brought up by G. Gaull)

Fernstrom: This is undoubtedly correct. We often confuse in the line between pharmacology and nutrition. Some investigators go to a pharmacologic extreme with precursors, and then attempt to supply the results with a nutritional relevance that is not there. This seems inappropriate. For example, tryptophan and tyrosine effects on blood pressure are pharmacologic, and likely to be of no nutritional relevance. However, there are some nutritional contexts in which precursor changes are known to alter transmitter synthesis and presumably release, since specific brain functions are altered, though to date these tend to be rather extreme (though not unknown) nutritional situations. For example, if rats are fed a corn-based diet for several weeks, brain tryptophan levels and serotonin synthesis rate fall (because corn is naturally low in tryptophan). As a consequence, the animals become very sensitive to painful stimuli, a condition that was predicted from other data on serotonin and pain, and that could be remediated immediately by administering tryptophan, or chronically by supplementing the diet with tryptophan (see Fernstrom and Lytle, 1976, Nutr. Rev. 34: 257—262). This dietary situation does have a human analog, in that certain Latin American populations subsist on corn as their main protein source, and thus could be experiencing low brain serotonin synthesis and heightened sensitivity to painful stimuli. What is presently unknown is whether normal changes in nutriture in normal animals can influence neuronal tryptophan or tyrosine levels and the syntheses of their respective transmitters sufficiently to modify particular brain functions. Hopefully, this issue will receive further attention in the future.

Young: Our difficulties in classifying tryptophan either as nutrient or as a drug is reflected by the confusion in the governmental food and drug administrations. In Britain tryptophan has been available for many years on prescription as an antidepressant. In the United States tryptophan is classified as a food supplement and is available over the counter in pharmacies, health food stores and even supermarkets. In Canada tryptophan was freely available as a dietary supplement until 1985 when it was reclassified as a drug. The reclassification was challenged legally by one of the companies selling tryptophan, but the trial judge's decision upheld the classification of tryptophan as a drug. In my opinion, if tryptophan is given for the sole purpose of influencing the metabolism of a neurotransmitter (5-hydroxytryptamine), which is usually the case, then it must be classified as a drug and not a nutrient.

Harper: For years we studied the requirements and nutritional needs of individual food components. If the intake of a particular nutrient is too low you develop some sorts of deficiencies. Above this critical minimal intake you have a certain safe level of nutrient intake, depending on the kind of nutrient. But if you rise the intake above this safe level, you may easily reach pharmaceutical or toxic effects caused by the respective nutrient. I think, with regard to our discussion, it is very important to make some efforts to distinguish the point from which a nutrient becomes a pharmacological agent.

Young: I would like to comment on the toxicity of tryptophan, a subject covered in recent reviews (Sourkes, 1983, Young, 1986). There are studies in the literature where doses up to 10 or even 20 g of tryptophan per day were given to patients or to normal subjects and no side effects were reported. Nevertheless, there are a few circumstances where tryptophan may have adverse effects. Supplementation of tryptophan in the diet of pregnant hamsters caused significant reductions in embryo and neonate survival and in neonatal weight of the pups (Meier and Wilson, 1983), so tryptophan should not be given to pregnant women. This is an important point because of a strong folklore that tryptophan is a "natural" treatment for insomnia, pain or depression. As such it might be preferred by pregnant women over compounds labelled as drugs. Tryptophan should also not be given to people with a source of physical irritation in the bladder or a history of bladder cancer. There is a large literature on association between bladder cancer and tryptophan metabolite 3-hydroxyanthranillic acid, but reviews of the topic indicates that 3-hydroxyanthranillic acid is only a carcinogen when there is a source of physical irritations in the bladder (Sourkes, 1983, Young, 1986). Recently it has been shown that the toxicity of tryptophan is very greatly elevated when rats are adrenalectomized (Trulson and Ulissey, 1987), so tryptophan should not be given to patients with adrenal insufficiency.

Gaull: Having initiated this discussion I would like to carry it one step further. If we all agree that there are pharmacological effects when normal nutrients are given in pharmacological doses, are we getting such effects because the systems normally protecting the brain are being overwhelmed or because physiological reactions have been simply pressed to their limits?

Harper: If you give more and more of one precursor amino acid, independent of whether you increase neurotransmission or not, you will proceed to a point where the system is overwhelmed and this is the point where you have toxic and lethal effect. But before this point is ever reached you have homeostatic control and various feedback regulations in the intact living system.

Neuhoff: I wonder if many of the effects you describe are only seen because you shifted the system out of its balanced steady state. But this is not physiology and this does not give us information about what happens in the intact system. So we should always have the buffer maintained when we study the effect of a certain nutrient or a drug, instead of knocking down homeostatic control mechanisms in our endeavour to get some sort of information out of the system.

Harper: I think it is very important to quantify the behavioural effects of tryptophan and to document it in a dose-effect relationship.

Leathwood: One must also take into account the composition of meals eaten at about the same time that tryptophan is consumed. Protein would be expected to decrease tryptophan's effects on the plasma TRP/LNAA ratio (and hence, on brain TRP and 5-HT) while a carbohydrate meal might well potentiate them (Leathwood, 1987, Proc. Nutr. Soc. 46: 143—156).

How good is the pharmacological evidence that the behavioural responses seen after tryptophan loading are indeed serotoninergic?

Bradberry: It seems very well established that tryptophan can have several behavioural effects on certain subtypes of the human population. But, as far as I see, nobody really cares about the possible pharmacological mechanism. Has it really been tested that the behavioural effects are indeed related to serotonin mediated phenomena? How good is the pharmacological characterization of the phenomena we are talking about?

Fernstrom: I can answer your question with regard to tryptophan's effects on blood pressure in hypertensive rats. You can block tryptophan's antihypertense action with metergoline, a serotonin anagonist, and enhance it with fluoxetine, a serotonin reuptake blocker. You can also prevent tryptophan's effects by pretreating rats with a dose of p-chlorophenylalanine (a tryptophan hydroxylase inhibitor) that by itself does not alter blood pressure (Sved et al., 1982, J. Pharmacol. Exp. Ther. 221: 329—333). Hence, for these experiments, care was taken to tie the biologic effect of tryptophan to an action on the serotonin pathway and synapse. But your question nonetheless makes an important point. Most investigators do not worry about the pharmacologic specificity of the effects they obtain with tryptophan. It is thus difficult to accept that all such effects are mediated via effects on serotonin synthesis and release. (For example, given sufficient tryptophan, it is conceivable that some could be converted to tryptamine in vivo, and lead to biologic effects).

Wurtman: This critizism is justified. In future we should intensify our efforts for demonstrating the pharmacologic evidence to support what we are suggesting. For instance, Peter Leathwood's study showing an effect of tryptophan on sleep, is difficult to interpret without such controls. For example, he should have seen whether drugs that also increase serotoninergic transmission (like fluoxetine, chlorimipramine or d-fenfluramine) also accelerated sleep onset, and compared their potency with that of tryptophan. Or, at least, he might have compared the potency of tryptophan with that of a mild hypnotic benzodiazepine (short-acting).

Leathwood: Richard Wurtman is quite right and we have already taken some of these points into account. Although we have not carried out systematic studies with other serotoninergic agents, fenfluramine is known to be a very effective sedative. In the sleep research community it is well known that the most consistent finding after treatments which can be expected to produce small increases in brain serotonin availability is an increase in sleepiness and a decrease in sleep latency (Koella, 1984, In: Sleep 1984, Fischer, Stuttgart, 6—10). We have not yet made a direct comparison between tryptophan and a benzodiazepine, but in doing so we need also to take into account other factors such as the frequency of side effects. I wonder, if the monkeys used by Michael Raleigh would offer much better models for studying the pharmacological evidence of the involvement of cerebral serotoninergic mechanisms in behavioural responses to the increased tryptophan supply. We have recently shown that the levels of tryptophan in the range he used are sufficient to increase brain tryptophan and brain stem serotonin in cyromolgus monkeys (Leathwood and Fernstrom, in preparation), so it is at least plausible that these manipulations are enhancing central serotoninergic function.

Raleigh: My colleagues, Gary Brammer, Art Yuwiler, and Michael McGuire and I have long been concerned with the question of whether the behavioral effects of tryptophan result from enhanced central serotonergic activity. In our view, if several criteria are met, then the effects of tryptophan can most parsimoniously be attributed to enhanced central serotonergic function. Further the larger the number of criteria that are met, the more likely alterations in serotonergic function underlie tryptophan's effects. As a minimal criterion we require that pharmacological treatments that diminish central serotonergic function lead to behavioral effects opposite to those following tryptophan. Thus the observation that PCPA or cyproheptadine treatments diminish while tryptophan administration increases approaching, grooming, and other affiliative behaviors suggests that tryptophan's effects may be due to alterations in central serotonergic function.

Secondly serotonergic systems may be considered as stronger mediators if 5-HTP reverses the effects of PCPA. Thus in vervet monkeys concurrent administration of 5-HTP and PCPA following PCPA pre-treatment restored the rates of grooming and approaching the baseline levels (Raleigh, et al., 1980, Exp. Neurology, 68: 322—334).

A third criterion is that if drugs that enhance serotonergic function by different mechanisms produce behavioral consequences similar to those of tryptophan, then the effects of tryptophan are likely to be due to its impact on serotonergic transmission. For example fluoxetine inhibits the reuptake of serotonin from the synaptic path (Fuller and Wong, 1977, Proc. Fed. Am. Soc. Exp. Bio. 36: 2154—2158) while quipazine is a receptor agonist (Fuller, 1982, In: Osborne Ed., Biol. of Ser. Transm. pp 221—247). Despite their different chemical structures and properties, tryptophan, fluoxetine, and quipazine produce strikingly similar effects on behavior at least in vervet monkeys. Thus all three drugs result in dose-dependent increases in approaching, grooming, resting, and eating, and decreases in locomoting, avoiding, being solitary, and being vigilant (Raleigh et al., 1985, Brain Res. 348: 274—282).

In combination with John Leathwood's and others data the observation that tryptophan treatment meets the criteria I have just described supports the possibility that tryptophan's effects are indeed due to enhancement of central serotonergic function.

Pratt: We are only talking about tryptophan as a precursor for the synthesis of serotonin. Has anybody considered the possibility that tryptophan has some other effects may be on membrane potentials or others not related to its precursor function for transmitter synthesis.

Leathwood: Researchers carrying out behavioral and clinical studies are not in a position to exclude the possibility that tryptophan influences behavior through mechanisms quite independent of serotonin metabolism. It should also be remembered that current knowledge of serotoninergic function is so fragmentary that attribution of behavioral effects of tryptophan (or of "serotoninergic" drugs) to changes in serotoninergic function is far more speculative than is usually admitted.

Huether: The other problem is, that people who study the effects of tryptophan on other processes than serotoninergic transmission are often not in the position to extrapolate from their findings to its possible significance for physiology and behaviour. There are several such findings that should concern us and that should at least be considered when we try to explain the behavioural effects of tryptophan.

For instance, it has been shown some years ago that tryptophan, at least in synaptosomal preparations, decreases the K_m-value of the tryptophan-hydroxylase for oxygen, and thereby accelerates the formation of serotonin independent from its precursor function (Katz, 1980, J. Neurochem. 35: 760)

We certainly must take into consideration the possibility that serotonin in addition to its transmitter-role may have a much broader function, being a hormon-like substance, an important mediator of intercellular communication, affecting not only neurons but also glial cells (astrocytes posses a high affinity uptake system for serotonin, see Kimelberg and Katz, 1985, Science 228: 889)) or endothelial cells (see the contribution of De Feudis, this volume). We do also know very little on the functional consequences of the largely increased formation of 5-HIAA in the brain and in the periphery after tryptophan-loading (hydroxyindole acetic acid affects, e. g. liver regeneration after partial hepatectomy, Yamada et al., 1986, Neurochem. Soc. Res. 11: 101).

Oral tryptophan administration increases not only the formation of serotonin in the brain, but also in the gut (own unpublished observations), and changes of intestinal function may affect other extraintestinal physiologic responses, including behaviour. We should also not forget that only about 1 % of the tryptophan is used to make indoles. What about possible physiologic and behavioural consequences of the other metabolites formed in much greater amounts after tryptophan is given.

Finally, the finding of increased blood serotonin in dominant monkeys seen by Michael Raleigh rises another important question: What does this measure mean physiologically? Are there more platelets or is the amount of serotonin stored in the platelet increased? Is the platelet-content of serotonin affected by tryptophan-loading and what are the implications then for the function of these platelets?

Raleigh: Prior to responding to Gerald Huether's interesting questions, I should note that under basal conditions the elevated serotonin present in dominant vervet and squirrel monkeys is almost entirely confined to platelets. Furthermore, irrespective of social status, in rats, squirrel monkeys, vervet monkeys, and humans in drug free conditions, more than 95 % of the serotonin in blood is found in platelets rather than plasma (Yuwiler, et al. 1981, Archives of General Psychiatry, 38: 619—626). For this reason we have focused our attention on either whole blood or platelet serotonin.

Until the physiological, behavioral, and psychological mechanisms that maintain the status-linked differences in blood serotonin have been specified, it may be premature to speculate about its significance. Presently it is possible that the differences arise from differences in precursor availability; tryptophan metabolism; platelet properties including their number, size, serotonin uptake and storage capacities; and serotonin degradation. We have observed that there are no status-linked differences in concentration of free, bound, and total plasma tryptophan (Raleigh et al., Psychopharmacology, 1981, 72: 241—246). We have not measured hepatic tryptophan pyrrolase activity. However this enzyme is induceable by stress and there are status-linked differences in the concentration of hormones (e. g. cortisol) which have been associated with stress. Thus, differential tryptophan metabolism is a viable possibility. Serotonin is largely degraded by plasma MAO and platelet MAO and there are no status-linked differences in the activity of these enzymes (Brammer et al., 1982, Int. J. Primatol., 3:337).

Some of our recent work speaks to Gerald Huethers second question. Studies of platelet properties have shown that whole blood serotonin levels, platelet serotonin content, and the serotonin uptake parameter V_{max} were stable within animals across repeated samplings. Blood serotonin level was strongly associated with platelet serotonin content and this in turn was strongly associated with V_{max}. These findings support the view that blood serotonin levels are a function of the number of platelet uptake sites (Brammer et al, 1987, Life Sciences, 41: 1539—1546).

The third question is important in part because the roles platelets play in blood clotting. Some time ago it was observed that in healthy human males a 50 mg/kg tryptophan load reliably increased blood serotonin (Yuwiler et al., 1981, Archives of General Psychiatry, 38: 619—626). This increase was transient and was almost entirely due to a rise in extraplatelet (e. e. plasma) 5-HT. Preliminary data from rats and monkeys suggest that this may be a general phenomenon. In none of these studies were alterations in blood clotting observed. This suggests that at least some aspects of platelet physiology may not be affected by precursor loading. This suggestion is supported obliquely by the observation that in carcinoid syndrome blood serotonin levels are unusually high but the incidences of thrombosis are not strikingly elevated.

In sum, the physiological significance of this status-linked difference remains to be determined but it is probably due to several factors. Until the mechanisms that translate different behavioral and environmental cues into altered blood serotonin levels are known the significance of this difference for peripheral and central monoaminergic function will remain difficult to specify.

Young: The evidence that the effects of tryptophan on mood and behavior in humans is mediated by serotonin is purely circumstantial. In general the effects of tryptophan are consistent with what is known about the role of 5-HT in the brain. However, other tryptophan metabolites including, tryptamine, tryptophol, 5-hydroxytryptophol, melatonin and quinolinic acid could possibly be responsible for part or all of the effects of tryptophan. There is also a report that tryptophan administration can increase brain protein synthesis (Jorgensen and Majumdar, 1976, Biochem. Med., 16: 37—46).

Chapter VI.

AMINO ACID AVAILABILITY AND BRAIN DYSFUNCTION

EXCITATORY AMINO ACIDS IN EPILEPSY AND IN ACUTE AND CHRONIC NEURONAL DEGENERATIVE DISORDERS

B. S. Meldrum

Department of Neurology
Institute of Psychiatry
De Crespigny Park,
London SE5 8AF,
U. K.

INTRODUCTION

Excitatory amino acids play a crucial role in the initiation and spread of epileptic activity (1). They are also excitotoxins that can cause selective neuropathology when present in concentrations moderately or substantially exceeding those normally found in the synaptic space. They probably contribute to selective neuronal loss occurring acutely after status epilepticus, transient cerebral ischaemia and hypoglycaemia (2). They may also be involved in a wide variety of chronic neurological disorders. These phenomena are of great interest in the context considered in this volume. Thus elevated concentrations of excitatory amino acids may arise in the brain as a result of peripheral abnormalities in amino acid availability or metabolism, abnormalities in the blood brain barrier, or abnormalities in the release, uptake or metabolism of the amino acids in the brain.

We shall review the endogenous excitatory amino acids and their specific receptor systems and their role in epilepsy. We shall also consider their role in acute neurotoxic disorders and their possible role in chronic degenerative disorders.

Endogenous Excitatory Amino Acids and Their Receptors

Amino acids with excitatory action that are found in the brain are listed in Table 1. There is ample evidence that L-aspartate and L-glutamate are released from nerve terminals upon stimulation. There are also powerful uptake systems providing transport for aspartate and glutamate into glia and into nerve terminals.

The sulphinic and sulphonic analogues of glutamate and aspartate are also released from hippocampal slices upon stimulation (3). Quinolinic acid is a metabolite on the kynurenine pathway for tryptophan (see Figure 1). It is synthesized in the liver and in the brain. The enzymes responsible for its synthesis and further metabolism occur predominantly in glial cells and there is no evidence for its stimulated release or for a specific uptake mechanism. Thus it is unlikely that quinolinate acts as a neurotransmitter. Nevertheless it may be involved in various pathological processes including epileptogenesis and acute and chronic neuronal degeneration. Thus we shall consider it in some detail.

NATO ASI Series, Vol. H20
Amino Acid Availability and Brain Function in Health and Disease. Edited by G. Huether

TABLE 1
Excitatory amino acids present in mammalian brain

		Concentration (μmol/g wet weight) (rat brain)	(human brain)
Asparte	$COOHCH(NH_2)CH_2COOH$	2.6	1.2
Glutamate	$COOHCH(NH_2)CH_2CH_2COOH$	11.3	10.2
Cysteine sulphinate	$COOHCH(NH_2)CH_2SOOH$	0.012	
Cysteate	$COOHCH(NH_2)CH_2SO_3H$	0.01—0.1	
Homocysteate	$COOHCH(NH_2)CH_2CH_2SO_3H$		
quinolinate	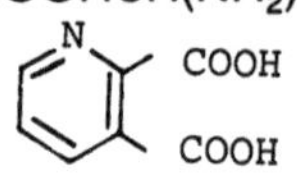		0.0008

Refs 38—40

There are also in the brain various di- and tri-peptides containing aspartate and glutamate that may have a role in neurotransmission, e. g. N-acetyl-aspartyl-glutamate (4). However, there is very little evidence of their involvement in the processes to be discussed and they will not be considered further.

Tryptophan
↓ (tryptophan pyrrolase)
Formylkynurenie
↓ (kynurenine formylase)
Kynurenine
↙ ↓ (kynurenine 3-hydroxylase)
Kynurenic acid ↓
3-Hydroxykynurenine
↓ (kynureninase)
3-Hydroxyanthranilic acid
↓ (3-hydroxyanthranilate oxidase)
QUINOLINATE
↙ ↓ (quinolinatephosphoribosyltransferase)
Picolinic acid ↓
Nicotinic acid
↓
Nicotinamide

FIGURE 1
The kynurenine pathway for tryptophan metabolism

Excitatory Amino Acid Receptors and Specific Antagonists

The post-synaptic receptors responding to glutamate and aspartate fall into at least three major subtypes, called N-methyl-D-aspartate (NMDA), kainate and quisqualate receptors after their preferred agonists (5). These receptors operate different receptor channels (or different states of similar channels) (6, 7) and different second messenger systems. The NMDA receptor opens channels permeable to Ca++ but subject to a voltage-dependent block by Mg++. This contributes to the pattern of intermittent shifts in resting potential associated with burst firing characteristically induced by NMDA (8).

Glutamate and quisqualate produce a steady depolarisation and sustained increase in firing rate.

Many synthetic analogues of glutamate have been tested as antagonists of glutamate and aspartate and of the selective agonists kainate, quisqualate and NMDA. Some highly potent and specific antagonists acting at the NMDA receptor are listed in Table 2.

TABLE 2
Selective antagonists at the NMDA receptor

Competitive antagonists		Non-competitive antagonists
D(−)2-amino-5-phosphonovalerate	(APV)	Phencyclidine
D(−)2-amino-7-phosphonoheptanoate	(APH)	(−)Cyclazocine
β-D-aspartylaminomethylphosphonate	(ASP-AMP)	Dextrorphan
γ-D-glutamylaminomethylphosphonate	(GLUAMP)	N-Allylnormetazocine
3(±)2-carboxypiperazin-4-yl)propyl-1-phosphonate	(CPP)	MK 801
	CGS 19755	Ketamine

Excitatory Amino Acids in Epilepsy

Excitatory amino acids applied focally to the cortex or hippocampus will induce focal seizure discharges (9). Administered systemically in high concentration they will induce generalized seizures. Release of excitatory amino acids into the extracellular space is enhanced during seizure activity. Spread of seizure activity from one brain region to another is dependent on excitatory amino acid neurotransmission; usually both NMDA and non-NMDA receptor mechanisms are involved.

Antagonists acting at the post-synaptic excitatory amino acid receptors are potent anticonvulsants. In particular NMDA antagonists, acting either as competitive inhibitors (such as 2-amino-7-phosphonoheptanoic acid and CPP) or as non-competitive inhibitors (such as MK 801) are potent anticonvulsants in a wide range of animal models of epilepsy (10, 2). The competitive inhibitors are remarkably potent when administered intracerebroventricularly in mice (exceeding the potency of benzodiazepines). They are however relatively much less potent when administered intraperitoneally to mice, as they penetrate the blood-brain barrier less efficiently than the benzodiazepines. The competitive antagonists are anticonvulsant when given intravenously to baboons, *Papio papio,* that show photosensitive epilepsy (11). 2-Amino-7-phosphonoheptanoic acid is approximately as potent as sodium valproate when tested in this way. A complete protection against photically-induced myoclonus is produced by a dose that does not produce sedation or motor side effects. Kynurenic acid, which blocks both NMDA and non-NMDA receptors is also effective in this model. None of the currently available competitive antagonists are orally active as anticonvulsants in rodents. However, compounds are now in development that do possess oral activity.

The non-competitive antagonists tend to produce alterations in motor activity at therapeutic doses. Confusional or psychotomimetic states are also seen after phencyclidine or ketamine. The extent to which MK 801 at therapeutic doses will alter cognitive or motor function is not yet clear.

The focal injection of excitatory amino acid antagonists into various brain regions can suppress seizure activity. Basal ganglia output pathways modulate seizure threshold in the limbic system and serve as a relay for some motor patterns of seizure activity (12, 13).

It is possible that abnormalities of the transport or metabolism of aspartate or glutamate are associated with particular forms of epilepsy. A rare and lethal syndrome of convulsions in infancy is associated with high CSF content of aspartate (14). Plasma concentration of glutamate and taurine is abnormal in patients with 3 per sec. spike and wave epilepsy and their first degree relatives (15).

Excitotoxic Mechanisms

That high levels of glutamate administered systemically to immature mice can damage retinal neurons was shown by Lucas & Newhouse (16). Subsequently Olney and colleagues administered glutamate and a variety of related amino acids to newborn mice, rats or monkeys and observed neuronal degeneration in various periventricular structures in which the blood-brain barrier was lacking (17). This raised the possibility that dietary glutamate or other excitatory amino acids could be responsible for neurodegenerative disorders, particularly in infants. The possibility that chronic neurodegenerative disorders may be due to excitatory amino acids (or their precursors) taken in the diet has been strengthened by recent studies relating to neurolathyrism and to the ALS-parkinsonism dementia complex of Guam (see below).

The characteristic feature of excitotoxic brain damage is its post-synaptic nature. Acutely this is manifest as focal swelling of dendrites and acute cytopathology (swelling of endoplasmic reticulum and of mitochondria) in neuronal somata. Chronically there is loss of intrinsic neurons with preservation of presynaptic elements and axons of passage. Excitotoxic damage can be induced with agents acting at any of the three receptor subtypes. However, there is very clear evidence for selectivity depending on receptor subtype density of distribution. Thus *in vivo* and *in vitro* pyramidal neurons in the CA1 zone of the hippocampus are most susceptible to the neurotoxic action of kainate, and the highest density of kainate receptors in the brain is around the mossy fibre terminals on the CA3 neurons. In cerebellar slices exposed to excitotoxins, Golgi cells are the most susceptible to kainate whereas granule cells are most susceptible to NMDA (18).

The excitotoxic action of NMDA is readily blocked by the co-injection of competitive antagonists or the systemic injection of a non-competitive antagonist. Quinolinic acid focally is a potent excitotoxin in the hippocampus or striatum. Its toxic effect can be blocked with direct antagonists such as APH or kynurenic acid. The ionic depenence of excitotoxic cell death has been studied using dissociated cultures of cortical or hippocampal neurones or tissue slices (19, 18). Acute swelling seen immediately after high concentrations of excitotoxins is dependent on entry of sodium and chloride apparently leading to an osmotic cell death (20). Cell death produced with lower concentrations or briefer exposures to excitotoxins, with a slower time course is dependent on extracellular calcium (18, 19). The mechanism may involve activation of proteases, phospholipases or calcium-calmodulin dependent protein kinases.

Excitotoxicity in Epileptic and Ischaemic Brain Damage

The selective neuronal loss that occurs in the hippocampus after status epilepticus is clearly excitotoxic in nature. Electron microscopy of the rat hippocampus after sustained seizure activity shows characteristic excitotoxic cytopathology even when the seizures are induced by agents that act on GABAergic transmission. Thus there is focal dendritic swelling in the stratum oriens and stratum radiatum and mitochondrial swelling, and eventually dark cell change or ischaemic cell change with loss of ultrastructural detail but with preserved presynaptic terminal morphology (21, 22, 23).

Not all the patterns of cytopathology occurring after cerebral ischaemia can be related to excitotoxic mechanisms. However, excitotoxicity appears to play an important part in the selective loss of hippocampal neurons occurring after transient forebrain ischaemia. Cytological changes observed 30—120 min after a period of ischaemia show a close resemblance to excitotoxic appearances both in terms of postsynaptic swelling and calcium overload in mitochondria (24). It is also possible to protect against acute cytological changes in hippocampal pyramidal neurons and against delayed neuronal loss by the focal injection of 2-amino-7-phosphonoheptanoic acid (25, 26, 27). Systemic injection of competitive or non-competitive antagonists also affords protection against delayed neuronal loss in the hippocampus (28, 26, 27). Occlusion of the middle cerebral artery in the rat provides a model of stroke. Kynurenic acid administered prior to and after the occlusion can reduce the volume of the cortical infarct by about 50 % (29). MK 801 can also decrease the pathology that follows either transient global ischaemia or permanent focal occlusion of the middle cerebral artery (30).

Quinolinic Acid and Huntington's Disease

The focal injection of kainic acid into the striatum in rats reproduces some of the biochemical features of Huntington's disease (loss of GABAergic and cholinergic markers, relative preservation of dopaminergic neurons) (31). Quinolinic acid provides a superior model in that the lesion it produces shows selective sparing or somatostatin and neuropeptide Y (32). Thus it has been suggested that excessive local production of quinolinic acid could be responsible for the pathology of Huntington's disease (33). In support of this is evidence that the activity of the quinolinic acid synthesizing enzyme (3-hydroxyanthranilic acid oxygenase) is increased in the striatum in Huntington's patients. The concentration of quinolinic acid is not, however, enhanced in the CSF in Huntington's patients (34).

Chronic Degenerative Disorders and Endogenous and Exogenous Excitotoxins

Exogenous excitotoxins appear to be responsible for at least two chronic degenerative neurological disorders. Lathyrism is a form of upper motoneurone disease found in Asia and Africa, among people consuming the *Lathyrus sativus* seed (chickling pea). This contains β-N-oxalylamino-L-alanine, an excitant amino acid acting on the quisqualate receptor. When fed to cynomolgus monkeys it produces motor signs of pyramidal damage (35).

The syndrome of amyotrophic lateral sclerosis-Parkinsonism-dementia occurring on Guam and some other Pacific islands can probably be related to consumption of the seeds of *Cycas circinalis* which contain an amino acid, β-N-methylamino-L-alanine, L-BMAA, that is excitotoxic in mouse brain and when fed chronically to macaque monkeys induces a degenerative syndrome with motor abnormalities resembling the syndrome found in Guam (36). In tissue culture the excitotoxic action of L-BMAA can be blocked by NMDA antagonists. Thus neuronal loss in the patients in Guam may be due to an excitotoxic action, mediated via the NMDA receptor, by L-BMAA or a metabolite of L-BMAA.

These studies showing that chronic degenerative disorders with varied pathology can result from the chronic consumption of excitotoxins raises the possibility that other degenerative disorders including amyotrophic lateral sclerosis, senile dementia of the Alzheimer type and various multiple system degenerations such as olivopontocerebellar atrophy could be due to either exogenous or endogenous excitotoxins. A sub group of patients with olivopontocerebellar atrophy, with recessive inheritance, have been shown to have an abnormality in the enzyme glutamate dehydrogenase. This is associated with a raised plasma glutamate level and a very high plasma glutamate response to an oral glutamate load (37). Thus accumulation of glutamate in the hindbrain might explain the olivopontocerebellar atrophy.

In these conditions the possibility has to be considered that chronic administration of excitatory amino acid antagonists could provide prophylaxis against neuronal degeneration (2).

REFERENCES

1) MELDRUM, B. (1986). Is epilepsy a disorder of excitatory transmission? In: *What is Epilepsy?* (Trimble, M. R., Reynolds, E. H., eds.). Churchill-Livingstone, Edinburgh, pp. 293—302.

2) MELDRUM, B. S. (1985b). Possible therapeutic applications of antagonists of excitatory amino acids. *Clin. Sci.* **68:** 113—122.

3) DO, K. Q., MATTENBERGER, M., STREIT, P., CUENOD, M. (1986). In: *In vitro* release of endogenous excitatory sulfur-containing amino acids from various rat brain regions. *J. Neurochem.* **46:** 779—786.

4) BERNSTEIN, J., FISHER, R. S., ZACZEK, R., COYLE, J. (1985). Dipepetides of glutamate and aspartate may be endogenous neuroexcitants in the rat hippocampal slice. *J. Neurosci.* **5:** 1429—1433.

5) DAVIES, J., EVANS, R. H., SMITH, D. A. S., WATKINS, J. C. (1982). Differential activation and blockade of excitatory amino acid receptors in the mammalian and amphibian central nervous system. *Comp. Biochem. Pharmacol.* **72C:** 211—224.

6) CULL-CANDY, S. G., USOWICZ, M. M. (1987). Multiple-conductance channels activated by excitytory amino acids in cerebellar neurons. *Nature* **325:** 525—528.

7) JAHR, C. E., STEVENS, C. F. (1987). Glutamate activates multiple single channel conductances in hippocampal neurons. *Nature* **325:** 522—525.

8) HERRLING, P. L., MORRIS, R., SALT, T. E. (1983). Effects of excitatory amino acids and their antagonists on membrane and action potentials of cat caudate neurones. *J. Physiol.* **339:** 207—222.

9) HAYASHI, T. (1952). A physiological study of epileptic seizures following cortical stimulation in animals and its application to human clinics. *Jap. J. Pharmacol.* **3:** 46—64.

10) CROUCHER, M. J., COLLINS, J. F., MELDRUM, B. S. (1982). Anticonvulsant action of excitatory amino acid antagonists. *Science* **216:** 899—901.

11) MELDRUM, B. S., CROUCHER, M. J., BADMAN, G., COLLINS, J. F. (1983). Antiepileptic action of excitatory amino acid antagonists in the photosensitive baboon, Papio papio. *Neurosci. Lett.* **39:** 101—104.

12) PATEL, S., CHAPMAN, A. G., MILLAN, M. H., MELDRUM, B. S. (1988). Epilepsy and excitatory amino acid antagonists. In: *Excitatory Amino Acids in Health and Disease* (Lodge, D., ed.). John Wiley and Sons Ltd., pp. 353—378.

13) MELDRUM, B. (1988). Initiation and neuroanatomical spread of seizure activity. In: *Recent Advances in Epilepsy* (Pedley, T., Meldrum, B. S., eds.). **Vol. 4,** Churchill-Livingstone, Edinburgh.

14) WEITZ, R., MERLOB, P., AMIR, J., REISNER, S. H. (1981). A possible role for aspartic acid in neonatal seizures. *Arch. Neurol.* **38:** 258—259.

15) VAN GELDER, N. M., ASLAM, J. N., METRAKOS, K., MacGIBBON, B., METRAKOS, J. D. (1980). Plasma amino acids in 3/sec spike-wave epilepsy. *Neurochem. Res.* **5:** 659—671.

16) LUCAS, D. R., NEWHOUSE, J. P. (1957). The toxic effect of sodium L-glutamate on the inner layers of the retina. *AMA Arch. Ophthalmol.* **58:** 193—204.

17) OLNEY, J. W., HO, O. C., RHEE, V. (1971). Cytotoxic effects of acidic and sulphur containing amino acids on the infant mouse entral nervous system. *Exp. Brain Res.* **14:** 61—76.

18) GARTHWAITE, G., GARTHWAITE, J. (1986). Neurotoxicity of excitatory amino acid receptor agonists in rat cerebellar slices: dependence on calcium concentration. *Neurosci. Lett.* **66:** 193—198.

19) CHOI, D. W. (1987). Ionic dependence of glutamate neurotoxicity. *J. Neurosci.* **7:** 369—379.

20) ROTHMAN, S. M. (1985). The neurotoxicity of excitatory amino acids is produced by passive chloride influx. *J. Neurosci.* **5:** 1483—1489.

21) EVANS, M. C., GRIFFITHS, T., MELDRUM, B. S. (1984). Kainic acid seizures and the reversibility of calcium loading in vulnerable neurons in the hippocampus. *Neuropathol. Appl. Neurobiol.* **10:** 285—302.

22) GRIFFITHS, T., EVANS, M. C., MELDRUM, B. S. (1983). Intracellular calcium accumulation in rat hippocampus during seizures induced by bicuculline of L-allylglycine. *Neuroscience* **10:** 385—395.

23) GRIFFITHS, T., EVANS, M. C., MELDRUM, B. S. (1984). Status epilepticus: the reversibility of calcium loading and acute neuronal pathological changes in the rat hippocampus. *Neuroscience* **12:** 557—567.

24) SIMON, R. P., GRIFFITHS, T., EVANS, M. C., SWAN, J. H., MELDRUM, B. S. (1984a). Calcium overload in selectively vulnerable neurons of the hippocampus during and after ischaemia: an electron microscopy study in the rat. *J. Cereb. Blood Flow Metab.* **4:** 350—361.

25) SIMON, R. P., SWAN, J. H., GRIFFITHS, T., MELDRUM, B. S. (1984b). Blockade of N-methyl-D-aspartate receptors may protect against ischaemic damage in the brain. *Science* **226:** 850—852.

26) MELDRUM, B., EVANS, M., SWAN, J., SIMON, R. (1987). Protection against hypoxic/ischaemic brain damage with excitatory amino acid antagonists. *Med. Biol.* **65:** 153—158.

27) SWAN, J. H., EVANS, M. C., MELDRUM, B. S. (1988). Long term development of selective neuronal loss and the mechanism of protection by 2-amino-7-phosphonoheptanoate in a rat model of incomplete forebrain ischaemia. *J. Cereb. Blood Flow Metabol.* (in press).

28) BOAST, C. A., GERHARDT, S. C., JANAK, P. (1986). Systemic AP7 reduces ischaemic brain damage in gerbils. In: *Excitatory Amino Acid Transmission* (Hicks, T. P., Lodge, D., McLennan, H., eds.), pp. 249—252.

29) GERMANO, I. M., PITTS, L. H., MELDRUM, B. S., BARTKOWSKI, H. M., SIMON, R. P. (1987). Kynurenate inhibition of cell excitation decreases stroke size and deficits. *Ann. Neurol.* (in press).

30) GILL, R., FOSTER, A. C., WOODRUFF, G. N. (1987). Systemic administration of MK 801 protects against ischemia-induced hippocampal neurodegeneration in the gerbil. *J. Neuroscience.*

31) COYLE, J. T., SCHWARCZ, R. (1976). Lesion of striatal neurones with kainic acid provides a model for Huntington's chorea. *Nature* **263:** 244—246.

32) BEAL, M. F., KOWALL, N. W., ELLISON, D. W., MAZUREK, M. F., SWARTZ, K. J., MARTIN J. B. (1986). Replication of the neurochemical characteristics of Huntington's disease by quinolinic acid. *Nature* **321:** 168—171.

33) SCHWARCZ, R., SHOULSON, I. (1987). Excitotoxins and Huntington's disease. In: *Animal Models of Dementia* (Coyle, J. T., ed.), Alan R. Liss, New York, pp. 39—68.

34) SCHWARCZ, R., TAMMINGA, C. A., KURLAN, R., SHOULSON, I. (1987). CSF levels of quinolinic acid in Huntingto's disease and schizophrenia. *Lancet.*

35) SPENCER, P. S., ROY, D. N., LUDOLPH, A., HUGUN, J., DWIVEDI, M. P., SCHAUMBURG, H. H. (1986). Lathyrism: evidence for role of the neuroexcitatory amino acid BOAA. *Lancet* **ii:** 1066—1067.

36) SPENCER, P. S., NUNN, P. B., HUGON, J., LUDOLPH, A. C., ROSS, S. M., ROY, D. N., ROBERTSON, R. C. (1987). Guam Amyotrophic lateral sclerosis-Parkinsonism-dementia linked to a plant excitant neurotoxin. *Science* **237:** 517—522.

37) PLAITAKIS, A., BERL, S., YAHR, M. D. (1982). Abnormal glutamate metabolism in an adult-onset degenerative neurological disorder. *Science* **21b:** 193—196.

38) MELDRUM, B. S. (1985a). GABA and other amino acids. In: *Handbook of Experimental Pharmacology* (Frey, H. H., Janz, D., eds.). Springer-Verlag, Berlin, pp. 153—188.

39) MORONI, F., LOMBARDI, G., MONETI, G., CORTESINI, C. (1984). The exitotoxin quinolinic acid is present and unevenly distributed in the rat brain. *Brain Res.* **295:** 352—355.

40) MORONI, F., LOMBARDI, G., CARLA, V., LAL, S., ETIENNE, P., NAIR, N. P. V. (1986). Increase in the content of quinolinic acid in cerebrospinal fluid and frontal cortex of patients with hepatic failure. *J. Neurochem.* **47:** 1667—1671.

CEREBRAL CAPILLARY AMINO ACID UPTAKE AND MEMBRANE FLUIDITY IN HEPATIC ENCEPHALOPATHY

G. Zanchin and P. Rigotti*

Departments of Neurology and Surgery*
University of Padova
Via Giustiniani 5
35128 Padova
Italy

Plasma and Brain Amino Acid Levels in Hepatic Encephalopathy

In recent years a great deal of attention has been devoted to the role played by plasma and brain amino acid (AA) derangements in the pathogenesis of hepatic encephalopathy (HE) (1, 2). Alterations in plasma AA levels in chronic liver failure in general involve those AA which the liver is responsible for clearing, i.e., aromatic AA, and those responsive to the higher concentrations of insulin and glucagon occurring in this condition, i.e.,

TABLE 1
Plasma ammonia and amino acids in rats after portacaval anastomosis

	Controls	PCA	PCA + MSO
Ammonia (μmol/ml)	0.10 ±	0.20 ± 0.01[a] (6)	0.22 ± 0.02 (6)
Amino acids (nmol/ml)			
Threonine	231 ± 14 (7)	138 ± 6[a] (7)	220 ± 7[b] (7)
Valine	199 ± 10	150 ± 10[a]	183 ± 15[b]
Methionine	56 ± 2	48 ± 2	72 ± 3[b]
Isoleucine	93 ± 4[c]	45 ± 4[a]	73 ± 8[b]
Leucine	169 ± 7[c]	89 ± 6[a]	134 ± 15[b]
Thyrosine	75 ± 3[c]	125 ± 8[a]	120 ± 7
Phenylalanine	62 ± 2	120 ± 6[a]	111 ± 6
Histidine	83 ± 4	99 ± 5	90 ± 5
Glutamine	650 ± 24[c]	967 ± 70[a]	289 ± 25[b]
Alanine	458 ± 33	441 ± 22	745 ± 42[b]
Lysine	473 ± 28	297 ± 14[a]	483 ± 15[b]
Arginine	206 ± 9[c]	147 ± 11[a]	164 ± 15

Results are given as means ± SEM with number of rats in parentheses.
[a] Significantly different from controls, $p < 0.01$
[b] Significantly different from PCA, $p < 0.01$
[c] Significantly different from PCA + MSO, $p < 0.01$
From Rigotti et al. (4).

NATO ASI Series, Vol. H20
Amino Acid Availability and Brain Function in Health and Disease. Edited by G. Huether

branched-chain amino acids (BCAA) (3). The typical pattern in HE (Table 1) is a decrease of BCAA (leucine, isoleucine, valine) and a large increase of aromatic AA (tyrosine and phenylalanine), but no change in tryptophan (4, 5). It is known that all these AA compete for entry into the brain across the blood-brain barrier (BBB) via a single carrier system specific for neutral AA; thus the decreased plasma ratio of the BCAA to the aromatic AA observed in HE should determine increased levels of the aromatic AA in the brain.

We measured brain AA levels in rats previously submitted to chronic portacaval shunt (PCS): indeed, the most prominent feature (Table 2) was a very large rise of aromatic AA (3—8), including tryptophan (5), and of histidine and glutamine, which was fairly uniform in the regions tested. However, BCAA were substantially unchanged (4) or were even moderately higher (5). Other AA, including the neurotransmitters glutamic acid and GABA (5), did not show significant changes. Similar alterations in cerebral AA levels have been observed in different experimental HE (1, 9, 10) and in CSF of patients with liver failure (11).

TABLE 2
Brain ammonia and amino acids in rats after portacaval anastomosis

	Controls	PCA	PCA + MSO
Ammonia (μmol/g	0.21 ± 0.02^{c} (6)	0.38 ± 0.03^{a} (6)	0.99 ± 0.08^{b} (6)
Amino acids (nmol/g)			
Threonine	515 ± 16^{c} (7)	578 ± 34 (7)	443 ± 11^{b} (7)
Valine	74 ± 4	87 ± 7	59 ± 3^{b}
Methionine	45 ± 1^{c}	60 ± 3^{a}	31 ± 4^{b}
Isoleucine	40 ± 1	33 ± 3^{a}	28 ± 1
Leucine	83 ± 3	87 ± 5	60 ± 3^{b}
Tyrosine	67 ± 3	276 ± 19^{a}	92 ± 7^{b}
Phenylalanine	42 ± 1^{c}	213 ± 10^{a}	72 ± 5^{b}
Histidine	69 ± 1	241 ± 18^{a}	81 ± 6^{b}
Glutamine	4824 ± 180^{c}	11493 ± 962^{a}	2644 ± 143^{b}
Alanine	611 ± 24^{c}	745 ± 19	1734 ± 84^{b}
Lysine	269 ± 14^{c}	247 ± 5	670 ± 23^{b}
Arginine	137 ± 4^{c}	128 ± 3	174 ± 10^{b}

Results are given as means ± SEM with number of rats in parentheses.
[a]Significantly different from controls, $p < 0.01$.
[b]Significantly different from PCA, $p < 0.01$.
[c]Significantly different from PCA + MSO, $p < 0.01$.
From Rigotti et al. (4).

The increased brain levels of aromatic AA could be relevant in the pathogenesis of HE since they are monoamine precursors and therefore could induce neurotransmitter impairment. It has been shown that a high brain concentration of tryptophan stimulates brain serotonin synthesis and turnover. Brain serotonin and its main metabolite 5-hydroxyindoleacetic acid are elevated after hepatectomy, NH_3 infusion, or PCS (12, 13), conditions in which cerebral levels of tryptophan are elevated. Another effect of cerebral AA imbalance could be impaired protein synthesis (8), although more recent observations point to an increased rate of protein breakdown instead (14). Furthermore, brain norepinephrine (NE) and dopamine (DA) depletion has been observed, probably secondary to inhibition of tyrosine-hydroxylase by the high levels of phenylalanine and tyrosine (15, 16,

6). The cerebral tyrosine that is not converted to the normal neurotransmitters, DA and NE, is decarboxylated to tyramine, and further to the weak neurotransmitter octopamine, and some of the excess phenylalanine is converted to phenylethanolamine. The overall effect of these metabolic events is the replacement of excitatory neurotransmitter by weaker ones (false neurotransmitters) or by those with inhibitory effects (17).

For these reasons, L-DOPA administration has been suggested as a therapeutic measure (18, 19); also, attempts are currently being made to normalize the plasma AA imbalance with intravenous infusion of AA mixtures rich in BCAA (2), which are able to compete with aromatics in the passage across the BBB and therefore prevent their accumulation in cerebral tissue.

Blood-Brain Barrier Amino Acid Transport in Hepatic Encephalopathy

Further investigation showed that the elevation of AA in the brain of animals following PCS could not be explained merely on the basis or their altered plasma concentrations. The marked cerebral increase of aromatic AA largely exceeds their rise in plasma; furthermore, BCAA either are unchanged or are slightly increased in brain, despite their decrease in the plasma and the increase of plasma aromatic AA, which by sharing the same neutral transport class at the BBB are uptake competitors.

Since these AA cannot be synthesized in the brain, the attention of investigators was directed to BBB permeability. We studied BBB AA transport in vivo by means of the double-label intracarotid injection technique described by Oldendorf (20). Significant differences were demonstrated between PCS rats and control animals (Table 3). BBB permeability was affected selectively: neutral AA transport was greatly increased in PCS animals, whereas basic AA showed a net decrease in their rate of passage from blood to brain and no changes were observed for GABA and glutamic acid (21—23).

TABLE 3

Brain Uptake Index of various amino acids in normal and shunted rats

Amino acids	Concentration of injections (μM)	Normal	Shunted	Difference (%)
L-tyrosine	13.1	26.3 ± 2.5 (6)	48.3 ± 2.9 (6)	+ 84**
L-phenylalanine	6.9	20.5 ± 1.5 (8)	33.9 ± 1.0 (9)	+ 65**
L-isoleucine	19.6	16.2 ± 1.8 (5)	28.7 ± 3.3 (5)	+ 77*
L-methionine	62.5	15.7 ± 1.5 (6)	25.5 ± 3.2	+ 63*
L-valine	23.5	13.5 ± 1.5 (4)	21.7 ± 8.7 (3)	+ 61*
L-tryptophan	41.2	10.3 ± 0.7 (5)	27.6 ± 1.9 (5)	+ 168**
L-arginine	58.8	15.3 ± 0.7 (5)	5.46 ± 0.7 (6)	- 64**
L-lysine	19.5	12.1 ± 1.1 (6)	5.76 ± 0.6 (5)	- 52**
L-glutamic acid	20.0	3.34 ± 0.8 (6)	3.98 ± 0.6 (4)	N.S.
GABA	10.8	2.48 ± 0.4 (5)	2.60 ± 0.5 (5)	N.S.

Values expressed as BUI ± SEM. Percentage differences given with statistical significance (Student's t test). Number of animals in parentheses.

* $p < 0.02$; ** $p < 0.001$.

From Zanchin et al. (23).

A possible increase of neutral AA transport in CNS cells, at least as measured in in vitro experiments, had previously been ruled out (1—10, 13—16, 18, 20—22, 24—29, 30—32). Uptake was studied in slices from brain regions of rats four weeks after portacaval anastomosis. No differences of the inulin compartment were observed between control and experimental animals. No detectable changes in uptake rates were found among portacaval shunted animals and controls during short-term experiments (10 minutes), whereas after 60-minutes incubation, uptake was decreased in some cases (Table 4). Unlike cellular uptake, the uptake of neutral AA by microvessels isolated from brain of PCS rats was significantly increased (33). These data suggest that the alteration of permeability does not involve cerebral cells, but involves BBB transport systems located in the cerebral capillary endothelial cells.

TABLE 4
Cerebral amino acid uptake in vitro after portacaval anastomosis

Amino acid		Forebrain	Cerebellum	Midbrain	Pons-medulla
Phenylalanine	C	2.31 ± 0.1 (4)	2.57 ± 0.2 (4) - 23 %	2.06 ± 0.1 (4)	1.40 ± 0.1 (3)
	PCS	2.38 ± 0.1 (4)	1.98 ± 0.1 (3) $p < 0.02$	2.10 ± 0.1 (4)	1.44 ± 0.1 (3)
Tryptophan	C	7.39 ± 0.4 (9)	7.00 ± 0.6 (9)	6.33 ± 0.3 (9)	3.99 ± 0.3 (9)
	PCS	7.26 ± 0.3 (8)	6.33 ± 0.3 (9)	5.78 ± 0.3 (9)	3.59 ± 0.2 (9)
Histidine	C	20.8 ± 1.4 (6)	17.7 ± 2.1 (5)	18.7 ± 1.4 (6) - 22 %	9.78 ± 1.5 (5)
	PC	20.0 ± 0.1 (6)	16.1 ± 1.4 (6)	14.6 ± 1.1 (6) $p < 0.05$	11.4 ± 1.9 (5)
Leucine	C	3.17 ± 0.2 (6)	3.21 ± 0.5 (5)	3.29 ± 0.2 (67	2.00 ± 0.2 (5)
	PCS	2.94 ± 0.2 (6)	2.72 ± 0.4 (6)	3.17 ± 0.2 (6)	2.30 ± 0.2 (6)
Valine	C	7.53 ± 0.3 (6) - 21 %	5.58 ± 1.1 (4)	7.04 ± 0.3 (3) - 34 %	6.19 ± 0.2 (4) - 46 %
	PC	5.92 ± 0.5 (4) $p < 0.05$	4.56 ± 1.0 (5)	4.67 ± 0.2 (4) $p < 0.001$	3.33 ± 0.3 (6) $p < 0.001$

C) control; PCS) portacaval shunted rats. Amino acid concentration 2 mM. Incubation time 60 minutes. Results expressed as a concentrative uptake (μmoles amino acid/ml intracellular water — μmoles amino acid/ml in medium at end of incubation). Averages ± SEM are given. Number of experiments in parentheses. Statistical significance obtained with the Student's t test. From Zanchin et al. (5).

With competitive inhibition experiments, we demonstrated that the increased brain permeability to neutral AA was due to the increased saturable component of transport (23). Afterwards, the kinetic parameters of neutral AA transport were studied. In rats with PCS, in comparison to controls, the phenylalanine V_{max} was approximately doubled, whereas the K_m was unchanged, and we did not observe any significant effect on the component of apparently nonsaturable uptake (Table 5) (32).

Thus, the increased BBB neutral AA transport after PCS was not related to changes in the affinity of the system or to its nonsaturable component, but presumably reflected a higher carrier availability, which was due either to increased carrier number or to increased carriers "cycling" across the membrane.

TABLE 5
Transport kinetic values in control and PCS rats

	Phenylalanine	
	Control	PCS
V_{max} (nmol/g/min)	14 ± 3	29 ± 6*
K_m (nmol/ml)	80 ± 17	116 ± 24
K_d µl/min/g)	69 ± 4	69 ± 8

* $p < 0.05$. From Rigotti et al. (32).

Possible Alterations Through Glutamine Formation

Once the changes in BBB AA permeability in HE have been demonstrated, the major question remains, what is the link between this alteration and liver failure.

To investigate the role of the altered AA profile in the plasma, the extraction of neutral AA of the brain in the presence of plasma of normal rats and of rats with PCS was studied (34). Increased neutral AA permeability after PCS had been previously demonstrated with intracarotid injection of the tested compound in an artificial bolus, therefore in a relatively unphysiological condition, that is, in the absence of the competing AA in the

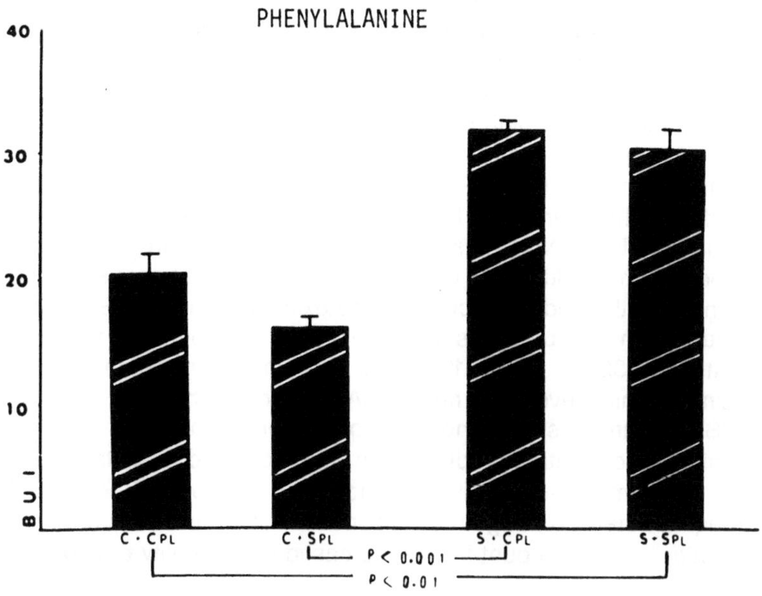

FIGURE 1
Brain Uptake Index of phenylalanine in "crossed plasma" experiments.
C + Cpl = control rats injected with plasma from controls; C + Spl = control rats injected with plasma from shunted rats; S + Cpl = shunted rats injected with plasma from controls; S + Spl = shunted rats injected with plasma from shunted rats. Results expressed as Brain Uptake Index ± SEM. N = 6—8. Significance at the Student's t test is reported. From Zanchin et al. (34).

plasma. This method allows minimal mixing of the injected solution with plasma, and therefore the passage of the studied AA takes place at the bolus/cerebral endothelium contact surfaces, without competition by substances contained in the blood (20).

In experiments with normal and PCS rats receiving intracarotid injections of a bolus made with the AA tested and their own plasma (Fig. 1), a higher uptake in PCS rats was also demonstrated; the tested neutral AA competed in the plasma with others sharing the BBB transport sites. In "crossed plasma" experiments, the difference in neutral AA BBB permeability between PCS and control animals was maintained, whether the injected bolus had been prepared with plasma of PCS or of control rats (Fig. 1). Therefore we concluded that the altered AA levels observed in the plasma do not seem to be a determining factor in the changes of BBB permeability after PCS (34—37).

Other experiments support this point. After PCS it is possible to improve liver perfusion by means of portal arterialization, that is, by connecting the empty portal bed to the arterial circulation. Plasma and brain AA levels were investigated in rats with PCS and arterialization of the portal stump with the right renal artery (31). The increased blood flow to the liver markedly normalized altered levels of aromatic AA in plasma and brain, whereas blood levels of ammonia and cerebral glutamine remained elevated, almost as in PCS rats. Interestingly, the known enhancement of BBB neutral AA transport in PCS rats was also maintained after liver arterialization (31).

Hyperammonemia is common in patients with liver cirrhosis and in laboratory animals with a PCS. The concept that hyperammonemia is closely related to HE is widely held; however, the mechanism whereby NH_3 may affect cerebral function remains unknown. Infusion of NH_4 salts in normal rats (Table 2) causes increase of brain neutral AA levels (28). AA transport of the BBB has been studied in these animals, and changes similar to those observed in PCS rats have been demonstrated (29). These changes did not seem to be a direct effect of high brain NH_3 concentration per se, but to be a metabolic consequence of high cerebral NH_3, that is, a large rise of glutamine in cerebral tissue. In fact, pre-treatment with methionine-sulfoximine (MSO), an inhibitor of glutamine synthesis, prevented both the changes in brain neutral AA concentration (28) and those in the AA transport in the BBB of NH_3-infused rats (27—29). In MSO-treated rats, brain NH_3 levels were higher, whereas glutamine concentration was lower, because of the inhibition of glutamine synthesis, the end product of NH_3 detoxification in brain. Similarly, the increase in brain concentration of neutral AAs and of their BBB transport was reversed by MSO also in PCS rats (27—32, 16, 18, 22, 10, 20, 4). Therefore there seems to be a correlation between brain glutamine levels and neutral AA transport at the BBB.

At the BBB, glutamine shares the transport system common to neutral AA. It is possible that a high concentration gradient from brain to blood of glutamine, as present in conditions of hyperammonemia, drives an increased transport of neutral AA from blood to brain through a heteroexchange mechanism with glutamine. This effect, referred to as "transacceleration", could account for the increased permeability to NAA observed in HE (25).

However, some arguments have been made against this hypothesis (24, 37). It has been stressed that the neutral AA transport system has such a low affinity for glutamine that it is unlikely that this metabolite could be an effective partner in an exchange mechanism. Moreover, an increased exchange by glutamine would obviously not account for the fact that transport of basic AA and of monocarboxylic acids is decreased after PCS. Furthermore, in experiments in which rats were chronically infused with NH_3, BBB transport of neutral AA was increased only many hours after brain glutamine content had already been raised; and after discontinuing the infusion, transport of tryptophan returned to normal despite the fact that brain glutamine remained elevated (30). Finally, an increased neutral AA uptake was demonstrated in vitro in isolated microvessels: in this experimental condition however glutamine levels were only slightly increased in brain capillaries from shunted animals and therefore could not account for an exchange mechanism (33).

As a whole, these experiments indicate that the BBB transport changes after HE are more dependent on a chronically high level of NH_3 than on the increase of brain glutamine. However, how hyperammonemia could accomplish this is still a matter of speculation.

Altered Membrane Fluidity in Brain Capillary Endothelial Membranes

Studies with autoradiography in rat have demonstrated regional heterogeneity of cerebral AA transport in normal conditions. The degree appears to be directly related to the relative regional vascularity and could mean that the number of AA carriers per unit of endothelial surface is similar in different cerebral regions (22). After PCS, the already known alterations of AA transport at the BBB — increases for the neutral AA and decreases for the basic AA — were observed in all regions tested , therefore the regional heterogeneity of transport rate for each amino acid (22) remained unchanged.

Changes in the density of the transport carriers at the cerebral endothelia and change in their mobility are both possible explanations for these phenomena, since they were also found in isolated capillaries from PCS rats (33).

The above mentioned studies (22) with autoradiography demonstrate that, in the same region, the patterns of permeability change are different among the amino acids tested, the extremes being for tryptophan (200 %) and leucine (30 %). This observation implies a selective change of BBB permeability, that is, a different modulation, within the same carrier system, of the transport of individual compounds, possibly meaning that a single transport class is made up of separate compounds that can be affected differently in chronic HE (24, 37).

The previously considered "crossed plasma" experiments showed that changes of the plasma AA profile after PCS do not account for the selective modification of AA transport at the BBB, stressing the importance of a "tissue factor" instead of a "plasma factor" (34). On the other hand, we have seen that the "glutamine exchange hypothesis" can at best only partially explain the complex changes in AA transport at the BBB observed after chronic PCS.

Alterations of the functional status of some neurotransmitter receptors has been reported in chronic and in acute models of HE — although with some contradictory findings (38, 35) — along with a modified composition of synaptic plasma membranes (26). For these reasons we looked for possible changes of brain capillary endothelial membrane fluidity, which could be relevant in explaining the above mentioned phenomena.

In brain capillaries isolated from the cerebral cortex of control and PCS rats, membrane fluidity was evaluated by measuring the fluorescence polarization of the lipid probe diphenylhexatriene (DPH), at 1°C intervals between 14 and 38°C. The DPH fluorescence anisotropy determined on PCS rats was significantly lower than that from control animals in the range 14—32°C. These data therefore gave evidence of an increased disorder of the capillary membrane lipid bilayer in PCS animals, which by modifying the function of protein molecules mediating transport, could contribute to the BBB permeability changes observed in this chronic model of HE (36). Thereafter, the question was raised whether this represents a phenomenon limited to brain capillaries, or more probably, a sort of "membrane disease" affecting other cells in the body and being somehow linked with chronic HE. However, experiments carried out on erythrocyte membranes of PCS rats failed to show any such changes in comparison with controls. Despite these results, we checked for possible structural differences between erythrocyte membranes of cirrhotic patients and healthy subjects. This study is still in progress, but from the limited number of cases investigated, it seems that red cell membranes from cirrhotics, even in the absence of clinically apparent HE, are altered. The slope calculated in terms of anisotropy decrease per degree of temperature was flattened in cirrhotic patients (-0.00221 ± 0.00025, mean $\pm$ SD, as compared to -0.00278 ± 0.00013 in controls) suggesting that the composition of their red cell membrane is different from that in normal subjects, originating an "intermediate fluid state" in the lipid bilayer matrix of erythrocyte membrane. The increased cholesterol/phospholipid ratio we found experimentally in cirrhotic patients was confirmatory; in fact, one of the best known effects of cholesterol is the dramatic elimination of the sharp phase transition in phospholipid bilayers.

CONCLUSIONS

In summarizing the above discussed data, we draw the following conclusions:

— in this model of chronic HE a selective modulation of the BBB occurs, different for different carrier-mediated compounds: for what AA are concerned, neutral AA permeability is greatly increased, whereas that of basic AA is decreased;

— these heterogeneous patterns cannot be explained by changes in passive diffusion, as it has been demonstrated, among other evidences, with competitive inhibition experiments;

— the so-called "unifying hypothesis", implicating a plasma neutral AA exchange with cerebral glutamine, would not fully account for the complex BBB changes observed;

— other factors must be considered: among these are the physical properties of brain capillary membranes, and their chemical composition;

— finally, selective modifications of BBB transport by hormonal, toxic, or pathologic modulations have been reported in recent years (Table 6): possibly the factors responsible are different in each case or some common mechanisms could be identified. A detailed comparative study in this sense is extremely important. Indeed, the existence of a selective modulation of brain capillary permeability is a very exciting acquisition: enlighting the mechanisms through which this is feasible for different compounds, could lead to a selective permeability control of BBB, allowing the manipulation for therapeutic purposes of the microenvironment of the brain.

ACKNOWLEDGEMENTS

This work was partly supported by a grant from the Italian Ministry of Public Education. We are indebted to Miss Annalisa Leone for kindly typewriting the manuscript.

TABLE 6
Reported changes in BBB nutrient transport in pathological conditions (37)

Transport system	Modulation	Condition	Reference
Hexose	Increased activity	Seizures	Chapman et al., 1977
		Insulin	Daniel et al., 1975; Hertz et al., 1981
	Decreased activity	Pentobarbital	Gjedde and Rasmussen, 1980
		Mercury	Pardridge, 1976
		Anoxia, ischemia	Betz et al., 1974, 1975
		Hyperglycemia	Gjedde and Crone, 1981; McCall et al., 1982
Neutral amino acid	Increased activity	Hepatic encephalopathy	James et al., 1978; Zanchin et al., 1979b; Mans et al., 1982
		Uremic encephalopathy	Jeppson et al., 1982
		Undernutrition	Freedman and Samuels, 1980
		Insulin	De Montis et al., 1978; Daniel et al., 1981
Basic amino acid	Decreased activity	Hepatic encephalopathy	Zanchin et al., 1979b; Mans et al., 1982
Monocarboxylic acid	Increased activity	Ketosis (neonatal, fasting, fatty diet)	Cremer et al., 1976; Gjedde and Crone, 1975; Howkins and Biebuyck, 1979
	Decreased activity	Hepatic encephalopathy	Sarna et al., 1979; Mans et al., 1980

REFERENCES

1) FISCHER, J. E., FUNOVICS, M., AGUIRRE, A., JAMES, J. H., KEANE, J. M., WESDORP, R. I. C., YOSHIMURA, N., WESTMAN, T. (1975). The role of plasma amino acids in hepatic encephalopathy. *Surgery* **78:** 276—290.

2) FISCHER, J. E., ROSEN, H. M., EBEID, A. M., JAMES, J. H., KEANE, J. M., SOETERS, P. B. (1976). The effect of normalization of plasma amino acids on hepatic encephalopathy. *Surgery* **80:** 77—91.

3) SOETERS, P. B., FISCHER, J. E. (1976). Insulin, glucagon, amino acid imbalance, and hepatic encephalopathy. *Lancet* **23:** 880.

4) RIGOTTI, P., JONUNG, T., PETERS, J. C., JAMES, J. H., FISCHER, J. E. (1985). Methionine sulfoximine prevents the accumulation of large neutral acids in brain of portacaval-shunted rats. *J. Neurochem.* **44 (3):** 929—933.

5) ZANCHIN, G., RIGOTTI, P., BETTINESCHI, F., VASSANELLI, P., BATTISITN, L. (1979a). Cerebral amino acid levels and uptake in rats after portocaval anastomosis: I. Regional studies in vitro. *J. Neurosci. Res.* **4:** 291—299.

6) TYCE, G. M., OWEN, C. A. Jr. (1978). Dopamine and norepinephrine in the brains of hepatectomized rats. *Life Sci.* **22:** 781—786.

7) VILLACARA, A., FRIES, W., AMBROSIN, V., MONTANARI, D., MARTINEZ, D., COLONNA, R., ZANCHIN, G. (1986). Membrane fluidity in hepatic encephalopathy. In: *Molecular Basis of Neural Function* (Tucek, S., Stipek, S., Stastuy, F. et al., eds.), Praha, p. 305.

8) WASTERLAIN, C. G., LOCKWOOD, A. H., CONN, M. (1978). Chronic inhibition of brain protein synthesis after portocaval shunting. A possible pathogenic mechanism in chronic hepatic encephalopathy in the rat. *neurology* **28:** 233—238.

9) IOB, V., MATTSON, W. J. Jr., SLOAN, M., COON, A. A., TURCOTTE, J. G., CHILD, C. G. (1970). Alterations in plasma-free amino acids in dogs with hepatic insufficiency. *Surg. Gynecol. Obstet.* **130:** 794—800.

10) MATTSON, W. J. Jr., JOB, V., SLOAN, M., COON, A. A., TURCOTTE, J. G., CHILD, C. G. (1970). Alterations of individual free amino acids in brain during acute hepatic coma. *Surg. Gynecol. Obstet.* **130:** 263—266.

11) CASCINO, A., CANGIANO, C., FIACCADORI, F., GHINELLI, F., MERLI, M., PELOSI, G., RIGGIO, O., ROSSI FANELLI, F., SACCHINI, D., STORTONI, M., CAPOCACCIA, L. (1982). Plasma and cerebrospinal fluid amino acid patterns in hepatic encephalopathy. *Dig. Dis. Sci.* **27:** 828—832.

12) CUMMINGS, M. G., JAMES, J. H., SOETERS, P. B., KEANE, J. M., FOSTER, J., FISCHER, J. E. (1976). Regional brain study of indoleamine metabolism in the rat in acute hepatic failure. *J. Neurochem.* **27:** 741—746.

13) CURZON, G., KANTAMANENI, B. G., FERNANDO, J. C., WOODS, M. S., CAVANAGH, J. B. (1975). Effects of chronic portocaval anastomosis on brain tryptophan, tyrosine, 5-hydroxytryptamine. *J. Neurochem.* **24:** 1065—1070.

14) DUNLOP, D. S., KAUFMAN, H., ZANCHIN, G., LAJTHA, A. (1984). Protein synthesis rates in rats with portacaval shunts. *J. Neurochem.* **43:** 1487—1489.

15) DODSWORTH, J. M., CUMMING, M. G., JAMES, J. H., FISCHER, J. E. (1974). Depletion of brain norepinephrine in acute hepatic coma. *Surgery* **75:** 811—820.

16) KAMATA, S., OKADA, A., WATANABE, T., KAWASHIMA, Y., WADA, H. (1980). Effects of dietary amino acids on brain amino acids and transmitter amines in rats with a portacaval shunt. *J. Neurochem.* **35:** 1190—1199.

17) BALDESSARINI, R. J., FISCHER, J. E. (1977). Substitute and alternative neurotransmitters in neuropsychiatric illness. *Arch. Gen. Psychiat.* **24:** 958—964.

18) LUNZER, M., JAMES, I. M., WEINMAN, J., SHERLOCK, S. (1974). Treatment of chronic hepatic encephalopathy with levodopa. *Gut* **15:** 555—561.

19) ZANCHIN, G., RIGOTTI, P., DUSSINI, N., BETTINESCHI, F., VASSANELLI, P., BATTISTIN, L. (1981). Effect of L-DOPA treatment on cerebral amino acid levels in rats after portocaval anastomosis. *Neurochem. Res.* **6:** 649—658.

20) OLDENDORF, W. H. (1970). Measurement of brain uptake of radiolabeled substances using a tritiated water internal standard. *Brain Res.* **24:** 372—376.

21) JAMES, J. H., ESCOURROU, J., FISCHER, J. E. (1978). Blood-brain neutral amino acid transport activity is increased after portacaval anastomosis. *Science* **200:** 1395—1397.

22) MANS, A. M., BIEBUYCK, J. F., SHELLY, K., HAWKINS, R. A. (1982). Regional blood-brain barrier permeability to amino acids after portocaval anastomosis. *J. Neurochem.* **38:** 705—717.

23) ZANCHIN, G., RIGOTTI, P., DUSSINI, N., VASSANELLI, P., BATTISTIN, L. (1979). Cerebral amino acid levels and uptake in rats after portocaval anastomosis: II. Regional studies in vivo. *J. Neurosci. Res.* **4:** 301—310.

24) HAWKINS, R. A., MANS, A. M., BIEBUYCK, J. F. (1987). Changes in brain metabolism in hepatic encephalopathy. *Neurochem. Pathol.* **6:** 35—66.

25) JAMES, J. H., JEPPSSON, B., ZIPARO, V., FISCHER, J. E. (1979). Hyperammonaemia, plasma amino acid imbalance and blood-brain amino acid transport: a unified theory of portal-systemic encephalopathy. *Lancet* **13:** 772—775.

26) JONES, E. A., SCHAFER, D. F., FERENCI, P., PAPPAS, S. C. (1984). Changes in the status of brain receptors for neurotransmitters in a rabbit model of hepatic coma: their potential significance. In: *Advances in Hepatic Encephalopathy and Urea Cycle Diseases* (Kleinberger, G., Ferenci, P., Riederer, P. et al., eds.), Karger, Basel, pp. 337—352.

27) JONUNG, T., RIGOTTI, P., JAMES, J. H., FISCHER, J. E. (1983). Methionine sulfoximine reduced the brain uptake index and the brain concentrations of neutral amino acids after portacaval anastomosis. *Surg. Forum* **34:** 38—40.

28) JONUNG, T., RIGOTTI, P., JEPPSSON, B., JAMES, J. H., PETERS, J. C., FISCHER, J. E. (1984). Methionine sulfoximine prevents the accumulation of large neutral amino acids in brain of hyperammonemic rats. *J. Surg. Res.* **36:** 349—353.

29) JONUNG, T., RIGOTTI, P., JAMES, J. H., BRACKETT, K., FISCHER, J. E. (1985). Effect of hyperammonemia and methionine sulfoximine on the kinetic parameters of blood-brain transport of leucine and phenylalanine. *J. Neurochem.* **45 (1):** 308—318.

30) MANS, A. M., BIEBUYCK, J. F., HAWKINS, R. A. (1983). Ammonia selectively stimulates neutral amino acid transport across blood-brain barrier. *Am. J. Physiol.* **245:** 74—77.

31) RIGOTTI, P., ZANCHIN, G., VASSANELLI, P., BETTINESCHI, F., DUSSINI, N., BATTISTIN, L. (1982). Cerebral amino acid levels and transport after portocaval shunt in rat: effect of liver arterialization. *J. Surg. Res.* **33:** 415—422.

32) RIGOTTI, P., JAMES, J. H., RIGGIO, O., FISCHER, J. E. (1985). Changes in the kinetics of phenylalanine and leucine transport across the blood-brain barrier after portacaval shunt. *Surg. Forum* **36:** 138—140.

33) CARDELLI-CANGIANO, P., CANGIANO, C., JAMES, J. H., JEPPSSON, B., BRENNER, W., FISCHER, J. E. (1981). Uptake of amino acids by brain microvessels isolated from rats after portacaval anasotmosis. *J. Neurochem.* **36:** 627—632.

34) ZANCHIN, G., RITOTTI, P., VASSANELLI, P., SALASSA, D., BATTISTIN, L. (1984). Plasma influence on the altered blood-brain barrier to amino acids after portacaval shunt: preliminary results. In: *Hepatic Encephalopathy in Chronic Liver Failure* (Capocaccia, L., Fischer, J. E., Rossi-Fanelli, F., eds.), Plenum Publishing Corporation, pp. 115—120.

35) ZANCHIN, G., MAGGIONI, F., SALASSA, D., VASSANELLI, P. (1984). GABA and dopamine receptors after chronic portacaval shunt in the rat. In: *Advances in Hepatic Encephalopathy and Urea Cycle Diseases* (Kleinberger, G., Ferenci, P., Riederer, P. et al., eds.), Karger, Basel, pp. 360—367.

36) ZANCHIN, G., VILLACARA, A., DA COL, C., COLONNA, R., MASSARI, S., VASSANELLI, P. (1985). Is fluidity of brain capillary endothelial membranes altered in chronic hepatic encephalopathy? *Neurology* **35 (Suppl.):** 250.

37) ZANCHIN, G., HUETHER, G. (1985). Features and functions of the blood-brain barrier. *Funkt. Biol. Med.* **4:** 95—105.

38) BARALDI, M. (1984). Portal-systemic encephalopathy in dogs: changes in brain GABA receptors and neurochemical correlates. In: *Advances in Hepatic Encephalopathy and Urea Cycle Diseases* (Kleinberger, P., Ferenci, P., Riederer, P. et al., eds.), Karger, Basel, pp. 353—359.

AMINO ACID THERAPY IN NEUROLOGICAL DISEASE

John H. Growdon

Department of Neurology
Massachusetts General Hospital
Boston, MA 02114

INTRODUCTION

Administration of purified amino acids to treat neurological disease rests upon the discovery that variations in the concentrations of amino acids circulating in the blood stream can influence rates of neurotransmitter synthesis and release in the central nervous system. There are three prominent examples in which individual amino acids that are precursors for neurotransmitter synthesis have been used to treat neurological disorders: tyrosine and L-dopa in Parkinson's disease (1, 2) tryptophan and L-5-hydroxytryptophan for posthypoxic intention myoclonus (3, 4) and threonine for treatment of spasticity (5). In each instance, the precursor amino acid is administered in pharmacological doses as a drug; there is no evidence that any of these diseases results from an amino acid dietary deficiency. Treatments of Parkinson's disease and posthypoxic intention myoclonus are well known and have been reviewed elsewhere (6); L-threonine administration in the treatment of spasticity is new and forms the basis of this chapter.

Anatomy and Physiology of Spasticity

Spasticity has been defined as "a motor disorder characterized by a velocity dependent increase in tonic stretch reflexes (muscle tone) with exaggerated tendon jerks, resulting from hyperexcitability of the stretch reflex as one component of the upper motor neurone syndrome" (7). Spasticity is a major cause of neurological disability and occurs in a wide variety of conditions with different etiologies: trauma to the spinal cord, multiple sclerosis, and stroke are common examples. Anatomically, spasticity results from loss of descending suprasegmental projections to motor neurons in the anterolateral part of spinal cord grey matter (8—10). Alpha motor neurons in the anterior horn of the spinal cord cause extrafusal muscle contraction, and under physiological conditions, maintain the reciprocal balance between agonist and antagonist that in essential for normal muscle tone and function (10, 11). Gamma motor neurons innervate intrafusal muscle fibers that serve to monitor and regulate extrafusal muscle contraction (12). Lengthening of the muscle as a whole is accompanied by stretching of spindle primary endings in intrafusal fibers; this information travels over Ia afferent nerves to the spinal cord, and modifies the discharge rate of the anterior horn cells (8, 11, 13—15). Physiologically, spasticity occurs because,

NATO ASI Series, Vol. H20
Amino Acid Availability and Brain Function in Health and Disease. Edited by G. Huether

alpha motor neurons are hyperexcitable. The local reflex arc (muscle spindle to mono- and polysynaptic spinal cord linkages to extrafusal muscle) remains intact but functions unchecked by suprasegmental influences that normally maintain a balance between excitatory and inhibitory signals (16, 17). Familial spastic paraparesis is the most pure example of spasticity because there is a single lesion affecting the corticospinal tract in the spinal cord (18, 19); there are no associated lesions to complicate the basis of spasticity. Patients with familial spastic paraparesis, therefore, constitute an ideal group for testing the safety and efficacy of L-threonine as a treatment for spasticity.

Biochemistry and Pharmacology of Spasticity

The body of knowledge describing the biochemical basis of spasticity is less systematized than information regarding anatomic and physiologic aspects of spasticity. Glycine and gamma-aminobutyric acid (GABA) have been identified as the major inhibitory transmitters in the spinal cord (20—22). Glycine is more plentiful in the cord than GABA, and is the major postsynaptic inhibitory transmitter (23—26); GABA is believed to mediate presynaptic inhibition (27). There is convincing evidence that glycine is the transmitter released by inhibitory interneurons located in the grey matter of the spinal cord and by Renshaw cells responsible for recurrent inhibition (24, 28—31). L-glutamate and L-aspartate have been proposed as excitatory transmitters released from terminals of large diameter myelinated primary afferent fibers and from excitatory interneurons, respectively (32, 33). Acetylcholine is the neurotransmitter released at the neuromuscular junction from axons whose cell bodies originate in the anterior horn of the spinal cord. An increase in serotonin neurotransmission produces extensor postures in rats, whereas an increase in norepinephrine transmission facilitates flexor reflexes (34). Although the anatomic location of neurons that release these various neurotransmitters is known, there is little information available regarding the biochemical changes that underlie specific aspects of spasticity in humans (35).

Drugs used in the treatment of spasticity (36, 37) have aimed at either enhancing presynaptic inhibition (benzodiazepines, such as diazepam; baclofen) or at diminishing muscular contraction to excitatory stimulation (dantrolene). Benzodiazepines increase presynaptic inhibition by increasing the affinity of GABA receptor sites for the endogenous ligand; at therapeutic doses, they do not normally affect synaptic events mediated by other transmitters such as glycine (38, 39). By increasing presynaptic inhibition, benzodiazepines decrease release of excitatory transmitters from afferent fibers with a resultant decrease in the gain of the stretch reflex. Baclofen was originally thought to function as a GABA agonist because it is a lipophilic derivative of GABA that crossed the blood-brain barrier and entered the CNS (37). Although baclofen does slow synaptic transmission in the cord, its electrophysiological and pharmacological profile is quite different from GABA (40—44). Baclofen reduces excitatory transmitter release, apparently by activating GABA b receptors and restricting calcium influx into presynaptic terminals (45, 46). As with benzodiazepines, the net effect of reducing excitatory transmitter release by afferent fibers (and presumably by interneurons) is to suppress reflex activity. Dantrolene's therapeutic effect occurs directly on contractile mechanisms within muscle, without any specific action on reflex pathways (36, 37). It reduces the depolarization-induced calcium efflux into the sarcoplasm caused by conducted muscle action potentials. Because sarcoplasmic calcium is necessary in order to activate myosin-ATPase and cause actin and myosin filaments to slide past one another, the force produced by muscle in response to its electrical activation is diminished by dantrolene, although the electromyogram is unchanged. These current modalities for treating spasticity are only partially effective in altering muscle tone and reflexes, and have significant side effects. Diazepam and baclofen do suppress flexor spasms in many patients, but neither improves walking (36, 37, 40). Side effects of these drugs include drowsiness, lightheadedness, and confusion; furthermore, additive CNS depression may occur when other agents, such as alcohol, are taken concomitantly (36). Dantrolene administration lessens muscle tone and hyperreflexia to the extent that it produces muscular paralysis. Thus, its use is counterproductive in those patients whose spasticity masks limb weakness; these individuals benefit from increased extensor tone that enables them to stand and walk on otherwise weak legs. Hepatic toxicity is a serious potential side effect that tempers physicians' inclination to prescribe dantrolene (47).

Rationale for L-Threonine Therapy in Spasticity

An alternate approach to the treatment of spasticity is to increase glycinergic inhibitory tone in spinal cord interneurons. This is a rational neuropharmacological strategy, but was impossible prior to the discovery that threonine increased glycine levels in the rat cord (48). Attempts at increasing brain or spinal cord levels of glycine by administering glycine itself were generally unsuccessful due to the limited passage of glycine across the blood-brain barrier (49). Glycine can be synthesized from serine in a reaction catalyzed by serine transhydroxymethylase (STHM); because this enzyme is fully saturated with serine under physiological conditions, exogenous administration of serine did not increase brain glycine levels. The finding that STHM was identical to threonine aldolase [EC 2.1.2.1.], and that the enzyme was unsaturated with threonine at normal brain concentrations, led Maher and Wurtman (48) to investigate the possibility that threonine availability might affect the biosynthesis of glycine. These investigators found that threonine administration to rats caused dose-dependent increases in spinal cord threonine and glycine levels. Seimers and Aprison (50) extended this finding by demonstrating a linear relationship between the concentrations of threonine and glycine in 10 sub-areas of the rat medulla, as well as in 8 major areas of the rat neuroaxis. Boyd *et al.* (51) studied the behavioral effects of threonine, and reported that it prevented the seizures in rats that usually follow intrathecally administered strychnine. Threonine's effect was mimicked by intrathecally administered glycine but not by serine. Taken together, these pre-clinical experiments indicate that administration of L-threonine increases CNS glycine levels, enhances glycinergic neurotransmission, and produces predicted behavioral effects in rats. If similar biochemical effects occurred in humans, L-threonine could be developed as a novel and effective treatment for CNS disorders, such as spasticity, in which increases in inhibitory tone are desired. Shortly after Maher and Wurtman's (48) original discovery, Barbeau et al. (52) reported that L-threonine suppressed spasticity in 6 of 6 patients with spinocerebellar degeneration; all showed improvement with lessening of tendon reflexes and muscle spasms. Our preliminary results (5) in patients with spasticity of a different etiology than subjects studied by Barbeau *et al.* indicate that this is a promising mode of therapy.

RESULTS OF A SINGLE DRUG OPEN LABEL PILOT STUCY OF L-THREONINE IN TREATING PATIENTS WITH FAMILIAL SPASTIC PARAPARESIS

Methods

Six patients with spastic paraparesis participated in a preliminary study to investigate whether L-threonine administration would increase CSF glycine levels and improve spasticity. The study was conducted according to a single drug open label protocol approved by the Human Studies Committee at the Massachusetts General Hospital (MGH). Patients with documented familial spastic paraparesis (Table 1) were admitted to the General Clinical Research Center at the MGH for 5 days. They related a complete history and underwent physical examinations and laboratory tests to monitor the safety and efficacy of L-threonine. Special emphasis was placed upon the detailed neurological examination, in which the cardinal features of spasticity were scored: muscle tone, strength, and tendon reflexes; abnormal reflexes; and ability to walk and run. Lumbar punctures were performed on the morning of the second hospital day after admission; blood samples were obtained at the same time, and plasma and CSF were frozen at −70° C until subsequent amino acid analysis. After completion of baseline clinical and laboratory examinations, patients began taking L-threonine capsules by mouth in three daily divided doses. Blood and CSF collections were repeated on the 4th admission day two hours after the morning L-threonine dose. All samples were stored at −70° C until submitted to amino acid analysis. Complete physical examination and laboratory tests were repeated on the 5th admission day. Patients were discharged taking L-threonine capsules in three divided daily doses; they returned as outpatients 2 and again 4 weeks later for neurological examination and assessment of drug safety and efficacy. Of the 6 patients, 5 received 500 mg tid, all 6 received 1000 mg tid, and 4 of them received 1500 mg tid of L-threonine. In each case, treatment with threonine lasted 1 month.

TABLE 1
Clinical Characteristics of Patients with Familial Spastic Paraparesis

Patient	Age	Sex	Extent of distability
1	47	M	Mild
2	66	M	Moderate
3	48	M	Severe
4	56	F	Severe, with urinary incontinence
5	65	M	Moderate
6	59	M	Severe

Mild = Walks without aids, Moderate = Walks with care, Severe = Walks with crutches

Results

L-threonine given as a single amino acid in capsules was well tolerated; there were no signs of clinical or laboratory toxicity, and all patients completed each trial they began. Spasticity scores before and after treatment were compared, and summarized results for each dose level are shown in Tables 2, 3, and 4. With doses of 1.5 g/day, only 1 patient (#3) showed clinical improvement; the others were either marginally improved or showed no benefit. Clinical improvement was more substantial with higher doses of L-threonine. With 3.0 g/day, 6 of 6 patients derived benefit in one or more aspect of spasticity: tone in the legs decreased in 2, urinary incontinence ceased in 1, proximal leg strength increased in 3, and walking improved in 4. In 1 of these 6 (#5), tone increased but there was improvement in three other aspects of spasticity (strength, walking, and less brisk tendon reflexes). With a dose of 4.5 g/day, all patients improved by at least 20 %: one patient had improvement in all aspects of the examination and all patients improved in at least four categories of spasticity.

TABLE 2
Clinical Response to L-Threonine in Patients with Spastic Paraparesis: 1.5 g/day

Name	Strength	Tone	Deep tendon reflexes	Running	Walking	Hopping R. Leg	Hopping L. Leg
1	S	S	S	S	S	S	S
2	S	I	I	S	S	S	S
3	I	S	S	S	S	I	I
4							
5	S	S	S	S	S	S	S
6	I	S	S	S	S	S	S

S = Stable, D = Deterioated, I = Improved

TABLE 3
Clinical Response to L-Threonine in Patients with Spastic Paraparesis: 3.0 g/day

Name	Strength	Tone	Deep tendom Reflexes	Running	Walking	Hopping R. Leg	Hopping L. Leg
1	S	S	I	S	I	I	I
2	S	I	I	I	I	I	I
3	I	I	I	S	I	I	I
4	S	S	I	S	S	I	I
5	I	D	I	S	I	S	S
6	I	S	S	S	S	S	S

S = Stable, D = Deteriorated, I = Improved

TABLE 4
Clinical Response to L-Threonine in Patients with Spastic Paraparesis: 4.5 g/day

Name	Strength	Tone	Deep tendon reflexes	Running	Walking	Hopping R. Leg	Hopping L. Leg
1	I	I	I	I	I	I	I
2							
3	I	S	S	S	I	I	I
4	I	I	S	S	S	I	I
5							
6	I	S	I	S	I	I	I

S = Stable, D = Deteriorated, I = Improved

To date, plasma and CSF amino acid analyses have been performed on samples obtained from 3 patients who received both the 1.5 g/day and 3 g/day doses. Neither plasma nor CSF levels of threonine or glycine increased significantly during administration of 1.5 g/day. In contrast to low doses of threonine, a dose of 3 g/day caused significant ($p < .05$) 2 to 4 fold increases in threonine and glycine levels in both plasma and CSF. Mean threonine levels in plasma rose from 78 $\pm$ 5.8 to 280 $\pm$ 11.1 nmol/ml and in CSF from 19.3 $\pm$ 2.3 to 70.2 $\pm$ 8.2 nmol/ml. The mean CSF glycine level was 3.2 $\pm$ 0.6 nmol/ml before L-threonine and 10.4 $\pm$ 0.6 nmol/ml during treatment.

Discussion

The results of this preliminary open label single drug study indicate that doses of L-threonine up to 4.5 g/day are well tolerated and free from toxicity. L-threonine produced a dose-related increase in plasma threonine, CSF threonine, and CSF glycine levels in patients with familial spastic paraparesis. These biochemical data support the hypothesis that L-threonine administration increases glycine levels in humans as in rats. Furthermore, there was a direct significant correlation between plasma threonine levels and CSF threonine and glycine levels. Additional data from future studies are needed in order to determine whether the relationship between plasma threonine and CSF glycine levels is a constant linear function or whether other variables influence eventual glycine levels. If the relationship is constant, CNS levels of glycine could be estimated simply by monitoring levels of threonine in a peripheral venous blood sample.

Doses of 3.0 and 4.5 g/day of L-threonine suppressed symptoms and signs of spasticity in all 6 patients. Most patients did not improve during treatment with 1.5 g/day; overall, CSF glycine levels did not increase with this dose of L-threonine. In the only patient with mild improvement during low dose treatment (#3), there was a slight rise in CSF glycine content. These observations are consistent with the hypothesis that L-threonine administration enhances glycine synthesis and release in the human spinal cord, and suggest that there is a causal relationship between inhibitory glycinergic transmission and spasticity suppression.

The magnitude of clinical benefit was similar to that reported in the only other study of L-threonine in spasticity. Barbeau *et al.* (52) gave 500 mg/day of L-threonine to patients with spasticity associated with familial spinocerebellar degeneration but did not measure glycine levels in CSF. In contrast to the previous report, patients in the present study with a different cause of spasticity did not improve until they received 3000 g/day of L-threonine. The higher dose required to suppress spasticity in familial spastic paraparesis than in those with familial spinocerebellar degeneration suggests that the dose of L-threonine needed to suppress spasticity may vary according to different etiologies of spasticity.

CONCLUSION

The increases in CSF glycine levels induced by L-threonine are consistent with expectations of increased glycine synthesis based upon preclinical data. The clinical benefits of L-threonine administration are consonant with current concepts of spinal cord anatomy, physiology, and biochemistry. Although additional placebo-controlled studies will be needed before accepting L-threonine as an effective treatment for familial spastic paraparesis, data from this preliminary study support the hypothesis that increases in glycinergic tone induced by L-threonine suppress symptoms and signs of spasticity.

REFERENCES

1) COTZIAS, G. C., PAPAVASILIOU, P. S., GELLENE, R. (1969). Modification of parkinsonism: Chronic treatment with L-dopa. *N. Engl. J. Med.* **280:** 337—345.

2) GROWDON, J. H. (1981). Tyrosine treatment in Parkinson's disease: clinical effects. *Neurology* **31:** 134.

3) VAN WOERT, M. H., SETHY, V. H. (1975). Therapy of intention myoclonus with L-5-hydroxythryptophan and a peripheral decarboxylase inhibitor — MK 486. *Neurology* **25:** 135—140.

4) GROWDON, J. H., YOUNG, R. R., SHAHANI, B. T. (1976). L-5-hydroxytryptophan in the treatment of several different syndromes in which myoclonus is prominent. *Neurology* **26:** 1135—1140.

5) GROWDON, J. H., GIBSON, C. J. (1982). Dietary precursors of neurotransmitters: treatment strategies. In: *Current Neurology* (Appel, S. H., ed.). New York: Wiley and Sons, **Vol. 4,** pp. 117—144.

6) NADER, T., GROWDON, J. H., MAHRE, T. J., WURTMAN, R. J. (1987). L-threonine administration increases CSF glycine levels and suppresses spasticity. *Neurology* **37:** 125.

7) LANCE, J. W. (1980). Symposium synopsis. In: *Spasticity: Disordered Motor Control* Feldman, R. G., Young, R. R., Koella, W. P., eds.). Chicago, Year Book, pp. 485—494.

8) WIESENDANGER, M. (1972). *Pathophysiology of Muscle Tone.* Neurology Series, **Vol. 9,** Springer Verlag, New York, pp. 1—46.

9) RUSHWORTH, G. (1960). Spasticity and rigidity: an experimental study and review. *J. Neurol. Neurosurg. Psychiatry* **23:** 99—118.

10) GRANIT, R. (1970). *The Basis of Motor Control.* Academic Press, New York.

11) MATTHEWS, P. B. C. (1972). *Mammalian Muscle Receptors and their Central Actions.* Camelot Press Ltd., London.

12) BURKE, D. (1983). A critical examination of the case for and against fusimotor involvement in disorders in muscle tone. In: *Motor Control Mechanisms in Health and Disease* (Desmedt, J. E., ed.). Progress in Clinical Neurophysiology, **Vol. 10,** Karger,Basel, pp. 133—215.

13) HAGBARTH, K.-E., WALLIN, G., LOFSTEDT, L. (1973). Muscle spindle response in normal and spastic subjects. *Scand. J. Rehabil. Med.* **5:** 156—169.

14) HAGBARTH, K.-E., WALLIN, G., LOFSTEDT, L. AQUILONIUS, S.-M. (1975). Muscle spindle activity in alternating tremor of Parkinsonism and in clonus. *J. Neurol. Neurosurg. Psychiatry* **38:** 636—641.

15) BURKE, D. (1980). Areassessment of the muscle spindle contribution to muscle tone in normal and spastic man. In: *Spasticity: Disordered Motor Control* (Feldman, R. G., Young R. R., Koella, W. P., eds.). Chicago, Year Book, pp. 261—328.

16) BISHOP, B. (1977). Spasticity: its physiology and management. Part I. Neurophysiology of spasticity: classical concepts. *Phys. Ther.* **57:** 371—376.

17) BISHOP, B. (1977). Spasticity: its physiology and management. Part II. Neurophysiology of spasticity: current concepts. *Phys. Ther.* **57:** 377—384.

18) HOLMES, G. L., SCHAYWITZ, B. A. (1977). Strumpell's pure familial spastic paraplegia: case study and review of the literature. *J. Neurol. Neurosurg. Psychiatry* **40:** 1003—1008.

19) RICHARDSON, E. P. Jr., BEAL, M. F., MARTIN, J. B. (1987). Degenerative disease of the nervous system. In: *Harrison's Principles of Internal Medicine* (Braunwald, E., Isselbacher, K. J., Petersdorf, R. G., et al., eds.). 11th edition, McGraw-Hill, New York, pp. 2011—2027.

20) CULLHEIM, S., KELLERTH, J. O. (1981). Two kinds of inhibition in cat spinal alphamononoeurones as differentiated pharmacologically. *J. Physiol.* **312:** 209—224.

21) NISTRI, A. (1983). Spinal cord pharmacology of GABA and chemically related amino acids. In: *Pharmacology,* **Vol. 1,** *Handbook of the Spinal Cord* (Davidoff, R. A., ed. Dekker). New York, pp. 45—104.

22) YOUNG, A. B., MACDONALD, R. L. (1983). Glycine as a spinal cord neurotransmitter. In: *Pharmacology,* **Vol. 1,** *Handbook of the Spinal Cord* (Davidoff, R. A., ed. Dekker). New York, pp. 1—43.

23) GRAHAM, L. T. Jr., et al. (1966). Distribution of some synaptic transmitter suspects in cat spinal cord. *J. Neurochem.* **14:** 465—47.

24) APRISON, M. H., DAVIDOFF, R. A., WERMAN, R. (1970). Glycine: its metabolic and possible transmitter roles in nervous tissue. In: *Handbook of Neurochemistry,* **Vol. 3.** Plenum Press, New York, pp. 381—397.

25) HOSLI, E., LJUNGDAHL, A., HOKFELT, T., HOSLI, L. (1972). Spinal cord tissue cultures — a model for autoradiographic studies on uptake of putative neurotransmitters such as glycine and GABA. *Experientia* **28:** 1342—1344.

26) EHINGER, B., FALCK, B. (1971). Autoradiography of some suspected neurotransmitter substances: GABA, glycine, glutamic acid, histamine, dopamine, and L-DOPA. *Brain Research* **33:** 157—172.

27) DAVIDOFF, R. A., HACKMAN, J. C. (1984). Spinal inhibition. In: *Anatomy and Physiology,* **Vol. 2 & 3,** *Handbook of the Spinal Cord* (Davidoff, R. A., ed. Dekker). New York, pp. 385—459.

28) LJUNGDAHL, A., HOKFELT, T. (1973). Autoradiographic uptake patterns of [^{3}H]GABA and [^{3}H]glycine in central nervous tissues with special reference to the cat spinal cord. *Brain Research* **62:** 587—595.

29) CURTIS, D. R., HOSLI, L., JOHNSTON, G. A. R., JOHNSTON, J. H. (1967). Glycine and spinal inhibition. *Brain Research* **5:** 112—114.

30) RIZZOLI, A. A. (1986). Distribution of glutamic acid, aspartic acid, gamma-aminobutyric acid and glycine in six areas of cat spinal cord before and after transection. *Brain Research* **11:** 11—18.

31) DALY, E. C., APRISON, M. H. (1974). Distribution of serine hydroxymethyltransferase and glycine transaminase in several areas of the central nervous system of the rat. *J. Neurochem.* **22:** 877—885.

32) PUIL, E. (1983). Actions and interactions of S-glutamate in the spinal cord. In: *Pharmacology,* **Vol. 1,** *Handbook of the Spinal Cord* (Davidoff, R. A., ed Dekker). New York, pp. 105—170.

33) SALT, T. E., HILL, R. G. (1983). Neurotransmitter candidates of somatosensory primary afferent fibers. *Neuroscience* **10:** 1083—1103.

34) GROWDON, J. H. (1978). Changes in motor behavior folloving the administration of serotonin neurotoxins. *Ann. New York acad. Sci.* **305:**510—523.

35) DELWAIDE, R. J., YOUNG, R. R. (1985). *Clinical Neurophysiology in Spasticity* (Restorative Neurology, **Vol. 1**), Elsevier Science Publishers.

36) YOUNG, R. R., DELWAIDE, P. J. (1981). Spasticity. *N. Engl. J. Med.* **304:** 28—34; 96—99.

37) DAVIDOFF, R. A. (1985). Antispasticity drugs: mechanisms of action. *Ann. Neurol.* **17:** 107—116.

38) CURTIS, R. D., LODGE, D., JOHNSTON, G. A. R., et al. (1976). Central actions of benzodiazepines. *Brain Res.* **118:** 344—347.

39) SCHMIDT, R. F., VOGEL, M. E., ZIMMERMANN, M. (1967). Die Wirkung von Diazepam auf die präsynaptische Hemmung und andere Rückenmarksreflexe. *Naunyn Schmiedebergs Arch. Pharmacol.* **258:** 69—82.

40) DELWAIDE, P. J. (1985). Electrophysiological analysis of the mode of action of muscle relaxants in spasticity. *Ann. Neurol.* **17:** 90—95.

41) CURTIS, D. R., LODGE, D., BORNSTEIN, J. C., PEER, M. J. (1981). Selective effects of (-)baclofen on spinal synaptic transmission in the cat. *Exp. Brain Res.* **42:** 158—170.

42) DAVIDOFF, R. A., SEARS, E. S. (1974). The effects of Lioresal on synaptic activity in the isolated spinal cord. *Neurology (Minneap)* **24:** 957—963.

43) FOX, S., KRNJEVIC, K., MORRIS, M. E., ET AL. (1978). Action of baclofen on mammalian synaptic transmission. *Neuroscience* **3:** 495—515.

44) KUDO, Y., JURACHI, M., FUKUDA, H. (1976). Action of B-(p-chlorophenyl)-y-aminobutyric acid (baclofen) on the isolated spinal cord of the frog. *Jpn. J. Pharmacol.* **26:** 99 p.

45) DESARMENIAN, M., SANTANGLO, F., OCCHIPINTI, G., et al. (1983). Electrophysiological study of $GABA_A$ versus $GABA_B$ receptors on excitation-secretion. In: *CNS Receptors From Molecular Pharmacology to Behavior* (Mandel, P., DeFeudis, F. V., eds.). Raven, New York, pp. 93—105.

46) DUNLAP, K., FISCHBACH, G. D. (1981). Neurotransmitters decrease the calcium conductance activated by depolarization of embryonic chick sensory neurones. *J. Physiol. (Lond)* **317:** 519—535.

47) RIDER, R. M., BROGDEN, R. N., SPEIGHT, T. M., AVERY, G. S. (1977). Dantrolene sodium; a review of its pharmacological properties and therapeutic efficacy in spasticity. *Drugs Jan.* **13:** 3—23.

48) MAHER, T., WURTMAN, R. J. (1980). L-threonine administration increases glycine concentrations in the rat central nervous system. *Life Sci.* **26:** 1283—1286.

49) PARDRIDGE, W. M. (1977). Regulation of amino acid availability to the brain. In: *Nutrition and the Brain,* **Vol. 1** (Wurtman, R. J., Wurtman, J. J., eds.). Raven Press, New York, pp. 141—204.

50) SIEMERS, E. R., DALY, E. C., APRISON, M. H. (1980). Is threonine a precursor of one of the central pools of glycine? *Amer. Soc. Neurochem. Abstr.* **11:** 69.

51) BOYD, D. K., BOXER, P. A., ANDESON, R. J. (1985). Antabonism of strychnine seizures by intrathecal administration of amino acids and anticonvulsants. *Soc. Neurosci. Abstr.* **11:** 1325.

52) BARBEAU, A., ROY, M., CHGOUZA, C. (1982). Pilot study of threonine supplementation in human spasticity. *J. Can. Sci. Neurol.* **2:** 141—145.

TRYPTOPHAN AND TYROSINE RATIOS TO NEUTRAL AMINO ACIDS IN DEPRESSED PATIENTS IN REGARD TO K_m: RELATION TO EFFICACY OF ANTIDEPRESSANT TREATMENTS

Svend E. Møller

Clinical Research Laboratory
Sct. Hans Mental Hospital
DK-4000 Roskilde
Denmark

INTRODUCTION

Formation in the brain of serotonin from tryptophan (Trp) and noradrenaline from tyrosine (Tyr) are important metabolic pathways related to neurotransmission and, thus, to brain function. Because the synthesis of these monoamines in the brain partly depends on the brain concentration of the respective precursor amino acids, it is of interest to identify the factors that normally influence brain Trp and Tyr levels.

The most important factor controlling brain Trp uptake from the periphery is its competition with other large neutral amino acids (LNAA), including valine (Val), isoleucine, leucine (Leu), Tyr and phenylalanine, for transport across the blood-brain barrier (BBB) (1, 2, 3). Therefore, the transport of Trp into the brain is favoured by a high plasma Trp concentration, or when the sum of the concentrations of the other LNAA is decreased. Conversely, a low plasma Trp level or a high level of the competing amino acids decreases the transport of Trp into the brain. Hence, the molar ratio in plasma of Trp to the sum of the other LNAA reflects the brain Trp concentration and serotonin synthesis. Implicitly, the BBB rate-limits the brain Tyr uptake, and the ratio Tyr/LNAA in plasma reflects similarly the brain Tyr concentration (4).

The decisive importance of competition in the LNAA transport to the brain has been established through studies on laboratory rats. Recent studies on isolated human brain capillaries, a model system of the human BBB, indicate that the Michaelis constant (K_m) of phenylalanine transport into human brain microvessels (22 μM (5)) is of the same size as that for uptake in cerebral hemisphere in conscious rats (26 μM (6)). At present time, it may be reasonable to assume that the K_m's of all the LNAA in the conscious rat are comparable to those in man, based on the findings for phenylalanine in the two species (7).

Information on the K_m of BBB amino acid transport provides not only the quantitative basis for understanding the importance of competition effects. The individual LNAA have

NATO ASI Series, Vol. H20
Amino Acid Availability and Brain Function in Health and Disease. Edited by G. Huether

different K_m's reflecting differences in the affinity of the various LNAA for the BBB neutral amino acid transport system. Thus, while the *regular* plasma ratios Trp/LNAA and Tyr/LNAA do not take into account these differences in affinity for the transport carrier, the *normalized* ratios do so by including the K_m's of individual LNAA in the ratios.

In a serial of clinical trials we have studied the association between the pretreatment plasma ratios Trp/LNAA and Tyr/LNAA and the antidepressant response to various drugs in depressed patients, who may suffer from a functional deficit of brain serotonin and/or noradrenaline. The aim of the present retrospective study was to evaluate whether the therapeutic response in 145 depressives from 6 separate trials associated stronger with the regular or the normalized plasma ratios Trp/LNAA and Tyr/LNAA.

Brain serotonin and noradrenaline possibly also take part in the regulation of the macronutrient specific appetite for carbohydrate and protein (8, 9, 10). Recently, a significant direct relationship has been shown between the sum of the plasma ratios Trp/LNAA and Tyr/LNAA in fasting individuals and the proportion of carbohydrate to protein in a freely chosen and consumed breakfast meal (11). This retrospective study has also evaluated whether the eating preferences of healthy females associated stronger with the regular or normalized plasma amino acid ratios.

METHODS

The following clinical studies were included in this retrospective investigation (antidepressant, number of depressed patients (reference)): L-tryptophan, n = 32 (12); imipramine, n = 39 (13); amitriptyline, n = 21 (14); nortriptyline, n = 26 (15); citalopram, n = 14, and maprotiline, n = 13 (16). Excluded from the study were two trials: clomipramine, n = 26 (13), and lithium plus L-tryptophan, n = 22 (14), due to weak correlations ($P > 0.2$, 2-tailed) between any one of the biochemical variables and the therapeutic outcome. The study of eating preferences included 31 healthy females (11).

Detailed information on the healthy subjects and the depressed patients, inclusion and exclusion criteria, dosage schedules, duration of treatment, analytical methods, serum steady-state drug levels, etc. are presented in the original reports. The severity of the depressive state was in all cases assessed by means of the Hamilton depression rating scale (HDRS) items 1—17 (17).

The normalized plasma ratios Trp/LNAA and Tyr/LNAA were calculated according to the formulas:

$$\frac{\text{Trp}}{\sum \frac{K_m(\text{Trp})}{K_m(i)} \times \text{LNAA}} \quad \text{and} \quad \frac{\text{Tyr}}{\sum \frac{K_m(\text{Tyr})}{K_m(i)} \times \text{LNAA}}$$

respectively, where $K_m(i)$ is the K_m of the individual competing LNAA. All the K_m's applied in the calculation of normalized ratios are from the conscious rat (6) (Table 1).

Relation between plasma amino acid ratios and therapeutic outcome was evaluated by means of the Spearman rank correlation coefficient (r_S) and 2-tailed significance test. A combined P value for a serial of studies was estimated by multiplying the sum of log P's with −4.605, which product follows a X^2 distribution with a number of df equal to twice the number of pooled P values (18). Relation between plasma amino acid ratios and the proportion of carbohydrate to protein consumed was evaluated by means of 2-tailed linear regression analysis.

TABLE 1

K_m's for the transport of various large neutral amino acids through the BBB of conscious rat

Amino acid	Val	Iso	Leu	Tyr	Phe	Trp
K_m (μM)	168	145	87	86	32	52

Each value is the mean of individual values for each of four dissected regions. Adapted from (6) with permission

RESULTS

The regular and normalized plasma ratios Trp/LNAA and Tyr/LNAA were correlated with the final HDRS scores and % reduction of HDRS scores. In the trial of amitriptyline both the ratio of free and total plasma Trp to 5 competing amino acids were applied, and in the

TABLE 2

Correlation coefficients (r_s) and P values for the correlations between various regular (R) and normalized (N) pretreatment plasma amino acid ratios in depressed patients and the final Hamilton depression rating scale (HDRS) scores and % reduction of HDRS Scores

				Final HDRS score		% Red. of HDRS score	
Study	(n)	R/N	Plasma ratio	r_s	P	r_s	P
L-Tryptophan	(32)	R	Trp/LNAA	0.30	0.100	−0.41	0.021
		N	Trp/LNAA	0.20	0.278	−0.30	0.097
		R	Trp/Val+Leu+Tyr	0.30	0.095	−0.40	0.022
		N	Trp/Val+Leu+Tyr	0.28	0.125	−0.38	0.033
Amitriptyline	(21)	R	free Trp/LNAA	0.13	—	−0.26	—
		N	free Trp/LNAA	0.05	—	−0.18	—
		R	Trp/LNAA	0.44	0.047	−0.43	0.054
		N	Trp/LNAA	0.41	0.065	−0.43	0.055
		R	Trp/Val+Leu+Tyr	0.47	0.036	−0.44	0.046
		N	Trp/Val+Leu+Tyr	0.46	0.039	−0.45	0.044
Citalopram	(14)	R	Trp/LNAA	0.51	0.064	−0.41	0.154
		N	Trp/LNAA	0.41	0.149	−0.34	0.232
		R	Trp/LNAA + Tyr/LNAA	0.57	0.036	−0.48	0.084
		N	Trp/LNAA + Tyr/LNAA	0.59	0.025	−0.50	0.071
		R	Trp/Val+Leu+Tyr	0.53	0.050	−0.43	0.129
		N	Trp/Val+Leu+Tyr	0.50	0.075	−0.40	0.162
Immipramine	(39)	R	Trp/LNAA + Tyr/LNAA	0.36	0.024	−0.31	0.055
		N	Trp/LNAA + Tyr/LNAA	0.29	0.076	−0.22	0.185
Nortriptyline	(26)	R	Tyr/LNAA	0.51	0.008	−0.53	0.005
		N	Tyr/LNAA	0.47	0.016	−0.49	0.011
Maprotiline	(13)	R	Tyr/LNAA	−0.53	0.065	0.59	0.034
		N	Tyr/LNAA	−0.51	0.077	0.58	0.039

Abbreviations: Tryptophan (Trp), tyrosine (Tyr), valine (Val), leucine (Leu), large neutral amino acids (LNAA)

trial of citalopram both the plasma ratio Trp/LNAA alone and the sum of the ratios Trp/LNAA and Tyr/LNAA were applied. Results are presented in Table 2.

Sixteen comparisons between the regular and the respective normalized plasma amino acid ratios were performed as regards which of the ratios that produced the numerically greater correlation coefficient. The sum of the normalized ratios Trp/LNAA and Tyr/LNAA associated stronger than the sum of the regular ratios with the final HDRS scores and % reduction of HDRS scores in the citalopram study, whereas the regular plasma ratios Trp/LNAA and Tyr/LNAA were superior to the normalized ratios in the remaining 14 comparisons ($P = 0.0018$, by the binomial distribution).

In addition to the qualitative, also a quantitative evaluation was performed. The combined 2-tailed P value for the 6 trials was calculated for each of the correlations between, on the one side, the regular and the normalized ratios Trp/LNAA and Tyr/LNAA, and, on the other side, the final HDRS scores and % reduction of HDRS scores. The amitriptyline study contributed with the plasma ratio total Trp/LNAA, and the citalopram study with the sum of ratios Trp/LNAA and Tyr/LNAA. The regular ratios associated stronger with the final HDRS scores ($X^2 = 39.95$, $P \ll 0.0005$) and % reduction of HDRS scores ($X^2 = 41.63$, $P \ll 0.0005$) than did the normalized ratios ($X^2 = 33.87$, $P = 0.0075$, and $X^2 = 34.68$, $P = 0.00054$, respectively, 12 df in all cases).

In a previous retrospective study the molar ratio in plasma of Trp to the sum of Val, Leu, and Tyr was shown to associate stronger with the therapeutic response to L-tryptophan in depressed patients than any other of the 31 combinations of plasma Trp ratios that can be formed from the 5 competitors (19). It was therefore of interest to evaluate which of the regular and normalized plasma ratios Trp/LNAA and Trp/Val + Leu + Tyr that associated strongest with the final HDRS scores and % reduction of HDRS scores in the trials of L-tryptophan, amitriptyline and citalopram, which all are serotonin-potentiating agents. The correlation coefficients and the exact P values are presented in Table 2 and the combined 2-tailed P values for the 3 trials in Table 3.

The X^2 values of the regular ratios Trp/LNAA and Trp/Val + Leu + Tyr were comparable and greater than those of the respective normalized ratios. However, while the X^2 values of the normalized ratio Trp/Val + Leu + Tyr were only slightly smaller than those of the respective regular ratio, the X^2 values of the normalized and regular ratio Trp/LNAA differed considerably (Table 3).

TABLE 3

Combined 2-tailed P values for correlations between regular (R) and normalized (N) plasma tryptophan ratios and therapeutic outcome in depressives treated with L-tryptophan, amitriptyline or citalopram

		Final HDRS score		% Red. of HDRS score	
R/N	Plasma ratio	χ^2 (6 df)	P	χ^2 (6 df)	P
R	Trp/LNAA	16.19	0.0139	17.33	0.0085
N	Trp/LNAA	11.84	0.069	13.40	0.039
R	Trp/Val+Leu+Tyr	17.36	0.0084	17.87	0.0070
N	Trp/Val+Leu+Tyr	15.82	0.0163	16.73	0.0155

Abbreviations: Tryptophan (Trp), large neutral amino acids (LNAA), valine (Val), leucine (Leu), tyrosine (Tyr), Hamilton depression rating scale (HDRS)

Ratio of Carbohydrate to Protein Consumed

The proportion of carbohydrate to protein in a freely chosen and consumed breakfast meal by 31 healthy females was correlated with the regular and normalized plasma amino acid ratios in the fasting individuals. The sum of the regular plasma ratios Trp/LNAA and Tyr/LNAA associated stronger with the proportion of carbohydrate to protein consumed than the sum of the respective normalized plasma amino acid ratios (Table 4).

TABLE 4
Linear correlation coefficients (r) for the relation between regular (R) and normalized (N) plasma amino acid ratios in fasting individuals and the proportion of carbohydrate to protein consumed in a freely chosen breakfast meal

R/N	Plasma amino acid ratios	r	P
R	Trp/LNAA + Tyr/LNAA	0.50	0.0048
N	Trp/LNAA + Tyr/LNAA	0.46	0.0091
R	Trp/Val+Leu+Tyr + Tyr/LNAA	0.46	0.0091
N	Trp/Val+Leu+Tyr + Tyr/LNAA	0.36	0.045

Abbreviations: Tryptophan (Trp), tyrosine (Tyr), valine (Val), leucine (Leu), large neutral amino acids (LNAA)

SUMMARY AND CONCLUSION

This retrospective study has shown that the clinical response in depressed patients to various antidepressants, thought to potentiate brain serotoninergic and/or noradrenergic function, associates stronger with the regular pretreatment plasma ratios Trp/LNAA and Tyr/LNAA than with the normalized ratios, in which the K_m's of individual amino acids are considered. Furthermore, the proportion of carbohydrate to protein consumed by healthy subjects associated stronger with the sum of the regular plasma ratios Trp/LNAA and Tyr/LNAA of the fasting individuals than with the sum of the normalized plasma ratios.

The smaller coefficients for the correlation between the therapeutic outcome in the depressives and the normalized plasma amino acid ratios compared with the respective regular ratios are most likely attributed to phenylalanine. Firstly, the K_m of phenylalanine is low as compared with the K_m of the other competing amino acids. This characteristic tends to give phenylalanine relatively more weight in the normalized plasma amino acid ratios. Secondly, while the normalized and regular plasma ratios Trp/Val + Leu + Tyr yielded comparable test statistics for the pooled P values from the trials of L-tryptophan, amitriptyline and citalopram, the normalized and regular ratios Trp/LNAA, which include phenylalanine, yielded test statistics that differed considerably in favour of the regular ratio.

A third piece of evidence, suggesting that phenylalanine impairs the association between the plasma Trp ratio and the therapeutic response in depressives, comes from a previous retrospective study. In that study all the 31 combinations of the plasma Trp ratio that can be formed from the 5 competing LNAA considered, were investigated in terms of which ratio that associated with the greatest number of responders to L-tryptophan treatment below the 15th percentile, and the smallest number of responders between the 15th and 30th percentiles, and above the 30th percentile. Phenylalanine consistently impaired the association between the various plasma Trp ratios and the therapeutic response (19).

Thus, there is a discrepancy between the present clinical findings and the results from studies of neutral amino acid transport across the BBB of the laboratory rat. In the latter studies, phenylalanine generally appeared as the neutral amino acid with the lowest K_m and as a very potent competitor of Trp transport at the BBB (1, 20). Because the K_m of phenylalanine transport into human brain microvessels (5) is of the same order as the K_m of phenylalanine transport across the BBB of the conscious rat (6), it might have been anticipated that phenylalanine strengthened rather than impaired the investigated relationships. On the other hand, some previous studies of laboratory animals have found support for the idea that plasma phenylalanine has relatively little to do with brain Trp (21, 22).

The controversy suggests that, in addition to the affinity for the transport carrier, there is a number of other factors, which must be integrated in order to describe a comprehensive mechanism for the regulation of neutral amino acid uptake by brain.

The correlation coefficients presented here from the trials of L-tryptophan and amitriptyline are somewhat smaller than those previously reported from the same material (23). In the previous study, the slight, but statistically significant correlations between the age and the plasma Trp and Tyr ratios were accounted for by calculation of residuals of the plasma Trp and Tyr ratio of the individual patient, which were subsequently correlated with the final depression score.

In conclusion, this retrospective study has shown that the regular plasma ratios Trp/LNAA and Tyr/LNAA associate stronger with brain processes in human individuals thought to implicate serotonin and noradrenaline than the respective normalized ratios, in which the difference in affinity of the various LNAA for the transport carrier at the BBB has been taken into account.

REFERENCES

1) PARDRIDGE, W. M. (1977). Kinetics of competitive inhibition of neutral amino acid transport across the blood-brain barrier. *J. Neurochem.* **28:** 103—108.

2) YUWILER, A., OLDENDORF, W. H., GELLER, E., BRAUN, L. (1977). Effect of albumin binding and amino acid competition on tryptophan uptake into brain. *J. Neurochem.* **28:** 1015—1023.

3) SMITH, Q. R. (1987). Regulation of neutral amino acid transport at the blood-brain barrier. *This volume.*

4) FERNSTROM, J. D., FALLER, D. V. (1978). Neutral amino acids in the brain: changes in response to food ingestion. *J. Neurochem.* **30:** 1531—1538.

5) CHOI, T. B., PARDRIDGE, W. M. (1986). Phenylalanine transport at the human blood-brain barrier. *J. Biol. Chem.* **261:** 6536—6541.

6) MILLER, L. P., PARDRIDGE, W. M., BRAUN, L. D., OLDENDORF, W. H. (1985). Kinetic constants for blood-brain barrier amino acid transport in conscious rats. *J. Neurochem.* **45:** 1427—1432.

7) PARDRIDGE, W. M. Personal communication.

8) ANDERSON, G. H., LI, E. T. S. (1987). Amino acids in the regulation of food intake. *This volume.*

9) BLUNDELL, J. E., HILL, A. J. (1987). Amino acids in appetite control. *This volume.*

10) LEIBOWITZ, S. F., WEISS, G. F., YEE, F., TRETTER, J. B. (1985). Noradrenergic innervation of the paraventricular nucleus: specific role in control of carbohydrate ingestion. *Brain Res. Bull.* **14:** 561—567.

11) MøLLER, S. E. (1986). Carbohydrate/protein selection in a single meal correlated with plasma tryptophan and tyrosine ratios to neutral amino acids in fasting individuals. *Physiol. Behav.* **38:** 175—183.

12) MøLLER, S. E., KIRK, L., HONORĚ, P. (1980). Relationship between plasma ratio of tryptophan to competing amino acids and the response to L-tryptophan treatment in endogenously depressed patients. *J. Affective Disord.* **2:** 47—59.

13) MøLLER, S. E., REISBY, N., ORTMANN, J., ELLEY, J., KRAUTWALD, O. (1981). Relevance of tryptophan and tyrosine availability in endogenous and "non-endogenous" depressives treated with imipramine or clomipramine. *J. Affective Disord.* **3:** 231—244.

14) MøLLER, S. E., HONORĚ, P., LARSEN, O. B. (1983). Tryptophan and tyrosine ratios to neutral amino acids in endogenous depression: relation to antidepressant response to amitriptyline and lithium + L-tryptophan. *J. Affective Disord.* **5:** 67—79.

15) MøLLER, S. E., ØDUM, K., KIRK, L., BJERRE, M., FOG-MøLLER, F., KNUDSEN, A. (1985). Plasma tyrosine/neutral amino acid ratio correlated with clinical response to nortriptyline in endogenously depressed patients. *J. Affective Disord.* **9:** 223—229.

16) MøLLER, S. E., DE BEURS, P., TIMMERMAN, L., TAN, B. K., LEIJNSE-YBEMA, H. J., STUART, M. H. C., PETERSEN H. E. H. (1986). Plasma tryptophan and tyrosine ratios to competing amino acids in relation to antidepressant response to citalopram and maprotiline. A preliminary study. *Psychopharmacology* **88:** 96—100.

17) HAMILTON, M. (1960). A rating scale for depression. *J. Neurol. Neurosurg. Psychiatry* **23:** 56—62.

18) SOKAL, R. R., ROHLF, F. J. (1969). Biometry. W. H. Freeman & Co., San Francisco, pp. 621—624.

19) MøLLER, S. E. (1980). Evaluation of the relative potency of individual competing amino acids to tryptophan transport in endogenously depressed patients. *Psychiatry Res.* **3:** 141—150.

20) DANIEL, P. M., MOORHOUSE, S. R., PRATT, O. E. (1976). Amino acid precursors of monoamine neurotransmitters and some factors influencing their supply to the brain. *Psycholog. Med.* **6:** 277—286.

21) FERNSTROM, J. D., LARIN, F., WURTMAN, R. J. (1973). Correlations between brain tryptophan and plasma neutral amino acid levels following food consumption in rats. *Life Sci.* **13:** 517—524.

22) MUNRO, H. N., FERNSTROM, J. D., WURTMAN, R. J. (1975). Plasma neutral amino acids and tryptophan in cirrhosis. *Lancet* **II:** 419.

23) MøLLER, S. E. (1985). Tryptophan to competing amino acids ratio in depressive disorder: relation to efficacy of antidepressive treatments. *Acta Psychiatr. Scand.* **72, Suppl. 325:** 1—31.

SERINE METABOLISM, beta-CARBOLINES AND PSYCHOSES

J. Bruinvels, L. Pepplinkhuizen and D. Fekkes

Biological Psychiatry Group
Departments of Pharmacology and Psychiatry
Erasmus University Rotterdam
P. O. Box 1738, 3000 DR Rotterdam
The Netherlands

INTRODUCTION

The striking potency of substances from exogenous sources — like hashish, mescaline or LSD — to evoke mental states similar to certain phenomena observed in psychotics, made many researchers speculate about the origin of psychoses. In this respect most attention was focussed on altered sensory perception of surroundings, distances, time, body, colours etc. and on subjective experiences of feelings of oneness and understanding of the universe. These so-called psychedelic experiences (1) have been noted in a variety of psychotic disorders including schizophrenia.

The first breakthrough of linking psychoses to the endogenous formation of psychotogenic substances was made by Osmond and Smythies in 1952 (2), introducing their transmethylation hypothesis. They postulated that schizophrenia resulted from the formation of mescaline-like substances by a faulty methylation of excessive catecholamines released during stress. In this they followed Harley-Mason's suggestion that 3,4-dimethoxyphenylethylamine (DMPEA) was a likely candidate as a psychotogenic substance, for it was reported to be highly potent in producing catatonia in animals. Thus instead of normal N- and O-methylation, abnormal O-methylation would occur in schizophrenia. As a result of their studies, it was realized that most of the hallucinogens had an indole nucleus. In order not to abandon the original hypothesis, it was assumed that the side chain of mescaline readily fused through its aminogroup forming an indole structure.

The chance observation that slightly detoriated adrenaline solutions gave unfavourable mental reactions resulted in the notion that this had to be due to the formation of adrenochrome which also has an indole structure.

Administration of adrenochrome did evoke typical psychedelic experiences (3, 4). However, the adrenochrome model for schizophrenia, assuming the formation of adrenochrome from adrenaline in schizophrenics, has since been abandoned as adrenochrome could not be detected in body fluids of schizophrenics or healthy controls (5, 6). When an-

NATO ASI Series, Vol. H20
Amino Acid Availability and Brain Function in Health and Disease. Edited by G. Huether

other class of hallucinogens was detected, namely the methylated derivatives of serotonin, N,N-dimethyltryptamine, 5-methoxy-N,N-dimethyltryptamine and 5-methoxytryptamine (7, 8, 9) the original hypothesis was extended to abnormal N-methylation of indoleamines. This suggestion found support by the enzymatic formation of N-methylated monoamines in vitro (10, 11, 12, 13).

Further support for a faulty methylation of catecholamines or indole amines was acquired through the experiments of Polin et al. (14), loading chronic schizophrenic patients with the methyldonor methionine combined with a MAO-inhibitor. Exacerbations of psychotic symptoms or a superimposed toxic psychotic state were noted in many chronic schizophrenic patients (approx. 58 %) by independent groups of investigators (15, 16, 17). In these experiments large doses of methionine (5-40 g/70 kg/day) were administered for periods of one week up till two months. Also when no MAO-inhibitor was used almost half of the patients had nonspecific psychotic symptoms (18, 19) and the re-occurrence of specific hallucinations and delusions reported by some patients (19) seems doubtful. Regrettably enough, research was not restricted to schizophrenia characterized by distorted sensory perceptions which has, according to us, seriously hampered progress up till now in this fascinating area. Too much interest was devoted to schizophrenia in general and it was blandly overlooked that impressive psychedelic phenomena had been observed in psychotic states known and classified in Europe as non-schizophrenic psychosis, e. g. "bouffēe dēlirante des dēgēnerēes", degeneration psychosis, anxiety-happiness psychosis (one of the cycloid psychoses) of Leonhard (see 20). These older concepts are still largely unknown to American psychiatry and perhaps the best way to approach these types of psychoses is to label them as acute "schizo-affective psychosis, manic-psychedelic type", thus extending RDC criteria with psychosensory symptoms and experiences.

Modification of Transmethylation Hypothesis

Searching for methyltransferases responsible for the formation of N-methylmonoamines Laduron (21) described in 1972 an N-methyltransferase converting dopamine into epinine (N-methyldopamine) in which N^5-methyltetrahydrofolic acid (CH_3FH_4) served as a methyl donor. Shortly afterwards it became clear that CH_3FH_4 was not acting as a methyl donor but is probably converted into N^5,N^{10}-methylenetetrahydrofolic acid (CH_2FH_4) which can dissociate into formaldehyde and tetrahydrofolic acid. The formaldehyde may react spontaneously with monoamines forming tetrahydro-isoquinolines and tetrahydro-beta-carbolines (THBC; 22, 23, 24, 25).

Since the patients studied by us were neuroleptic resistent, it was postulated that beta-carbolines are endogenously formed in these patients by an increased conversion of serine into glycine — a situation which may occur in porphyric patients — which concommitantly increases the synthesis of CH_2FH_4 (26). From Figure 1 it can be seen that a one-carbon unit is transferred from serine to FH_4 with the simultaneous formation of glycine and CH_2FH_4 respectively. The dissociation of CH_2FH_4 into FH_4 and formaldehyde may lead to the formation of beta-carbolines as mentioned above. An increased demand for glycine in the body will therefore result in an increased formation of CH_2FH_4 which normally will be converted into CH_3FH_4. This latter compound will transfer its CH_3 group to homocysteine via vitamin B_{12} forming methionine. If CH_2FH_4 accumulates due to some pathological condition, formation of beta-carbolines will occur of which several are known to be potent psychotogenic substances (9).

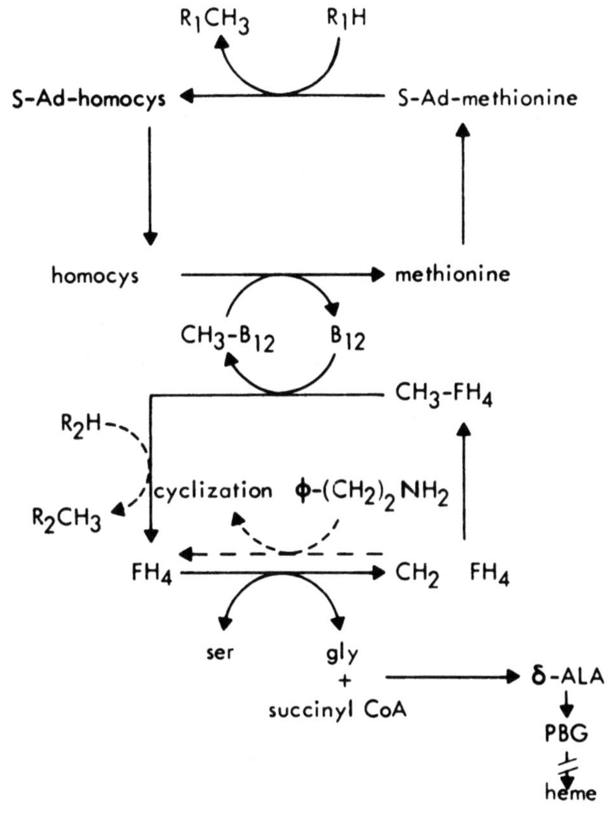

FIGURE 1

One-carbon transfer in relation to serine-glycine conversion and the formation of beta-carbolines. δ-ALA = δ-aminolaevulinic acid; PGB = porphobilinogen; ser = serine; gly = glycine; FH_4 = tetrahydrofolic acid; S-AD = S-adenosyl; Φ-$(CH_2)_2NH_2$ = indoleamine or catecholamine; homocys = homocysteine; CoA = coenzyme A.

CLINICAL STUDIES

When to patients suffering from an episodic "schizo-affective psychosis, manic psychedelic type", after their recovery, a single dose of serine (2 mmol/kg, p. o.) was administered, the characteristic symptoms — especially the sensory perceptual distortions — were evoked 2—4 hours after ingestion of the amino acid. Oral administrations of serine was without effect in another group of patients who also suffered from acute psychotic episodes but without disturbances of sensory perceptions (20, 27, 28). These experiments not only suggest that the group of episodic patients with disturbances of sensory perceptions may indeed represent a separate entity of psychotic illness but also showed that in this group of patients a disturbance in serine metabolism exists. The occurrence of perceptual distortions fits very well into the proposed accumulation of CH_2FH_4 and the subsequent formation of beta-carbolines.

BIOCHEMICAL STUDIES

Plasma Amino Acids

In order to find a disturbance in serine metabolism, blood samples from a group of patients who reacted on serine were analysed for their amino acid composition and compared with blood samples of three other groups of patients, namely a group of patients who reacted on glycine, a small group of manic depressed patients and a group of healthy controls.

The results showed that in serine positive (serpos) patients the fasting plasma level of serine was decreased and that of taurine was increased. These two changes were not found in the plasma of the other groups of patients nor in plasma obtained from healthy controls (29). Calculation from plasma concentrations of serine and glycine after loading with each of the amino acids, indicated an increased conversion of serine into glycine and a decreased formation of serine from glycine indicating that the equilibrium reaction is biased towards the formation of glycine. This shift cannot be caused by an altered activity of the enzyme serine hydroxymethyltransferase (SHMT) since both reactions are changed in opposite direction. The observed inhibition of glycine conversion, however, may be caused by a diminished availability of the co-substrate CH_2FH_4 suggesting that CH_2FH_4 is used for other reactions like the formation of beta-carbolines. The impaired formation of serine probably contributes to the decreased plasma level of serine found in serpos patients.

The increased plasma levels of taurine have probably to be ascribed to an enhanced synthesis of this amino acid. This means that less homocysteine is available for accepting the methylgroup of CH_3FH_4 for the formation of methionine. This diminished transfer of CH_3 groups will thus result in an accumulation of CH_3FH_4 which subsequently will dissociate into THF and formaldehyde via the intermediate CH_2FH_4. It has indeed been shown that the formation of THBC from tryptamine occurs if serine, in the presence of tryptamine and FH_4 is incubated with purified serine hydroxymethyltransferase obtained from ox liver (30). Since this incubation mixture did not contain CH_3FH_4-homocysteine methyltransferase, an accumulation of CH_3FH_4 and CH_2FH_4 occurs and the latter will dissociate into FH_4 and formaldehyde as discussed above. In fact this reaction system may represent a model for the pathological condition where most of the homocysteine is used for the biosynthesis of taurine and is therefore not available as a methyl acceptor in the synthesis of methionine.

For the conversion of homocysteine into cystathionine by the enzyme cystathionine beta-synthase serine is needed as a co-substrate. An increased formation of taurine is therefore a second factor contributing to the decreased plasma level of serine.

Beta-Carbolines

Porphyric rats can be used as a model where an increased consumption of glycine occurs and thus enhancing the conversion of serine into glycine. If serine is administered to these rats, the animals become cataleptic (31). After screening the plasma of these rats for beta-carbolines, a positive correlation was found between the plasma concentration of norharman and the duration of catalepsy (32). These *in vivo* results clearly support the hypothesis that, under certain conditions, an increased conversion of serine into glycine will give rise to the formation of beta-carbolines via the simultaneous conversion of THF into CH_2FH_4 and decomposition of the latter substance into FH_4 and formaldehyde as described above.

For the determination of beta-carbolines a solid phase extraction method was developed excluding artifactual formation of these compounds (33).

To see whether beta-carbolines can also be detected in serpos patients, plasma obtained from acute psychotic patients was analysed. Norharman was found in plasma of these patients in an at least five-fold higher concentration than in plasma of nonsymptomatic patients or healthy controls (34). The beta-carboline norharman has been reported to possess anxiogenic properties (35) and also inhibits the enzyme monoamineoxidase (36). These properties indicate that norharman is a likely candidate for an endogenously formed "psychotic" substance.

Fibroblast Experiments

Because of the fact that serine metabolism is disturbed in these patients, experiments were initiated to find out whether any serine-related enzyme is responsible for this disturbance. Therefore it was decided to measure the activities of serine hydroxymethyltransferase (SHMT; reaction 1 in Fig. 2), cystathionine beta-synthase (CBS; reaction 5), CH_2FH_4 dehydrogenase (reaction 2) 5,10-methenyl-FH_4 cyclohydrolase (reaction 3) and 10-formyl-FH_4 synthetase (reaction 4). A defect in one of the latter three enzymes of the

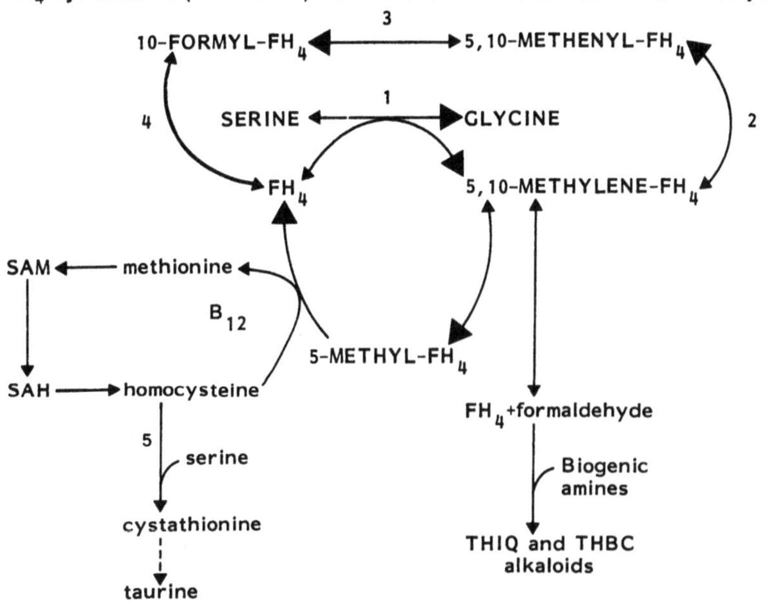

FIGURE 2
Main routes of serine metabolism. FH_4 = Tetrahydrofolic acid; SAM = S-adenosylmethionine; SAH = S-adenosylhomocysteine; THIQ = tetrahydroisoquinoline;THBC = tetrahydro-beta-carboline.

folic acid cycle may lead to an accumulation of CH_2FH_4 and thus favouring the formation of beta-carbolines. For this purpose fibroblasts were used because preliminary evidence suggested that this psychosis may be a familial disorder. An important advantage of using fibroblasts is that measured enzyme activities are not influenced by drug treatment of the patients, while this is not the case when enzyme activities are measured in blood cells. However, none of the measured enzymes in the fibroblasts obtained from serpos patients showed a deviating activity (37). It can therefore be concluded that the psychotic symptoms occurring after the administration of serine do not seem to be due to a malfunctioning of one of the investigated enzymes. These findings seem to be in contrast with those of Carl *et al.* (38), who found that the activity of SHMT in blood cells of schizophrenics is decreased. Another group of investigators (39) also found abnormal SHMT activity in plasma of patients with a diagnosis of major psychosis. However, in both studies not only the group of patients studied was different but also the specimens in which the enzyme activities were measured differed. Moreover, these investigators used blood or blood cells in which the presence of drugs may have influenced SHMT activity as pointed out above. Thus from the experiments with fibroblasts obtained from serpos patients one may conclude that no differences in enzyme activities as compared to those in control fibroblasts exist.

Another possibility is the presence of one or more factors in the plasma of our patients which may influence enzymes involved in the metabolism of serine. Therefore, it was decided to subculture fibroblasts from patients as well as from controls in the presence of 10 % serum obtained from a serpos patient during an acute psychotic episode or in the presence of 10 % serum obtained from a healthy individual. It was found that after 3 days of subculturing in the medium containing 10 % patient serum, the specific activities of SHMT and CBS in the fibroblasts were significantly higher than those subcultured in the medium containing 10 % control serum (40). In addition the concentration of both serine and methionine in the fibroblasts as well as in the medium were significantly decreased upon subculturing with medium containing 10 % patient serum. This indicates that not the transport into the fibroblasts but the metabolism of these amino acids is altered by some factor present in the serum of the patients. Moreover, the concentration of taurine in the fibroblasts was increased while its concentration in the medium was unchanged. These results fit very well with the above reported data on the fasting plasma concentrations of serine and taurine in serpos patients. However, the plasma concentration of methionine was not significantly decreased in serpos patients while it was significantly decreased in both the fibroblasts and the culture medium. This difference can be explained by the fact that human fibroblasts do not contain betaine-homocysteine methyltransferase (41), an enzyme which is also capable of synthesizing methionine.

CONCLUSION

From the presented experiments it can be concluded that there is one (or more) factor(s) present in the serum of serpos patients which is responsible for the reported changes in amino acid pattern by activating SHMT and CBS. These changes may subsequently create an accumulation of CH_2FH_4 resulting ultimately in the formation of the beta-carboline norharman. The latter substance may be responsible for evoking one or more of the psychotic symptoms.

REFERENCES

1) KELM, H. (1973). The Hoffer-Osmond diagnostic test. In: *Orthomolecular Psychiatry* (Hawkins, D., Pauling, L., eds.), Freeman & Co, San Francisco, pp. 327—341.

2) OSMOND, H., SMYTHIES, J. (1952). Schizophrenia: a new approach. *J. Ment. Sci.* **98:** 30.

3) HOFFER, A., OSMOND, H., SMYTHIES, J. (1954). Schizophrenia: a new approach II. Results of a year's research. *J. Ment. Sci.* **100:** 29—45.

4) HOFFER, A., OSMOND, H. (1959). The adrenochrome model and schizophrenia. *J. Nerv. Ment. Dis.* **128:** 18—35.

5) SZARA, S., AXELROD, J., PERLIN, S. (1958). Is adrenochrome present in the blood? *Amer. J. Psychiat.* **115:** 162—163.

6) FELDSTEIN, A. (1959). On the relationship of adrenaline and its oxidation products to schizophrenia. *Amer. J. Psychiat.* **116:** 454—456.

7) SZARA, S. (1956). Dimethyltryptamine: its metabolism in man; The relation of its psychotic effect to the serotonin metabolism. *Experientia* **12:** 441—442.

8) SZARA, S., ROCKLAND, L. H., ROSENTHAL, D., HANDLON, J. H. (1966).Psychological effects and metabolism of N,N-diethyltryptamine in man. *Arch. Gen. Psychiat.* **15:** 320—329.

9) BRIMBLECOMBE, R. W., PINDER, R. M. (1975). Hallucinogenic agents. *Wright Scientechnica,* Bristol.

10) AXELROD, J. (1961). Enzymatic formation of psychotomimetic metabolites from normally occuring compounds. *Science* **134:** 343.

11) AXELROD, J. (1962). The enzymatic N-methylation of serotonin and other amines. *J. Pharmacol. Exp. Ther.* **138:** 28—33.

12) SAAVEDRA, J. M., AXELROD, J. (1972). Psychotomimetic N-methylated tryptamines: formation in brain and in vitro. *Science* **175:** 1365—1366.

13) WYATT, R. J., SAAVEDRA, J. M., AXELROD, J. (1973). A dimethyltryptamine — forming enzyme in human blood. *Amer. J. Psychiat.* **130:** 754—760.

14) POLLIN, W., CARDON, P. V., KETY, S. S. (1961). Effects of amino acid feedings in schizophrenic patients. *Science* **133:** 104—105.

15) COHEN, S. M, NICHOLS, A., WYATT, R., POLLIN, W. (1974). The administration of methionine to schizophrenic patients: a review of ten studies. *Biol. Psychiat.* **8:** 209—225.

16) WYATT, R. J., TERMINI, B. A., DAVIS, J. (1971). Biochemical and sleep studies of schizophrenia. A review of the literature 1960—1970. Part I. Biochemical studies. *Schizophrenia Bull.* **4:** 10—66.

17) NESTOROS, J. N., BAN, T. A., LEHMANN, H. E. (1977). Transmethylation hypothesis of schizophrenia. Methionine and nicotinic acid. *Int. Pharmacopsychiat.* **12:** 215—246.

18) HAYDU, G. G., DHRYMIOTIS, A., KORENYI, C., GOLDSCHMIDT, L. (1965). Effects of methionine and hydroxychloroquine in schizophrenia. *Amer. J. Psychiat.* **122:** 560—564.

19) ANTUN, F. T., BURNETT, G. B., COOPER, A. J., DALY, R. J., SMYTHIES, J. R., ZEALLY, A. K. (1971). The effects of L-methionine (without MAO I) in schizophrenia. *J. Psychiat. Res.* **8:** 63—71.

20) BRUINVELS, J., PEPPLINKHUIZEN, L. (1985). Disturbances in serine-glycine metabolism in relation to acute psychoses with psychedelic symptoms. In: *Pathochemical Markers in Major Psychoses* (Beckmann, H., Riederer, P., eds.), Springer-Verlag, Berlin, pp. 59—73.

21) LADURON, P. (1972). N-methylation of dopamine to epinine in brain tissue using N-methyltetrahydrofolic acid as the methyl donor. *Nature New Biol.* **238:** 212—213.

22) LEYSEN, J., LADURON, P. (1974). Characterization of an enzyme yielding formaldehyde from 5-tetrahydrofolic acid. *FEBS Lett.* **47:** 299—303.

23) MANDEL, L. W., ROSEGAY, A., WALKER, R. W., VANDENHEUVEL, W. J. A. (1974). 5-Methyltetrahydrofolic acid as a mediator of the formation pyridoindoles. *Science* **187:** 741—743.

24) LAUWERS, W., LEYSEN, J., VERHOEVEN, H., LADURON, P. (1975). The condensation products of biogenic amines with formaldehyde, enzymatically formed from 5-methyltetrahydrofolic acid. *Biomed. Mass. Spectr.* **2:** 15—22.

25) MELLER, E., ROSENGARTEN, H., FRIEDHOFF, A J., STEBBIUS, R. D., SILBER, R. (1975). 5-Methyltetrahydrofolic acid is not a methyldonor for biogenic amines: enzymatic formation of formaldehyde. *Science* **187:** 171—173.

26) BRUINVELS, J., PEPPLINKHUIZEN, L., VAN TUIJL, H. R., MOLEMAN, P., BLOM, W. (1980). Role of serine, glycine and the tetrahydrofolic acid cycle in schizoaffective psychosis. A hypothesis relating porphyrin biosynthesis and transmethylation. In: *Enzymes and Neurotransmitters in Mental Disease* (Usdin, E., Sourkes, T. L., Youdim, M. B. H., eds.), John Wiley & Sons Ltd., Chichester, pp. 139—154.

27) PEPPLINKHUIZEN, L., BRUINVELS, J., BLOM, W., MOLEMAN, P. (1980). Schizophrenia-like psychosis caused by a metabolic disorder. *The Lancet* **I:** 454—456.

28) WUNDERINK, A., PEPPLINKHUIZEN, L., BRUINVELS, J. (1986). Nutrition and psychosis. In: *Progr. Brain Res.* **vol. 65** (Van Ree, J. M., Matthysse, S., eds.), Elsevier Science Publisher's, pp. 49—57.

29) BRUINVELS, J., PEPPLINKHUIZEN, L. (1984). Impaired glycine-serine conversion and increased plasma taurine levels in episodic psychotic patients with psychedelic symptoms. *J. Psychiat. Res.* **18:** 307—318.

30) PEARSON, A. G., TURNER, A. J. (1979). The formation of beta-carboline alkaloids mediated by serine hydroxymethyltransferase. *FEBS Lett.* **98:** 96—98.

31) SCHOUTEN, M. J., BRUINVELS, J., PEPPLINKHUIZEN, L., WILSON, J. H. P. (1983). Serine and glycine-induced catalepsy in porphyric rats. *Pharmacol. Biochem. Behav.* **19:** 245—250.

32) SCHOUTEN, M. J., BRUINVELS, J. (1986). Endogenously formed norharman (beta-carboline) in platelet rich plasma obtained from porphyric rats. *Pharmacol. Biochem. Behav.* **24:** 1219—1223.

33) SCHOUTEN, M. J., BRUINVELS, J. (1985). High-performance liquid chromatography of tetrahydro-beta-carbolines extracted from plasma and platelets. *Anal. Biochem.* **147:** 401—409.

34) SCHOUTEN, M. J., (1986). The occurrence of beta-carbolines in man and rat, putative biochemical substrates responsible for psychosis. *Thesis Erasmus University Rotterdam,* The Netherlands.

35) SKOLNICK, P., CRAWLEY, J. N., GLOWA, J. R., PAUL, S. M. (1984). Beta-carboline-induced anxiety states. *Psychopathol.* **17:** 52—60.

36) BUCKHOLZ, N. S., BOGGAN, W. O. (1977). Monoamineoxidase inhibition in brain and liver produced by beta-carbolines: structure-activity relationships and substrate specificity. *Biochem. Pharmacol.* **26:** 1991—1996.

37) FEKKES, D., BRUINVELS, J. (1986). Serine and folate metabolism in fibroblasts from episodic psychotic patients with psychedelic symptoms. *Biol. Psychiat.* **21:** 951—959.

38) CARL, C. F., CREWS, E. L., CARMICHAEL, S. M., BENESH, F. C., SMYTHIES, J. R. (1978). Four enzymes of one-carbon metabolism in blood cells of schizophrenics. *Biol. Psychiat.* **13:** 773—776.

39) WAZIRI, R., WILCOX, J., SHERMAN, A. D., MOTT, J. (1984). Serine metabolism and psychosis. *Psychiat. Res.* **12:** 121—136.

40) BRUINVELS, J., FEKKES, D., PEPPLINKHUIZEN, L. (1986). Disturbances of serine metabolism and endogenous formation of beta-carbolines in a group of episodic psychotic patients. *Brit. J. Pharmacol.* **89:** 476P.

41) MUDD, S. H., UHLENDORF, B. W., HINDS, K. R., LEVY, H. L. (1970). Deranged B_{12} metabolism: studies of fibroblasts grown in tissue culture. *Biochem. Med.* **4:** 215—239.

BRAIN MONOAMINE OXIDASE B ACTIVITY AND AMINO ACID PRECURSOR TREATMENT OF PARKINSON'S DISEASE, DEPRESSION AND EPILEPSY

M. B. H. Youdim and P. Riederer*

Rappaport Familiy Research Institute
and Department of Pharmacology
TECHNION, Haifa, Israel
and
* Clin. Neurochem., Department of Psychiatry
Univ. Würzburg, Würzburg, FRG

ABSTRACT

The importance of monoamine oxidase (MAO) A and B in maintaining the cytoplasmic levels of monoamines (dopamine, noradrenaline, serotonin and phenylethylamine) low within the aminergic neurones and their role in cytoplasmic release mechanism are well established. Animal behavioural and biochemical studies have shown that MAO in the brain is in excess and inhibition of enzyme by more than 80 % is required for amino release into "function". The alteration in brain concentrations of dopamine and serotonin after L-dopa (L-dihydroxyphenylalamine) and L-tryptophan treatments respectively, are not as obvious as those seen in animals pretreated with selective MAO inhibitors. Therefore, enzyme inhibition and amino acid precursor loading would appear to be obvious and logical choice for treatment of Parkinson's disease and depressive illness, where alterations in dopamine and serotonin metabolism have been implicated in their pathophysiology. The therapeutic success of l-deprenyl (MAO-B inhibitor) in combination with L-dopa or L-5-hydroxytryptophan for the management of Parkinson's disease and depression has manifested itself in the predominance of MAO-B in the extrapyramidal regions and raphe nucleus of human brain. The apparent potentiation of the pharmacological activities of the neurotransmitters, derived from amino acids precursors, by MAO-B inhibitor is supported by the clinical data, since neither treatment alone appears to be as effective. MAO-B is an enzyme capable of oxidizing a variety of primary, secondary and tertiary aliphatic and aromatic monoamines. This property of MAO-B lends itself to the oxidation of inert amines, including the Parkinson inducing compound, MPTP (N-methyl-4-phenyl-1, 2,3,6-tetrahydropridine), and the anticonvulsant, milacemide (2-n-pentylaminoacetamide) to neuroactive metabolites. Precursor amino acid treatment of neurological disorders, where inhibitory neurotransmitters GABA and glycine are implicated have not been successful, since, these amino acids hardly cross the blood brain barrier (BBB). However, the potent glycine-derivative anticonvulsant, milacemide, readily crosses the BBB where it is actively converted to glycineamide, followed by glycine. *In vitro* and *in vivo* studies have

NATO ASI Series, Vol. H20
Amino Acid Availability and Brain Function in Health and Disease. Edited by G. Huether

demonstrated that the major (80—90 %) route of its metabolism is solely mediated by MAO-B and is the only one so far identified in the brain. The oxidation of milacemide is selectively inhibited by MAO-B rather than MAO-A inhibitors. Milacemide, therefore, behaves as a prodrug and opens a new avenue for the development of similar drugs derived from other amino acids. Thus, just as in the case with "dopamine replacement" therapy of Parkinson's disease by L-dopa, similar therapeutic approaches can be envisaged for the treatment of neurological disorders involving alterations in inhibitory amino acid neurotransmitters.

INTRODUCTION

Monoamine oxidase (MAO) occupies an important historic as well as academic place in the brain biochemistry and function (1—4). It was among the first of the monoamine neurotransmitter enzymes to be discovered and for which a clear function described, namely oxidative inactivation of aliphatic pentylamine and aromatic (noradrenaline, tyramine) amines. The identification of the antitubercolosis drug, iproniazid, as an irreversible inhibitor of this enzyme together with the discovery of antipsychotic activity of chlorpramazine in the early 1950s were the hallmark of the new directions in psychopharmacology, psychotherapy and an insight into brain function and neurotransmitter metabolism. Extensive studies have demonstrated the ability of MAO to (A) detoxify (inactivate or oxidize) a whole variety of monoamines among which are the well known neurotransmitters (adrenaline, nor-adrenaline, dopamine (DA) and serotonin (5-HT)); false neurotransmitters (octopamine) and indirectly acting (tyramine and phenylethylamines) amines (4, 2), (B) convert the inert tertiary amine, MPTP, (N-methyl-4-phenyl-1,2,3,6-tetrahydropyridine), to the selective Parkinson inducing dopamine neurotoxin MPP+ (N-methyl 6-phenyl 1,2 dihydopsridinium ion) (5), and (C) oxidize the secondary amine, such as the anticonvulsant milacemide (2-n-pentylaminoacetamide) to glycine a potent major inhibitory amino acid neurotransmitter (6—8). The above three, conventionally different, actions of MAO would indicate that modulation of activity of this enzyme and amino acid precursor loading can bring about an improvement or be deleterious in brain function. Therefore, assigning a single function to such a complex enzyme as MAO would be enormous. In this paper we shall discuss the role of MAO in relation to the disorders of Parkinson's disease and depressive illness where L-dopa (l-dihydroxyphenylalanine) and serotonin precursor loading respectively have been employed in the treatment of these disorders. Furthermore, and unconventional aspect of brain MAO activity will be discussed, where the enzyme can oxidize a prodrug, readily accessible through blood brain barrier (BBB) to an important inhibitory neurotransmitter which itself does not do so.

Parkinson's Disease L-Dopa, Loading and Brain MAO-B Inhibition

The recognition in early 1970 (9, 10) that human brain MAO existed as multiple forms, and MAO-B predominates (~ 80 %) in extrapyramidal (basal ganglia) regions and that dopamine was oxidized by the latter enzyme (11) in human brain (9) lends support to the notion that a selective MAO-B inhibitor would be useful as adjunct to the L-dopa therapy of Parkinson's disease (12). Such considerations were based on the inability of MAO-A inhibitor, clorgyline, to inhibit human brain MAO using dopamine as a substrate and on animal experimental data showing that L-dopa, even at high doses (> mg/kg), alone in rats does not profoundly affect brain dopamine levels or animal behaviour (13). However, L-dopa in combination with an MAO inhibitor (pargyline or tranylcypramide) significantly increased dopamine in the striatum and induced an overt dopamine behavioural syndrome of hyperactivity and stereotypity, which could be blocked by dopamine antagonists, haloperidol and chlorpramazine (13). Furthermore, 10 years of therapy with high doses of L-dopa Parkinsonian patients had shown increased loss of therapeutic response and side effects. Therefore, an MAO-B inhibitor, which would selectively inhibit MAO-B rapidly and be devoid of any side effects (e. g. cheese effect) of classical MAO inhibitors, might be useful in reducing the high conventional doses of L-dopa therapy and potentiate the pharmacological activity of dopamine formed from it. l-Deprenyl, a selective MAO-B inhibitor devoid of cheese effect (14) was chosen (12, 15, 16). The success and acceptance of l-deprenyl-L-dopa combination as a therapeutic approach to Parkinson disease is affirmed by the confirmatory reports from other groups (17, 18, 19). The rapid clinical response and diminished incidents of L-dopa side effects, especially the on-off phenomenon, is one feature of this drug combination. However, the mechanism action of l-deprenyl remains in doubt even though the addition of l-deprenyl to conventional L-dopa therapy can result in daily reduction of L-dopa dosage by as much as 50 % (20). These data would suggest that the L-dopa effects are tentiated in a manner similar to the observations in animals given the combination of MAO inhibitor and L-dopa (13), namely where brain levels of DA are substantially increased. The only study so far on brains from Parkinsonian patients treated with l-deprenyl-L-dopa combination do show that DA and not 5-HT is increased significantly more in caudate nucleus, s. nigra, globus pallidus and putamen than in the same regions from patients on conventional L-dopa therapy only (21). The MAO-B activity of the above brain regions after l-deprenyl therapy is almost completely inhibited while substantial MAO-A activity remains (21). This is considered as one reason why 5-HT concentrations were unaltered. Thus, it is apparent that the degree of MAO-A or MAO-B inhibition is most crucial for determination of 5-HT on DA accumulation and release. Indeed, tranylcypramine treated rats given L-dopa or L-tryptophan show behavioural hyper-activity when MAO activity is inhibited more than 85 % and where brain DA or 5-HT is respectively increased (22, 13). In animals where MAO activity is fully inhibited by tranylcypramine the behavioural responses to L-dopa or L-tryptophan can be observed until 18—24 hours after the tranylcypramine injection. Thereafter, the responses disappear. This phenomenon can readily be explained by the recovery of MAO activity by more than 15 % within the 24 hour period (13, 23). The essentiality of MAO inhibition followed by precursor loading for the management of PD can also be found in cases where treatment with l-deprenyl only failed to show clinical response (15, 20).

From the studies of Riederer and Youdim (21) it is apparent that l-deprenyl does inhibit MAO-B with a resultant increase of DA. However, *in vitro* studies of O'Carroll *et al.* (24) have shown that in human brain caudate nuclear DA is oxidized by MAO-A and MAO-B so that selective inhibition of MAO-B would result in a smaller increase of brain DA. Indeed, this seems to be the case since the increase of DA in brains from l-deprenyl treated subjects is not substantial (21) whereas phenylethylamine, a selective MAO-B substrate, is increased in the extrapyramidal regions by more than 3000 % (25). Therefore, the MAO-A activity remaining after l-deprenyl may have access to DA. The effect of l-deprenyl on DA metabolism and function *in vivo* may be a resultant of compartmentation of MAO-A and B between extra and intraneuronal sites (24). It is possible that using brain homogenates only the MAO in extraneuronal sites are being expressed since uptake would not be operative. Thus, the enrichment of MAO-B in DA neurons might very well lead to its dominant role in the metabolism of the neurotransmitter in the human brain and this may be one reason why the pharmacological activity of L-dopa is potentiated by l-deprenyl in Parkinson subjects. Whether MAO-B inhibition is the sole mechanism by which l-deprenyl initiates its anti-Parkinson activity (19) needs to be investigated using either another MAO-B inhibitor or other selective MAO-A inhibitors in combination with L-dopa. The only results available with clinical doses of clorgyline have shown that although this inhibitor can initiate a significant increase of 5-HT in the human brain regions, DA concentrations were unaltered (26).

Treatment of Seizure and Spasticity Disorders with Glycine-Prodrug; the Role of Brain MAO-B Activity

It is well recognized that GABA (γ-amino butyric acid) and glycine are the major inhibitory neurotransmitters in the central nervous system, and together they form the largest densities of neurons within the central nervous system. The conventional approach to the treatment of seizure related disorders has been the use of drugs which act on central GABA neurons. Although these drugs have advanced the understanding of GABA neurons, in not all cases have these approaches been therapeutically successful, especially in the case of myoclonus (27). One feature of GABA and glycine which has prevented their greater understanding and interaction has been the inability of the amino acids to cross the BBB. However, both amino acids exert anticonvulsant activity when injected directly into the brain. Difficulties still exist for glycine since there are no available agonists or metabolic inhibitors of this amino acid. However, the ability to modulate brain glycine may now be possible using the novel anticonvulsant milacemide (28).

Milacemide (2-n-pentylaminoacetamide) is a pentylamine derivative of glycine which unlike the latter amino acid readily crosses the BBB where it accumulates and is primarily metabolized to glycineamide (G) and glycine (28). Milacemide has anticonvulsant activity equivalent to valproate in many animal models. It is particularly effective in protecting against bicuculline, pentylenetetrazol and picrotoxin induced convulsion as well as cobalt-induced epileptogenic foci and supramaximol electro shock-induced tonic extension (28). Although the mechanism of its action is unresolved the formation and accumulation of glycine in the brain is thought to be the basis of its action. Thus, oral or i. p. administration of milacemide results in an increase of glycine concentration in forebrain, cerebellum and medulla-pons, raphe and striatum but not in other tissues such as heart, liver and kidney.

N-dealkylation to G is the major (90 %) pathway of milacemide metabolism in the brain. G is transformed, probably by hydrolytic reaction, to glycine and where the latter accumulates and is incorporated into the protein. It was apparent, therefore, that milacemide acts as a prodrug and its transformation as a secondary amine to G involved an oxidative reaction of MAO. This metabolic route of metabolism is the only one so far identified in the brain. The possibility of MAO involvement in the metabolism of milacemide to G has recently been investigated *in vitro and in vivo* (6—8). Such a phenomenon was not considered far fetched since not only primary (DA, 5-HT, and noradrenaline) but tertiary (MPTP, and 5-methoxy N-N'dimethyltryptamine (5MeODT)), as well as secondary N-methyltyramine, adrenaline are substrate of MAO.

Using mitochondria prepared from various animal and human tissue sources (rat liver and brain; human brain, placenta and platelets and isolated bovine chromaffin cells know to contain MAO-A-B, MAO-A and MAO-B) (4), the metabolism of milacemide was investigated. These studies indicated that MAO-B rather than MAO-A could be involved in the oxidization of milacemide to glycine. The apparent K_m values of MAO-A and MAO-B for milacemide oxidation were calculated to be in the range of $\approx$ 1200 μM and $\approx$ 50 μM respectively. The latter K_m values with MAO-B are very close to the K_m values of MAO-B for non-hydroxylated monoamines benzylamine, phenyl-ethylamine, pentylamine and MPTP (7, 8). The role of MAO-B in the oxidation of milacemide is further strengthened by the *in vitro, ex vivo* and *in vivo* effects of selective MAO inhibition. The latter studies have clearly demonstrated a role for MAO-B since the selective inhibitors of this enzyme (l-deprenyl and AGN 1135) (29) rather than inhibitor of MAO-A selectively blocks the oxidation of milacemide *in vitro.* Furthermore, while injection of selective *in vivo* doses of l-deprenyl resulted in increased urinary excretion of unchanged ^{14}C-milacemide and lower values of ^{14}C-glycineamide, as compared to control, clorgyline (MAO-A inhibitor) was without effect until doses at which it also inhibited MAO-B (7, 8). Therefore, milacemide acts a prodrug and its conversion *in vivo* in the brain is handled by MAO-B. This unconventional ability of brain MAO-B to metabolize an unusual substrate is not uncommon, since MAO-B also oxidizes methylhistamine (30) and n-acetylputrescine (K_m 39 μm (Youdim, unpublished data)). However, both histamine and putrescine are substrates for diamine oxidase, an enzyme absent from human and rat brain. The question, therefore, remains as to what is the function of MAO-B in the brain for which no clear cut endogenous substrate, except phenylethylamine, has been described. The possibility arises that MAO-B might be involved in the processing of methyl-histamine to an as yet unknown compound or oxidation of n-acetylputrescine to GABA as suggested by Seiler and Al-Therib (31) for a second pool of this neurotransmitter (32).

Glycine as the Anticonvulsant Component of Milacemide

A number of well controlled studies have demonstrated the anticonvulsant activity of milacemide (33, 28). If glycine formation, via the action of MAO-B, is responsible for its action the use of selective MAO inhibitors should highlight this. Hyperbaric oxygen toxicity (HBO) of the central nervous system develops at partial pressure of oxygen greater than 2.0 atmosphere and manifests itself in generalized seizures. Various antipletic agents, except magnesium sulphate, have in the past failed to induce latency in HBO induced seizures. We have examined the effects of milacemide (400—600 mg/kg i. p.) on HBO induced (4—5 atms) seizures in rats. The studies (34) so far show that milacemide significantly increases the latency of seizures to HBO by 400—500 %, and the latency can be reversed by selective doses of l-deprenyl (MAO-B inhibitor) but not by MAO-A inhibitor, clorgyline (34). Therefore, it can be concluded that the formation of glycine could in part be involved in the preventation of seizures since milacemide has the ability of also increasing brain GABA (28).

Interaction of Glycine with 5-HT Neurons

The brain stem and upper spinal cord contain the highest densities of glycine and 5-HT neurons (27). The close anatomical distribution of the two neurotransmitters may also indicate a physiological interaction between them. Animal behavioural and clinical human studies show involvement of 5-HT in oxygen deprivation myoclonus seizures. Thus, drugs that increase 5-HT functional activity either increasing its synthesis (L-tryptophan plus MAO inhibitor) or act as 5-HT agonists are thought to be antimyoclonic (35, 36). However, as of now no adequate or successful therapeutic agents have been described for this disorder. The sensitivity of 5-HT synthesis to the lack or excess of oxygen is a well known aspect of this neurotransmitter (37). We chose to study the interaction of milacemide with 5-HT in two animal behavioural models. Milacemide in increasing doses (10—300 mg/kg) significantly potentiated, rather than inhibited, the behavioural responses to (a) tranylcypromine plus L-tryptophan and (b) 5-MeODMT (Youdim, unpublished data). At present, little is known of the mechanisms involved in this reaction but it is more than possible that glycinergic neurones may be indirectly disinhibitory on post-synaptic 5-HT neurons. The potentiation of 5-HT function is not via the 5-HT_2 receptor sites, but would indicate importance in myoclonus. Thus, our preliminary have clearly demonstrated that milacemide (100 mg/kg) inhibits myoclonus induced by p,p-DDT (1,1,1-trichloro-2,2-bis (p-chlorophenyl) ethane) in rats (Youdim, unpublished results; 38). These data together with those described earlier for the increased latency of seizure by milacemide in HBO would indicate a new pharmacological and possibly clinical tool to examine seizure due to oxygen toxicity and deprivation.

Glycine Therapy and Spinal Spasticity

Crowdon (see this volume) has outlined the role of the inhibitory neurotransmitter glycine, in the upper spinal cord and brain stem. They have suggested and presented evidence that glycine derived from threonine therapy can increase inhibitory tone in the spinal cord and be useful for treatment of spinal spasticity. The preliminary open study using 4—5 g/day of threonine in six patients with familial spastic paraparesis are encouraging enough to support a role for glycine in this disease. However, the extremely large doses of threonine used could have adverse side effects due to its metabolism to other metabolites. If glycine is indeed involved in the therapeutic response, the prodrug milacemide would be an excellent choice for delivery of glycine as outlined in the previous sections of this paper. Furthermore its interaction with the newly discovered NMDA (N-methyl-D-aspartate) receptor (39) needs investigation.

SUMMARY

From the studies reported here it is apparent that central MAO has an important role as a functional regulator of dopamine formed from L-dopa in clinical therapy of Parkinson's disease. Its inhibition can result in a better utilization of dopamine, and lead to a lower therapeutic dosage of L-dopa. Whether the dopamine formed is mainly inactivated by MAO-A or B has not been established since selective MAO-A inhibitors without the cheese effect are not available clinically (Youdim, unpublished results). What is however clear is that MAO-B inhibitors have found a true place in the management of Parkinson's disease when combined with L-dopa. Whether MAO-B activity is involved in the pathogeneses of the disease and its inhibitors would be useful as agents for retarding the progressive dopamine neuron degeneration (40, 41) remains open for further investigation. What is known is that MAO-B inhibitors do present the Parkinsonian syndrome induced by the neurotoxin MPTP in primates and some rodents (5).

On the other hand brain MAO-B activity can be utilized to deliver the amino acid glycine, normally not able to cross the BBB, into the brain via its coupling with an MAO-B substrate. Milacemide is the first example of such a prodrug. In the future the use of this drug and development of similar drugs from taurine and β-alanine will, no doubt, greatly enhance our understanding of the functional purpose of such amino acids in the brain, for which no clear agonists or antagonists have been described. Therefore, it can be envisaged that certain other neurological disorders with abnormal amino acid pathology may be corrected or treated in a similar fashion to Parkinson's disease via the use of amino acid neurotransmitter replacement therapy.

REFERENCES

1) FINBERG, J. P. M., YOUDIM, M. B. H. (1984). Monoamine oxidase. In: *Handbook of Neurochemistry* (Lajtha, A., ed.), Plenum Press, New York, **4:** 293—330.

2) YOUDIM, M. B. H., FINBERG, J. P. M. (1985). Monoamine oxidase inhibitor antidepressants. In: *Psychopharmacology* **2/1** (Grahame-Smith, D. G., Hippius, H., Winokur, G., eds.) Excerpta Medica Amsterdam, pp. 35—71.

3) TIPTON, K. F. (1986). Enzymology of monoamine oxidase. *Cell Biochem. Function* **4:** 79—88.

4) YOUDIM, M. B. H., FINBERG, J. P. M., TIPTON, K. F. (1987b). Monoamine oxidase. In: *Catecholamine II. Handbook of Experimental Pharmacology* (Weiner, N., Tredelenburg, U., eds.). Springer-Verlag, Berlin (in press).

5) MARKEY, S. P., CASTAGNOLI, N. JR., TREVOR, A. J., KOPIN, I. J. (Eds.) (1986). MPTP: *A Neurotoxin Producing a Parkinson Syndrome.* Academic Press, New York.

6) YOUDIM, M. B. H. DE VAREBEKE, P. J., ROBA, J., CAVALIER, R., GILBERT, P. (1987a). Formation of glycine from the novel anticonvulsant milacemide is mediated by brain monoamine oxidase B. In: *Epilepsy* (Manleis, J., ed.), Raven Press, New York (in press).

7) DE VAREBEKE, P. J., CAVALIER, R., DAVID-REMAELE, M., YOUDIM, M. B. H. (1987a). Formation of the neurotransmitter glycine from the anticonvulsant milacemide is mediated by brain monoamine oxidase B. *J. Neurochem.* (in press).

8) DE VAREBEKE, P. J., CAVALIER, R., DAVID-REMAELE, M., YOUDIM, M. B. H. (1987b). Conversion of the novel anticonvulsant milacemide to glycine in brain is mediated by monoamine oxidase B. *9th Catecholamine Conference* (Sandler, M., Dahlstrom, A., Belmaker, H., eds.), Alan R. Liss, New York (in press).

9) COLLINS, G. G. S., SANDLER, M., WILLIAMS, E. D., YOUDIM, M. B. H. (1970). Multiple forms of human brain mitochondrial monoamine oxidase. *Nature* **225:** 817—820.

10) YOUDIM, M. B. H., COLLINS, G. G. S., SANDLER, M. (1972). Human brain monoamine oxidases: multiple forms and selective inhibitors. *Nature* **236:** 225—228.

11) JOHNSTON, J. P. (1968). Some observation upon a new inhibitor of monoamine oxidase in brain tissue. *Biochem. Pharmacol.* **17:** 1285—1297.

12) BIRKMAYER, W., RIEDERER, P., YOUDIM M B. H., LINAUER, W. (1975). The potentiation of the antiakinetic effect of L-dopa treatment by an inhibitor of MAO-B deprenyl. *J. Neural Transm.* **36:** 303—326.

13) GREEN, R. A., MITCHELL, D. B., TORDOFF, F. C., YOUDIM, M. B. H. (1977). Evidence of dopamine deamination by both type A and type B monoamine oxidase in rat brain *in vivo* and for the degree of enzyme inhibition necessary for functional activity of dopamine and 5-hydroxytrypramine. *Br. J. Pharmacol.* **60:** 343—349.

14) KNOLL, J., MAGYER, K. (1972). Some puzzling effects of monoamine oxidase inhibitors. *Adv. Biochem. Psychopharmacol.* **5:** 393—408.

15) BIRKMAYER, W., RIEDERER, P., AMBRAZI, L., YOUDIM, M. B. H. (1977). Implications of combined treatment with madopar and l-deprenil in Parkinson's disease. *Lancet* **I:** 439—444.

16) LEES, A. J., SHAW, K. M., KOHOUT, L., STERN, G. M., ELSWORTH, J. D., SANDLER, M., YOUDIM, M. B. H. (1977). Deprenyl in Parkinson's disease. *Lancet* **II:** 791—795.

17) BIRKMAYER, W., RIEDERER, P., YOUDIM, M. B. H. (1982). (-) Deprenyl in the treatment of Parkinson's disease. *Clin. Neuropharmacol.* **5.** 195—230.

18) RINNE, U. K. (ed.) (1983). A new approach to the treatment of Parkinson's Disease. *Acta Neurol. Scand. Suppl.* **95 68:** 7—144.

19) YOUDIM, M. B. H. (1986). Pharmacology of MAO-B inhibitors: mode of action of (-) deprenyl in Parkinson's disease. *J. Neural Transm. Suppl.* **22:** 91—106.

20) YAHR, M. W. (1978). Overview of present day treatment of Parkinson's disease. *J. Neural Transm.* **43:** 227—238.

21) RIEDERER, P., YOUDIM, M. B. H. (1986). Brain monoamine oxidase activity and monoamine metabolism in Parkinson patients treated with (-) deprenyl. *J. Neurochem.* **46:** 1359—1365.

22) GREEN, A. R., YOUDIM, M. B. H. (1975). Effect of Monoamine oxidase inhibition by clorgyline, deprenyl or tranylcypomine on 5-HT concentration in brain and hyperactivity following subsequent tryptophan administration. *Br. J. Pharmacol.* **55:** 415—422.

23) YOUDIM, M. B. H., ASHKENAZI, R. (1982). Regulation of 5-HT catabolism. In: *Serotonin in Biological Psychiatry* (Schooler, J. C., Usdin, E., eds. B. T. H. O.), Raven Press, New York, pp. 35—60.

24) O'CARROL, A. M., FOWLER, C., PHILLIPS, J. P., TOBIA, I., TIPTON, K. F. (1983). The deamination of dopamine by human brain monoamine oxidase. *Arch. Pharmacol.* **322:** 198—222.

25) RIEDERER, P., KONRADI, C., SCHAY, V., KIENTL, E., BIRKMAYER, W., DANIELCZYK, W., YOUDIM, M. B. H. (1986). Localization of MAO-A and MAO-B in human brain. A step in understanding the therapeutic action of l-deprenyl. *Adv. Neurol.* **43:** 111—118.

26) BEVAN JONES, A. B., PARE, C. M. B., NICHOLSON, W. J., PRICE, K., STACEY, R. S. (1972). Brain amine concentrations after MAO inhibitor administration. *Br. Med. J.* **1:** 17—19.

27) MELDRUM, B. S. (1986). Drugs acting on amino acid neurotransmitters. *Adv. Neurol.* **43:** 687—706.

28) ROBA, J., CAVALIER, R., CORDI, A., GORISEN, H., HERIN, M., DE VARBEKE, P. J. (1986). Milacemide. In: *New Anticonvulsant Drugs* (Meldrum, B. S., Porter, H. H., eds.), J. Libbey and Co., London, pp. 179—180.

29) KALIR, A., SABBACH, A., YOUDIM, M. B. H. (1981). Selective acetylenic "suicide" and reversible inhibitors of monoamine oxidase type A and type B. *Br. J. Pharmacol.* **73:** 55—64.

30) SUZUKI, O., YOSHIANAO, K., OYA, M. (1979). Characterization of some biogenic monoamines as substrates of type A and type B monoamine oxidase. In: *Monoamine Oxidase Structure, Function and Altered Functions* (Singer, T. P., Von Korff, R. W., Murphy, D. L., eds.), Academic Press, New York, pp. 197—204.

31) SEILER, N., AL-THERIB, M. J. (1974). Puterescine catabolism in mammalian brain. *Biochem. J.* **144:** 29—35.

32) NOTO, T., HASHIMOTO, H., NAKAO, J., KAMINURA, H., NAKAJIMA, T. (1986). Spontaneous release of GABA formed from putrescine and its enhanced CA^{2+}-dependent release by high K^{+} stimulation in the brains of freely moving rats. *j. Neurochem.* **46:** 1877—1880.

33) HOUTKOOPR, M. A., OORSHOL, C. A. E. H., RENTNEESTER, T. W., HOEPPENER, P. S. E. A., ONKELINX, C. (1986). Double blind study of milacemide in hospitalized therapy-resistant patients with epilepsy. *Epilepsia* **27:** 255—262.

34) YOUDIM, M. B. H., KEREM, D., DUVDOVAN, Y. (1987c). L-Deprenyl reverses the increase of seizure threshold induced by milacemide in hyperbar ic oxygen toxicity. *Eur. J. Pharmacol.* (Submitted).

35) FAHN, S., MARSDEN, C. D., VAN WOERT, M. (eds.) (1986). Myoclonus.*Adv. Neurol.* **43:** Raven Press, New York.

36) PRATT, J. A., ROTHWELL, J., JENNER, P., MARSDEN, C. D. (1986). P,P'-DDT-induced myoclonus in the rat and its application as an animal model of 5-HT-sensitive action myoclonus. *Adv. Neurol.* **43:** 577—588.

37) CARLSSON, A. (1974). The *in vivo* estimation of rated of tryptophan and tyrosine hydroxylation: effect of alterations in enzyme environment and neuronal activity. In: *Aromatic Amino Acids in the Brain.* Ciba Foundation Symposium **22** (New Series), Elsevier, Amsterdam, pp. 117—134.

38) CHUNG HWANG, E., VAN WOERT, M. (1978). P,P'-DDT-induced neurotoxic syndrome experimental myoclonus. *Neurology* **28:** 1020—1025.

39) JOHNSON, J. W., ASCHER, P. (1981). Glycine potentiates the NMDA response in cultured mouse brain neurons. *Nature* **325:** 522—525.

40) BIRKMAYER, W., KNOLL, J., RIEDERER, P., YOUDIM, M. B. H. (1983). (-) Deprenyl leads to prolongation of L-dopa efficacy in Parkinson's disease. *Mol. Probl. Pharmaco-Psychiatry* **19:** 170—177.

41) BIRKMAYER, W., KNOLL, J., RIEDERER, P., YOUDIM, M. B. H., HARS, V., MARTON, H. (1985). Increased life expectancy resulting from addition of l-deprenyl to madopar treatment of Parkinson's disease. A long term study. *J. Neural. Trans.* **64:** 113—117.

SUMMARY AND DISCUSSION-REPORT OF CHAPTER VI

Brian S. Meldrum

SUMMARY

The six rather disparate presentations of this session all concerned neurological or psychiatric disorders. We may ask how far these presentations contribute to the theme of this workshop, namely how manipulations of central neurotransmitters (by altering precursor availability or transport) increase our understanding of behavior, and of the causes and therapy of diseases.

Firstly in the contribution that I gave on glutamate and aspartate as excitatory amino acid transmitters, possible manipulations by precursor availability were not discussed because in the brain the dietary precursor for glutamate and aspartate is glucose or carbohydrate. You *can* manipulate brain levels of glutamate and aspartate by this means. If you produce hypoglycaemia you will decrease the brain glutamate content and increase aspartate content. This is probably due to changing the balance in their entry into and exit from the TCA-cycle. It seems to be a mechanism for conserving carbon units. The increase in aspartate probably does contribute to the symptoms of hypoglycaemia. It does not provide a therapeutic approach. And the other approach would be to increase brain glucose by increasing plasma glucose supply. This changes the pathological outcome of ischemia and also some effects of epilepsy. A high blood glucose makes either focal brain ischaemic damage or generalized brain ischaemic damage worse. This is not thought to be dependent on changes in brain aspartate and glutamate. It is instead dependent on brain lactate levels and changes in intracerebral pH-values. So neither of these possibilities of manipulating cerebral excitatory amino acid transmitter levels is a useful approach for therapy.

The presentation by Giorgio Zanchin on the effects of portocaval anastomoses in rats provided clear evidence for changes in the brain uptake index for plasma amino acids. He related this effect to changes in the properties of the endothelial cell membranes, although it might be argued that in the chronic preparations he uses there could be some involvement of astrocytes as well. He did not provide any specific evidence that the changes in the brain uptake indices for aromatic amino acids produce changes in neurotransmitters that could be responsible for any of the features of hepatic encephalopathy. Perhaps this is a very difficult topic to approach, but evidently it can and should be done.

The contribution of John Growdon firstly reminded us of the two classical examples which provide the basis for his approach, firstly that you can treat Parkinsonism with L-dopa and this is a very major therapy in neurology. Secondly, there is a rather minor

NATO ASI Series, Vol. H20
Amino Acid Availability and Brain Function in Health and Disease. Edited by G. Huether

syndrome, post-hypoxic intention myoclonus, a proportion of which can be treated by L-tryptophan or 5-HTP. But this phenomenon does not apply to other forms of epilepsy. Thus only a very restricted type of myoclonus responds to serotonin precursors. John Growdon proposed that a third neurologic syndrome might be treated with an amino acid precursor. He described a syndrome in which glycine is deficient and enhanced glycine would help. In familial spastic paraparesis he gave L-threonine and produced a therapeutic effect that correlated with the increase in CSF-glycine.

This presentation linked with the final presentation of Moussa Youdim who showed that treatment with milacemide produced a large increase in brain glycine content in animals. He was proposing milacemide as a treatment for some forms of myoclonus and other disorders associated with seizures. Of course one would ask instantly whether these two therapeutic (L-threonine and milacemide) approaches would operate in the same cases. Their apparent mechanism is the same, namely to raise glycine levels in the brain. In both cases the changes in glycine levels in the brain should have a physiological effect. Further studies to document the mechanism are required.

Svend Møller in his presentation described a very sophisticated study of the ratios of plasma amino acids in patients with depression. He tried to correlate these plasma amino acid profiles to the response to therapy. There were two problems in this presentation. One was the fact that the correlation did not come out as expected. The predictions made from the model of competitive uptake inhibition regulating serotonin and noradrenaline synthesis and turnover did not fit with the therapeutic outcome. This weakened the argument for a causal link between precursor availability and therapeutic success. In addition, it became clear, that, if one wished to establish such a relationship, one would need independent tests of serotoninergic function or noradrenergic function in the patient groups and show that these correlated with the therapeutic outcome.
the correlation did not come out as expected. The predictions made from the model of competitive uptake inhibition regulating serotonin and noradrenaline synthesis and turnover did not fit with the therapeutic outcome. This weakened the argument for a causal link between precursor availability and therapeutic success. In addition, it became clear, that, if one wished to establish such a relationship, one would need independent tests of serotoninergic function or noradrenergic function in the patient groups and show that these correlated with the therapeutic outcome.

Finally Jaques Bruinvels presented a specific group of patients with periodic psychosis. He described an effect of serine administration to nonsymptomatic patients evoking the psychosis again. However, here it turned out that the formation of serine from its precursor glycine was decreased, while there was an increased formation of taurine, for which serine is a precursor. An increased conversion of serine into glycine may also lead to the formation of beta-carbolines in these patients. Indeed the beta-carboline, norharman, is identificable in plasma during an acute psychotic episode in these patients. Also in porphyric rats serine administration results in an increased formation of horharman. In these rats glycine consumption is increased and consequently more serine is converted into glycine. These rats also show catalepsy under these circumstances. Jaques Bruinvels found a clear correlation between the norharman level and the degree of catalepsy. The increased norharman levels which he found in acute psychotic patients were absent in nonsymptomatic patients or in healthy controls. Here again some details are missing, but it is a hypothetic sequence of considerable interest that merits further investigation. I have presented a rather crude caricature of the issues raised under the main theme by the presentations of this session in order to provoke discussion.

DISCUSSION-REPORT

Cell death in Parkinsonism and in ischaemia: May changes in the release and degradation, or in the compartmentation of neurotransmitter have neurotoxic effects?

Pratt: These presentations, and in particular that of Brian Meldrum, raise an important consideration, namely that we must not neglect to look for possible toxic effects of certain agents for instance excitatory amino acids. If we are able to identify substances which cause cell death, it may be easier to solve what is happening, rather then looking for slight, small behavioral changes. To look for these relevant effects is at least as relevant as looking at rather small changes in precursor availability and transmitter synthesis.

Youdim: I think the considerations made by Oliver Pratt are rather important as a similar phenomenon may operate in Parkinson's disease. It was suggested that dopamine itself and certain metabolic changes which occur during ageing are involved in the pathogenesis of the disease. We know that during ageing monoamine oxidase B activity increases. It is possible that the increased turnover of dopamine itself has a neurotoxic effect. In 1982 we published a retrospective study on the outcome of DOPA-treatment of Parkinson patients with and without deprenyl. The deprenyl-treated patients appeared to live longer and we came to the conclusion that deprenyl may prevent the degeneration of dopaminergic neurons. At that time, no other drug had been shown to have this effect. Deprenyl also blocked the MPTP-Parkinson inducing effect. So I suggested that the neurotoxic effect of MPTP or of a naturally occurring MPTP like substance or of dopamine itself may be involved in the pathogenesis of the disease. Several groups are now trying to replicate our clinical result and the preliminary data reported last year from the clinical study in the United States supports the idea that deprenyl improves the outcome of L-dopa therapy and is therefore useful in the treatment of Parkinsonism.

Balazs: In contrast to monoamines, amino acid transmitters are always available in great excess in nerve cells. However, the relevant question is whether they are available in the right site. Furthermore it is of interest that for example as a result of metabolic stress (anoxia, hypoglycaemia) there is a selective vulnerability of certain types of nerve cells in particular parts of the nervous system. Moreover the different structures even within a single cell seem to be differentially vulnerable. Thus it is seldom recognized that nerve terminals are apparently the most resistent structures under adverse conditions. Some time ago we studied the effect of anoxia on the cerebellum using slices which retained structural preservation when incubated under aerobic conditions (Bosley et al, 1983, J. Neurochem, 40: 189—201). We were particularly interested in the influence of anoxia on presynaptic functions. These were assessed by examing the calcium-dependent stimulus coupled transmitter release. In the cerebellar cortex glutamate (Glu) and GABA resepectively are the major excitatory and inhibitory transmitter. To our surprise we found that evoked GABA release was unaffected by periods up to 35 min of anoxia. It was even more surprising that induced excitatory amino acid release was markedly stimulated during this period of anoxia. Combined biochemical and morphological studies indicated that a selective vulnerability of different structures may account for these findings. In anoxia the earliest alterations were detectable in glia, while nerve terminals were the structures by far the most resistant to damage. As a result there is a redistribution of amino acids: while GABA is preferentially formed in nerve cells, Glu is also present in glial cells in relatively high concentration. Therefore the leakage of amino acids from the compromized glia will lead to preferential accumulation of Glu or GABA by nerve terminals which are the structures relatively undamaged under these conditions. This may account for the unaltered evoked release of GABA and increased release of Glu during anoxia. These observations have been confirmed and extended by Beneriste et al. (1984, J. Neurochem. 43: 1369—1374), who showed a selective increase in the extracellular content of Glu, Asp and Tau in the hippocampus in ischemia *in vivo.* There is now strong evidence indicating that excitatory amino acids may be involved in the pathogenesis of various brain disorders, such as resulting from metabolic stress. With respect of therapeutic implication it is important to learn more about the ways how the metabolism, release and compartmentation of the excitatory amino acids can be influenced, in addition to attempting to antagonize their effect at the receptor level.

Meldrum: We are confident that excitatory amino acids are involved in the brain damage occurring after status epilepticus. They are at least partially involved in ischaemic brain damage. This does not, however, prove that we shall get a major clinical effect by using antagonists in stroke or in other symptomes associated with cerebral ischaemia. There have been many disappointments in clinical trials of protective agents. People have looked at barbiturates, opiates and a whole range of compounds in different animals, but no major break-through has been made. With respect to clinical applications we have not yet got the optimal antagonists for cerebral protection in stroke. The competitive antagonists do not seem to enter the brain in sufficient amounts. Compounds like phencyclidine and MK8O1 may have acute toxic side effects.

Van Gelder: How confident are you that it is the seizures in status epilepticus that damage the brain cells and that it is not the peripheral breakdown of the brain's supply by cardiac and pulmonary distress which causes brain ischaemia or anoxia?

Meldrum: It is certainly right that death from status epilepticus does depend on the systemic side effects and the unfavourable neurologic outcome also correlates closely with such factors, as the increase in body temperature and changes in the cerebral circulation and oxygen supply. This is one thing. But there is clear evidence from experimental studies that even if you keep all these parameters in close physiological limits, you can still get damage on hippocampal neurons that relates to the excessive neuronal activity and not to any of the systemic stresses.

Van Gelder: Certainly seizures may cause brain damage; nevertheless, a large number of seizure patients are still alive until 60 or 80 years. They have a very well defined focus. The focus remains at a fixed location. They will have their seizures perhaps only once a month and certainly once a year. This indicates that the cell population which is the origin of the seizures remains the same. It cannot be that these neurons are progressively dying off during successive seizures, because then the focus would disappear. So there is a discrepancy between the theory saying that the neurons are dying off during the seizure and the fact that the focus remains constant over a long period in spite of repeated seizures.

Meldrum: In order to cause neuronal death during status epilepticus you need sustained excitation going on for more than 40—90 minutes. So there is no reason to expect that, a simple focus that discharges from time to time for a rather short period, will produce an increased rate of cell death. It is true that many patients remain stable in terms of their pathology throughout their lifetime. Other patients, however, do show some progressive neuronal loss particularly in the hippocampus. This is very clearly seen, if you look at the increase in hippocampus neuronal loss, with age in epileptic patients. So both phenomena exist.

What are the causes of the altered amino acid transport in hepatic encephalopathy?

Lajtha: The change in transport activity for aromatic amino acids in portocaval shunted rats reported by Giorgio Zanchin is a rather exciting observation. We have been talking about amino acid loading and precursor availability affecting transmitter synthesis, but in addition to competition for the carrier and metabolic changes, other changes may also occur. In this context the question arises whether the carrier for amino acid transport at the blood-brain barrier is induceable or suppressable by long lasting alterations of the brain's amino acid supply.

Qu.Smith: Two hypotheses have been proposed to explain the changes that occur in neutral amino acid transport into brain in hepatic encephalopathy. James et al. (1979, Lancet 2: 772—775) have proposed that brain glutamine, which is elevated in hepatic encephalopathy, increases capillary amino acid transport by an exchange mechanism (trans-stimulation). In contrast, Mans et al., (1983, Am. J. Physiol. 245: C74—77) have challanged the trans-stimulation hyptothesis and suggested that transport is induced in some manner, possibly involving ammonia. Recently, we have done some experiments that suggest that the trans-stimulation hypothesis may not be correct (Smith et al., 1987, J. Cereb. Blood Flow Metabol. 7:S501). In one set of animals we elevated brain tyrosine by 10—20 fold for one week by feeding animals a high tyrosine diet. At the end of the week we examined the kinetics of amino acid transport into brain and found no change in the V_{max} or K_m of the neutral amino acid carrier. In a second set of animals we acutely raised brain concentrations of tyrosine (10X), leucine (20X) or glutamine (3X), and again found no change in the blood-brain barrier transport activity. These results suggest that the neutral amino acid carrier is not sensitive to trans-stimulation in vivo. The changes that occur in transport activity with hepatic encephalopathy probably involve induction of the transport carrier.

Balazs: I am not aware of studies demonstrating the induction of amino acid transport-systems in structures associated with the blood brain barrier. However, we had observed recently (Balazs et al., 1987, Dev. Brain Res., in press) that in cerebellar granule cells developing in culture the expression of the acidic amino acid (AAA) carrier is controlled by environmental conditions. AAA uptake is hardly detectable when granule neurons are grown in a serum-free medium. However, chronic depolarization of the cells, by elevating the concentration of K^+ in the medium results in the induction of the AAA carrier.

Pratt: To decide if induction is involved in the changes of the transport activity for aromatic amino acids in portocaval shunted rats it would be very important to establish the minimum time needed to produce the effects. A transstimulation effect of glutamine would occur immediately. At the other extreme, induction by analogy with, for example, the ketone body carrier (Love et al., Krebs et al.) might well need days. The possibility of induction must not be excluded. We should also take into consideration metabolic changes in the endothelial cells affecting their properties not only at the luminal, but also at the abluminal surface including their interaction with astrocytes.

Rogers: We also have used animal models in order to examine the effects of amino acids on hepatic encephalopathy. As you know, infusion of ammonia will produce changes in plasma amino acids similar to that of hepatic disease. We have found that pancrealectomy in dogs prevents these amino acid changes and that, although adrenalectomy did not prevent the changes, the infusion of epinephrine or epinephrine and norepinephrine caused a similar change in plasma amino acids. These changes appear to be mediated at least in part by increased plasma insulin and glucagon. The addition of ammonia to the infusate of epinephrine and norepinephrine had little effect, excepted that the brain tryptophan increased about 2-fold (Strombeck, D.R. et al., 1984, Am. J. Physiol. 247:, E 276) It would appear to me that an increase in ammonia per se is an essential factor that contributes to abnormal brain amino acids and the ensuing encephalopathy.

Is the modulation of brain glycine a promising strategy for treatment of neurologic dysfunctions?

Commissiong: The two glycinergic interneurons involved in the observed clinical effectiveness of threonine in treating spasticity are probably the Renshaw cell and the Ia reciprocal inhibitory interneuron. Both the alpha and gamma interneurons are hyperactive during most forms of spasticity. The Renshaw cell provides a direct inhibitory input to the alpha motoneurons. However, the Renshaw cell also inhibits the Ia reciprocal inhibitory interneuron, thereby disinhibiting (i.·e, providing a net excitation) to the alpha motoneuron. Therefore, on the basis of what is currently known about the physiology and the microcircuity of the spinal cord, it is difficult to predict a net inhibitory effect of threonine (acting via glycine) on motoneuron hyper-excitability during spasticity. The results provided from this clinical trial are therefore of great interest from a basic science point of view as well.

Growdon: There are more than 20 putative neurotransmitter compounds including classical transmitters, neuropeptides, and amino acids in the spinal cord that could affect the reflex arc. Information regarding the way in which these chemicals interact to control normal muscle function, or are altered in various spastic states, is lacking. Electrophysiological studies have been limited because there are no human reflexes known to be subserved primarily by glycine, or any other transmitter that is segmentally localized within interneurons of the spinal cord. Given the current state of uncertainty regarding the biochemical bases of human spasticity, it is easy to formulate compelling reasons why a treatment is destined to succeed, or absolutely doomed to failure. Generally, such opinions in either direction are wrong: pharmacologists are plagued by failed drugs that were once considered promising. Examples also abound in which a drug worked, but not according to the mechanisms originally proposed; treatment of spasticity with baclofen is a case in point. Fortunately, the question of a treatment's clinical efficacy can be settled by data and not according to theoretical discussions. We have begun a double blind study of L-threonine in patients with familiar spastic paraparesis in order to provide more data on this treatment modality. If we are able to confirm our preliminary results that L-threonine administration suppressed aspects of spasticity, every effort should be made to uncover its mode of action.

Youdim: Of course with threonine loading one cannot say what would be the concentration of glycine in the brain, since there will be substantial peripheral metabolism. Therefore a drug such as milacemide may give a more dramatic clinical response since the first step in its metabolism in the brain is the formation of glycine. Milacemide is a highly potent anticonvulsant. The ability of this drug to potentiate the postsynaptic 5-HT receptor function is fascinating and intriguing. We have no doubt that it will provide a tool for examining the interaction of serotonergic and glycinergic systems in the brain stem and upper spinal cord. The implications for hyperbaric oxygen toxicity, hypoxia induced myoclonus and spasticity are obvious since an alteration of 5-HT neurotransmission has been implicated.

What is the cause and the significance of changed plasma amino acid ratios in depressed patients?

Møller: With regard to L-tryptophan and depression, one study performed on mildly depressed patients from 5 general practices showed that the antidepressive effect of L-tryptophan was comparable with that of amitriptyline and amitriptyline plus L-tryptophan, while all 3 treatments were superior to placebo (J. Thomson et al., 1982, Psycholog. Med. 12: 741—751). On the other hand, there seems to be a consensus that L-tryptophan does not possess antidepressant properties, in general, in severely depressed hospitalized patients. However, two studies on in-patients have found support for the suggestion that there may be a small biochemical subgroup of patients with major affective disorder who respond favourably to L-tryptophan. In those studies, the depressives with low pretreatment levels of the plasma ratio tryptophan to the sum of the other large neutral amino acids (LNAA) showed significantly greater clinical response than the remaining depressives (S. E. Møller et al., 1976, Psychopharmacology; 49: 205—213: S.E. Møller et al., 1980, J. Affective Disord. 2: 47—59). However, since the putative biochemical subgroup comprised only about 15 % of the patient sample, the results may from a practical point of view be of limited importance.

A more interesting finding is that the clinical response in the depressives failed to associate with the pretreatment levels of the free and total plasma tryptophan, but correlated statistically significant with the pretreatment plasma ratio tryptophan/LNAA. As presented in my contribution further 5 studies have shown statistically significant correlations between the pretreatment plasma ratios tryptophan/LNAA or tyrosine/LNAA and the clinical response in depressives to treatment with a variety of antidepressant drugs. (The range of the plasma ratios tryptophan/LNAA and tyrosine/LNAA in the 6 studies was from mean minus 28 ± 9 (SD) % to mean plus 31 ± 11 (SD) %). These consistent findings suggest an association between amino acid availability from plasma to brain, as reflected by the plasma amino acid ratios, and neuronal function in the human brain, which is of great theoretical importance. From a practical point of view, the plasma amino acid ratios may be a tool for increasing the efficacy of antidepressant treatment.

Pratt: The interesting findings of Svend Møller comparing raw ratios of tryptophan or tyrosine to other large neutral amino acids do not need elaborate statistical treatment. Even, if the K_m-values differed somewhat in the human from those in the rat, the use of even somewhat inappropriate K_m-values would still be expected to improve the efficiency of the model for transport across the blood-brain barrier. The fact that he does not find such an improvement in either instance suggests to me that a blood brain barrier transport model alone is not adequate to explain his very important findings in relation to the efficiency of these drugs. We have to look for some other mechanism related to these amino acids.

Møller: I agree with Oliver Pratt's comments. The findings suggest that competition plays an important role in the neutral amino acid transport to the human brain, yet, the application of K_m-values impaired the association between the biochemical and clinical findings in the depressed patients. Estimation of K_m-values is based on unidirectional influx, and they may therefore not describe the amino acid movements across the blood-brain barrier comprehensively. The measurement of the net uptake by the brain of individual amino acids could possibly provide better information on this issue and thereby strengthen the association between the plasma ratios of tryptophan or tyrosine to other large neutral amino acids and the clinical improvement. As judged from the present findings, the interaction effect of valine and leucine on the tryptophan transport seems to be underestimated and that of phenylalanine overestimated when using K_m-values alone.

Leathwood: According to my opinion, Svend Møller found a very interesting approach to a difficult problem. I would like to ask if you have tried using stepwise discriminant analysis of all clinical variables, to see (a) which seem to be the most powerful predictors of depression and (b) where the TRP/LNAA and TYR/LNAA ratios fall in this spectrum.

Møller: There are no useful clinical predictors of depression, although twin studies of manic-depressive disorders have shown evidence of a strong genetic factor with a concordance rate 3 to 4 times higher among monozygotic than dizygotic twins (e. g. A. Berthelsen et al., 1977, Br. J. Psychiatry; 130: 330—351). As regards biochemical predictors, I can think of only one study which found support for a predisposing (rather than a causative) factor to the occurence of depression. The frequency of depression in patients with low levels of the serotonin metabolite 5-hydroxyindoleacetic acid in the cerebrospinal fluid was increased compared with that seen in patients with normal levels (H.M. van Praag & S. de Haan, 1979, Psychiatr. Res. 1: 219—224). Concerning amino acids, most studies have shown nearly the same distributions of the plasma ratio tryptophan/LNAA or tyrosine/LNAA in samples from depressed patients and healthy controls. However, drugs that decrease the plasma ratio tyrosine/LNAA, e. g. combined oral contraceptives, may possibly predispose to the occurrence of depressive symptoms in some individuals. (S. E. Møller, 1981, Neuropsychobiology; 7: 192—200).

Gliosis, Ageing and Parkinsonism: What is the role of monoamine oxidase in astrocytes?

Vernadakis: It was very interesting to hear from Moussa Youdim that monoamine oxidase B is contained in astrocytes. Do you have an idea on the role of this enzyme in glial cells? Is it possible that it contributes significantly to the degradation of monoamines released from neurons and taken up by astrocytes, as has been suggested from our studies, and those of Hertz, Schousboe, Kimelberg (see contribution of Vernadakis for complete refs.)?

Youdim: Until now we have no idea on the functional importance of monoamine oxidase B in astrocytes. Its presence there, however, rises several important implications. Under conditions of gliosis, in particular in the aged brain, we find a general increase of monoamine oxidase B activity which correlates with the increase in the number of astrocytes. So it seems justified to conclude that the increase of monoamine oxidase B in aged brain is in fact due to the increase of astrocytes containing monoamine oxidase B. Another implication is related to the pathogenesis of Parkinsonism and treatment with l-deprenyl. As you all know dopamine can also be degraded by monoamine oxidase B to DOPAC and H_2O_2. With ageing, due to the increased activity of monoamine oxidase B, the formation of H_2O_2 could rise. l-Deprenyl inhibits monoamine oxidase B, and therefore, causes a decreased formation of H_2O_2 . H_2O_2 plays a very important role in lipid peroxidation of membranes since it can interact with iron to form free radicals, many of which are cytotoxic. Interestingly enough, two areas in the brain, the substantia nigra and the globus pallidum, contain tremendous amounts of iron , about as much as the liver (g/wet weight). In the brain of Parkinson patients where most of the dopaminergic neurons have degenerated, only 5 to 8 % of total monoamine oxidase B activity is found in neurons. The bulk of monoamine oxidase B is located in astrocytes and it has been suggested that L-DOPA can be converted to dopamine and then to DOPAC and H_2O_2 in the astrocyte. If radicals are involved in the pathogenesis of Parkinsons disease, by giving a monoamine oxidase B inhibitor we block the formation of H_2O_2 and free radicals. In this context it is interesting to note that in the brains of Parkinson patients the amount of ferritin and free iron is increased in the substantia nigra and globus pallidum. On the other hand, the amount of glutathione reductase and of ascorbate, two important antioxidant agents, is decreased. I should mention here that in mice and monkeys which have been given MPTP, a similar depletion of these antioxidants has been observed. Another interesting finding is that the younger the animal, i. e. the less the degree of glioses in their brains, the less effective is MPTP neurotoxity in these animal models.

Chapter VII.

AMINO ACID AVAILABILITY AND BRAIN DEVELOPMENT

DEVELOPMENTAL CHANGES IN CEREBRAL AMINO ACIDS AND PROTEIN METABOLISM

A. Lajtha, H. Sershen and D. Dunlop

Center for Neurochemistry
Ward's Island
New York, NY 10035

INTRODUCTION

This chapter will examine developmental changes in the free amino acid pool and in protein metabolism in the brain, and the possible relationship of the changes in amino acids and the proteins. An important aspect of this subject, the developmental changes in amino acid supply to the brain via capillary transport through the blood-brain barrier, will not be discussed in detail, since it will be treated in the next chapter by Dr. Lefauconnier.

Much is known about genetic control of protein synthesis, and nutritional influences on protein metabolism have been studied at some detail. Little study has been made of influences on and controls of brain protein metabolism, and even less is known about the influences or controls of protein breakdown.

Developmental Changes in the Cerebral Free Amino Acid Pool

The composition of the free amino acid pool in brain is different from that in other organs, and also from that in plasma. The high level of glutamic acid and the specific presence of GABA are examples of its unique characteristics. Although the composition is fairly stable in most circumstances, significant regional heterogeneity in the distribution of several amino acids has been found (1, 2). This problem was recently investigated in more detail in 40—50 discrete brain areas and in nuclei, and severalfold differences in level were found between areas, especially for glutamate, aspartate (3), and taurine (4); differences in essential amino acids were also found (Banay-Schwartz, Palkovits, and Lajtha, in preparation). The significance of this regional heterogeneity in terms of amino acid function, and in terms of protein metabolism, is not well understood. Heterogeneity in distribution of glutamate and GABA in synaptosomal preparations has also been studied (5, 6). Synaptosomal levels can be significantly above or below the level of the amino acid in the cytoplasm. This heterogeneous distribution is stable under most conditions, although it can be altered by drugs, hypoxia, and other influences.

There are significant changes in the amino acid pool during development. Just as in the regional distribution pattern, the developmental changes are complex. In general, the levels of non-essential amino acids such as glutamic acid, glutamine, and aspartic acid increase while those of essential amino acids decrease. This may indicate a decrease

NATO ASI Series, Vol. H20
Amino Acid Availability and Brain Function in
Health and Disease. Edited by G. Huether

with decreasing rate of protein turnover in the levels of the amino acids primarily used for protein synthesis, while amino acids utilized for other functions increase with increasing function. Of the neurotransmitter or neurotransmitter precursor amino acids, glycine, phenylalanine, serine, and taurine decrease and GABA increases (1). The pattern is still more complex in that the developmental period of rapid change in level not only differs for the various amino acids (7, 8), but is also different in different species; aspartate increases in most species but not in guinea pigs, for example, and the absolute level and degree of change in taurine vary greatly in the different species. It is likely that developmental patterns depend on the specific brain structure assayed. In our study of changes during aging the patterns were different in various brain areas and nuclei (in preparation). Examples of perinatal and postnatal changes in amino acid levels are given in Table 1.

TABLE 1
Perinatal Changes in the Free Amino Acid Pool in Mouse Brain

	μmoles amino acid per g brain tissue				
	Foetus		Newborn		Adult
	15 d	19 d	0 h	24 h	
Taurine	14.1	12.2	14.1	**15.6**	**8.01**
Glutamic acid	7.54	5.71	4.81	**5.02**	**11.7**
Threonine	**4.28**	**0.90**	0.93	0.90	0.56
Proline	0.89	0.52	0.66	**0.57**	**0.15**
Alanine	5.08	3.01	**4.29**	**0.80**	0.56
Lysine	0.86	0.88	**0.91**	**0.41**	0.29

(From 1, 7, 8) Mouse brain free amino acid content was measured in different development stages.

DEVELOPMENTAL CHANGES IN TRANSPORT

When changes in transport are discussed, it is important to emphasize that amino acids are transported by several transport systems, and that the properties of the various systems differ. It is likely that a change in the environment alters only a few of the systems, or that the change may be to a different degree or in a different direction. For example, since not all the systems require Na or Na gradients, changes in Na distribution would affect only some of the systems.

In looking at inhibition patterns of amino acid uptake by structurally related analogs, we could distinguish at least ten transport systems for amino acids in brain (9) (Table 2). It is likely that more than ten systems are present and will later be identified. We found some indications that the high- and low-affinity transport of an amino acid may represent different systems (10). If this is true for all the high-affinity systems, there may be 16 or more systems in brain. Not all systems are present in all structures — brain capillaries contain considerably fewer systems than brain cell membranes do (11), and the high-affinity transport systems are primarily present at nerve-ending membranes (12). The substrate specificity of the high-affinity uptake is similar to that of the low-affinity systems (13).

With the presence of a fairly large number of transport systems with heterogeneous distribution, it is not surprising that the developmental changes in transport also vary, in

TABLE 2
Classes of Amino Acid Transport in Brain

Gly, Pro, Sar
Leu, Phe, Val, Met
Ala, Ser, Cys
Pro
Gly
GABA
Tau
Asp, Glu
Lys, Arg, Orn
Lys

(From 9) Representative amino acids transported by the same system are indicated in each group.

different structures and for different amino acids. Although several amino acid transport systems are present in the brain, there is not a specific system for each amino acid; rather, each system has affinity for a number of structurally related amino acids, and an amino acid in turn has affinity to more than one system. It seems that the major change during development in transport systems is quantitative rather than qualitative; that is, the level and activity of some of the systems change rather than their properties.

It is well-known that the blood-brain barrier is less restrictive in the immature brain than in the mature organ. In our studies, all 10 amino acids administered in vivo penetrated the brain to a greater extent in the newborn than in the adult (14). Although in these studies some amino acids reached the same level in the brain as was present in the plasma, some stayed at lower levels and a few increased above plasma levels (14), indicating both transport and a barrier. When capillary transport was measured either in vivo or in vitro, transport activity, somewhat unexpectedly, was higher in young than in adult brain (11, 15). Tissue transport showed a different developmental change in that cellular uptake increased during development, as measured with brain slices (16), while capillary uptake decreased. The change in slices was complex, with uptake increasing initially, then decreasing to adult levels (17); in general, essential amino acid uptake developed at earlier fetal stages than neurotransmitter amino acid uptake (16).

Changes in Transport Activity Versus Changes in Amino Acid Levels in the Brain

Since data are now available on the differences in the regional levels of amino acids and on developmental changes in amino acid levels in the brain on one hand, and on regional and developmental changes in amino acid transport activity on the other, the composition of the free amino acid pool and alterations in the pool can be compared with the transport activity to determine to what degree the level and distribution of an amino acid are determined by its transport.

Although it is generally assumed that blood-brain barrier transport, that is, capillary transport, is the rate-limiting and level-determining component, this does not seem to be true for brain amino acids. Capillary transport of essential amino acids (such as phenylalanine and leucine) is much greater than for non-essential amino acids (such as glutamate and aspartate), and for some non-essentials is negligible (such as glycine and GABA) (18). Nevertheless, the level of non-essential amino acids (such as glutamate, aspartate, glycine, and GABA) is 10—100 times as high as the level of essential amino acids. Clearly, capillary transport activity is not a major determinant in this case. Cellular uptake and retention are more important: even if an amino acid such as glutamate enters the brain slowly, if it is then taken up and retained by cells it will accumulate to a greater degree than an amino acid which, although it penetrates rapidly, is not retained by the tissue.

Passage through the capillaries is rapid — in general, more rapid for essential amino acids, but significant for the non-essentials as well. With tracer doses we estimated the half lives of essential amino acids in minutes, of the non-essentials in hours (19). With such rapid uptake other processes such as exit, cellular and particulate uptake, metabolism, and binding are likely to play important roles.

We previously compared changes in cellular transport, as measured by uptake in brain slices, with the levels of amino acids. In particular, we compared regional heterogeneity of uptake in slices and in vivo with regional heterogeneity in levels (2, 20), and also compared developmental changes in uptake with developmental changes in levels of amino acids (1, 17). In each case there was some correlation, in that increased or decreased transport was reflected in increased or decreased levels respectively, but there were also a number of exceptions, in that the changes in level and in uptake were in opposite directions. We concluded that transport, as measured by in vivo uptake rates or by rates of uptake in slices, is not the sole control of levels. It is likely that cellular transport plays an important role, but that with slices in vivo cellular transport activity can not be measured precisely. We found indications that some transport activity may be inactive under normal conditions, but is activated by tissue damage during preparation of brain slices (21). Influx and steady state also did not always change in a similar manner, indicating differences between heterogeneity of uptake and that of exit (17).

Since transport activity is complex a) passage through capillaries differs from that through cells and particulates, b) structurally related compounds influence each other's transport depending on affinity to the carrier, c) metabolism and d) exit are also of influence — it is not surprising that it is difficult to find close correlation between transport activity and metabolite levels.

TABLE 3
Penetration of Nonessential Amino Acids Through the Blood Brain Barrier

		μmol/g tissue		
		Taurine	Glutamate	Glycine
Plasma	Control	0.6	0.04	0.3
	Experiment	12	6.2	3.5
Brain	Control	10	11	1.0
	Experiment	14	15	2.2

(From 22) Plasma levels were increased by intragastine administration of the amino acid for a few hours in mice.

Although under most conditions it is likely that passive diffusion does not play a major role, the barrier to diffusion is not absolute. Even the levels of the usually non-penetrating amino acids such as aspartate, glutamate, glycine, and taurine could be increased in the brain if their levels in the plasma were significantly increased (22). In these experiments (Table 3) glutamate levels did not have to be higher in plasma than in brain to cause increases in the brain; presumably there was a large concentration gradient from plasma to brain extracellular fluid, since glutamate is rapidly removed from the extracellular fluid and accumulated in cells.

Developmental Changes in Brain Protein Synthesis

Labeled amino acids administered in vivo are rapidly incorporated into brain proteins, indicating a high rate of synthesis. The rate of replacement for whole brain is about 0.5—0.7 % per hour; of course, all proteins are not metabolized at this average rate. As a crude estimate, in adult brain we could detect a fast and a slow pool for protein turnover: a small (5 % of tatal) rapidly metabolized pool with an average half-life of 7 hours, and a large (95 % of total) pool with a half-life of about 8 days (23) (Table 4). There is a significant degree of heterogeneity within the pools, and even possibly in the metabolic rate of a simple protein at different sites. In myelin protein turnover there is significant heterogeneity, with indications that most myelin proteins on the outer regions are metabolized differently (24), possibly more rapidly, while some portion may be metabolically stable (25). A recent example is the report that actin and tubulin in the synaptic plasma membrane and synaptic junction fraction turn over about twice as slowly as in the cytosolic fraction (26).

TABLE 4
Half-Lives of Tissue Proteins

Liver	100 %	26 h
Kidney	41 %	18 h
	63 %	63 h
Brain		
adult	3.5 %	7 h
	96.5 %	9.7 days
young	99 %	2 dys

(From 23)

Although some small fraction of brain proteins may be stable or metabolized slowly, more than 95 % seem to be replaced several times during the life span (27). A similar 20-fold difference in level (5 % versus 95 %) and in turnover rate (7 hours versus 9 days) in the two metabolic pools of proteins in adult brain (23) results initially about 50 % of the incorporation occurring in each of the pools — as much in the small active pool as in the large less active pool.

In the developing brain, protein synthesis was shown to occur at a higher rate than in the slow pool of adult tissue, but the rapidly metabolized adult pool was not detectable (28). Since the slow pool in the adult contains the major portion of proteins (95 %) there must be a change in the metabolic rate of most proteins during development. Recently we compared the developmental changes in amino acid incorporation into brain proteins in various fractions — neuronal versus glial-enriched fractions (29), particulate fractions (30), and fractions according to molecular weight (31). The developmental changes in metabolic rate were similar in each fraction, again indicating that the decrease in turnover rate with maturation seems to be similar for most brain proteins. The ratios of protein synthesis rates to RNA content were constant throughout development (32), indicating nucleic acids as the rate-determining substrate.

Developmental Changes in Protein Breakdown

In adult brain, with no net change in protein content, total protein synthesis and breakdown are equal, while during growth and net deposition of protein, synthesis is greater than breakdown. Rates of incorporation in the adult (0.5—0.7 %/hour) are close to the rate of growth of brain proteins in the first postnatal days. Hence, if protein synthesis rate remained constant during growth, inhibition of breakdown would be sufficient for growth.

When in vivo breakdown was measured a significantly higher breakdown rate was found in the immature brain (33, 34). This increased breakdown during the active growth phase may be related to alterations in brain composition or to cell death.

When breakdown of proteins was measured in vitro in brain slices (35), the rates were similar to those in vivo and they decreased during development. Slices may be a suitable system for studying developmental change in breakdown and its control factors in detail. Slices, however, may not be as suitable for studying changes in synthesis rates, since synthesis is greatly reduced in slices, especially in adult tissue, where incorporation in slices is only 6—10 % of that in vivo (36).

Amino Acid Effects on Protein Metabolism

Although the composition of the amino acid pool is fairly stable in the brain, a number of factors, nutritional, pharmacological, or physiological, can alter its composition and distribution. Amino acid distribution within the brain can also be influenced, for example by synaptosomal uptake inhibitors that alter neurotransmitter amino acid levels (37). The changes can be complex; protein-free diet results in the increase of some amino acids and the decrease of others, and changes their uptake (19, 38), and the effects of total food withdrawal differs from that of protein restricition (39). Drugs, such as morphine, psychotropic drugs, and especially insulin, influence the levels of specific amino acids (1, 40). Brain protein composition and metabolism are more stable than the free amino acid pool, but a number of factors have been shown to alter brain protein metabolism and content (41). Amino acids have been shown, in a number of tissues, to influence protein metabolism. In a recent report, for example, starving increased and refeeding decreased myofibrillar protein breakdown; after starving, proteolysis decreased only when protein or amino acids were added to the meal, whereas either protein-containing or protein-free diet enhanced muscle protein synthesis (42).

Because of the effect of amino acids on protein metabolism in other tissues, it is of interest to establish whether similar effects can be observed in the brain.

As a rule, brain proteins are more stable than proteins in other organs; for example, the effects of malnutrition in brain, if any can be observed, are much milder (43). This may be because in brain malnutrition results in simultaneous reduction in protein synthesis and breakdown, while in other organs such as muscle, synthesis decreases but breakdown increases (44).

Of the amino acid effects on brain protein metabolism the effect of increased phenylalanine levels were studied in the most detail because of the possibility that the mental retardation in phenylketonuria is the result of an abnormal protein metabolism caused by the elevated levels of phenylalanine. Mild hyperphenylalaninemia does not result in changes in protein content in brain, although in acute experiments a slight decrease in incorporation was observed (45), but myelination in the young was impaired (46). The content of neural polyribosomes (as determined in vitro) is decreased in hyperphenylalaninemia, but the injection of the amino acids that are transported with phenylalanine restored this decrease and restored the acylation levels of the methionyl-tRNA initiator species, which may be the mechanism of the changes (47). Ribosomal protein phosphorylation may also be affected (48), although some of the changes may be artifactual and the result of increase by phenylalanine of in vitro ribonuclease activity (49). Some of the observed decrease in incorporation may have been due to inhibition of uptake by the brain of the labeled precursor amino acid, rather than direct inhibition of protein synthesis (50). This problem has been recently reviewed (51).

We tested the effects of elevated amino acids on brain protein synthesis in three different sets of experiments. In one set of experiments amino acid incorporation was measured in immature brain explants incubated up to ten days with elevated amino acids. The protein turnover rate was similar to that in vivo, but was not altered by elevated amino acids, including phenylalanine (52). In another more recent set of experiments, following the finding that brain levels of taurine, glycine, or GABA can be elevated if large doses are administered, we measured the effect in adult mice. Increased taurine caused inhibition of amino acid incorporation in muscle and liver, but not in brain — increased GABA or glycine had no effect in any organ (53). In the third set of experiments the incorporation of tracer doses and that of large doses of several amino acids into brain proteins were compared in immature animals; and no differences were found (54). Experimental circumstances need to be carefully controlled, since amino acid administration can alter body temperature and the access of the labeled precursor to the site of protein synthesis, but it seems that brain protein synthesis under most circumstances is not sensitive to elevated amino acids.

This lack of sensitivity of brain protein turnover to alterations in amino acid levels is in contrast to muscle protein metabolism, which has been shown in a number of experiments to be altered when amino acid levels are altered.

In perfused rat heart protein synthesis was increased 40 % when amino acids were elevated to 5 times that in plasma (55). Starvation decreased muscle protein synthesis rates, while protein degradation remained unchanged, in adults, but increased in young (56). This change with starvation, decreased synthesis and increased breakdown of proteins, was found in the extensor digitorium but not in the soleus muscle (57). Increase in amino acids promoted protein synthesis and inhibited degradation in diaphragm muscle (58).

These findings indicate that protein synthesis and breakdown are controlled in different ways; not only is the response in brain different from that in muscle, but different muscles also show differences in sensitivity to amino acids.

REFERENCES

1) HIMWICH, W. A., AGRAWAL, H. C. (1969). Amino acids. In: *Handbook of Neurochemistry* (Lajtha, A., ed.). **Vol. 1**, pp. 33—52.

2) BATTISTIN, L., LAJTHA, A. (1970). Regional distribution and movement of amino acids in the brain. *J. Neurol. Sci.* **10:** 313—322.

3) PALKOVITS, M., LANG, T., PATTHY, A., ELEKES, I. (1986). Distribution and stress-induced increase of glutamate and aspartate levels in discrete brain nuclei of rats. *Brain Res.* **373:** 252—257.

4) PALKOVITS, M., ELEKES, I., LANG, T., PATTHY, A. (1986). Taurine levels in discrete brain nuclei of rats. *J. Neurochem.* **47:** 1333—1335.

5) GEDDES, J. W., NEWSTEAD, J. D., WOOD, J. D. (1980). Stability of the glutamate content of synaptosomes during their preparation. *Neurochem. Res.* **10:** 1107—1125.

6) WOOD, J. D., JOHNS, K. L. (1985). Stability of the amino acid content of nerve endings after administration of γ-aminobutyrate agonists and antagonists. *Neuropharmacology* **24:** 685—687.

7) LAJTHA, A., TOTH, J. (1973). Perinatal changes in the free amino acid pool of the brain in mice. *Brain Res.* **55:** 238—241.

8) PERRY, T. L. (1982). Cerebral amino acid pools. In: *Handbook of Neurochemistry* (Lajtha, A., ed.). **Vol. 1**, pp. 151—188.

9) SERSHEN, H., LAJTHA, A. (1979). Inhibition pattern by analogs indicates the presence of ten or more transport systems for amino acids in brain cells. *J. Neurochem.* **32:** 719—726.

10) DEBLER, E. A., SERSHEN, H., LAJTHA, A., GENNARO, J. F., Jr. (1986). Superoxide radical-mediated alteration of synaptosome membrane structure and high-affinity -(^{14}C)aminobutyric acid uptake. *J. Neurochem.* **47:** 1804—1813.

11) SERSHEN, H., LAJTHA, A. (1976). Capillary transport of amino acids in the developing brain. *Exp. Neurol.* **53:** 465—474.

12) LEVI, G., RAITERI, M. (1973). GABA and glutamate uptake by subcellular fractions enriched in synaptosomes: critical evaluation of some methodological aspects. *Brain Res.* **57:** 165—185.

13) DEBLER, E. A., LAJTHA, A. (1987). High-affinity transport of γ-aminobutyric acid, glycine, taurine, L-aspartic acid, and L-glutamic acid in synaptosomal (P_2)tissue: A kinetic and substrate specificity analysis. *J. Neurochem.* **48:** 1851—1856.

14) SETA, K., SERSHEN, H., LAJTHA, A. (1972). Cerebral amino acid uptake in vivo in newborn mice. *Brain Res.* **47:** 415—425.

15) CREMER, J. E., BRAUN, L. D., OLDENDORF, W. H. (1987). Changes during development in transport processes of the blood-brain barrier. *Biochim. Biophys. Acta* **448:** 633—637.

16) SERSHEN, H., LAJTHA, A. (1976). Perinatal changes of transport systems for amino acids in slices of mouse brain. *Neurochem. Res.* **1:** 417—428.

17) PICCOLI, F., GRYNBAUM, A., LAJTHA, A. (1971). Developmental changes in Na^+, K^+ and ATP and in the levels and transport of amino acids in incubated slices of rat brain. *J. Neurochem.* **18:** 1135—1148.

18) OLDENDORF, W. H., SZABO, J. (1976). Amino acid assignment to one of three blood-brain barrier amino acid carriers. *Am. J. Physiol.* **230:** 94—98.

19) TOTH, J., LAJTHA, A. (1977). Rates of exchange of free amino acids between plasma and brain in mice. *Neurochem. Res.* **2:** 149—160.

20) BATTISTIN, L., GRYNBAUM, A., LAJTHA, A. (1969). Distribution and uptake of amino acids in various regions of the cat brain in vitro. *J. Neurochem.* **16:** 1459—1468.

21) BRACCO, F., GENNARO, J. Jr., LAJTHA, A. (1982). Relationship of morphologic damage and amino acid uptake in incubated slices of brain. *Exp. Neurol.* **76:** 606—622.

22) TOTH, E., LAJTHA, A. (1981). Elevation of cerebral levels of nonessential amino acids in vivo by administration of large doses. *Neurochem. Res.* **6:** 1303—1320.

23) LAJTHA, A., LATZKOVITS, L., TOTH, J. (1976). Comparison of turnover rates of proteins of the brain, liver, and kidney in mouse in vivo following long-term labeling. *Biochim. Biophys. Acta* **425:** 511—520.

24) LAJTHA, A., TOTH, J., FUJIMOTO, K., AGRAWAL, H. C. (1977). Turnover of myelin proteins in mouse brain in vivo. *Biochem. J.* **164:** 323—329.

25) SHAPIRA, R., WILHELMI, M. R., KIBLER, R. F. (1981). Turnover of myelin proteins of rat brain, determined in fractions separated by sedimentation in a continuous sucrose gradient. *J. Neurochem.* **36:** 1427—1432.

26) SEDMAN, G. L., JEFFREY, P. L., AUSTIN, L., ROSTAS, J. A. P. (1986). The metabolic turnover of the major proteins of the postsynaptic density. *Mol. Brain Res.* **1:** 221—230.

27) LAJTHA, A., TOTH, J. (1966). Instability of cerebral proteins. *Biochem. Biophys. Res. Commun.* **23:** 294—298.

28) LAJTHA, A., DUNLOP, D., PATLAK, C., TOTH, J. (1979). Compartments of protein metabolism in the developing brain. *Biochim. Biophys. Acta* **561:** 491—501.

29) SHABAZIAN, F. M., JACOBS, M., LAJTHA, A. (1986). Regional and cellular differences in rat brain synthesis in vivo and in slices during development. *Int. J. Dev. Neurosci.* **4:** 209—215.

30) SHABAZIAN, R. M., JACOBS, M., LAJTHA, A. (1986). Rates of amino acid incorporation into particubrain. *Biochem. J.* **208:** 659—666.

31) SHABAZIAN, F. M., JACOBS, N., LAJTHA, A. (1986). Amino acid incorporation in relation to molecular weight of proteins in young and adult brain. *Neurochem. Res.* **11:** 647—660.

32) DUNLOP, D. S., BODONY, R., LAJTHA, A. (1984). RNA concentration and protein synthesis in rat brain during development. *Brain Res.* **294:** 148—151.

33) BANAY-SCHWARTZ, M., BRACCO, F., DUNLOP, D., LAJTHA, A. (1983). Alterations and heterogeneity of protein breakdown in the nervous system. In: *Molecular Aspects of Neurological Disorders* (Austin, L., Jeffrey, P. L., eds.). Academic Press, Sidney, pp. 159—171.

34) DUNLOP, E. S., McHALE, D. M., LAJTHA, A. (1982). The rate of protein degradation in developing brain. *Biochem. J.* **208: 659—666.**

35) DUNLOP, D. S., Van ELDEN, W., PLUCINSKY, I., LAJTHA, A. (1981). Brain slice protein degradation and development. *J. Neurochem.* **36:** 258—265.

36) DUNLOP, S., VAN ELDEN, W., LAJTHA, A. (1977). Developmental effects on protein synthesis rates in regions of the CNS in vivo and in vitro. *J. Neurochem.* **29:** 939—945.

37) WOOD, J. D., SCHOUSBOE, A., KROGSGAARD-LARSEN, P. (1980). In vivo changes in the GABA content of nerve endings (synaptosomes) induced by inhibitors of GABA uptake. *Neuropharmacology* **19:** 1149—1152.

38) TOTH, J., LAJTHA, A. (1980). Effect of protein-free diet on the uptake of amino acids by the brain in vivo. *Exp. Neurol.* **68:** 443—452.

39) ENWONWU, C. O. (1987). Differential effect of total food withdrawal and dietary protein restriction on brain content of free histidine in the rat. *Neurochem. Res.* **12:** 483—487.

40) TOTH, E., LAJTHA, A. (1981). Drug-induced changes in the composition of the cerebral free amino acid pool. *Neurochem. Res.* **6:** 3—12.

41) LAJTHA, A., DUNLOP, D. (1981). Turnover of protein in the nervous system. *Life Sci.* **29:** 755—767.

42) GOODMAN, M. N., DEL PILAR GOMEZ, M. (1987). Decreased myofibrillar proteolysis after refeeding requires dietary protein or amino acids. *Am. J. Physiol.* **253:** E52—E58.

43) WINICK, M. (1974). Malnutrition and the developing brain. In: *Brain dysfunction in metabolic disorders* (Plum, F., ed.). Res. Publ. Assoc. Nerv. Ment. Dis. **Vol. 53,** pp. 253—261.

44) BANAY-SCHWARTZ, M., GIUFFRIDA, A. M., DE GUZMAN, T., SERSHEN, H., LAJTHA, A. (1979). Effect of undernutrition on cerebral protein metabolism. *Exp. Neurol.* **65:** 157—168.

45) GEISON, R. L., SIEGEL, F. L. (1975). Tolerance of protein and lipid synthesis to mild hyperphenylalaninemia in developing rat brain. *Brain Res.***92:** 431—441.

46) SHAH, S. N., JOHNSON, R. C. (1978). Effect of postweaning hyperphenylalaninemia on brain development in rats: myelination, lipid, and fatty acid composition of myelin. *Exp. Neurol.* **61:** 370—379.

47) HUGHES, J. V., JOHNSON, T. C. (1978). Experimentally induced and natural recovery from the effects of phenylalanine on brain protein synthesis. *Biochim. Biophys. Acta* **517:** 473—485.

48) ROBERTS, S., MORELOS, B. S. (1980). Cerebral ribosomal protein phosphorylation in experimental hyperphenylalaninemia. *Biochem. J.* **190:** 405—419.

49) ROBERTS, S., MORELOS, B. S. (1976). Role of ribonuclease action in phenylalanine-induced disaggregation of rat cerebral polyribosomes. *J. Neurochem.* **26:** 387—400.

50) ANTONAS, K. N., COULSON, W. F. (1975). Brain uptake and protein incorporation of amino acids studied in rats subjected to prolonged hyperphenylalaninemia. *J. Neurochem.* **25:** 309—314.

51) HUETHER, G. (1987). Regulation of the free amino acid pool of the developing brain. A lesson learned from experimental phenylketonuria. In: *Amino Acids in Health and Disease: New Perspectives.* Alan R. Liss, Inc., pp. 107—122.

52) SCHNEIDER, J. F., DE MARTINI, J. E., TOTH, J., LAJTHA, A. (1978). The effect of amino acids on protein metabolism as measured in long-term experiments in immature brain explants. *J. Neurobiol.* **9:** 29—41.

53) TOTH, E., LAJTHA, A. (1984). Brain protein synthesis rates are not sensitive to elevated GABA, taurine, or glycine. *Neurochem. Res.* **9:** 173—179.

54) DUNLOP, D., LAJTHA, A., TOTH, J. (1977). Measuring brain protein metabolism in young and adult rats. In: *Mechanisms, Regulation and Special Functions of Protein Synthesis in Brain* (Roberts, S., Lajtha, A., Gispen, W. H., eds.). Elsevier, Amsterdam, pp. 79—96.

55) MORGAN, H. E., EARL, D. C. N., BROADUS, A., WOLPERT, E. B., GIGER, E., JEFFERSON, L. S. (1971). Regulation of protein synthesis in heart muscle. I. Effect of amino acid levels on protein synthesis. *J. Biol. Chem.* **246:** 2152—2162.

56) LI, J. B., HIGGINS, J. E., JEFFERSON L. S. (1979). Changes in protein turnover in skeletal muscle in response to fasting. *Am. J. Physiol.* **236:** H222—H228.

57) LI, J. B., GOLDBERG, A. L. (1976). Effects of food deprivation on protein synthesis and degradation in rat skeletal muscles. *Am. J. Physiol.* **231:** 441—448.

58) FULKS, R. M., LI, J. B., GOLDBERG, A. L. (1975). Effects of insulin, glucose, and amino acids on protein turnover in rat diaphragm. *J. Biol. Chem.* **259:** 290—298.

BLOOD-BRAIN AMINO ACID TRANSPORT DURING DEVELOPMENT

J. M. Lefauconnier

INSERUM U 26 (Director: J. M. Bourre)
Hôpital Fernand Vidal
200 rue du Faubourg Saint Denis
75475 Paris Cedex 10
France

INTRODUCTION

Protein synthesis in the developing brain is higher than in adult brain (1) while the capillary density is lower (2). It thus seems probable that there is a difference in blood-brain amino acid transport across the blood-brain barrier mainly by three system: one system for large neutral amino acids (the system L), one system for basic amino acids, one system for acidic amino acids (3) and perhaps a fourth system for small neutral amino acids which is referred to as the ASC system (4). The ASC and acidic systems have a small activity and amino acids are mainly transported by the L and basic amino acid systems. The problem is, that if there is a high blood-brain transport of amino acids in young animals, does it come through a high transport of all amino acids or through specific modifications of some transport systems as seen in other tissues (5).

The techniques used to measure blood-brain amino acid transport were mainly: the brain uptake index technique (6) and an intravenous technique (7). Smith, Takasato, Sweeney and Rapoport proposed two years ago an in situ brain perfusion technique which gives reliable results (8). In immature animals, the same techniques have been used as in adults with some particular difficulties. I will describe the techniques we have successively used and the results we have obtained by using these techniques, then I will discuss them with those obtained by other investigators.

The Brain Uptake Index Technique

This technique was described by Oldendorf in 1971 (6). It consists in the arterial injection of a mixture of the ^{14}C amino acid and tritiated water in the right common artery. The animal is decapitated 5—15 sec. later. The right hemisphere is removed and the tissue sampled is digested in 1 ml Soluene in a water bath at 50 °C. After cooling scintillation mixture is added for the measurement of radioactivity. 10 μl of the injection solution are also counted.

The brain uptake index (BUI) is defined as:

NATO ASI Series, Vol. H20
Amino Acid Availability and Brain Function in Health and Disease. Edited by G. Huether

$$\frac{\frac{^{14}C}{^{3}H}\ \text{in brain}}{\frac{^{14}C}{^{3}H}\ \text{in injectate}} \times 100 \text{ or: } \frac{E_{\text{amino acid}}}{E_{H_2O}} \times 100$$

Where E is the extraction of the amino acid or of water.

In adult animals, this technique has given a lot of information about blood-brain transport of amino acids in particular the competition of large neutral amino acids for the transport system (9). However it gives only the extraction value (E) and it must be associated with cerebral blood flow measurements in order to obtain the influx (J)

$J = FEC$

With:

F : Cerebral blood flow

C : Arterial concentration of the substance studied

In young animals, this technique cannot be used as such because the smallest needle completely occludes the carotid artery and the technique works only if a free flow past the needle persists throughout the procedure. Several modifications have been made to overcome this problem: i) to compare only 15-day-old and adult animals (10), ii) to compare newborn rabbits and adult rats (11), iii) to inject the solution into the left ventricle of the heart (12). We modified this technique by performing a retrograde injection into the

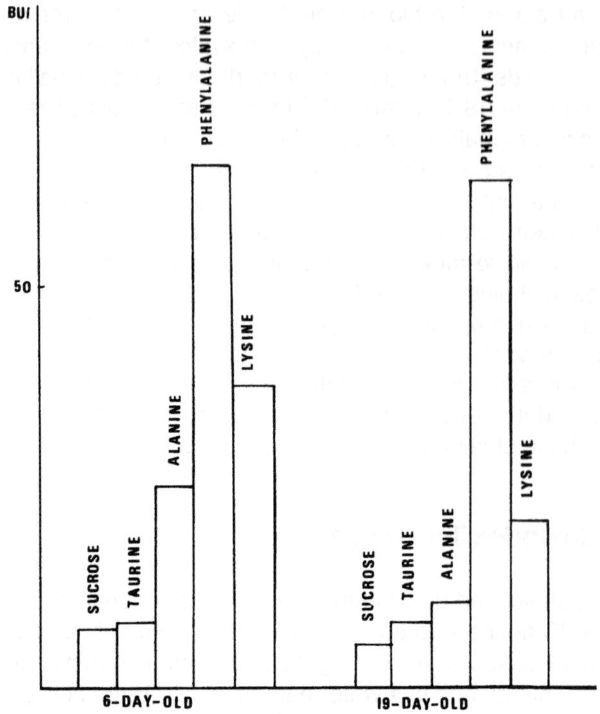

FIGURE 1
Evolution of the BUI of sucrose and 4 amino acids
(taurine, alanine, phenylalanine and lysine) during development

right brachial artery of ether anesthetized rats (13). This resulted in a clear bolus in the axillary and subclavian artery in the retrograde direction and a bolus in the common carotid artery in the orthograde direction. The BUI could thus been estimated. We found that two groups of amino acids were clearly differentiated in 5-day-old rats: one in which amino acids had a BUI close to that of sucrose (8): proline, glycine, glutamic acid, taurine; another in which amino acids had BUIs higher than 25. In 19-day-old rats, the first group had a BUI always similar to that of sucrose but in the second group, 3 subgroups could be further differentiated: neutral amino acids with a BUI greater than 40 (phenylalanine, tryptophane, leucine, methionine), neutral amino acids with a BUI between 10 and 20 (alanine, serine, cysteine, threonine) and the basic amino acid (lysine) with a BUI less than 20 (Fig. 1).

We also studied the self and cross inhibition process. The BUI of all amino acids with a higher BUI than sucrose was decreased by 10 mM concentration of the corresponding amino acid. In cross inhibition studies: MeAIB did not significantly inhibit the transport of any amino acid. Alanine and serine both inhibited the transport of alanine, serine and threonine but not that of leucine. Leucine inhibited the transport of alanine, serine and threonine but this was much less than its self inhibition. Finally, when we replaced sodium by mannitol, the BUI of alanine was decreased by 34 % and that of leucine was not modified.

These results suggest 2 types of transport for neutral amino acids:

- the first, transporting phenylalanine, leucine, methionine and tryptophane among the neutral amino acid studied, is characterized by:
  - a high BUI throughout the development
  - the absence of inhibition by the methylated amino acids
  - the absence of inhibition by the amino acids of the other group
  - a sodium independency
- the second group includes alanine, serine, cysteine and threonine; it is characterized by:
  - a significant decrease in activity during development
  - the absence of inhibition by the methylated amino acids
  - the presence of inhibition via amino acids of the same group, likewise but to a lesser extent via amino acids of the other group
  - a substantial decrease in activity in the absence of sodium

Therefore the first group is similar to the L system described in adult animals, while the second group looks like the ASC system. The data obtained by using this technique suggest that an ASC system of transport is much more active in immature than in 19-day-old animals. However, this technique is not very good as it gives a rather high BUI of sucrose, while this value is low in adult animals. The reason for this high BUI of sucrose is that the retrograde radioactive injectate is into the brachial artery and distributes partly to the general circulation. One part of the radioactive sucrose rapidly goes to brain with the clear bolus, but another part mixed with blood goes to brain later, while the tritiated reference leaves the vessels by entering the tissues. This can be corrected by removing the ^{14}C vascular radioactivity which is the radioactivity measured in arterial blood multiplied by the vascular space. In addition, two problems have been evoked for the use of this technique in adult animals, these problems exist also for young animals: 1) the injection into the carotid artery can change the blood flow, 2) the perfusion fluid can be mixed together with circulating blood. This would alter the concentration of perfusate before the fluid reaches the brain capillary.

The Intravenous Technique

The technique consists in the iv injection of radioactive amino acids at a various rate to obtain a constant radioactive concentration in blood (14). We have used a modification of this method by injecting rapidly in ether anesthetized rats a radioactive bolus and measuring the radioactivity in plasma during the whole period between the injection and decapitation (15). The animal was decapitated at a time varying from 20 to 40 seconds after injection. The brain was then removed from the skull, placed on a ice-chilled filter paper and dissected in several regions. These regions were placed in tared vials that were reweighed. They were digested as described above.

The influx is calculated according to the formula: J = FEC

$$FE = \frac{Q(T)}{\int_0^T Ca.\,dt} \qquad \text{FE is the clearance}$$

Where

Q(T): radioactivity in 1 g brain at time T minus vascular radioactivity

$$\int_0^T Ca.\,dt = \overline{C}a\,T$$

Ca is the mean arterial concentration between time t = 0 and t = T

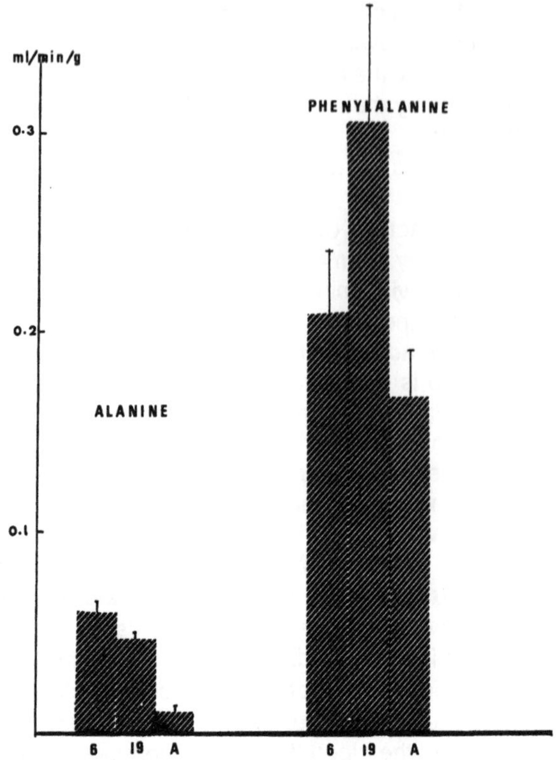

FIGURE 2

Evolution of the clearance of alanine and phenylalanine during development

By this technique we measured the evolution of phenylalanine and alanine transport between 6-day-old and adult animals (Fig. 2). The clearance of phenylalanine was 0.206 ± 0.028 ml/min/g in the parietal cortex of 6-day-old rats, 0.307 ± 0.064 ml/min/g in 20-day-old rats, and 0.171 ± 0.025 ml/min/g in adult rats. The clearance of alanine decreased from 0.063 ± 0.005 ml/min/g in parietal cortex of 6-day-old rats to 0.012 ± 0.005 ml/min/g in the same region of adult rats.

We also compared the information given by this technique and the BUI technique in blood-brain transport of amino acids in hypothyroid animals (15). This transport was studied as capillary density is decreased in immature hypothyroid rats (16). Thyroid hormones have also been shown to alter amino acid transport in brain tissue (17). Daniel, Love and Pratt (18) have shown that the blood-brain influx of leucine, valine and lysine was lower in adult rats made hypothyroid at birth than in control rats. Hypothyroidism was induced by injections of propylthiouracile from day 1 after birth. Blood-brain transport and cerebral blood flow were studied on day 16 after birth.

Cerebral blood flow was greatly decreased in hypothyroid animals (more than 50 %). The decrease was higher in cerebellum (—67 %) than in other regions (—50 % in cortex). The measurement of amino acid transport was made by:

- The brain uptake index technique
  A large increase in extraction (+ 200 %) was observed for alanine, serine and systeine in hypothyroid rats, a much smaller increase was observed for phenylalanine and leucine.
- The intravenous technique, which allowed the measurement of clearance of phenylalanine and alanine. A decrease was observed for phenylalanine and an increase for alanine. The influx was decreased for phenylalanine, and not modified for alanine because the concentration of alanine was slightly decreased in hypothyroid animals.

This experiment shows that the iv technique gives results which are closer to physiological conditions than the brain uptake index technique. In this technique we obtain the influx by multiplying the clearance obtained by the arterial concentration of the amino acid, while in the brain uptake index technique, we had to measure the blood flow and multiply the extraction by the blood flow and then the arterial concentration. However, the iv technique has a great disadvantage: as the radioactive products are injected in blood, it is not possible to obtain the k_m of transport because competition with other amino acids takes place. It seems also possible that even on a very short period of time, a small part of the tracer can be metabolized.

The in Situ Brain Perfusion Technique

This technique has been described about 2 years ago by Smith, Takasato, Sweeney and Rapoport. Recently, it has been applied for studies during development (19). The right cerebral hemisphere of a pentobarbital anesthetized rat was perfused with saline or plasma containing the ^{14}C-labelled neutral amino acid (NAA), ^{3}H-inulin and 0—10 μmol/ml of unlabelled NAA through the common carotid artery. After 20—60 seconds of perfusion, the rat was decapitated and samples from brain and perfusion fluid were analyzed for tracer content. Cerebrovascular NAA influx was calculated from the unidirectional uptake of ^{14}C NAA into brain. Influxes were measured for three NAA's: leucine, cycloleucine and methylaminoisobutyric acid (MeAIB). L-leucine, an amino acid of the L-system was transported from blood to brain with a 2 fold greater rate in immature than in adult rats. This difference was due to a greater maximal transport capacity (V_{max}) of the blood-brain barrier in infant rats. V_{max} for saturable L-leucine influx in 7-day-old rats was not L-leucine specific, and was demonstrated for the nonmetabolized analog, cycloleucine. There were no significant developmental differences in the transport K_m for either L-leucine (0.026 μmol/ml) or cycloleucine (0.29 μmol/ml). At both ages, L-leucine transport was sodium independent and was inhibited by BCH, the model substrate for the L-system but not by MeAIB.

There was a significant saturable component of MeAIB transport into the brain of 7-day-old rats. Calculated values for V_{max} and K_m equaled 5.10^{-4} μmol/s/g and 0.1 μmol/ml respectively. Saturable MeAIB influx was sodium dependent and it was competitively inhibited by L-alanine but not by BCH, which is consistent with transport by the A system. These data demonstrate an increased transport capacity for both small and large neutral amino acids at the blood-brain barrier of infant rats. This is the first in vivo demonstration of A transport system at the blood-brain barrier.

DISCUSSION

Two different results have been observed concerning the blood-brain transport of amino acids during development.

1) A greater influx of amino acids from blood to brain has been obtained in young animals, as compared with adult ones. This has been shown by:

— Sershen and Lajtha (11) who used a modification of the brain uptake index technique by injecting the radioactive products in the heart left ventricle.

— Pardridge and Mietus (10) who compared the uptake in the brain of newborn rabbits with that in the brain of adult rats. Later, Pardridge measured blood-brain transport in adult rabbits and suggested that the differences were due to differences in species and not related to development (9).

— Baños, Daniel and Pratt (13) used a variable intravenous extraction technique. They showed that the influx was much higher in immature animals than in adult animals.

— We also showed that influx from plasma to brain by the L-system was 2 fold higher in 7-day-old rats than in adult rats (19).

2) The existence of a greater number of transport systems in immature than in adult animals:

— We have shown that small neutral amino acids: alanine, serine, cysteine and threonine were transported in immature rats by the ASC system, the efficacy of which decreases during development (15).

— We have also shown that MeAIB was transported into the brain of 7-day-old rats by a saturable system, probably system A (19).

The existence of these systems in immature animals probably means that the young brain is not able to synthesize enough amino acids for protein synthesis. A small part of the amino acids must then be brought to the brain by plasma transport systems.

REFERENCES

1) LAJTHA, A., DUNLOP, D. (1981). Turnover of proteins in the nervous system. *Life Sci.* **29:** 755—767.

2) BÄR, T., WOLFF, J. R. (1972). On the vascularization of the rat's cerebral cortex. In: *Bibliotheca Anatomica No. 11* (Ditzel, J., Lewis, D. H., eds.). S. Karger, Basel.

3) OLDENDORF, W. H., SZABO, J. (1976). Amino acid assignment to one of three blood-brain barrier amino acid carriers. *Am. J. Physiol.* **230:** 94—98.

4) SERSHEN, H., LAJTHA, A. (1979). Inhibition pattern by analogs indicates the presence of ten or more transport systems for amino acids in brain cells. *J. Neurochem.* **32:** 719—726.

5) CHRISTENSEN, H. N. (1973). On the development of amino acid transport systems. *Federation Proc.* **32:** 19—28.

6) OLDENDORF, W. H. (1971). Brain uptake of radiolabeled amino acids, amines and hexoses after arterial injection. *Am. J. Physiol.* **221:** 1629—1639.

7) BAÑOS, G., DANIEL, P. M., MOORHOUSE, S. R., PRATT, O. E. (1975). The requirement of the brain for some amino acids. *J. Physiol* **246:** 539—548.

8) SMITH, Q. R., TAKASATO, Y., SWEENEY, D. J., RAPOPORT, S. I. (1985). Regional cerebrovascular transport of leucine as measured by the in situ brain perfusion technique. *J. Cereb. Blood Flow Metab.* **5:** 300—311.

9) PARDRIDGE, W. M. (1983). Brain metabolism: a perspective from the blood-brain barrier. *Physiol. Rev.* **63:** 1481—1535.

10) CREMER, J. E., CUNNINGHAM, V. J., PARDRIDGE, W. M., BRAIN, L. D., OLDENDORF, W. H. (1979). Kinetics of the blood-brain barrier transport of pyruvate, lactate and glucose in suckling, weaning and adult rats. *J. Neurochem.* **33:** 439—445.

11) PARDRIDGE, W. M., MIETUS, L. J. (1982). Kinetics of neutral amino acid transport through the blood-brain barrier of newborn rabbit. *J. Neurochem.* **38:** 955—962.

12) SERSHEN, H., LAJTHA, A. (1976). Capillary transport of amino acids in the developing brain. *Exp. Neurol.* **53:** 465—474.

13) LEFAUCONNIER, J. M., TROUVÊ, R. (1983). Developmental changes in the pattern of amino acid transport at the blood-brain barrier in rats. *Develop. Brain Res.* **6:** 175—182.

14) BAÑOS, G., DANIEL, P. M., PRATT, O. E. (1978). The effect of age upon the entry of some amino acids into the brain and their incorporation into cerebral proteins. *Develop. Med. Child Neurol.* **20:** 335—346.

15) LEFAUCONNIER, J. M., LACOMBE, P., BERNARD, G. (1985). Cerebral blood flow and blood-brain influx of some neutral amino acids in control and hypothyroid 16-day-old rats. *J. Cereb. Blood Flow Metab.* **5:** 318—326.

16) EAYRS, J. T. (1954). The vascularity of the cerebral cortex in normal and cretinous rats. *J. Anat. 88:* 164—173.

17) RIGGS, T. R., POTE, K. G., IM, H.-S., HUFF, D. W. (1984). Thyroxine-induced changes in the development of neutral α-amino acid transport systems of rat brain. *J. Neurochem.* **42:** 1260—1268.

18) DANIEL, P. E., LOVE, E. R., PRATT, O. E. (1975). Hypothyroidism and amino acid entry into brain and muscle. *Lancet* **2:** 872.

19) NAGASHIMA, T., LEFAUCONNIER, J. M., SMITH, Q. R. (1987). Developmental changes in neutral amino acid transport across the blood-brain barrier. *J. Cereb. Blood Flow Metab.* **Suppl. 1:** 524.

NUTRITION AND SELECTIVE ACCUMULATION PROCESSES ESTABLISH THE FREE AMINO ACID POOL OF DEVELOPING CHICK EMBRYO BRAIN

N. M. van Gelder and F. Bélanger

Centre de Recherche en Sciences Neurologiques
U. de Montréal, et Département de neurobiologie
Institut de Recherche Clinique de Montréal
110 Avenue des Pins ouest, Montréal, (Québec)
Canada H2W 1R7

INTRODUCTION

Control of the fetal amino acid supply resides in a series of interlocking metabolic and environmental factors which are difficult to isolate experimentally. Such factors typically include the diet and alimentary efficiency of the mother, the dynamic influences of her health and hormonal status on these processes, as well as the placental circulation and metabolism. Further control may be exercised by the fetus itself in the form of regulatory entry and/or transport mechanisms governing the transfer of specific (or all) amino acids from blood to embryo, and within the tissues proper. The manner in which these complex variables interact, and their relative importance for fetal brain development, remains uncertain (1, 2). Nevertheless, several lines of evidence seem to suggest a dominant role of the maternal metabolism in regulating the amino acid requirements for the fetus, with the latter only exercising a more selective control over this phenomenon.

Inborn errors of amino acid metabolism (e. g., phenylketonuria) causing amino acedemias and, secondarily, specific amino acid deficiencies, often do not seem to have a decisive impact on brain maturation until after birth (cf. 3). Secondly, the seminal investigations by Stein *et al.* (4) but also by many others (review: 5), indicate that severe malnutrition or Kwashiorkor in pregnant women, similarly, does not cause permanent brain damage provided the diet of the underweight neonate or of nursing mothers is corrected soon after birth (see o. a. 6—11). Many of the data seem to indicate that important maternal compensatory mechanisms become activated when the fetus is threatened by an imbalanced or insufficient amino acid supply (cf. 5). Especially in cases of inborn errors of metabolism in the fetus it can be assumed that in some manner the maternal circulation provides the protection, so that intrauterine brain development continues in (almost) normal manner. This in turn strongly suggests that the free amino acid pool in embryonic tissue is largely governed by the amino acid composition of its nutritional supply. If this is indeed the case, then it would seem imperative to determine when this embryonic dependency on a proper balance of amino acids in the fetal food reserves becomes replaced or

NATO ASI Series, Vol. H20
Amino Acid Availability and Brain Function in Health and Disease. Edited by G. Huether

begins to be modified by endogenous tissue metabolism. The relevance of this issue to the prevention of the potentially damaging effects of dietary or pathological amino acid disturbances during pregnancy is quite clear.

Over the past year this laboratory (12, 13, 14) has begun to address the problem by studying the relationship, in the developing chick, between the amino acid composition of the food supply (the yolk), the intervening extra-embryonic circulation (the vitelline plexus), and three differentiating excitable organs (brain, heart and eye). Unlike the mammalian fetus, the chick embryo is totally dependent on non-renewable and fixed amounts of the individual amino acids present in the yolk. Furthermore, from day four onwards (E_4), the vitelline circulation represents the only means by which the developing organism can access this nutritional pool of amino acids. Finally, by using free valine in the various organs as an internal reference, it was possible to obtain a direct measure of the changing amino acid balance of different tissues in relation to their stage (age) of development, as well as compared to the dynamic alterations of the circulating amino acid composition. Here we present a small portion of the large amount of data which these studies generated, and confine our report to brain only. The methods employed have been fully described in the references quoted.

Amino Acid Content in Yolk, E_2 Embryo

In Table 1 the amino acid balance relative to free valine is shown for the nutritional supply, the still very primitive brain region, and the developing blood plexus which does not yet demonstrate a circulation connected to the embryo. With reference to the essential amino acids, valine in the yolk together with leucine is present in the highest concentrations: 70 and 100 μmoles in 35 g of yolk. Taurine (6 μmoles) and phosphoethanolamine (1—2 μmoles) represent the least abundant amino acids, while among other metabolic, i. e. the so-called non essential, amino acids only glutamic acid (180 μmoles) is found in higher amounts than valine.

During this same period the abundance of amino acids relative to each other in the still immature extra-embryonic blood system (Table 1, row 2), such as LEU vs ILE and TYR vs PHE in roughly 2:1 proportion, remains similar to that in yolk. With reference to valine, however, all ratios are somewhat lower in the blood organ than in the yolk, suggesting a preferential entrance of valine into the developing vitelline plexus. Also, three notable exceptions are apparent. Both taurine and phosphoethanolamine, despite their very low yolk content, have become highly concentrated in the primitive vitelline circulation. Glutamic acid, high in the yolk, is halved in the plexus while an almost stochiometric reciprocal increase of glutamine is seen. Thus, amidation of glutamic acid appears already established in the plexus as well as selective capture mechanisms for two amino acids present in very low amounts in the embryonic food reserves.

In the two day old cephalic region (Table 1, row 3), again a selective increase of phosphoethanolamine and taurine occurs but the remaining valine ratios are very close to those seen in the yolk, with perhaps only a preferential accumulation of isoleucine. For the most part then, and with few exceptions, the brain amino acid profile at the onset of differentiation very closely resembles its food source. This suggests that, until the extra-embryonic circulation has started to function, the embryo obtains most of its amino acids directly by passive diffusion from the yolk. Further confirmation for this was obtained by injections of 100 μmoles valine or up to 200 μmoles of taurine into the yolk after one day

TABLE 1
Amino acid levels relative to valine (± .S.D.) in two day chick embryonic tissues

Embryo E_2 Injected E_1		Essential amino acids										Metabolic acids					
	n	THR	ILE	LEU	PHE	TYR	MET	CTH	TAU	POE	ASP	GLU	GLN	GABA	SER	GLY	ALA
SHAM																	
Yolk (E_0)	3	0.85 (0.12)	0.82* (0.03)	1.54* (0.02)	0.44* (0.01)	0.73* (0.08)	0.37 (0.08)	n. d.	0.08* (0.03)	0.08* (0.01)	0.02* (0.14)	2.59* (0.28)	0.74* (0.07)	n. d.	0.93 (0.10)	0.67* (0.07)	0.81 (0.03)
Vitelline plexus	8	0.72 (0.08)	0.68 (0.03)	1.14 (0.08)	0.29 (0.04)	0.50 (0.06)	0.29 (0.03)	0.07 (0.01)	0.39 (0.10)	1.00 0.23)	1.15 (0.08)	1.35 (0.14)	1.91 0.23)	n. d.	1.03 (0.33)	0.79 (0.18)	0.80 (0.10)
Head region	8	0.86 (0.22)	*0.99 (0.22)	1.32 (0.22)	*0.52 (0.15)	*0.91 (0.23)	0.28 (0.06)	*0.10 (0.02)	*3.84 (1.00)	*5.79 (1.67)	1.13 (0.58)	*2.44 (0.52)	1.78 (0.44)	n. d.	1.07 (0.33)	0.86 (0.18)	0.87 (0.30)
VALINE 50 μm																	
Vitelline plexus	10	0.38† (0.15)	0.43† (0.17)	0.71† (0.29)	0.16† (0.06)	0.31† (0.10)	0.18† (0.07)	0.05 (0.03)	0.21† (0.08)	0.59† (0.22)	0.66† (0.27)	0.81† (0.29)	0.98† (0.39)	n. d.	0.59† (0.23)	0.48† (0.20)	0.48† (0.19)
Head region	13	*0.54† (0.25)	*0.61† (0.23)	0.84† (0.30)	*0.25† (0.13)	*0.43† (0.18)	0.16† (0.06)	0.08† (0.08)	*2.14† (1.20)	*2.92† (1.30)	0.58† (0.26)	*1.56† (0.81)	1.05† (0.52)	n. d.	*0.85 (0.33)	*0.68† (0.23)	0.63 (0.26)
TAURINE 200 μm																	
Vitelline plexus	10	0.70 (0.12)	0.69 (0.05)	1.20 (0.11)	0.29 (0.04)	0.54 (0.09)	0.29 (0.03)	0.07 (0.01)	4.85† (3.74)	1.09 (0.40)	1.37† (0.22)	1.65† (0.46)	1.86 (0.47)	n. d. (0.30)	1.02 (0.16)	0.77 (0.10)	0.89 (0.25)
Head region	13	*0.96 (0.20)	1.11 (0.36)	*1.42 (0.33)	0.38 (0.20)	0.95 (0.43)	*0.22 (0.04)	*0.11 (0.04)	*10.51† (7.36)	*6.15 (1.92)	1.15 (0.58)	*3.75† (1.39)	*2.48† (0.92)	n. d.	*1.46† (0.38)	*1.20† (0.30)	*1.13 (0.35)

† against proper sham control
* against vitelline plexus
$p \leqslant 0.05$ using Standard Error of the Difference between two different means as a significance criteria rather than Student's t test. For small groups, the latter method is less reliable and more likely to lead to false suggestions of significant differences.

of incubation (Table 1; 12). Following such manipulations of the amino acid balance in the nutritional reserve, both the plexus and the brain region exhibit corresponding increases of, respectively, either valine or taurine (Table 1). With valine, all ratios decrease but no other modifications occur, indicating that an increase of valine as such has no effect on the amino acid balance except to increase tissue valine levels. In contrast, an additional amount of taurine in the yolk causes glutamic acid and aspartic acid to increase in the plexus, while in the brain an elevation of glutamic acid and glutamine is observed, besides the further accumulation of taurine; and increase of serine and glycine is also noted (see 15). Hence, neither in the brain nor in the circulation can an excessive elevation of amino acids be prevented when the nutritional supply becomes altered. Moreover, the work at present in progress clearly reveals that early metabolic changes, such as are produced by excessive taurine, can cause damage to the central nervous system.

Excessive Taurine During Embryogenesis: Ataxia

When after 100 μmoles of valine or taurine injections on days 1, 7, or 15 the embryos are allowed to develop and hatch, more then 85 % of the 30 valine treated chicks appeared normal. The only most consistent difference between these and saline treated, control chicks (50 μmoles NaCl) was that they hatched more often 24—12 h earlier. On the other hand, taurine, especially when injected between E_7—E_{15}, caused in well over 75 % of the hatchlings (40 chicks) from moderate to severe hypotonia, staggering gate and crooked or paralysed toes. Among the eggs injected with 150 μmoles or more of taurine, only approximately 20 % hatched successfully, with the majority of the chicks appearing seemingly too weak to peck out of the egg. The biochemical and anatomical details implicated in this phenomenon are as yet under study and it appears premature to discuss these here. However, from some of the results it is already quite certain that at least until day 9 of chick embryogenesis, the developing organism has few mechanisms in place to protect itself against an amino acid excess. This is further corroborated by injections of 50 μmoles glycine or GABA, and even less, which proved practically 100 % lethal to the embryos when introduced into the nutritional supply before day 7.

Amino Acid Extraction into the Blood Plexus and Brain

Most amino acids in the yolk are extracted by the circulation in proportion to their concentration between E_2—E_9 of embryogenesis, with circulating levels showing a peak between E_7—E_{10} (Figs. 1, 2). Using free valine content of a tissue as an index of net protein synthesis (16), Figure 1A suggests that the rate of its utilisation in the developing mesencephalon is faster up to day 7 than the amino acid can be obtained from the blood organ, which extracts valine in increasing amounts from the non-renewable nutritional supply (Fig. 1A). At the time that peak levels are present in the circulation (E_7—E_{10}), only slightly more free valine accumulates in the tissue; as blood concentrations begin to fall, tissue content resumes its decrease. Throughout the 15 days of embryogenesis, the amount of valine used by the developing brain seems to exceed the amount which can be replaced from the circulation; tissue free valine levels are therefore lower than in the blood. This pattern of change may be contrasted with that observed for glutamic acid (Fig. 1B). Similar to valine, the tissue content at E_4 is higher than for the circulation, but

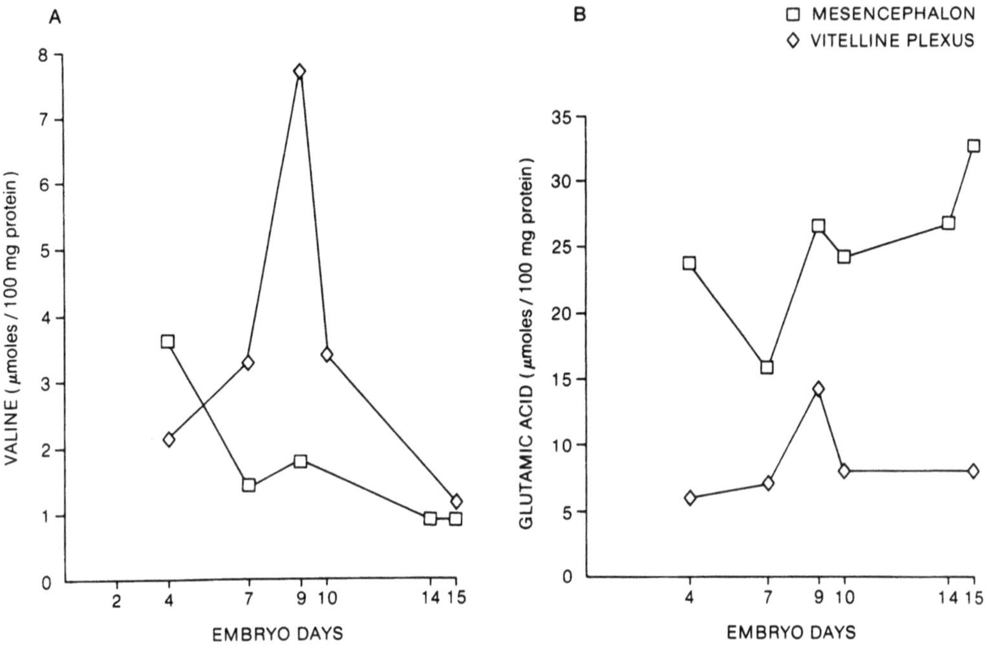

FIGURE 1
Valine and glutamic acid levels, standardised to 100 mg protein, in the extra-embryonic circulation and neural tissue during 15 days of chick embryo development. Note that even during peak blood levels of valine (A), at best only a small increase of brain valine occurs. Overall, from E_4 onwards, valine assimilation into proteins exceeds its rate of replenishment by the blood, resulting in a gradual decrease of the free valine levels in neural tissue. By day 15, exhaustion of essential amino acids, such as valine, in the non-renewable yolk supply may place a limit on further brain development, until the post-hatching period. In contrast, tissue glutamate (B) always stays more elevated than the blood levels. Between E_4—E_7, rapid protein accrementition accounts for the apparent fall in brain glutamic acid. As circulating levels increase (E_7—E_{10}), the balance between protein and glutamic acid content is redressed and becomes restored to what it was on day 4. Only around day 14 does it seem that endogenous synthesis or further active accumulation of glutamic acid begins to occur.

this difference is much more pronounced than for valine. The brain to blood glutamate difference is 5 fold on day 4, whereas for valine it is at most only twice as much. Subsequently (E_4—E_7), during the phase when the protein content of the differentiating neural tissue increases rapidly, an apparent fall in free glutamic acid is seen, although the content stays above blood levels; as the glutamate levels in the blood rise, the balance between free glutamic acid and protein content is redressed to what it was on day 4. Finally, when blood glutamate begins to decrease and then steadies, the amino acid is retained by the brain. By day 14, endogenous synthesis or (further) active extraction of glutamate form the circulation begins to occur.

If instead of changing protein content, the free amino acid pool is expressed directly with reference to valine, one can demonstrate that the embryonic circulation for the most part maintains an essential amino acid balance which reflects the amino acid composition of the yolk. Their tissue levels, in turn, closely follow the fluctuations observed in the blood plexus (13, 14). On the other hand, a few amino acids are actively extracted from the yolk by the tissues even before E_4, when the vitelline circulation is not yet fully developed. This is exemplified in Figure 2, starting with, as point of comparison, glutamic acid (2A, compare with 1B).

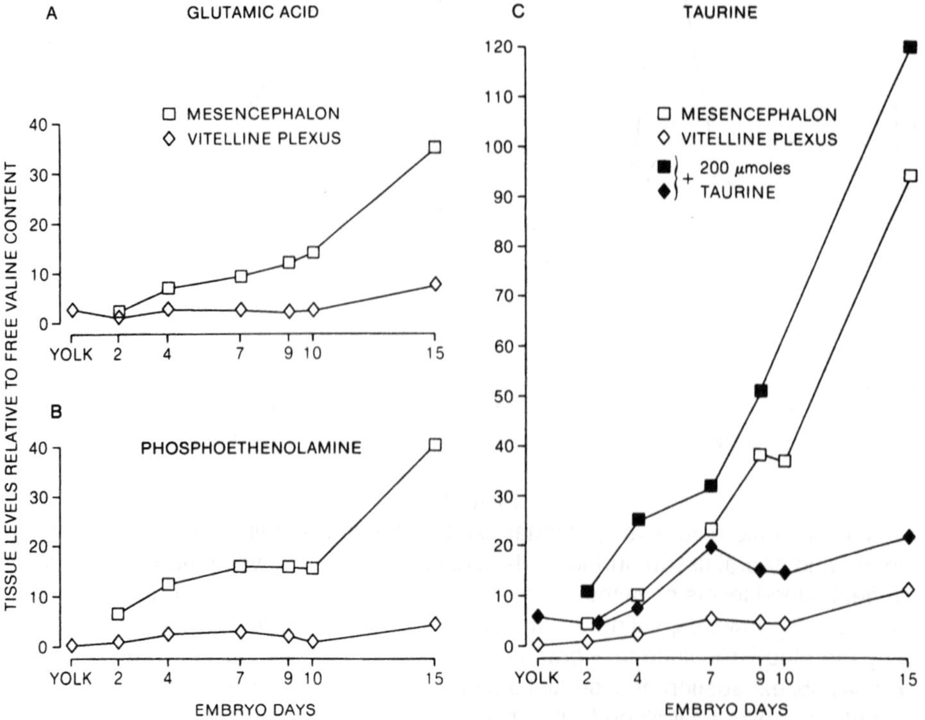

FIGURE 2

Amino acid accumulation with reference to free valine, used as an index of embryonic maturation (protein accrementition). A. The two fold ratio of glutamic acid to valine found in yolk (Table 1 and y-variate) is more or less maintained in the blood plexus, with some relative glutamate increase occurring (E_9—E_{15}) as blood valine declines. In the brain, starting with a similar ratio on day 2 glutamic acid accumulates steadily until E_{10}, when a steep rise is seen; this is in part, but not entirely, accounted for by the gradual decrease in tissue free valine. B. Phosphoethanolamine is present at only 10 % of the valine concentration in yolk but the metabolite is nevertheless more actively concentrated in neural tissue between E_0—E_4. The blood supply to the embryo also maintains a higher valine ratio than is found in the yolk. C. Taurine accumulates even more rapidly and for a longer period (E_9). The amounts which neural tissue extracts, first directly from yolk (E_0—E_4) and then from the blood supply, is determined by the quantities available in the nutritional source. When extra taurine is injected into the yolk (black symbols), the brain in the early stages of development contains at least twice the usual amounts of taurine (E_2—E_7). However, the rate of accumulation during the second week of development does not increase, despite the much higher blood taurine levels. This is but one demonstration that the early embryo, especially, is most susceptible to an unbalanced amino acid supply.

Amino Acid Content Relative to Valine

Throughout the first 10 days of embryogenesis the glutamic acid content of the extra-embryonic circulation stays constant relative to valine and at roughly two fold higher levels (see YOLK, y-variate). As valine decreases in the blood (1 A), glutamic acid and practically all other amino acids begin to represent a larger percentage of the total circulating amino acid pool. The glutamic acid content of brain gradually increases in relation to valine until day 9, when this amino acid becomes one of the dominant constituents of the free amino acid pool in neural tissue. As tissue valine decreases (E_{10}—E_{15}) from 2 μmoles/100 mg protein to 1 μmole (Fig. 1 A), glutamic acid content increases 3 fold relative to valine.

Even though phosphoethanolamine (1 μmole total yolk) and taurine (6 μmoles total) are present in yolk at much lower concentrations than glutamic acid (180 μmoles) or valine (70 μmoles), their blood levels are as high or higher than for these last mentioned amino acids (Fig. 2B, 2C). The amounts which accumulate in the brain are also much higher than for glutamic acid, especially in the case of taurine. These amino acids are therefore selectively extracted from the yolk by the plexus, as well as from the circulation by the brain. Note that much of the accumulation occurs before day 7 for phosphoethanolamine and until day 9 for taurine, the remaining three fold relative increase being accounted for in part by the decline of valine during that same period (E_{10}—E_{15}). At least in terms of taurine, the maximum is determined by the amount of the amino acid available. Following an injection of taurine (200 μmoles) both blood and tissue levels increase their taurine content, especially during the early embryonic period, and even before the extra-embryonic circulation becomes functional (E_2—E_4, Fig. 2C).

Whether this accumulation during the early period of development might be by passive diffusion (glutamic acid, valine) or by a dynamic process (taurine, phosphoethanolamine), the evidence indicates that it is limited and determined by the supply of an amino acid to which the embryo is exposed, first directly and later via the blood. One suspects that the success of the mother to maintain a balanced amino acid composition in the blood supply to the fetus plays an equally dominant role in preventing irreparable brain damage during development.

High Taurine During Embryogenesis: A Genetic Trait?

As one example one may consider Friedreich's Ataxia which represents a recessive autosomally inherited neurological disorder, marked by cardiac myopathy and a gradual development of ataxia in association with peripheral sensory failure (17, 18). Such patients have been shown to demonstrate enhanced alimentary uptake of taurine when presented with high dietary taurine levels (Fig. 3). In these individuals the rate of disappearance of taurine from the blood (Fig. 3) as well as its rate of excretion (19) remain proportional to the circulating levels. It is possible therefore that the high taurine content found in the heart and cerebellum of such patients (18, 20) reflect a genetic inability of tissues to restrict the entrance of taurine, thereby causing an excessive accumulation of the amino acid when dietary exposure is high. Unaffected individuals, in contrast, demonstrate remarkably steady and tissue specific taurine levels, which are usually independent of fluctuating blood concentrations (cf. 21). A genetic mutation of the F.A. type, in combination with a natural tendency of pregnant and lactating women to conserve taurine (22), would thus expose the fetus of F.A. mothers to unnaturally high blood levels of taurine during early development as well as during the neonatal nursing period. Alternatively, the fetus or especially, the neonate, having inherited from the father a propensity for increased taurine uptake from maternal blood or milk (23), could run a similar risk. These possibilities remain to be investigate but the analogy between the symptoms observed in chick hatchlings exposed embryonically to higher than normal taurine levels, and those gradually appearing in maturing F.A. patients, may not be fortuitous. It is evident, however, that much further work will be needed before one can state unequivocally whether a nutritional model of (very) excessive taurine exposure during development will be useful in the study of this rather unpredictable neurological disease.

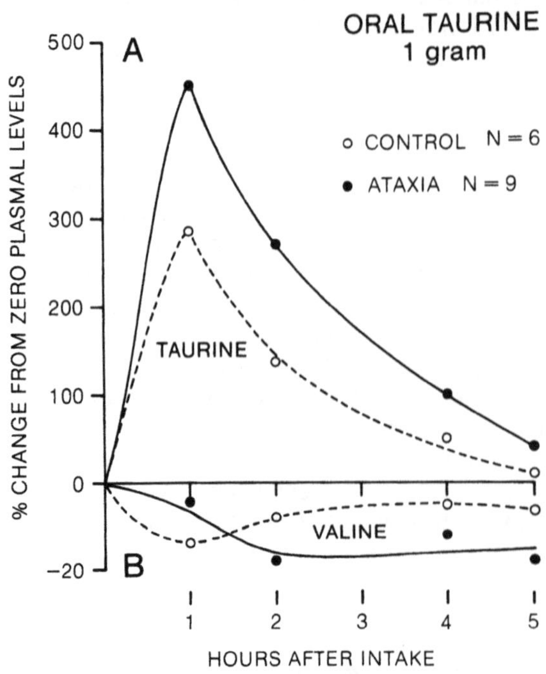

FIGURE 3

Friedreich's Ataxia patients demonstrate closely similar blood amino acid concentrations after 18 hrs of fasting. However, when challenged with a 1 g dietary load of taurine, the patients show higher blood taurine levels. This represents enhanced alimentary absorption, because the five hour rates of decline in the blood (cf. Fig.) and of excretion (19) remain proportional to blood levels, and normal. Since taurine content in heart and cerebellum are also higher in this genetic disorder, the affected individuals may tend to absorb more than the usual amounts of taurine into their cells, either because during fetal life the mother's blood has elevated taurine levels (cf. 24) or they inherited this tendency from the father. The mother's milk would also contain more taurine (cf. 22, 23) and the maturing cerebellum and heart of the nursing neonate may thus be further exposed to unnaturally elevated taurine levels; the developing symptoms are not unlike those seen in chicks presented with extra taurine during embryo-genesis (from 20 with permission of Alan R. Liss, Inc.).

ACKNOWLEDGEMENT

This work, carried out at the Institut de Recherche Clinique de Montréal, was made possible by a generous donation from D. Bloom to N. van Gelder with additional funding provided by Dr. M. Chrétien, Director of the Institute.

REFERENCES

1) ANDERSON, G. H. (1979). Control of protein and energy intake: Role of plasma amino acids and brain neurotransmitters. *Can. J. Physiol. Pharmacol.* **57:** 1043—1057.

2) WINICK, M. (1975). Effects of malnutrition on the maturing nervous system. *Adv. Neurol.* **13:** 193—246.

3) HSIA, Y. E., WOLF, B. (1981). Disorders of amino acid metabolism In: *Basic Neurochemistry* (Siegel, G. J., Albers, R. W., Agranoff, B. W., Katzman, R., eds.). Little, Brown and Co., Boston, pp. 563—600.

4) STEIN, Z., SUSSER, M., SAENGER, G., MAROLLA, F. (1972). Nutrition and mental performance. *Science* **178:** 708—713.

5) VAN GELDER, N. M. (1984). Malnutrition, cerebral excitability and intelligence. *Microscopia Electr. Biol. Celular* **8:** 227—243.

6) CHUGANI, H. T., PHELPS, M. E. (1986). Maturational changes in cerebral function in infants determined by ^{18}FDG positron emission tomography. *Science* **231:** 840—843.

7) FREEDMAN, L. S., SAMUELS, S., FISH, I., SCHWARTZ, S. A., LANGE, B., KATZ, M., MORGANO, L. (1980). Sparing of brain in neonatal undernutrition: amino acid transport and incorporation into brain and muscle. *Science* **207:** 902—904.

8) LATHAM, M. C. (1974). Protein-calorie malnutrition in children and its relation to psychological development and behavior. *Physiol. Rev.* **54:** 541—565.

9) RAKIC, P., BOURGEOIS, J.-P., ECKENHOFF, M. F., ZECEVIC, N., GOLDMAN-RAKIC, P. S. (1986). Concurrent overproduction of synapses in diverse regions of the primate cerebral cortex. *Science* **232:** 232—235.

10) SHAHBAZIAN, F. M., JACOBS, M., LAJTHA, A. (1986). Regional and cellular differences in rat brain protein synthesis in vivo and in slices during development. *Int. J. Devel. Neuroscience* **4:** 209—215.

11) WINICK, M., MEYER, K. K., HARRIS, R. C. (1975). Malnutrition and environmental enrichment by early adoption. *Science* **190:** 1173—1176.

12) VAN GELDER, N. M., BÉLANGER, F. (1987a). Methods to study changing free amino acid pools during chick development. *J. Neuroscie. Res.* In press.

13) VAN GELDER, N. M., BÉLANGER, F. (1987b). The development of the amino acid pools in chick embryo brain, heart and eye: taurine, valine, glutamine, phosphoethanolamine. *J. Neuroscie. Res.* In press.

14) VAN GELDER, N. M., BÉLANGER, F. (1987c). The accumulation of essential and metabolically derived amino acids in developing chick embryo organs. *J. Neuroscie. Res.* In press.

15) VAN GELDER, N. M., DRUJAN, B. D. (1978). Interrelated changes of amino acids in the retina and optic tectum of a marine fish with alterations of illuminating conditions. *Brain Res.* **159:** 137—148.

16) DUNLOP, D. S., VAN ELDEN, W., LAJTHA, A. (1975). A method for measuring brain protein synthesis rates in young and adult rats. *J. Neurochem.* **24:** 337—344.

17) AIRAKSINEN, E. M., PARTANEN, J. (1985). Muscle taurine in chicken and diseased muscle. *Prog. Clin. Biol. Res.* **179:** 407—412.

18) BARBEAU, A. (1984). The Quebec Cooperative Study of Friedreich's Ataxia: 1974—1984 — 10 years of research. *Can. J. Neurol. Scie.* **11:** 646—660.

19) FILLA, A., BUTTERWORTH, R. F., BARBEAU, A. (1979). Pilot studies on membranes and some transport mechanisms in Friedreich's Ataxia. *Can. J. Neurol. Scie.* **6:** 285—289.

20) VAN GELDER, N. M., ROY, M., BÉLANGER, F., PARIS, S., BARBEAU, A. (1987). Subtle defects in the regulation of the free amino acid balance in Friedreich Ataxia: A relative deficiency of histidine in combination with a mild but chronic hyperammonemia. In: *Basic and Clinical Aspects of Nutrition and Brain Development* (Rassin, D. K., Haber, B., Drujan, B. D., eds.). Alan R. Liss, Inc., New York. In press.

21) VAN GELDER N. M., PARENT, M. (1982). Protein and taurine of maternal diets: permanent effects on cerebellar-brain stem amino acid levels in mature offspring. *Neurochem. Res.* **7:** 987—998.

22) ARMSTRONG, M. D. (1973). Decreased taurine excretion in relation to child birth, lactation and progestin-estrogen therapy. *Clin. Chim. Acta.* **46:** 253—256.

23) RASSIN, D. K. (1984). Nutritional requirements for the fetus and the neonate. In: *Neonatal Infections — Nutritional and Immunological Interactions* (Ogra, P. L., ed.). Grune and Stratton, Inc., pp. 205—227.

24) CHESNEY, R. W., FRIEDMAN, A. L., ALBRIGHT, P. W., JAX, D. K., GINGERY, R., GUSOWSKI, N. (1981). Studies on the renal handling of taurine: changes during maturation and after altered dietary intake. *Adv. Expr. Med. Biol.* **139.** 47—63.

CONSEQUENCES OF EXPERIMENTAL MODULATIONS OF AMINO ACID AVAILABILITY DURING BRAIN DEVELOPMENT

Gerald Huether

Forschungsstelle Neurochemie
Max-Planck-Institut für experimentelle Medizin
Göttingen, Germany

GENERAL CONSIDERATIONS

There is no question that amino acid imbalances, as many other nutritional imbalances, can affect the developing brain, can alter the process of its structural and functional maturation. The question for us is: How can we exactly assess, identify and prevent the circumstances under which nutritional imbalances will cause irreversible alterations of normal brain development? There is no simple answer to this question. It took us quite a while to realize that one and the same nutritional imbalance may have rather different consequences, that its outcome will depend on several other factors.

The first factor is the state of maturity of the developing brain when it is confronted with a particular nutritional imbalance. There are certain periods during brain development when the brain, or better: the developmental processes that take place in certain areas of the brain, are particularily vulnerable to changes of their nutritional supply. We have learned a lot about the role of the time of onset of nutritional imbalances on the neurologic outcome in children with severe protein malnutrition or in PKU-children whose dietary treatment was started too late.

The second factor interfering with the outcome of nutritional imbalances during brain development is already much less understood: It is the interaction of nutritional inadequacies with other extrinsic or intrinsic prerequisites of normal brain development. Some of these extrinsic factors can be other nutrients which may enhance or suppress the effects of an inadequate supply of certain nutrients on the developing brain, but they can also be other environmental factors, reaching from external temperature and daylight exposure to the intoxication by environmental noxes or the kind and degree of emotional and sensory stimulation or stress. Intrinsic factors which may enhance or suppress the consequences of nutritional imbalances on the developing brain may be individual variations in the efficiency and the state of maturation of the peripheral metabolic pathways and mechanisms involved in the control of the brain's nutrient supply.

And, as if this would make the picture not yet sufficiently complicated, a third factor additionally contributes to the outcome of nutritional imbalances on the developing brain. The role of this important factor is even less understood, because it acts not prior to, not together with, but after the nutritionally caused defect of the developing brain. It is plas-

NATO ASI Series, Vol. H20
Amino Acid Availability and Brain Function in Health and Disease. Edited by G. Huether

ticity. Only in rare, exceptional cases we get an impression of the tremendous plasticity exerted by a damaged developing brain. In all other cases we have no idea to what extent plasticity, by means of structural and functional rearrangements does contribute to remodel and to rearrange the brain's connectivity, and to make the original defect vanish — at least in terms of subsequent brain function. Or, the other way around, without plasticity the effects of nutritional imbalances on the developing human brain may be much more dramatic than they — from a functional point of view — seem to be.

The brain's plasticity to compensate for certain developmental defects creates an additional problem which is of equal theoretical and practical importance, and which is related to the reversibility and irreversibility of developmental processes in general, and of developmental failures in particular: If the proper formation of a certain structure or a specific neuronal pathway is disturbed, we end up with structural aberrations, such as deficits in cell number and connectivity. All structural deficits of the developing brain are irreversible in later life. If, however, other neurons, by increasing their dendritic arborization or by additional axonal sprouting and synapse formation will compensate for the original defects we may not see any functional deficit in later life. The conclusion that the consequences of early nutritional deprivation, for example, can be reversed in later life, is wrong. There is no way to reverse any process during brain development, once it became manifested in terms of structural changes. It is the change of structure that, on the one hand, represents an irreversible final stage of a developmental process and that, on the other hand, becomes the starting point — the basis of all future developmental events.

In order to differentiate between primary causes and secondary consequences of nutritionally induced developmental failures, one has to extrapolate from findings that have been made in animal models designed to investigate the effects of well defined, e. g., amino acid imbalances on the developing brain. Nevertheless, one has to be aware that these extrapolations may easily overemphasize the actual contribution of an individual factor, in particular, if the original findings were obtained under well described experimental conditions which exaggerate the role of specific factors, and which do not apply to other, more complex situations. These are some of the most pressing problems we are confronted with in our search for consequences of an altered amino acid availability during brain development.

FACTORS AFFECTING THE AVAILABILITY OF AMINO ACIDS TO THE DEVELOPING BRAIN

To some extent, the free amino acid pool of the developing brain can be used as a mirror of the brain's amino acid availability. Data on concentrations of free amino acids in the brain tissue are easily obtained, and characteristic differences have been reported between species, and within one and the same species, between various brain regions. The concentrations of free amino acids in the brain or in individual brain regions are affected by the metabolic state of the organism and by its state of activity. Metabolic diseases and inborn errors of amino acid metabolism, hormonal imbalances, age and sex have all been shown to affect the size and the composition of the free amino acid pool of the developing brain.

Nevertheless, our understanding of the metabolic causes and of the physiologic significance of these changes is still very poor. Alterations of the concentrations of individual amino acids observed in the brain tissue under various experimental conditions have often been attributed to an altered amino acid transport at the blood-brain barrier. The concept of carrier-mediated amino acid transport systems at the blood-brain barrier has been very useful to explain some effects on brain uptake of amino acids, e. g., by competition. However, the contribution of pure transport phenomena to the size and composition of the free amino acid pool of the brain should not be overemphasized: Failures of amino acid metabolism (phenylketonuria or experimental hyperphenylalaninaemia (1)), alterations of nutritional amino acid supply (feeding of amino acid enriched diets (2)), metabolic strain (saline injections (3)), or sensory stimulation (light exposure (4)) and stress (immobilization (5) visual isolation (4)) have all been shown to affect the free amino acid pool and the availability of individual amino acids in the developing brain. In each case, the initial acute effects are followed by long-term responses, i. e., metabolic adaptation and counterregulation. The free amino acid pool appears to be regulated, very similar to an ecologic system, by a subtle steady state between a variety of extrinsic and intrinsic factors.

Intrinsic determinants are the rates of amino acid uptake and efflux, their metabolization, their utilization for protein synthesis, and their liberation from proteolysis. Although no experimental data are available at present, it seems very likely that each of these factors can be modified by activators and inhibitors of neuronal activity, by the state of activity of individual pathways, and maybe even by the release of transmitters and modulators. Extrinsic determinants of the regulation of the free amino acid pool of the developing brain are all those factors that affect the availability of amino acids from blood circulation, e. g., shifts of the metabolic state and the amino acid supply of the body, the ecologic supersystem in which the brain is embedded.

Alterations of the availability of individual amino acids may therefore be considered as signals, in the sense that they transmit information on shifts of the metabolic state of the body or of individual pathways, and that they provoke specific responses to restore the original equilibrium. The adult brain is able to activate humoral or neuronal feedbacks to restore such a shifted steady state. In the developing brain, these feedbacks are not yet elaborated. Therefore, its process of growth and maturation must adapt itself to the altered amino acid supply. Protein synthesis, cell proliferation and cell death, axonal outgrowth, pathway formation, transmitter metabolism, e. g., may become shifted in response to an altered availability of amino acids during brain development.

CONSEQUENCES OF AN INADEQUATE AMINO ACID AVAILABILITY ON PROTEIN SYNTHESIS AND BRAIN DEVELOPMENT

During brain growth, protein anabolism is greater than protein catabolism. The amino acid requirement is therefore exceptionally high and neuronal protein synthesis is particularily dependent on an adequate supply of at least those amino acids which are essential to brain growth. There is a good deal of evidence that short term as well as moderate variations of the brain's amino acid supply do not affect the process of protein accretion in the developing brain significantly (see contribution of A. Lajtha, this volume). But what about the huge and chronic alterations of the brain's amino acid supply encountered in certain diseases, in inborn errors of metabolism or in protein undernutrition?

An excellent and paradigmatic model to study the influences of such rather severe amino acid imbalances on the developing brain is experimental hyperphenylalaninaemia (hyper-Phe). Suckling rats can be rendered hyperphenylalaninaemic by the daily injection of phenylalanine together with alpha-methyl-phenylalanine, an in vivo inhibitor of phenylalanine hydroxylase (6). Under these conditions, blood Phe level is elevated constantly by 20—30 fold, whereas almost all other amino acids are depleted in serum (about 50 % of control values) and in the brain (about 70—80 % of control values (7)).

The observation that protein synthesis is disturbed under hyper-Phe conditions has been made by several other groups (8, 9, 10, 11). Concomitantly with the decreased incorporation of labelled amino acids, a disaggregation of polyribosomes due to an increased RNAase activity has been found (12). Supplementation of other large neutral amino acids reversed these effects (13). Obviously, the disturbed protein synthesis is primarily due to the inadequate availability of other amino acids in hyper-Phe.

Two features of this irregular protein synthesis are most remarkable:

(a) The synthesis of the more complex protein species seems to be more affected than that of the simpler, short-chained ones. This has nicely been demonstrated in respect to the various myelin-associated proteins some years ago by Hommes et al. (14). They found a selective increase in the turnover rate of the smaller myelin proteins in hyper-Phe rats. The rate of synthesis of the larger myelin protein species, however, was decreased, and so was myelin deposition.

(b) In general, protein synthesis of neural cells seems to be most vulnerable to an inadequate supply of amino acids during the phases of most rapid growth of individual brain structures, when net protein anabolism, and therefore the amino acid requirement is particularily high. In suckling rats, the cerebellum is the most rapid growing brain region. Between day 5 and day 15 post partum, a massive proliferation of neuroblasts in the external granular layer generates all the granule cells settling down in the internal granular layer. It is the proliferation of these granule cell precursors which is severely affected under hyper-Phe conditions (15). Between day 10 and day 20, a huge dendritic tree is grown out by the Purkinje cells, and again, it is this event of rapid growth which is retarded in hyper-Phe rats (16). The outgrowth of axons is another process requiring massive protein anabolism. The effects of hyper-Phe on this process can easily be monitored in cultured chick spinal ganglia. Rising Phe-concentrations in the medium from 0.1 to 2.0 mM cause a dose-dependent reduction of axonal outgrowth during 24 hrs. This growth retardation is abolished when other amino acids are supplemented to the medium (unpublished results).

A decreased rate of protein synthesis during these phases of proliferation and outgrowth of processes in the developing hyper-Phe brain is sufficient enough to affect the tight sequential timing of interdependent neurodevelopmental events, and will therefore result in an increased rate of cell death (15). Such deficits in cell number, synaptic contacts and connectivity are irreversible structural changes that limit the further developmental potencies of the brain.

Untreated PKU is only one example where the developing brain is confronted with an inadequate availability of amino acids essential to its growth and development. Other inborn errors of amino acid metabolism are characterized by a similar elevation of certain amino acids and a concomitant depletion of other amino acids in the blood and in the brain. Severe protein undernutrition will also restrict the amino acid supply to the developing brain, and in undernourished suckling rats, very similar to hyper-Phe rats, cerebellar development is most affected, and within the cerebellum, similar morphogenetic deficits are found (15).

These effects of undernutrition, of course can be prevented by an increased amino acid supply. The effects of hyper-Phe can also be prevented, when only the depletion of other amino acids in the blood is restored. In dietarily treated PKU-children, the depletion of other amino acids disappears, when their Phe intake is reduced. And it is this effect, rather than the prevention of the accumulation of Phe, that appears to be the reason for the success of the dietary treatment: In suckling hyper-Phe rats, a situation can be created where the depletion of other amino acids in the blood and in the brain is restored, but the Phe-elevation is maintained, simply by the injection of an additional dose of lysine (17). In spite of the persisting elevation of Phe, brain development is then no longer affected. The growth retardation and the morphogenetic deficits are no longer seen, and even the depletion of serotonin and HIAA in the brain is restored (18, 19).

CONSEQUENCES OF AN ALTERED AMINO ACID AVAILABILITY ON NEUROTRANSMITTER FORMATION AND DEVELOPMENTAL SIGNALLING

Even though severe amino acid imbalances can affect protein anabolism and brain development, a possible role of the availability of amino acids as signals involved in the regulation of protein synthesis remains speculative unless future experiments will show that the rate of the accretion of certain proteins is selectively affected by an increased or reduced availability of individual amino acids. At present, much more data have been published supporting the idea that the availability of amino acids may be involved in the regulation of developmental events via another mechanism, i. e., by precursor control of the synthesis of important mediators of intercellular communication. Among other "neurotransmitters" serotonin and catecholamines have been found at early, prenervous stages of ontogenetic and phylogenetic development. The interaction of these "prenervous" transmitters with their receptors affects a variety of morphogenetic events during embryogenesis, including, at least in sea urchin embryos, the rate of cell division and of protein synthesis (see contribution of J. A. Wallace, this volume). At early phylogenetic stages, e. g. for the free living cells of Tetrahymena, the term "neurotransmitter" does not make much sense anyhow. Here, as well as during early ontogenesis, these informational substances are involved in the control of transscription and translation of genetic information via the activation of second messenger systems and protein phosphorylation, or they modulate other basic cell functions, like contractility, migration and intracellular transport (for references see 20).

For these reasons, several investigators have proposed that these monoamines may have some pre-transmission (trophic) function prior to their use as mediators of chemical neurotransmission at the synapse (21, 22, 23). A summary of such proposed functions of, e. g., serotonin is given elsewhere (20). Specifically, during neurogenesis and brain development, these monoamines play a part in the regulation of, e. g., neural tube closure, neuronal and glial differentiation, cell migration, outgrowth of processes and synaptogenesis (24, 25, 26, 27).

Given this evidence for monoamines as coordinating mediators of intercellular communication during neural ontogeny, the possible control of their formation by the availability of their respective precursor amino acids would be of outstanding importance: (a) The normal occurring alterations of the availability of Trp and Tyr during embryogenesis and brain development may be part of the control-mechanism for appropriate amounts of amines being synthesized and released at appropriate developmental ages by the appropriate cell populations. (b) Any experimental alteration of the availability of these precursors would interfere with this normal signalling pattern, and may thereby affect normal development.

There is little doubt that the formation of serotonin is affected by the availability of Trp not only in the adult brain, but also during early development. In the two day old chick embryo, for instance, administration of Trp will intensify the serotonin-related fluorescence in the chick notochord and in the cells of the ventral roof of the neural tube (28). Similarily, at later developmental stages, an acute injection of Trp will cause a prompt elevation of not only brain serotonin, but also of 5-HTP and 5-HIAA (20). These results indicate that the amount of serotonin produced by a certain cell population is dependent on the availability of Trp. The availability of Trp to the developing embryo is primarily dependent on the amount of Trp that can be released from the protein stores of the egg, or in mammals, on the maternal Trp-supply. The availability of Trp to individual developing cell populations is dependent on what is left by the others, i. e. on the rate of Trp-consumption for protein anabolism, of Trp-metabolism and of Trp-release by proteolysis in other cells of the developing embryo. An increasing availability of Trp to the whole embryo throughout embryogenesis affect this steady state. The injection of additional Trp into the unincubated chicken egg does not cause significant changes of concentrations of serotonin and 5-HIAA in the developing brain. However, the ability of the brain of these chick embryos to produce a certain amount of serotonin within a limited time in response to a rise of its Trp-supply is remarkably increased by this treatment. This increased serotonin-synthesizing capacity is measurable throughout embryogenesis and can still be detected at the 20th day after hatching (20).

One cannot deny the possiblility that shifts in the nutritional supply in a given population will cause shifts in the amount and the proportions of essential substrates (e. g. Trp) made available to the descendants by their parents. It is possible that these shifts during early ontogenesis will influence and modify all those processes of metabolic and phenotypic maturation which are dependent on a well defined availability of a particular amino acid.

To test this assumption, we bred rats for three consecutive generations on a Trp-enriched diet (3.8-fold increase in Trp-content) and measured the concentrations of Trp, serotonin and 5-hydroxyindoleacetic acid in the brain of each generation's offspring at days 5, 10, 15 and 20 post partum.

In the first generation (offspring of normal rats placed on the Trp-enriched diet two weeks before mating), the concentration of Trp in the brain was significantly reduced until 20 days of age (i. e., during the whole suckling period), and so was the brain concentration of serotonin. In the offspring of the F 2 generation a recovery from the depletion of Trp and of serotonin in the brain was observed. In the third generation the serotonin concentration in the forebrain was no longer different from controls. However, in this generation, the initial brain weight and the postnatal growth of the body and the brain was significantly decreased. The adult rats of the third generation were significantly smaller than the control rats bred on normal diets, and so was their daily food intake relative to body weight (29, 30).

These results indicate that an altered nutritional supply of tryptophan during the development of an individual may induce a sequence of metabolic adaptations. Such responses seem to last and to become continuously modified through several generations — until a new steady state is finally achieved. In the course of this adjustment, the expression of certain phenotypic parameters may be either activated or suppressed. In nature, an increased availability of Trp or other amino acids to the individuals of a certain population would probably never occur. Here, the nutritional changes would be much more complex, involving simultaneous changes of several substrates and cofactors, generally superimposed by other environmental alterations.

SUMMARY

An increased or decreased availability of individual amino acids to the adult brain can modulate certain functional capacities of adult nerve cells, and therefore brain function. Similarily, an altered availability of amino acids can affect certain functional capacities of developing nerve cells, too, and therefore brain development. Instead of the function, it is then the formation of functional circuits which is modulated by the amino acid availability.

Two principal possibilities of such modulations and their effects on the developing brain can be differentiated:

(a) Rather severe imbalances of the brain's amino acid supply. They may occur under pathological conditions, e. g. in inborn errors of metabolism, in animal models of such diseases or in severe protein undernutrition. An inadequate availability of essential amino acids in respect to their enormous requirements for the rapid protein anabolism of the developing brain effects a variety of developmental processes, e. g., the rate of cell proliferation, outgrowth of processes, formation of synapses, survival of cells etc. Due to the general interference with normal protein anabolism, the effects are not very specific for the kind of the amino acid imbalance. Different imbalances may have similar consequences. Which developmental event becomes most affected will primarily depend on the time of onset and the duration of the imbalance.

(b) Rather mild changes in the supply of amino acids. Such changes seem to be a general feature of normal metabolic maturation. They can be enhanced or suppressed by experimental environmental modifications of the developing organism. The possible consequences of such changes are discussed in respect to Trp-availability and serotonin formation during development, because serotonin is an important modulator and intercellular signal during embryonic development. Its production and release by developing serotoninergic cells is dependent on their Trp-supply. Sudden rise in the availability of Trp from the embryonic circulation leads to an increased serotonin formation and will enhance the serotonin-mediated effects on developing cells. Permanent changes of the availability of Trp during embryonic life, however, are counterregulated by metabolic adaptations involving long lasting changes of the metabolization of Trp and serotonin.

REFERENCES

1) HUETHER, G., (1987). Regulation of the free amino acid pool of the developing brain. A lesson learned from experimental phenylketonuria. In: *Amino Acids in Health and Disease: New Perspectives* (Kaufman, S., eds.) Alan R. Liss, Inc., pp. 107—122.

2) THOEMKE, F., HUETHER, G. (1984). Breeding rats on amino acid imbalanced diets for three consecutive generations affects the concentrations of putative amino acid transmitters in the developing brain. *Int. J. Dev. Neurosci.* **2:** 567—574.

3) HUETHER, G., THOEMKE, F., SPROTTE, U., STEINHAUS, U., NEUHOFF, V. (1985). Altered precursor availability and metabolism of monoamines in the brain caused by the injection of physiologic saline to suckling rats. *Neurochem. Int.* **7:** 725—730.

4) HUETHER, G., NEUHOFF, V. (1987). The effects of sensory stimulation on the free amino acid pool of the developing brain. *Neurochem. Res.* (in press).

5) KENNETT, G. A., CURZON, G., HUNT, A., PATEL, A. J. (1986). Immobilization decreases amino acid concentrations in plasma but maintains or increases them in brain. *J. Neurochem.* **46:** 208—212.

6) GREENGARD, O., YOSS, M. S., DELVALLE, J. A. (1976). a-Methyl-phenylalanine, a new inducer of chronic hyperphenylalaninaemia in suckling rats. *Science* **192:** 1007—1008.

7) HUETHER, G., SCHOTT, K., SPROTTE, U., THOEMKE, F., NEUHOFF, V. (1984). Regulation of the amino acid availability in the developing brain. No physiological significance of amino acid competition in experimental hyperphenylalaninaemia. *Int. J. Dev. Neurosci.* **2:** 43—54.

8) ANTONAS, K. N., COULSON, W. F. (1975). Brain uptake and protein incorporation of amino acids studied in rats subjected to prolonged hyperphenylalaninemia. *J. Neurochem.* **25:** 309—314.

9) GEISON, R. L., SIEGEL, F. L. (1975). Tolerance of protein and lipid synthesis to mild hyperphenylalaninemia in developing rat brain. *Brain Res.* **92:** 431—441.

10) HUGHES, J. V., JOHNSON, T. C. (1976). The effects of phenylalanine on amino acid metabolism and protein synthesis in vitro. *J. Neurochem.* **26:** 1105—1113.

11) HUGHES, J. V., JOHNSON, T. C. (1978). Abnormal amino acid metabolism and brain protein synthesis during neutral development. *Neurochem Res.* **3:** 381—399.

12) ROBERTS, S.,MORELOS, B. S. (1976). Role of ribonuclease action in phenylalanine-induced disaggregation of rat cerebral polyribosomes. *J. Neurochem.* **26:** 387—400.

13) HUGHES, J. V., JOHNSON, T. C. (1978). Experimentally induced and natural recovery from the effects of phenylalanine on brain protein synthesis. *Biochim. Biophys. Acta* **517:** 473—485.

14) HOMMES, F. A., ELLER, A. G., TAYLOR, E. H. (1982). Turnover of the fast components of myelin and myelin proteins in experimental hyperphenylalaninemia. *J. Inher. Metab. Dis.* **5:** 21—27.

15) HUETHER, G., KAUS, R., NEUHOFF, V. (1983). Brain development in experimental hyperphenylalaninaemia: Disturbed proliferation and reduced cell numbers in the cerebellum. *Neuropediatrics* **14:** 12—19.

16) HUETHER, G., NEUHOFF, V. (1981). Use of alpha-methylphenylalanine for studies of brain development in experimental phenylketonuria. *J. Metab. Dis.* **4:** 67—68.

17) HUETHER, G., KAUS, R., NEUHOFF, V. (1985). Amino acid depletion in the blood and the brain tissue of hyperphenylalaninaemic rats is abolished by the administration of additional lysine. A contribution to the understanding of the metabolic defects in phenylketonuria. *Biochem. Med.* **33:** 334—341.

18) HUETHER, G. (1986). The depletion of serotonin in the brain of hyperphenylalaninaemic rats is caused by the exhaustion of tryptophan from the blood. *Neurochem. Res.* **11:** 1663—1668.

19) HUETHER, G., SCHOTT, K., NEUHOFF, V. (1984). Decreased concentrations of tryptophan and serotonin in the brain of developing hyperphenylalaninaemic rats. A contribution to the understanding of the metabolic defects in phenylketonuria. In: *Progress in Tryptophan and Serotonin Research* (Schlossberger, H. G., Kochen, W., Linzen, B. & Steinhart, H., eds.), Walter de Gruyter & Co., Berlin-New York, pp. 429—435.

20) HUETHER, G., and REIMER, A. (1987). Availability of tryptophan and formation of indoleamines during embryogenesis. In: *Progress in Tryptophan and Serotonin Research* (Bender, D. A., Joseph, M. H., Kochen, W., and Steinhart, H. eds.), Walter de Gruyter & Co., Berlin-New York, pp. 237—244.

21) GUSTAFSON, T., TONEBY, M. (1970). On the role of serotonin and acetylcholine in sea urchin morphogenesis. *Exp. Cell Res.* **62:** 102—117.

22) BUZNIKOW, G. A., SHMUKLER, Y. B. (1980). Possible role of "prenervous" neurotransmitter in cellular interactions of early embryogenesis: a hypothesis. *Neurochem. Res.* **6:** 55—68.

23) LAUDER, J. M. (1985). Roles for neurotransmitter in development: possible interactions with drugs during the fetal and neonatal periods. *Progr. Clin. Biol. Res.* **163:** 375—380.

24) LAUDER, J. M., WALLACE, J. A., KREBS, M., PETRUSZ, P., MC CARTHY, K. (1982). In vivo and in vitro development of serotoninergic neurons. *Brain Res. Bull.* **9:** 605—625.

25) HAYDON, P. G., MC COBB, D. P., KATER, S. B. (1984). Serotonin selectively inhibits growth cone motility and synaptogenesis of specific identified neurons. *Science* **226:** 561—564.

26) LAUDER, J. M. (1985). Hormonal and humoral influences on brain development. *Psychopharmacol.* **8:** 121—155.

27) CHUBAKOV, A. R., GROMOVA, E. A., KNOVALO, G. V., SARKISON, E. F., CHUMASOV, E. J. (1986). The effects of serotonin on the morpho-functional development of rat cerebral neocortex in tissue culture. *Brain Res.* **369:** 285—297.

28) WALLACE, J. A. (1982). Monoamines in the early chick embryo: Demonstration of serotonin synthesis and the regional distribution of serotonin-concentrating cells during morphogenesis. *Am. J. Anat.* **165:** 261—276.

29) THOEMKE, F., HUETHER, G., KAUS, R., NEUHOFF, V. (1983). Altered concentrations of serotonin and dopamine in the forebrain of rats bred for three successive generations on a tryptophan-enriched diet. *Hoppe-Seyler's Physiol. Chem.* **364:** 1282—1283.

30) HUETHER, G. (1984). The influence of increased availability of tryptophan on the formation of tryptamine and serotonin during early ontogenesis. In: *Progress in Tryptophan Research* (Schlossberger, G. G., Kochen, W., Linzen, B. & Steinhart, H., eds.), Walter de Gruyter & Co., Berlin-New York, pp. 613—622.

NEUROTRANSMITTERS AND EARLY EMBRYOGENESIS

J. A. Wallace

Department of Anatomy,
School of Medicine
University of New Mexico
Albuquerque, New Mexico
U. S. A.

INTRODUCTION

Substances acting as neurotransmitter agents within the adult nervous system may also have additional roles during early phases of embryonic development. These substances, including primarily acetylcholine and monoaminergic compounds, have been suggested to influence various stages of embryogenesis from the earliest cell divisions of fertilized ova, to periods of gastrulation and subsequent morphogenesis, and to the initial phases of neurogenesis and synaptogenesis in the early forming central nervous system (CNS). Discussed here is a short review of studies involving some of these developmental roles for neurotransmitter substances, with emphasis given to investigations concerning the period of morphogenesis and organogenesis. Detailed reviews occur elsewhere concerning neurotransmitters in relation to early cleavage divisions (1), neurogenesis (2) and growth cone motility and synaptogenesis (3).

In leading to a discussion of what roles neurotransmitters may have in embryogenesis, some background is necessary to understand the phylogenetic history of these substances in relation to their proposed ontogenetic functions during what has been called the "prenervous" period of embryogenesis (1), the stage *prior* to when neurotransmitters act as a means of communication between neurons. Evolutionarily, the appearance of transmitters precedes that of the nervous system, with several of these substances occurring within single cell organisms, such as Protozoa, as well as in simple multicellular organisms (1). In these instances, the "neurotransmitters" may act as intracellular regulators of metabolism (4). With further evolutionary progression, it has been proposed that these same substances advanced from intracellular to intercellular messengers, culminating eventually with their participation in the rapid intercellular communication mediated by nerves (5).

In a similar ontogenetic pattern, several neurotransmitters are found in sea urchin embryos at the one cell stage (fertilized ova), and during early mitotic divisions, these substances appear to participate in control over cleavage divisions and macromolecular synthesis (6). Slightly later in embryogenesis, these same compounds have been shown to be involved in intercellular communication between neighboring blastomeres (7). Further-

NATO ASI Series, Vol. H20
Amino Acid Availability and Brain Function in Health and Disease. Edited by G. Huether

more, during subsequent development in sea urchins, transmitters such as acetylcholine and serotonin (5-HT; or related indoleamines, tryptamine or 5-methoxytryptamine) regulate morphogenetic cell movements involved in the process of gastrulation and in the formation of the primitive gut (8, 9).

Since, as described, above, neurotransmitter substances may function intracellularly as well as intercellularly during different phases of ontogenesis, this raises the issue of the mechanism by which neurotransmitters influence embryonic cells. Must transmitters be synthesized within responding cells if the substances act intracellularly? Alternatively, if these compounds can be released between neighboring cells, can the neurotransmitters act on cell surface receptors which in turn alter levels of secondary intracellular messengers? Or, can transmitters from the extracellular environment be taken-up within cells, where they affect metabolic functions, cell motility or cell shape changes? During the embryonic period of morphogenesis, we will see: 1) that synthesis of neurotransmitters occur within certain embryonic tissues, 2) that many embryonic cells possess efficient uptake mechanisms for various transmitters, and 3) that receptors exist for certain transmitters that may influence either the synthesis or uptake of *different* neurotransmitter agents.

MONOAMINES AND MORPHOGENESIS

In avian and amphibian species, monoaminergic compounds [catecholamines (CA) and indoleamines (such as 5-HT)] occur endogenously within early embryos during the stages of morphogenesis (10—14). The embryonic structure most often associated with the synthesis of these amines is the notochord (12, 15). However, it has also been shown that CA and 5-HT occur within yolk material, which can then serve as an additional source of these substances to the embryos (16, 17). In this context, many embryonic tissues possess uptake mechanisms for these amines, yet as will be discussed further below, the uptake mechanisms for CA and 5-HT are found *neither:* within the same embryonic tissues, at the same locations within a single tissue, nor within the same cells at any one stage of embryogenesis (18).

Catecholamines and Embryogenesis

By biochemical assays, a sequential appearance of CA compounds, including dopamine (DA), norepinephrine (NE) and epinephrine (E), has been shown in chick embryos from 1 to 3 days of incubation (16). Using the formaldehyde-induced fluorescence (FIF) technique, endogenous CA FIF appears primarily in the notochord (10, 19). Following the treatment of chick embryos with the CA precursor, L-DOPA, the intensity of FIF for CA is greatly enhanced within the notochord (15, 18, 20). This increase in the notochordal fluorescence is markedly diminished by the DOPA decarboxylase inhibitor, RO4-4602 (15). Besides the chick embryo's ability to synthesize CA during the period of morphogenesis, several embryonic structures demonstrate the capacity to accumulate NE, including the notochord, neural tube, myotomes, scleratomes and gut mesenchyme (10, 15, 18, 20). The specific locations of FIF resulting from NE uptake varies within these tissues over the period of morphogenesis (15, 18). Furthermore, this FIF is decreased by desmethylimipramine, an inhibitor of adult neuronal CA uptake, but not by corticosterone, a blocker of adult *extra*neuronal CA uptake (15).

In an attempt to examine the potential role for CA in chick morphogenesis, Lawrence and Burden (10) altered the embryonic levels of CA by treating embryos with an inhibitor of monoamine oxidase (MAO; an enzyme that degrades CA), or with an inhibitor of CA biosynthesis (α-methyl-p-tyrosine). Either treatment resulted in abnormal embryonic flexure and rotation, severe malformations of the forebrain, and failure of the neural tube to close in the spinal cord (spina bifida). However, the results of this teratologic study are difficult to interpret because the drugs were delivered at high concentrations, within the mM range. On the other hand, another recent study has re-examined this issue by injecting low doses of NE or DA beneath the blastoderm of early chick embryos (21). After 48 hours of exposure of embryos to NE or DA at doses of 0.65 μM, the development of the neural tube, notochord and somites were severely affected (again involving the absence of formation of forebrain structures, and the presence of spina bifida). It was also noted that there was impaired yolk degradation concomitant with a fivefold decrease in glycogen levels in treated embryos compared to control embryos. From these findings, Sarasa and Climent (21) proposed that CA may stimulate glycogenolysis (mediated by cAMP and/or by calcium), thus reducing available stores of energy to embryonic cells. In support of this proposal was the finding that the observed teratogenic effects of CA administration could be countered by the simultaneous injections of low doses of either glucose or a calcium chelating agent (EDTA). These results suggest that a role for CA in chick morphogenesis may be related to the regulation of the metabolic activity of certain embryonic cells, involving calcium-dependent yolk degradation, thereby influencing the utilization of energy stores.

Serotonin and Embryogenesis

Besides the presence of CA in early chick embryos, 5-HT also occurs endogenously in these embryos during the stages of morphogenesis, as mentioned earlier. 5-HT has been detected biochemically in preparations of isolated chick notochords (11). However, when chick embryos are examined by anti-5-HT immunocytochemistry, immunostaining of endogenous 5-HT is detected only within caudal portions of the notochord and within the adjacent underlying endoderm of the embryos, at these early ages (14, 22). Synthesis of 5-HT within chick embryos has been demonstrated by an increase in the intensity of 5-HT immunostaining after the *in vitro* treatment of the embryos with the 5-HT precursors, L-tryptophan or 5-hydroxytryptophan (14). (The sites with enhanced immunostaining, after the treatment of embryos with these 5-HT precursors, correspond with the embryonic locations of 5-HT uptake, as described below.) Confirming evidence of 5-HT synthesis was provided by a marked reduction in the immunosatining obtained with either precursor in the presence of 5-HT synthesis inhibitors (RO4-4602 or parachlorophenylalanine).

As with the ability of chick embryos to take up CA, the embryos also have the capacity for 5-HT uptake. Following the *in vitro* incubation of embryos with 5-HT, the localization of sites possessing 5-HT uptake capabilities can be demonstrated by 5-HT FIF, as shown in Figure 1. These sites include: restricted portions of the developing brain (within the floor plate of the mesencephalon and myelencephalon), scleratome cells of somites, and the floor plate of caudal (or posterior) portions of the early forming spinal cord and subjacent notochord (14). With respect to the location of 5-HT uptake in the vicinity of the caudal spinal cord, this region is where the neural tube is actively closing, and is termed the caudal neuropore. Of interest to the process of closure, the uptake of 5-HT occurs in a

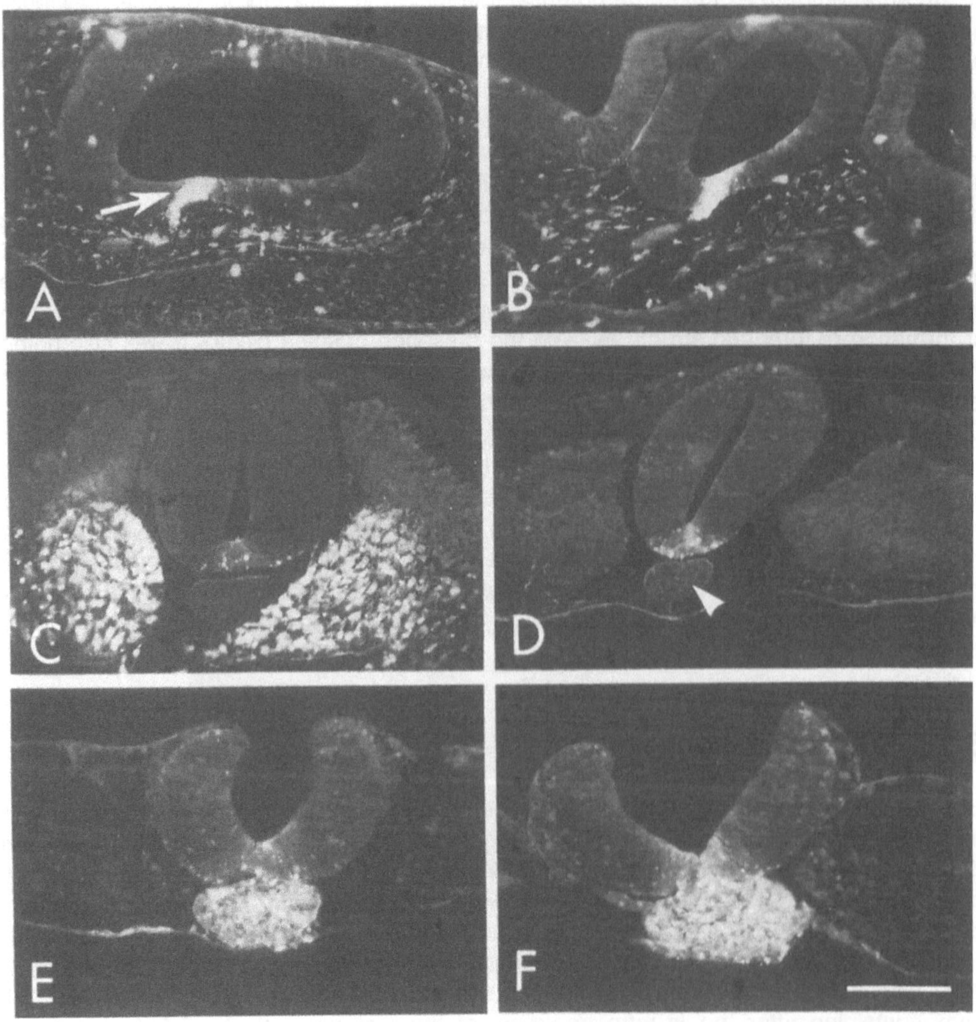

FIGURE 1

After the *in vitro* incubation of a 2 day old chick embryo with 5-HT, histofluorescence for 5-HT is found at several embryonic levels. (A) Within the mesencephalon, 5-HT FIF is restricted to the floor plate (arrow) of the early forming brain. (B) A similar location of FIF is observed at the level of the otic discs in the myelencephalon. (C) At the level of the cervical neural tube, intense 5-HT FIF is localized to scleratome cells of somites advanced in differentiation. (D) Approaching the caudal end of the spinal cord, floor plate FIF increases markedly, yet no FIF is noted in the adjacent notochord (arrowhead). (E and F) At the level of the caudal neuropore where the neural tube is actively involved in closure, notochordal and floor plate 5-HT FIF becomes very intense. This represents the sole location where *notochordal* FIF occurs. Bar = 100 μm.

very small portion of the spinal cord and notochord (in terms of the rostro-caudal extent of these structures) and is found progressively further caudally with the increasing age of the embryos in conjunction with posterior advance in development of closure of the caudal neuropore. Thus, there appears to be a spatio-temporal relationship between the changing sites of 5-HT uptake caudally in chick embryos, and the continuing progression of neural tube closure at the end of the developing spinal cord. This relationship raises the question of whether the uptake or presence of 5-HT in the region of the caudal neuropore is physiologically significant to the process of neural tube closure, and thus, if altered levels or metabolism of 5-HT in early embryos is relevant to the occurrence of spina bifida.

Teratologic studies have examined the effects of changes in the levels of 5-HT within chick embryos during the stages of morphogenesis by exposing the embryos either to L-tryptophan (23), to 5-hydroxytryptophan (24), to an MAO inhibitor, or to 5-HT itself (25). The resulting malformations were similar to those observed in experiments where chick embryos were exposed to 5-HT synthesis inhibitors or to 5-HT receptor antagonists (25). The common abnormalities consisted primarily of defects in brain formation, somitogenesis, and in neurulation (often seen as failure of closure in the caudal neural tube). Nonetheless, these studies share the same problems that were encountered in some of the teratologic experiments described earlier that investigated the role of CA in chick morphogenesis. Specifically, the concentrations of the compounds to which the embryos were exposed were excessively high (often within the mM range, even in *in vitro* whole embryo culture experiments). Furthermore, with respect to the various drugs used as 5-HT antagonists or inhibitors of 5-HT synthesis, these neuropharmacologic agents have been characterized for their actions on adult neuronal systems, while their specificity and/or action on early embryonic tissues (including non-neuronal structures such as the notochord) have not been evaluated.

Taking into consideration these problems inherent in many teratologic studies, we sought to examine the possibility that the uptake of 5-HT within the neural tube and notochord may be necessary for normal morphogenesis in the chick embryo, particularly involving closure of the neural tube in the region of the caudal neuropore. To initiate these investigations, we first characterized the 5-HT uptake mechanisms in chick embryos with respect to their similarities to adult neuronal 5-HT uptake systems. Utilizing 5-HT immunocytochemistry, we demonstrated that the uptake of exogenously supplied 5-HT was specific, since the induced 5-HT immunostaining was not decreased in the presence of a hundredfold excess of NE (14). Furthermore, we also found that embryonic 5-HT uptake mechanisms were sodium and temperature dependent. In addition, the responsiveness of the embryonic 5-HT uptake systems to various well-known 5-HT uptake inhibitors was examined (Fig. 2). Utilizing low concentrations of exogenously supplied 5-HT (0.1 μM) in *in vitro* studies, all tricyclic uptake inhibitors investigated were shown to almost completely eliminate 5-HT immunostaining, thus the uptake of 5-HT. The doses of the uptake inhibitors required to obtain this degree of blockade of 5-HT uptake is within the range of concentrations (< 10 μM) used in experiments involving inhibition of 5-HT uptake within synaptosomal fractions isolated from adult neurons. Quantification of the varying effectiveness of these uptake inhibitors is currently being examined by image analysis, utilizing microdensitometry.

Pilot studies have also been initiated to examine the potential teratologic effects of 5-HT uptake inhibitors when used in the range of concentrations shown to be effective in blocking 5-HT uptake in early chick embryos. At present, utilizing an *in ovo* paradigm for

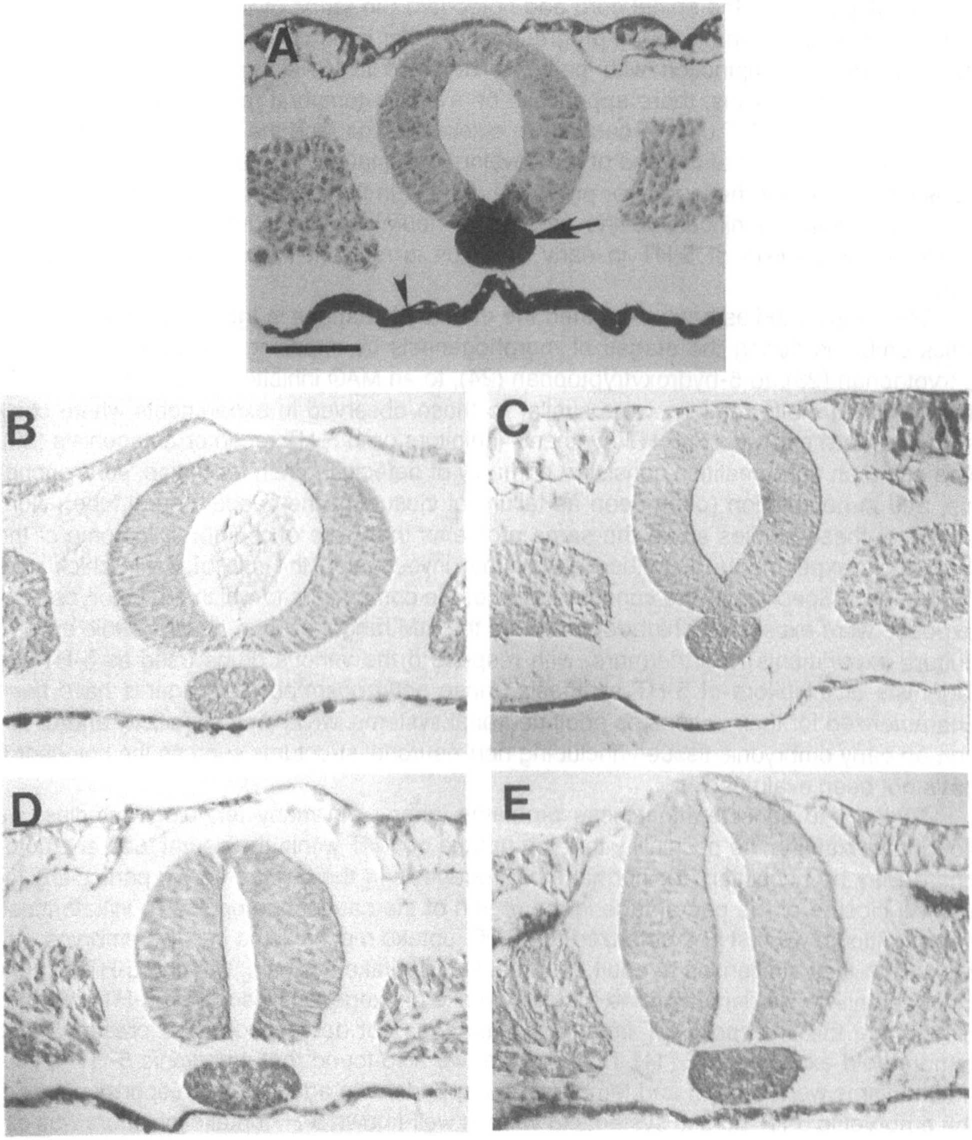

FIGURE 2

Examination of the efficacy of various tricyclic antidepressants to inhibit 5-HT uptake (and the resulting 5-HT immunostaining) in embryos exposed *in vitro* to 0.1 μM 5-HT and 10 μM pargyline. (A) Intense immunostaining occurs in the notochord (arrow) and underlying endoderm (arrowhead) in control embryos treated with only 5-HT and pargyline. (B—E) Shows the marked reduction in 5-HT immunostaining in the notochord and endoderm when the embryos are simultaneously cultured with 10 μM (B) chlorimipramine, (C) fluoxetine, (D) amitriptyline, and (E) imipramine. Bar = 40 μM.

the administration of the drugs (followed by a survival period of 24 hours), we have unfortunately found relatively high numbers of malformations in control saline-treated animals, as well as in drug-treated embryos. However, to date, a consistent finding seen *only* in drug-treated chicks is a change in the cytoarchitecture of embryonic cells that occur caudally, solely within those regions where 5-HT is normally taken-up (Fig. 3). The floor plate and notochord in the vicinity of the caudal spinal cord in drug-treated embryos is filled with numerous dark vacuoles. Furthermore, the boundaries of cells within the floor plate and notochord are indistinct, and cell debris is common. Clarification of these cellular changes as involving necrosis within these specific regions, will require electron micrographic examination.

Lastly, further compounding the complexity of understanding the potential means by which neurotransmitter substances can be interfered with during embryogenesis, is the realization that there are possible interactions between different transmitter systems. An

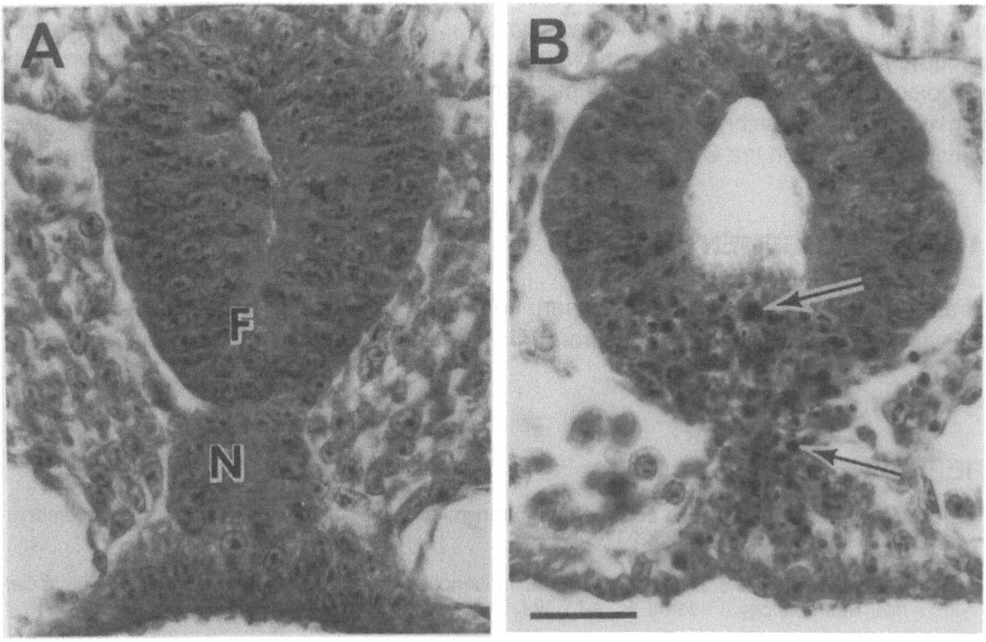

FIGURE 3
A comparison of the morphology in the spinal cord and notochord at the caudal end of chick embryos at 3 days of incubation. The embryos were exposed *in ovo* for the prior 24 hour period to either (A) sterile saline or (B) to the 5-HT uptake inhibitor, fluoxetine, at an administered dose of 10 μM. Notice in (A) that the floor plate (F) of the spinal cord and the notochord (N) appear normal when stained with hematoxylin and eosin. However, the fluoxetine-treated embryo shows abnormal development (arrows) in the floor plate and notochord, where 5-HT is normally taken-up. The cells in this region contain many dark staining inclusions and have indistinct cell borders and possess considerable cell debris. Bar = 25 μM.

example of this involves the potential inter-relationship between acetylcholine and monoamines. Strudel and Gateau (26) have shown that acetylcholine receptors exist in early chick embryos. Of specific interest is their finding that the stimulation of these receptors with nicotine sulfate or carbachol, not only increases endogenous levels of CA and 5-HT in the notochord (11), but also enhances the capability of the notochord to take up NE (19). Therefore, as the existence of additional transmitter substances are demonstrated in early embryos, the total possible interactions between these agents further complicates our ability to investigate the role of individual neurotransmitters acting during the period of morphogenesis.

CONCLUSIONS

From the above descriptions of the presence of monoamines and of the highly specific distribution of uptake mechanisms for these compounds in early chick embryos undergoing morphogenesis, it should be considered that factors that influence levels of CA or 5-HT in adult neuronal systems must also be assumed to have the potential of altering these same transmitter systems in early embryos. Changes in neurotransmitter levels induced by variations in amino acid availability, or by exposure of embryos to neuropharmacologic agents that alter the synthesis, metabolism and/or action of any transmitter substance, may possibly have long range effects on the morphology, survival or differentiation of cells within the CNS, as well as within other non-neuronal embryonic organs.

ACKNOWLEDGEMENTS

Thanks are given to Isabel Lopez-Colberg and Rolanda R. Maez for their excellent technical assistance. Supported by funds from NIH grants NS19712 and RR08139.

REFERENCES

1) BUZNIKOV, G. A. (1984). The action of neurotransmitters and related substances on early embryogenesis. *Pharmac. Ther.* **25:** 23—59.

2) LAUDER, J. M., KREBS, H. (1984). Hurmoral influences on brain development. In: *Advances in Cellular Neurobiology* (Federoff, S., ed.). **Vol. 5.** Academic Press, New York, pp. 3—51.

3) KATER, S. B. (1985). Dynamic regulators of neuronal form and connectivity in the adult snail *helisoma.* In: *Model Neural Networks and Behavior* (Selvevstor, A. I., ed.). Plenum Publishing Corp., New York, pp. 191—109.

4) TOMKINS, G. (1975). The metabolic code. *Science* **189:** 760—763.

5) McMAHON, D. (1974). Chemical messengers in development. *Science* **185:** 1012—1021.

6) BUZNIKOV, G. A., KOST, A. N., KUCHEROVA, N. F., MNDZHOYAN, A. L., SUVOROV, N. N., BERDYSHEVA, L. V. (1970). The role of neurohumors in early embryognesis. III. Pharmacological analysis of the role of neurohumors in cleavage divisions. *J. Embryol. Exp. Morph.* **23:** 549—569.

7) BUZNIKOV, G. A., SHMUKLER, Y. B. (1980). Possible role of "prenervous" neurotransmitters in cellular interactions of early embryogenesis: a hypothesis. *Neurochem. Res.* **6:** 55—68.

8) GUSTAFSON, T., TONEBY, M. (1971). How genes control morphogenesis. *Amer. Scientist* **59:** 452—462.

9) BUZNIKOV, G. A., SAKHAROVA, A. V. MANUKHIN, B. N., MARKOVA, L. V. (1972). The role of neurohumors in early embryogenesis. IV. Fluorometric and histochemical study in cleaving eggs and larvae of sea urchins. *J. Embryol. Exp. Morph.* **27:** 339—351.

10) LAWRENCE, I. E., Jr., BURDEN, H. W. (1973). Catecholamines and morphogenesis of the chick neural tube and notochord. *Amer. J. Anat.* **137:** 199—208.

11) STRUDEL, G., RECASENS, M., MANDEL, P. (1977). Identification de cãtecholamines et de sērotonin dans les chordes d'embryons de Poulet. *C. R. Acad. Sc. Paris* **284:** 967—969.

12) GODIN, I., GIPOULOUX, J. D. (1986). Notochordal catecholamines in exogastrulated *xenopus* embryos. *Develop. Growth and Differ.* **8:** 137—142.

13) SIMS, T. J. (1977). The development of monoamine-containing neurons in the brain and spinal cord of the salamander, *Ambystoma mexicanum. J. Comp. Neurol.* **173:** 319—336.

14) WALLACE, J. A. (1982). Monoamines in the early chick embryo: Demonstration of serotonin synthesis and the regional distribution of serotonin-concentrating cells during morphogenesis. *Am. J. Anat.* **165:** 261—276.

15) NEWGREEN, D. F., ALLAN, I. J., YOUNG, H. M., SOUTHWELL, B. R. (1981). Accumulation of exogenous catecholamines in the neural tube and non-neural tissues of the early fowl embryo. *Wilhelm Roux's Archiv.* **190:** 320—330.

16) IGNARRO, L. J., SHIDEMAN, F. E. (1968). Appearance and concentrations of catecholamines and their biosynthesis in the embryonic and developing chick. *J. Pharmacol. Exp. Ther.* **159:** 38—48.

17) HUETHER, G., REIMER, A. (1987). Availability of tryptophan and formation of indoleamines during embryogenesis. In: *Progress in Tryptophan and Serotonin Research* (Bender, D., Joseph, D., Kochen, W., Steinhart, H., eds.). Walter de Gruyter and Co., Berlin and New York, pp. 237—243.

18) WALLACE, J. A. (1979). Biogenic amines in the development of the early chick embryo. *Ph. D. Thesis,* University of California, Davis.

19) STRUDEL, G., MEINIEL, R., GATEAU, G. (1977). Recherches d'amines fluorigènes dans les chordes d'embryons de Poulet traitēs par des cholinergiques. *C. R. Acad. Sc. Paris* **284:** 1097—1100.

20) KIRBY, M. L., GILMORE, S. A. (1972). A fluorescence study on the ability of the notochord to synthesize and store catecholamines in early chick embryos. *Anat. Rec.* **173:** 469—478.

21) SARASA, M., CLIMENT, S. (1987). Effects of catecholamines on early development of the chick embryo: Relationship to effects of calcium and cAMP. *J. Exper. Zool.* **241:** 181—190.

22) SAKO, H., TOKUZO, K., OKADO, N. (1986). Immunocytochemical study on the development of serotonergic neurons in the chick: Distribution of cell bodies and fibers in the brain. *J. Comp. Neurol.* **253:** 61—78.

23) EMANUELSSON, H., PALĒN, K. (1975). Effects of L-tryptophan on morphogenesis and growth in the early chick blastoderm. *Wilhelm Roux' Archiv.* **117:** 1—17.

24) SCHOWING, J., SPRUMONT, P. VAN TOLEDO, B. (1977). Influence of L-5-hydroxytryptophan on the development of the chick embryo. *C. R. Acad. Sci. Paris* **171:** 1163—1166.

25) PALĒN, K., THORNEBY, L., EMANUELSSON, H. (1979). Effects of serotonin and serotonin antagonists on chick embryogenesis. *Wilhelm Roux's Archiv.* **187:** 89—103.

26) STRUDEL, G., GATEAU, G. (1977). Sur l'existence de rēcepteurs cholinergiques chez le très jeune embryon de Poulet. *C. R. Acad. Sci. Paris* **284:** 469—472.

ROLE OF EXCITATORY AMINO ACIDS IN THE DEVELOPMENT OF CEREBELLAR GRANULE NEURONES

R. Balázs*, O. S. Jørgensen, N. Hack*, V. Gallo, A. Kingsbury and C. Cotman+**

MRC, Developmental Neurobiology Unit,
Institute of Neurology, London WC1N 1PJ, U. K.
* Present address: Netherlands Institute for Brain Research,
1105 AZ Amsterdam, The Netherlands
** Psychochemistry Institute, University of Copenhagen,
Copenhagen, Denmark
+ Department of Psychology,
University of California,
Irvine, California, USA

INTRODUCTION

It is now established that acidic amino acids, such as glutamate (Glu) and aspartate (Asp), which are associated with the tricarboxylic acid cycle and play a metabolic role in the brain, are also abundant excitatory transmitters in the CNS (e. g., 1). It is also well documented that by virtue of their excitatory action, these amino acids and their analogues in excess can exert toxic effects on nerve cells (e. g., 2). Here we are reporting new observations which indicate that in addition during development, excitatory amino acid (EAA) transmitters may have a trophic action on nerve cells which depends on their maturational stage (3).

Cerebellar Granule Cell Cultures

The discovery of the trophic effect of EAA receptor stimulation was facilitated by the unique properties of the culture system we have been studying (e. g., 4 for review). This is a culture enriched (about 95 %) in interneurones predominantly granule cells, obtained from the cerebellum of 6—8 day-old (P6—8) rats. The dissociation of the tissue together with manipulation of the culture conditions result in an initial selection for immature granule cells. Many of these cells would continue to proliferate *in vivo,* but *in vitro* they are unable to do so and thus are prematurely induced to differentiate. Differentiation is characterized by neurite emission soon after plating, the production of a fine network of fibres and the migration of neurones along the fibres to form clumps. After the first few days, the aggregates increase progressively in size and fibres fasciculate to form large interconnecting bundles. The morphological appearance of the cultures does not change much after the end of the first week. Biochemical indices of maturation show that after a

NATO ASI Series, Vol. H20
Amino Acid Availability and Brain Function in
Health and Disease. Edited by G. Huether

few days' lag period there is a rapid phase of development reaching a plateau by about 8 days *in vitro* (DIV), i. e., quicker than the adult level of maturation is reached *in vivo*. This is understandable since, due to neurogenesis during the first three weeks after birth, cerebellar development *in vivo* is protracted.

The properties of the culture which have been of particular importance for our work are (a) relatively homogeneous cellular composition (predominance of excitatory granule cells) and (b) the more or less synchronous differentiation of these cells *in vitro*.

Depolarization and Granule Cell Survival in Culture

The question which led us to the recognition of the trophic role of EAA was why does K^+-induced depolarization promote granule cells survival (5). This effect has been described by Lasher and Zagon (6), but there is evidence that chronic depolarization in cultures also promotes the survival and/or affects the differentiation of immature nerve cells other than the cerebellar granule cells (for reference see 5). It should be noted, however, that not every class of neurones is dependent on elevated K^+ for survival; thus foetal nerve cells from the hippocampus or the cerebral cortex survive well in media containing physiological concentrations of K^+ ions (7, 8). In similar foetal calf serum (FCS)-containing media cerebellar granule cells die, usually within the first week unless K^+ levels are increased to > 20 mM (Table 1). The dependence of granule cells on depolarization develops within a narrow time-span. When cells grown in 5 mM K^+ containing medium (5K) for 2 days were exposed to 25 mM K^+ (25K) they survived as well as the cells cultured from the time of seeding in 25K. On the other hand, when cells were exposed to elevated K^+ from the fourth day in culture they died more or less at the same time as the nerve cells which were continuously grown in 5K. It should be noted that *in vivo* after their last mitosis differentiating granule cells need about the same length of time, about two days, to reach the internal granular layer (IGL), where they first receive an afferent input, primarily from the mossy fibres (9).

TABLE 1
Medium K^+ levels and granule cells survival

		Cell numbers per dish (as a percentage of those in K25 at 2 DIV)		
K^+ (mM)		2 DIV	5 DIV	12 DIV
5		112	87	(20)
10		93	88	(20)
15		93	91	(37)
20		100	100	84
25		100	114	92
5 —> 25	at 2 DIV		83	84
	at 4 DIV		74	(22)

Granule cells were grown in 10 % FCS containing medium on poly-L-lysine coated plastic as described (19). Data are from Gallo et al. (5). Bracketed values indicate that very few nerve cells were detectable, if at all, under the conditions specified.

During the first few days in culture the morphological and biochemical development of the granule cells was similar in the presence and the absence of elevated K^+. Nevertheless, after about 5 days the low K^+ neurones died abruptly. It would appear, therefore, that granule cells become dependent on elevated K^+ in parallel with the expression of their differentiated characteristics and that high $[K^+]_e$ may mimick influences mediated *in vivo* by the synaptic input received by granule neurones at this developmental stage.

In the following experiments we examined whether the exposure to high K^+ imitates influences that are normally elicited by the bioelectric activity and/or the depolarization of the cells. There is evidence that neuronal activity is necessary for the survival of certain populations of nerve cells (10). However, the inhibition of bioelectric activity by means of tetrodotoxin or xylocaine did not prevent the survival promoting effect of elevated K^+. On the other hand, depolarization induced calcium entry into granule cells was essential for their survival. Interference with this process, by blocking Ca^{2+} entry with 10 mM Mg^{2+} or by chelating Ca^{2+} with EGTA, resulted in the death of granule cells grown in K25 more or less at the same time — about a week after seeding — as in K5.

The most convincing findings were, however, obtained when we observed (11) that, in contrast to nerve terminal preparations from the CNS, cultured granule cells are endowed with voltage-sensitive calcium channels (VSCC) at which dihydropyridine (DHP) calcium effectors are as potent functional ligands as in muscle tissue. Supplementation of the culture medium with DHP calcium agonists at a concentration of K^+ (15 mM) at which the survival of granule cells is compromised, resulted in the rescue of the cells. The potency of these agents was high: 1×10^{-8} M BAY K8644 or 5×10^{-8} M of the Sandoz compound (+)-(S)-202 791 gave full protection. Moreover, calcium antagonists including the enanthiomer of the Sandoz compound, (−)-(R)-202 791, completely blocked the effect of elevated K^+ on cell survival. Further studies indicated that the effect of Ca^{2+} is mediated through reactions involving calmodulin, since, although the specificity of calmodulin inhibitors is limited, trifluoperazine and calmidazolium blocked the survival promoting effect of elevated K^+ at low concentration (1 μM).

These findings clearly showed, therefore, that the survival-promoting effect of chronic depolarization is mediated through Ca^{2+} entry, but the vital question was — what is the physiological event which is mimicked by high K^+ *in vitro?* The observations described above on cultured granule cells — such as the narrow time-frame of the development of K^+ dependence which is associated with the initial differentiation of granule cells and the critical involvement of Ca^{2+} entry in the effect — when compared with the *in vivo* development of postmitotic granule cells, led to the proposal that depolarization *in vitro* mimicks influences exerted *in vivo* by the afferent input that is received by these cells at the end of their migration primarily from the mossy fibres (5). Many of these fibres are glutamatergic (12) and granule cells possess excitatory amino acid receptors, including the N-methyl-D-aspartate (NMDA)-preferring subtype, which gates channels permeable not only to Na^+ and K^+ but also to Ca^+ (13, 14). Thus the hypothesis proposes specifically that 1) depolarization induced Ca^{2+} entry *in vitro* mimicks the NMDA receptor-induced Ca^{2+} entry evoked by mossy fibre stimulation *in vivo* and 2) only granule cells at a specific maturational state require this influence for survival. Furthermore, it is proposed that the initial stage of maturation is reached by granule cells after about 2 days in culture and *in vivo* by the time the migrating cells arrive into the IGL. The hypothesis predicts therefore that granule cell survival can be secured in culture by NMDA receptor stimulation.

NMDA Promotes the Survival of Cerebellar Granule Cells in Culture

Our recent observations supported the hypothesis outlined above by showing that NMDA has a dramatic effect in rescuing granule cells grown in media containing K^+ at concentrations at which the survival of these cells is compromised (3). Light microscopic examination of the cultures suggested that the effect was dependent on both the concentration of NMDA and the membrane potential of the cells. However, in order to screen more objectively the influence of these variables we adopted a method involving the oxidation by viable cells only of a tetrazolium salt (MTT) to a blue reaction product (15). In this and the following studies the effect of different conditions on cell survival was assessed in comparison with estimates obtained in cultures in K25 (control). The findings, using the MTT method, indicated that both the efficiency and the potency of NMDA was a function of the medium K^+ concentration. Maximal viability was, as a percentage of the control (K25), about 50 % in K5, more than 60 % in K10, and over 90 % in K15. Furthermore, the half maximal effective concentration of NMDA (μM) was about 50, 35, and 20 at 5, 10 and 15 mM K^+ respectively. It is noteworthy that the K_i for the inhibition by NMDA of labelled 2-amino-5-phosphonovalerate (APV) binding to cerebral membranes preparations is of similar magnitude (11 μM, 16).

The observation that the effect of NMDA depended on the degree of depolarization of the cells is consistent with the properties of the NMDA receptor-ionophore complex (14). It is well established that the excitatory action of NMDA receptor agonists is sensitive to the membrane potential. This is a consequence of the voltage-dependence of the blockade by Mg^{2+} of the receptor-linked ion channels (17). The culture medium contained physiological concentrations of Mg^{2+}, which at the normal resting potential would inhibit responses to NMDA (18). Partial depolarization of the cells would therefore alleviate the inhibition.

The findings with the MTT method were substantiated by quantitating cell survival using DNA estimations (Table 2). The effect of NMDA in promoting cell survival was completely blocked by the NMDA receptor antagonist, AVP. It was a consistent finding that

TABLE 2
Effect of NMDA on granule cell survival

K^+ (mM)	10	10	10	25
NMDA (μM)	—	140	140	—
D, L-APV (μM)	—	—	280	—
DNA	31 ± 2.6^{a}	$82 \pm 2^{a,b}$	$16 \pm 2.9^{a,c,d}$	100 ± 4.0
Protein	55 ± 3.2^{a}	$82 \pm 2.7^{a,b}$	$40 \pm 8.4^{a,c,d}$	100 ± 4.0
N-CAM	43 ± 6.0^{a}	$75 \pm 11^{a,b}$	$41 \pm 8.2^{a,c}$	100 ± 7.6

Estimates are means ± S. E. (n = 3—9) and are expressed as a percentage of the control (K25) values per dish. Granule cells were cultured for 7—8 days. Content of DNA and protein was estimated by the method of Burton (32) and Lowry et al. (33) respectively and that of N-CAM as described by Jørgensen (34). Results are unpublished observations by Balázs, Hack and Jørgensen. Data were analysed by one-way analysis of variance ($P < 0.01$ in each case), followed by Duncan's multiple range test. The following significant differences ($P < 0.05$) are indicated:
a — in comparison with controls (K25) reduction was significant
b — addition of NMDA to D10 caused significant elevation
c — APV blocked the effect of NMDA
d — in comparison with K10 supplementation with APV (+NMDA) resulted in a significant reduction

AVP significantly reduced cell survival to below that seen in K10 in the absence of NMDA. This observation suggested that NMDA receptor stimulation must have occurred in K10, although EAA were not added to the medium. However, small amounts of Glu may have been present in FCS and formed from glutamine. Glu estimates were below the detection limit of the spectrophotometric method used ($< 20\ \mu M$), but it is known that Glu in low concentration preferentially stimulates NMDA receptors (14).

Table 2 also shows that compared with the number of cells the amount of protein was less reduced in K10 or in the presence of APV. This may relate to the fact that although their total number is kept low by cytosine arabinoside, non-neuronal cells constitute in K10 a greater proportion of the total number of cells and their size is larger than in K25 (19). Since DNA estimation cannot distinguish between neuronal and non-neuronal cells certain parameters, which are relatively good nerve cell markers, were also estimated. These included tetanus toxin receptors and proteins, such as N-CAM, D3 and synaptin. We observed previously that in cerebellar neuronal cultures tetanus toxin or N-CAM antibodies bind predominantly to nerve cells (20, 19), while Nybroe *et al.* (21) have found that the concentration of N-CAM is much greater in nerve cells than in astrocytes. In Table 2 only N-CAM estimates are given, but the other 'neuronal' indices gave comparable results. Thus the dramatic rescue of cells by NMDA involved primarily nerve cells.

CONCLUSIONS

The results show that NMDA promotes the survival of cerebellar granule cells in culture. It seems therefore that EAA are not only accredited transmitters, which in excess can exert a toxic action on nerve cells, but they are also trophic agents at certain stages of neuronal development.

The maturational stage dependence is highlighted by the changes in the responsiveness of granule cells to NMDA in the cerebellum during development. Garthwaite and Garthwaite (22) studying the acute toxic effect of EAA in cerebellar slices have observed that only differentiating granule cells are vulnerable to NMDA, although those in the upper part of the IGL are not adversely affected. These are the cells which have just arrived in the IGL, and according to our hypothesis the survival-promoting effect of depolarization or of NMDA in culture mimicks the influence exerted *in vivo* by the new afferent innervation of these cells. The short-term experiments of Garthwaite and Garthwaite (22) cannot provide information concerning whether or not NMDA and mossy fibres exert a trophic influence on this cohort of granule cells. However, the observations of Rakic and Sidman (23) are consistent with such ar role of mossy fibres *in vivo*. These authors found that in weaver mutant mice most of the postmitotic granule cells die, in part as a result of defective migration and a failure to reach their proper position and thus of receiving their appropriate innervation. The granule cells which survive in ectopic positions are those which are innervated by aberrant mossy fibres.

As the maturation of the cells progresses, the stage of differentiation when NMDA is apparently a trophic factor for the granule cells, is superseded by a stage when the cells become vulnerable to NMDA (22). It is tempting to suggest that both the trophic and the toxic effect of NMDA are mediated though the same mechanism, notably increased Ca^{2+} influx, and that the maturational stage dependent difference in the response may relate to the effectiveness of regulatory mechanisms which keep Ca^{2+} levels within non-toxic limits. Finally, fully differentiated granule cells are not sensitive to NMDA (22, 24): they are not vulnerable and, in contrast to the younger differentiating cells which are destroyed by NMDA, this EAA elicits only very weak depolarizing responses from the adult granule cells.

It has been suggested that NMDA receptors play a role in plasticity in the mature CNS (see chapters in Ref. 25) and — although the evidence is rather limited — in the developing brain (26, 27). It seems plausible therefore, that trophic influences mediated by NMDA receptor stimulation by synaptically released EAA are not unique to cerebellar granule cells. During development nerve cells, including the cortical pyramidal cells which are believed to be glutamatergic, make abundant projections even to areas which they do not innervate in the adult. These redundant projections may fulfil a functional role by providing, at the right time, a trophic influence on their transient synaptic partners.

Various transmitters have been implicated previously as trophic agents in the developing CNS (e.g., 28—31). It is an attractive possibility that signalling systems which in evolution prove to be successful, should be used — depending on the developmental state of the cell which receives the stimulus — for purposes other than transmission.

ACKNOWLEDGEMENTS

V. Gallo was supported by a Long-Term EMBO Fellowship, O. S. Jørgensen received support for this work from the Danish Medical Research Council. We are indebted to J. C. Watkins, W. W. Habig and R. O. Thompson respectively for the gift of AVP, the labelled and unlabelled tetanus toxin. Helpful discussions with John Garthwaite, Jeff Watkins and Silvio Varon are acknowledged.

REFERENCES

1) CURTIS, D. R., JOHNSTON, G. A. R. (1974). Amino acid transmitters in the mammalian central nervous system. *Ergebn. Physiol.* **69:** 97—188.

2) ROTHMAN, S. M., OLNEY, J. W. (1986). Glutamate and the pathophysiology of hypoxic-ischemic brain damage. *Ann. Neurol.* **19:** 105—111.

3) BALÁZS, R., JØRGENSON, O. S. (1987). A new trophic function of excitatory transmitter amino acids. *Neurosci.* **22:** S41.

4) BALÁZS, R., GALLO, V., KINGSBURY, A., THANGNIPON, W., SMITH, R., ATTERWILL, CH., WOODHAMS, P. (1987). Factors affecting the survival and maturation of nerve cells in culture. In: *Glial-Neuronal Communication in Development and Regeneration* (Althaus, H. H., Seifert, W., eds.), Springer, Berlin, pp. 286—302.

5) GALLO, V., KINGSBURY, A., BALÁZS, R. (1987). The role of depolarization in the survival and differentiation of cerebellar granule cells in culture. *J. Neurosci.* **7:** 2203—2213.

6) LASHER, R. S., ZAGON, I. S. (1972). The effect of potassium on neuronal differentiation in cultures of dissociated newborn rat cerebellum. *Brain Res.* **41:** 482—488.

7) BANKER, G. A., COWAN, W. M. (1979). Further observations on hippocampal neurons in dispersed cell culture. *J. Comp. Neurol.* **187:** 469—494.

8) DICHTER, M. A. (1978). Rat cortical neurons in cell culture: culture methods, cell morphology, electrophysiology and synapse formation. *Brain Res.* **149:** 279—293.

9) ALTMAN, J. (1982). Morphological development of the rat cerebellum and a source of its mechanisms. In: *The Cerebellum: New Vistas* (Chan-Palay, V., Palay, S., eds.), Springer, Berlin, pp. 8—49.

10) HARRIS, W. A. (1981). Neural activity and development. *Ann. Rev. Physiol.* **43:** 689—710.

11) KINGSBURY, A., BALÁZS, R. (1987). Effect of calcium agonists and antagonists on cerebellar granule cells. *Eur. J. Pharmacol.* **140:** 275—283.

12) SOMOGYI, P., HALASY, K., SOMOGYI, J., STORM-MATHIESEN, J., OTTERSEN , O. P. (1986). Quantification of immunogold labelling reveals enrichment of glutamate in mossy and parallel fiber terminals in cat cerebellum. *Neurosci.* **19:** 1045—1050.

13) CULL-CANDY, S. G., OGDEN, D. C. (185). Ion channels activated by L-glutamate and GABA in cultured cerevellar neurones of the rat. *Proc. Roy. Soc. London* **224B:** 367—373.

14) MAYER, M. L., WESTBROOK, G. L. (1987). The physiology of excitatory amino acids in the verbrate central nervous system. *Progr. Neurobiol.* **28:** 197—276.

15) MANTHORPE, M., FAGNANI, R., SKAPER, S. D., VARON, S. (1986). An automated colorimetric microassay for neurotrophic factors. *Dev. Brain Res.* **25:** 191—198.

16) OLVERMAN, H. J., JONES, A. W., WATKINS, J. C. (1984). L-Glutamate has higher affinity than other amino acids for [3H]-D-AP5 binding sites in rat brain membranes. *Nature* **307:** 460—462.

17) NOWAK, L., BREGESTOVSKI, P., ASCHER, P., HERBERT, A., PROSCHIANTZ, A. (1984). Magnesium gates glutamate-activated channels in mouse central neurons. *Nature* **307:** 462—465.

18) EVANS, R. H., FRANCIS, A. A., WATKINS, J. C. (1977). Selective antagonism by Mg^{2+} of amino acid-induced depolarization of spinal neurones. *Experientia* **33:** 489—491.

19) THANGNIPON, W., KINGSBURY, A., WEBB, M., BALÁZS, R. (1983). Observations on rat cerebellar cells in vitro: influence of substratum, potassium concentration and relationship between neurones and astrocytes. *Dev. Brain Res.* **11:** 177—189.

20) MEIER, E., REGAN, C. M., BALÁZS, R. (1984). Changes in the expression of a neuronal surface protein during development of cerebellar neurones *in vivo* and in culture. *J. Neurochem.* **43:** 1328—1335.

21) NYBROE, O., ALBRECHTSEN, M., DAHLIN, J., LINNEMANN, D., LYLES, J. M., MOLLER, C. J., BOCK, E. (1985). Biosynthesis of the neuronal cell adhesion molecule: Characterization of polypeptide C. *J. Cell Biol.* **101:** 1—6.

22) GARTHWAITE, G., GARTHWAITE, J. (1986). *In vitro* neurotoxicity of excitatory amino acid analogues during cerebellar development. *Neurosci.* **17:** 755—767.

23) RAKIC, P., SIDMAN, R. L. (1973). Organization of cerebellar cortex secondary to deficit of granule cells in weaver mutant mice. *J. Comp. Neurol.* **152:** 133—162.

24) GARTHWAITE, G., YAMINI, B. Jr., GARTHWAITE, J. (1987). Selective loss of Purkinje and granule cell responsiveness to N-methyl-D-aspartate in rat cerebellum during development. *Dev. Brain Res.*, in press.

25) COTMAN, C. W., IVERSEN, L. L. (eds.) (1987). Excitatory amino acids in the brain — focus on NMDA receptors. *Trend. Neurosci.* **7:** 263—302.

26) RAUSCHECKER, J. P., HAHN, S. (1987). Ketamine-sylazine anaesthesia blocks consolidation of ocular dominance changes in kitten visual cortex. *Nature* **326:** 183—185.

27) SINGER, W., KLEINSCHMIDT, A., BEAR, M. F. (1986). Infusion of an NMDA receptor antagonist disrupts ocular dominance plasticity in kitten striate cortex. *Soc. Neurosci. Abstr.* **12:** 786.

28) BALÁZS, R., PATEL, A. J., LEWIS, P. D. (1977). Metabolic influences on cell proliferation in the brain. In: *Biochemical Correlates of Brain Structure and Function* (Davison, A. N., ed.), Academic Press, New York, pp. 43—83.

29) BUZNIKOV, G. A. (1984). The action of neurotransmitters and related substances on early embryogenesis. *Pharmacol. Ther.* **25:** 23—59.

30) LAUDER, J. M. (1987). Neurotransmitters as morphogenes. In: *Neurochemistry of Functional Neuroteratology* (Boer, G. J., ed.), Elsevier, Amsterdam, in press.

31) REDBURN, D. A., SCHOUSBOE, A. (eds.) (1987). *Neurotrophic Activity of GABA during Development.* Alan R. Liss, New York.

32) BURTON, K. (1956). A study of the conditions and mechanisms of the diphenylamine reaction for the colorimetric estimation of DNA. *Biochem. J.* **62:** 315—323.

33) LOWRY, O. H., ROSEBROUGH, N. J., FARR, A. L., RANDALL, R. J. (1951). Protein measurement with the Folin phenol reagent. *J. Biol. Chem.* **193:** 265—275.

34) JØRGENSEN, O. S. (1981). Neuronal membrane D2-protein during rat brain ontogeny. *J. Neurochem.* **37:** 939—946.

GAMMA-AMINOBUTYRIC ACID AS A NEUROTROPHIC AGENT

A. Schousboe[a], B. Belhage[a], E. Meier[b], R. Hammerschlag[c], G. H. Hansen[a]

[a] Department of Biochemistry A, Panum Institute
University of Copenhagen
DK-2200 Copenhagen N, Denmark
[b] Present Address:
Department of Pharmacology, H. Lundbeck & Co., LTD.
DK-2500 Valby, Denmark
[c] Division of Neurosciences, Beckman Research Institute
City of Hope National Medical Center
Duarte, CA 91010, USA

INTRODUCTION

There is accumulating evidence that different classical neurotransmitters such as serotonin, catecholamines and the amino acid GABA may serve as epigenetic factors during neuronal development and differentiation (1, 2, 3, 4). Since a variety of macromolecular compounds, mainly peptides, are also known to act as neurotrophic factors (5) it is clear that the extremely complex processes which govern the development and differentiation of the neuronal network and communication system in the central nervous system are subject to regulation at a number of different levels. The present review will concentrate on a summary of the recent progress in studies concerned with the possible role of GABA and GABA agonists as agents stimulating neuronal development *in vitro* and *in vivo.*

Effect on Cell Morphology

Since the original observation by Wolff and coworkers (6) that GABA enhances synaptogenesis in rat superior cervical ganglia, a number of studies mainly on neurons in tissue culture have been concerned with a characterization of the effect of GABA on the morphological development of neurons. In cerebellar granule cells (7) as well as in retinal neurons (8) and neuroblastoma cells (9) addition of GABA to the culture media has been shown to promote neurite formation. At the electron microscope level this effect seems to be associated with increases in the cytoplasmic densities of e. g. rough endoplasmic reticulum, Golgi apparatus and coated vesicles whereas the densities of mitochondria and smooth endoplasmic reticulum are unaffected (Table 1). These effects of GABA on the morphological differentiation of the neurons were found to be induced *via* an interac-

NATO ASI Series, Vol. H20
Amino Acid Availability and Brain Function in Health and Disease. Edited by G. Huether

TABLE 1

Summary of effects of GABA on the ultrastructure of cerebellar granule cells

Organelle	Effect of GABA	
Mitochondria	none	—
Golgi apparatus	increased	$p<0.0005$
Neurotubules	increased	$p<0.025$
Ribosomes	none	—
Rough endoplasmic reticulum	increased	$p<0.025$
Smooth endoplasmic reticulum	none	—
Coated vesicles	increased	$p<0.05$
Other vesicles	increased	$p<0.05$

Cerebellar granule cells were cultured for 1—7 days in the absence or presence of 50 μM GABA and the temporal development of the cytoplasmic densities of the different organelles was followed. Electron micrographs were taken at random and densities of the structures were determined double-blind using the point method for morphometric analysis. Summary from Hansen *et al.* (7).

tion between GABA and high affinity $GABA_A$ receptors since the action of GABA could be completely blocked (10) by the specific GABA receptor antagonist (11) bicuculline and mimicked (10) by the specific GABA receptor agonist (12) 4,5,6,7-tetrahydroisoxazolo[5,4-c]pyridin-3-ol (THIP).

The temporal development of these GABA or GABA receptor agonist-induced morphological changes has recently been studied (13). It was found that a one hour exposure to THIP was sufficient to induce the increase in the cytoplasmic density of the different subcellular structures. Moreover, it has been shown that exposure of the cerebellar granule cells to THIP only enhances the morphological development provided the cells are exposed during the first week in culture (Table 2). This indicates that a developmental window exists during which neurons are particularly sensitive to the trophic actions of GABA (14).

TABLE 2

Effect of THIP treatment on morphological development in cerebellar granule cells as a function of the culture period

Culture period (days)	Effect on cytoplasmic density of organelles
4	Increased
7	Increased
11	No effect

Cerebellar granule cells were cultured for the periods indicated above and subsequently exposed to 150 μM THIP for 6 hours. After fixation and preparation of the cultures for electron microscopy the cytoplasmic densities of organelles such as Golgi apparatus, rough endoplasmic reticulum and coated vesicles were determined. Results from Hansen *et. al.* (13) and unpublished experiments of G. H. Hansen, B. Belhage and A. Schousboe.

Effect on Receptor Expression

GABA receptors in CNS normally consist of high and low affinity entities (15, 16, 17). On the basis of this it was surprising that cerebellar granule cells in culture only expressed the high affinity GABA receptors (16). It was subsequently shown that treatment of the cultures with GABA (50 μM) during the culture period led to induction of low affinity GABA receptors in addition to high affinity GABA receptors (18, 19). Simultaneously it was reported that GABA or GABA agonists injected intraocularly induced GABA receptors in retina (20, 21) and that elevation of brain GABA levels by GABA-transaminase inhibitors led to an increase in GABA receptors in the brain (22, 23). All of these findings strongly indicate that GABA is able to regulate the expression of its own receptors by formation of additional receptor sites which at least in some of the models studied exhibited different affinities for GABA.

The mechanism by which GABA induces low affinity GABA receptors has been studied using GABA receptor agonists and antagonists. It was found (Table 3) that the effect of GABA could be mimicked by the GABA receptor agonist THIP and blocked by the receptor antagonist bicuculline (24) or picrotoxin (25), a blocker (26) of GABA gated chloride channels. These results strongly imply that the action of GABA is mediated *via* the preexisting high affinity GABA receptors on the cells. This appears to be the same mechanism by which GABA acts as a neurite promoting agent (cf. above).

The temporal development of the induction by GABA of low affinity GABA receptors has recently been studied. It was found that low affinity GABA receptors can be demonstrated in membranes prepared from cerebellar granule cells 3 h after addition of THIP to the culture media (13). Moreover, it appears that this induction is only possible at early developmental stages of the neurons since cultures older than 7 days did not respond to THIP treatment (14). Similar developmental patterns have also been reported for induction of GABA receptors in retina (27).

TABLE 3
Effects of GABA-mimetics, GABA-antagonists and protein synthesis inhibitors on GABA-receptor expression

Culture condition	GABA-binding K_D (nM) High aff.	Low aff.
Control	7.0	n. d.
50 μM GABA	7.0	546
150 μM THIP	6.8	476
150 μM THIP + 150 μM BIC	14.3	n. d.
150 μM THIP + 150 μM Picrotoxin	4.9	n. d.
150 μM THIP + 20 μg/ml Actinomycin D	8.3	n. d.
150 μM THIP + 50 μg/ml Cycloheximide	11.0	n. d.

Cells were cultured for 7—12 days in plain culture media (controls) or media to which the agonists or antagonists were added. In case of the protein synthesis inhibitors, exposure of the cells to THIP and inhibitor was restricted to 6 h after 4 days of culture (cf. Table 2). K_D values were determined from Scatchard plots by computer analysis (28, 29).
BIC: Bicuculline methobromide.
n.d.: means not detectable.
From Meier *et al.* (19, 25), Belhage *et al.* (24) and unpublished results of B. Belhage and A. Schousboe.

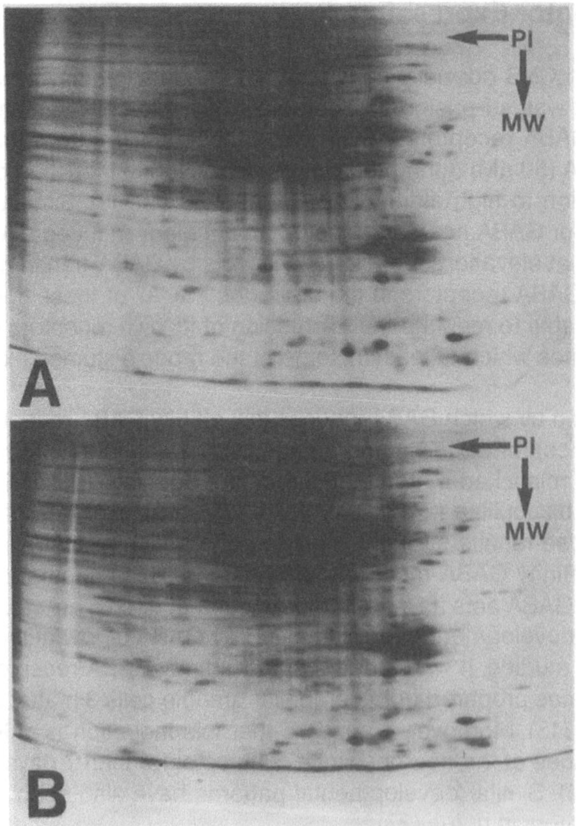

FIGURE 1

Fluorographic patterns of two-dimensional gel electrophoresis of extracts of cerebellar granule cells cultured for 4 days in control culture media and an additional 8 h in either similar serum-free media (A) or serum-free media containing 150 μM THIP (B). In both cases cold methionine was removed from the culture media and replaced by ^{35}S-methionine (20 μCi per culture). Lyophilized cell samples were resuspended in 30 μl of a mixture of 1 % SDS, 10 % β-mercaptoethanol, and 9.5 M urea and the pH was adjusted to 10 with NaOH. To this homogenate was added 10 μl of a mixture of 10 % NP-40, 8 % 5—7 ampholines, 2 % 3—10 ampholines (v/v) (Bio-Rad). The homogenizer was rinsed twice with 20 μl each of O'Farrell's sysis buffer (38). Samples were then spun at 100,000 × g for 60 min (22°C) before loading the supernatant on the first dimension gel. Also, duplicate 2 μl aliquots of the homogenates were added to 500 μl 10 % TCA, and the precipitate was counted to determine the amount of radioactive material applied to the gel. The first dimension isoelectric focusing tube gel (pH ~ 4.5 to 8.5) was run for 19 h at 400 V. Ten percent SDS slab gels run at 18 mA separated proteins in the second dimension according to molecular weight (~ 20,000 to 200,000 daltons). Following SDS electrophoresis, the gels were fixed and stained with Coomassie blue, impregnated with PPO (39) and dried prior to fluorographic exposure at −70°C for up to 2 weeks. Gels were compared by keeping the product of radioactivity added to the gel and exposure time constant.

Unpublished results of R. Hammerschlag, E. Meier and A. Schousboe.

The appearance of low affinity GABA receptors after treatment of the cells with GABA could conceivably be the result of either a conversion of preexisting dormant receptor sites or it may require *de novo* synthesis of low affinity receptors. The observation (Table 3) that the effect of THIP could be completely blocked by protein synthesis inhibitor, cycloheximide, and the RNA synthesis inhibitor, actinomycin D, strongly favor the latter hypothesis thus supporting the notion that the expression of low affinity receptors is indeed the result of an induction.

Effect on Protein Synthesis

Based on early reports that GABA may enhance brain protein synthesis (30, 31, 32, 33) and the finding that enhancement of expression of GABA receptors by GABA requires protein synthesis (cf. above) it might be anticipated that treatment of cerebellar granule cells with THIP would enhance incorporation of amino acids into proteins. Cells were cultured in control media or THIP-containing media and protein synthesis was monitored by measuring the incorporation of ^{35}S-methionine into proteins which were subsequently separated by two dimensional gel electrophoresis. Figure 1 shows fluorograms prepared from such gels and there are several spots in which the incorporation of radioactivity appears enhanced after treatment with THIP. It should, however, be noted that the overall incorporation of ^{35}S-methionine was similar in the two sets of cultures. The finding that the level of several proteins is increased by THIP is in agreement with the observation that such treatment increases the number of GABA receptors (cf. above) and enhances the expression of the neuron associated proteins N-CAM and $\gamma\gamma$-enolase (34). That the total incorporation of ^{35}S-methionine was unaffected by THIP treatment is consistent with the observation that overall protein synthesis in brain is unaffected by elevated brain GABA levels (35).

Effects In Vivo

The findings that elevated GABA levels induce GABA receptors *in vivo* both in retina and in the brain (22, 20, 21, 23) might suggest that other neuronal functions could also be influenced by GABA or GABA agonists during development *in vivo.* In order to obtain such information rats were treated with THIP during rearing. It was found (Fig. 2) that the specific activity of glutamate decarboxylase which is associated with GABAergic neurons (36) was increased during development subsequent to THIP treatment (37). In addition, the maturation of N-CAM was enhanced in the THIP treated rats reflecting a similar observation in cultured cerebellar granule cells (34). It may accordingly be concluded that *in vivo* neuronal development and differentiation in brain is enhanced by GABA and GABA agonists.

ACKNOWLEDGEMENTS

The work was supported by the Danish Medical Research Council and the NOVO Foundation. The expert technical assistance by Mrs. Inge Damgaard, Hanne Fosmark, Grete Rossing and Judy Bobinski is gratefully acknowledged.

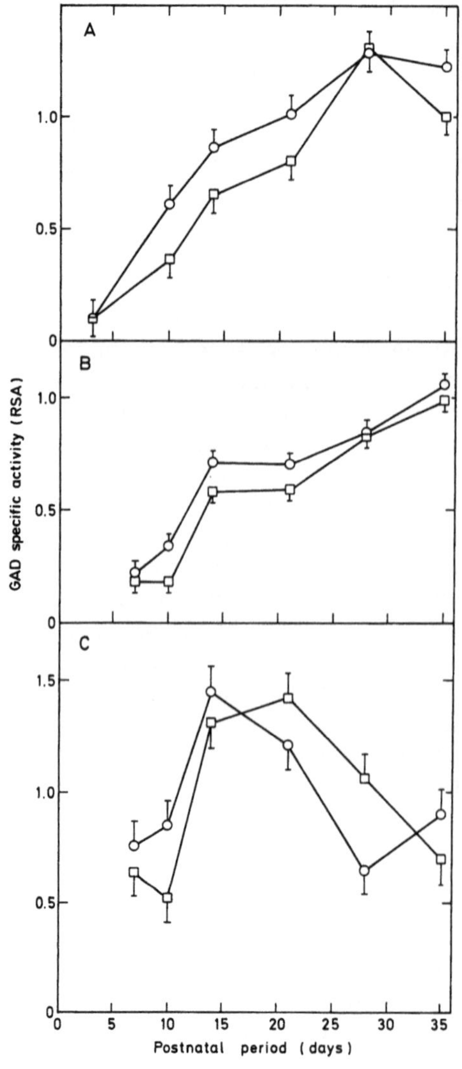

FIGURE 2
Relative specific activity of GAD in residual forebrain (A), hippocampus (B), and cerebellum (C) in rats treated with 60 nmol/g body weight i. p. THIP (○) and control rats (□). Each point represents the average of 5 animals and SEM are indicated by vertical bars. One RSA = specific activity of GAD in residual forebrain of 35-day-old control animals (4.25 nmol × min^{-1} × mg^{-1} protein). A and B: Residual forebrain and hippocampus in THIP-treated animals expressed significantly higher specific activity of GAD than in control animals (A: $p = 0.003$ and B: $p = 0.002$). C: Cerebellum in THIP-treated animals and control animals displayed no significant difference as a whole. However, a significant interaction between the two criteria, postnatal day and ± THIP, was observed ($p = 0.01$) due to significantly higher specific activity of GAD at 7—14 days ($p = 0.03$) and a significantly lower specific activity at 21—28 days ($p < 0.001$).
From Meier *et al.* (37).

REFERENCES

1) HAYDON, P. G., McCOBB, E. P., KATER, S. B. (1984). Serotonin selectively inhibits growth cone motility and synaptogenesis of specific identified neurons. *Science* **226:** 561—564.

2) KÖNIG, N., DRIAN, M.-J., PRIVAT, A., LAMANDE, N., PARES-HERBUTE, N., SCHACHNER, M. (1986). Dissociated cells of foetal rat pallium grown in culture medium supplemented with noradrenaline: Effects of the expression of neuron-specific enolase and cell adhesion molecule L1. *Neurosci. Lett.* **66:** 67—72.

3) LAUDER, J. M. (1983). Hormonal and humoral influences on brain development. *Psychoneuroendocrinology* **8:** 121—155.

4) REDBURN, D. A., SCHOUSBOE, A. (eds.) (1987). *Neurotrophic Activity of GABA During Development.* Alan R. Liss, Inc., N. Y., pp. 1—277.

5) MANTHORPE, M., RUDGE, J. S., VARON, S. (1986). Astroglial cell contributions to neuronal survival and neuritic growth. In: *Astrocytes* (Fedoroff, S., Vernadakis, A., eds.), **Vol. 2,** Academic Press, N. Y., pp. 315—376.

6) WOLFF, J. R., JOO, F., DAMES, W. (1978). Plasticity in dendrites shown by continuous GABA administration in superior cervical ganglion of adult rat. *Nature Lond.* **274:** 72—74.

7) HANSEN, G. H., MEIER, E., SCHOUSBOE, A. (1984). GABA influences the ultrastructure composition of cerebellar granule cells during development in culture. *Int. J. Devl. Neurosci.* **2:** 247—257.

8) SPOERRI, P. E. (1987). GABA-mediated alterations in a neuronal cell line and in cultures of cerebral and retinal neurons. In: *Neurotrophic Activity of GABA During Development* (Redburn, D. A., Schousboe, A., eds.). Alan R. Liss, Inc., N. Y., pp. 189—220.

9) SPOERRI, P. E., WOLFF, J. R. (1981). Effect of GABA-administration on murine neuroblastoma cells in culture. *Cell Tiss. Res.* **218:** 567—579.

10) MEIER, E., HANSEN, G. H., SCHOUSBOE, A. (1985). The trophic effect of GABA on cerebellar granule cells is mediated by GABA-receptors. *Int. J. Devl. Neurosci.* **3:** 401—407.

11) CURTIS, D. R., DUGGAN, A. W., FELIX, O., JOHNSTON, G. A. R., McLENNAN, H. (1971). Antagonism between bicuculline and GABA in the cat brain. *Brain Res.* **33:** 57—73.

12) SCHOUSBOE, A., LARSSON, O. M., KROGSGAARD-LARSEN, P. (1985). Lack of a high affinity uptake system for the GABA agonists THIP and isoguvacine in neurons and astrocytes cultured from mouse brain. *Neurochem. Int.* **7:** 505—508.

13) HANSEN, G. H., BELHAGE, B., SCHOUSBOE, A., MEIER, E. (1987). Temporal development of GABA-agonist induced alterations in ultrastructure and GABA receptor expression in cultured cerebellar granule cells. *Int. J. Devl. Neurosci.* **5:** 263—269.

14) BELHAGE, B., HANSEN, G. H., SCHOUSBOE, A., MEIER, E. (1987). GABA-agonist promoted formation of low affinity GABA-receptors on cerebellar granule cells is restricted to early development. *Int. J. Devl. Neurosci.* (in press).

15) FALCH, E., KROGSGAARD-LARSEN, P. (1982). The binding of the GABA agonist ^{3}H-THIP to rat brain synaptic membranes. *J. Neurochem.* **38:** 1123—1129.

16) MEIER, E., SCHOUSBOE, A. (1982). Differences between GABA receptor binding to membranes from cerebellum during postnatal development and from cultured cerebellar granule cells. *Dev. Neurosci.* **5:** 546—553.

17) WANG, Y.-J., SALVATERRA, P., ROBERTS, E. (1979). Characterization of ^{3}H-muscimol binding to mouse brain membranes. *Biochem. Pharmacol.* **28:** 1123—1128.

18) MEIER, E., DREJER, J., SCHOUSBOE, A. (1983). Trophic actions of GABA on the development of physiologically active GABA receptors. In: *CNS-Receptors from Molecular Pharmacology to Behaviour* (Mandel, P., DeFeudis, F. V., eds.). Raven Press, N. Y., pp. 47—58.

19) MEIER, E., DREJER, J., SCHOUSBOE, A. (1984). GABA induces functionally active low-affinitive GABA receptors on cultured cerebellar granule cells. *J. Neurochem.* **43:** 1737—1744.

20) MADTES, P. C., REDBURN, D. A. (1983). GABA as a trophic factor during development. *Life Sci.* **33:** 979—984.

21) MADTES, P. C., REDBURN, D. A. (1983). Synaptic interactions in the GABA system during postnatal development in retina. *Brain Res. Bull.* **10:** 741—745.

22) BEART, P. M., SCATTON, B., LLOYD, K. (1985). Subchronic administration of GABAergic agonists elevates [^{3}H]-GABA binding and produces tolerance in striatal dopamine catabolism. *Brain Res.* **335:** 169—173.

23) SYKES, C., PRESTWICH, S., HORTON, R. (1984). Chronic administration of the GABA-transaminase inhibitor ethanolamine-O-sulphate leads to up-regulation of GABA binding sites. *Biochem. Pharmacol.* **33:** 387—393.

24) BELHAGE, B., MEIER, E., SCHOUSBOE, A. (1986). GABA-agonists induce the formation of low affinity GABA-receptors on cultured cerebellar granule cells via preexisting high affinity GABA-receptors. *Neurochem. Res.* **11:** 599—606.

25) MEIER, E., BELHAGE, B., DREJER, J., SCHOUSBOE, A. (1987). The expression of GABA receptors on cultured cerebellar granule cells is influenced by GABA. In: *Neurotrophic Activity of GABA During Development* (Redburn, D. A., Schousboe, A., eds.). Alan R. Liss, Inc., N. Y., pp. 139—159.

26) OLSEN, R. W. (1981). GABA-benzodiazepine-barbiturate receptor interactions. *J. Neurochem.* **37:** 1—13.

27) MADTES, P. C., BASHIR-ELAHI, R. (1986). GABA-receptor binding site 'induction' in rabbit retina after nipecotic acid treatment: Changes during development. *Neurochem. Res.* **11:** 55—61.

28) McPHERSON, G. A. (1983). A practical computer-based approach to the analysis of radioligand binding experiments *Comp. Prog. Biomed.* **17:** 107—114.

29) MUNSON, P. I., RODBARD, D. (1980). A versatile computerized approach for characterization of ligand binding systems. *Anal. Biochem.* **107:** 220—239.

30) CAMPBELL, M. K., MAHLER, H. R., MOORE, W. J., TEWARI, S. (1966). Protein synthesis systems from rat brain. *Biochemistry* **5:** 1174—1184.

31) GOERTZ, B. (1979). Effect of GABA on cell-free protein synthesizing systems from mouse brain. *Exp. Brain Res.* **34·** 365—372.

32) SNODGRASS, S. R. (1973). Studies of GABA and protein synthesis. *Brain Res.* **59:** 339—348.

33) TEWARI, S., BAXTER, C. F. (1969). Stimulatory effect of GABA upon amino acid incorporation into protein by a ribosomal system from immature rat brain. *J. Neurochem.* **16:** 171—180.

34) MEIER, E., JØRGENSEN, O. S. (1986). Gamma-aminobutyric acid affects the developmental expression of neuron-associated proteins in cerebellar granule cell cultures. *J. Neurochem.* **46:** 1256—1262.

35) TOTH, E., LAJTHA, A. (1984). Brain protein synthesis rates are not sensitive to elevated GABA, taurine, or glycine. *Neurochem. Res.* **9:** 173—179.

36) ROBERTS, E. (1979). New directions in GABA research I: Immunocytochemical studies of GABA neurons. In: *GABA-Neurotransmitters: Pharmacochemical, Biochemical and Pharmacological Aspects* (Krogsgaard-Larsen, P., Scheel-Krüger, J., Kofod, H., eds.). Munksgaard, Copenhagen, pp. 28—45.

37) MEIER, E., JØRGENSEN, O. S., SCHOUSBOE, A. (1987). Effect of repeated treatment with a GABA receptor agonist on postnatal neural development in rats. *J. Neurochem.* **49** (in press).

38) O'FARRELL, P. H. (1975). High resolution two-dimensional electrophoresis of proteins. *J. Biol. Chem.* **250:** 4007—4021.

39) LASKEY, R. A., MILLS, A. D. (1975). Quantitative film detection of ^{3}H and ^{14}C in polyacrylamide gels by fluorography. *Eur. J. Biochem.* **56:** 335—341.

ROLE OF TAURINE IN BRAIN DEVELOPMENT AND VISION

C. E. Wright and G. E. Gaull

New York State Institute for Basic Research
in Developmental Disabilities
Staten Island, New York
and
The Nutrasweet Company
Skokie, Illinois, USA

INTRODUCTION

The sulfonic acid containing amine, taurine, is an intriguing ubiquitous simple amino acid. Unlike many of the neuroactive amino acids discussed at this NATO Workshop, a specific biological role of taurine to either the central of peripheral nervous system remains unresolved. Taurine is one of the most abundant amino acids present in the pool of animal tissues. For some tissues such as developing brain (1—3), and mature retina (4—6) it is present in relatively high concentrations. In brain, concentrations of taurine are equal to or slightly higher than those of the neuroactive gamma-aminobutyric acid; and glutamate is the only amino acid present in greater amounts. Unlike brain tissue, taurine in the retina constitutes 40—50 % of the total free amino acids. This concentration is the highest of any region of the central nervous system. Because of its ubiquitous distribution, its abundance in the central nervous system, and its many physiological actions on the brain and eye it is apparent that taurine has fundamental, but undetermined function(s) on these organs. Whether taurine acts as a neurotransmitter (7—9), a neuromodulator (8), an osmotic regulator (10—11), a membrane stabilizer (12), an antioxidant (13), some other undiscovered function, or is multifunctional and encompasses some or all of these functions, remains to be determined.

Chemistry and Biochemistry of Taurine

In order to understand better a biological role of taurine in brain and retina it is essential to present some of the chemical and biochemical properties of this molecule. Taurine is a beta amine possessing a sulfonate. It is this structural combination that gives taurine its unique biological characteristics and physiological role(s). A well known and often ignored capacity of free carboxylic-containing amino acids is to form stable metal complexes.

NATO ASI Series, Vol. H20
Amino Acid Availability and Brain Function in Health and Disease. Edited by G. Huether

Taurine forms fewer stable metal complexes with various transition metals than other amino acids (14). The low or unmeasurable stability constants of taurine-metal complexes are the result of the sulfonate ion, which is a poor ligand. Thus, it fails to form a stable five-membered ring complex of metal and amino acid in which both the amino and the acidic carboxylic groups participate in chelate formation. Since taurine fails to form chelates like most other amino acids, it is truly the only "free" amino acid present in tissues. This chemical property may be critical to its biological function.

Taurine, like all amino acids, behaves as an amphoteric electrolyte; however, some physiochemical differences are expected as a result of the presence of the sulfonate ion. Taurine possesses a more acidic acid group (pK1' > 1.50) as well as a more acidic ammonium group (pK2' = 8.74) than other amino acids (1). As a result taurine maintains a zwitterionic form within a wide range around physiological pH. This characteristic of taurine may also be a necessity for its biological function.

Traditionally, taurine has been considered to be mainly the biochemical inert end product of methionine and cysteine catabolism. Certainly, taurine is not utilized for production of energy, is not a major source of free sulfate or sulfur, and is not translationally incorporated into protein. Even though taurine is conjugated with several different bile acids, the amount consumed by this well-recognized reaction consumes less than one per cent of the total pool (15).

It is the reaction of the amino group that reflects the biochemical reactivity of taurine. The amino group of taurine, under enzymatic conditions, can react with carboxylic acid-containing molecules to form amide linkages. In addition to bile acids, many different xenobiotics react in a similar manner (16). Retinoic acid reacts with taurine to form retinotaurine in biliary metabolites (17). Taurine also forms peptide bonds in some low-molecular-weight acidic peptides in brain synaptosomal and vesicular preparations (18, 19). These taurine-containing peptides have not been detected in non-neuronal tissue, and their physiological function in the central nervous system are unknown. Amide linkages of taurine are not limited to small brain peptides; in certain prokaryotes and lower eukaryotes several compounds have been found showing amide linkages of taurine to polysaccharides and lipids (20—22). For all of the above taurine-containing compounds, specific enzymes are involved in their synthesis.

An example of the chemical reactivity of the amino group of taurine is its ability to react with several halogen oxidants to form stable and non-toxic taurine haloamines (23). The nitrogen of taurine can react with either the hypohalite anion or the halinium cation. As a result of this reaction taurine can effectively scavenge and attenuate toxic halide derived oxidants which are formed during the enzymatic and chemical breakdown of peroxides (23); this is especially true for tissues which generate oxidants, which are excitable and are rich in membranes (1, 13).

It is this antioxidant capability of taurine which may be its central biochemical mechanism, namely, to protect cells and tissues by attenuating and eliminating toxic compounds, such as oxidants, secondary bile acids, and xenobiotics. Taurine is abundant in most tissues, however, it is more abundant where oxidants and potentially toxic substances are found. The chemical reactivity of the amino group allows taurine to perform this protective function, while it is its sulfonic acid end of the molecule that is critical in its performance. As a result of these unique molecular characteristics, attenuation and elimination by taurine spares tissues from adverse effects and results in their normal function. Thus, taurine mediates the continuous threat of cellular disorganization and self destruction.

Taurine in the Nervous System

For nearly twenty years there has been a burgeoning interest in the neurochemistry of taurine. Numerous biochemical and electrophysiological studies have led to the suggestion that taurine may have a role as a neuromodulator or neurotransmitter in neural tissues. Most aspects of the neurochemistry of taurine have been authoritively reviewed by Oja and Kontro (24).

In most mammalian species, taurine concentrations are high in fetal neural tissues, gradually decreasing during postnatal development, and reaching adult levels at approximately the time of weaning (2, 25, 26). This change in taurine concentration is striking when compared with the increasing levels of glutamic acid, gamma aminobutyric acid, and aspartic acid during ontogeny of the central nervous system.

Recently, Sturman (27, 28) has shown conclusive evidence that taurine plays an important role in brain development. In prenatal and postnatal deprivation of taurine in kittens gross physical and morphological dysfunctions occur, including cellular aberrations in the brain. Kittens born to deficient mothers and nursed on such mothers' taurine-limited milk are smaller than normal; at weaning (around 8 weeks) being 60 % of normal weight. Similar kittens given oral supplements of taurine reach full size, demonstrating directly that taurine is necessary for growth. The deficient kittens display other abnormalities as well: abnormal development of the hind legs, a peculiar gait with excessive abduction and paresis, and thoracic kyphosis. These symptoms are indicative of cerebellar dysfunction.

It had been found earlier (29) that, in addition to the retinal degeneration in taurine-deficient cats, there is disorganization and disruption of the cells of the tapetum lucidum, a reflective layer behind the retina of cats. Study of the cerebellum then disclosed an alteration in its morphology. Many cells of the external granule cell layer were still in evidence at 8 weeks in cerebellum from deficient kittens, but had still largely migrated to the internal molecular layer in normal cerebellum, as expected. Mitotic figures in many cells of the deficient cerebellar external granule cell layer indicated that cell division was still going on, although this process normally is complete by about 3 weeks of age. No mitotic figures were seen in the normal tissue. Supplementation of deficient kittens with taurine resulted in an increase in the cerebellar taurine content to normal and the occurrence of normal migration of cells. The delay in cell migration in the absence of adequate taurine must upset the strict timetable of cerebellar development, resulting in a failure to form many synaptic connections.

In addition, Sturman's group has found that the visual cortex of deficient kittens also is affected by lack of taurine (30). At birth, neuroblasts at both the ventricular and pial zones have failed to complete their differentiation and have not migrated into the molecular layer; subsequent arborization is poor and much organization must be lost.

Taurine in the Retina

Taurine is exceptionally abundant as a free amino acid in the vertebrate retina (4, 5). Unlike the brain where taurine decreases gradually after birth, taurine concentrations increase during neonatal development (5). Adult retinal concentrations of taurine range from 10 mM in the frog to approximately 50 mM in the rat and rabbit (6). Intermediate amounts are found in other species (4, 6). Of any region derived from the central nervous system, the retina possesses the highest concentration of taurine. Taurine is concentrated in the retinal pigmented epithelium and photoreceptor cell layer, with maximal amounts (50—80 mM) present in the inner segments and outer nuclear layer (6, 31).

Taurine may be supplied to the retina via the vitreous humor (32, 33), the circulating plasma (34) or may be synthesized in situ. Activities for cysteine oxidase, 3-sulfinoalanine decarboxylase and hypotaurine oxidase have been observed in retinal tissues from several animal species (35—37). The amount of taurine synthesized by these enzymes is species dependent. Although it is apparent that the retinas of some animals can synthesize taurine, the millimolar concentrations present in the retina can be accounted for the ability of the retinal pigmented epithelium to accumulate taurine (38). A temperature-sensitive, sodium-dependent, energy-dependent, high-affinity transport system (6, 39) allows retinal pigmented epithelium to rapidly concentrate taurine against a 400- to 500-fold gradient (6). This accumulation occurs prior to being slowly transferred to the photoreceptors and neural retina (33). This slow passage of taurine is not well understood and probably reflects the action of the blood-retinal barrier which excludes free entry of taurine and other neuroactive amino acids into the subretinal space (40). The presence in the retina of two binding proteins specific for taurine (41, 42) may be important in the accumulation and transport of taurine into the inner retina.

Although functions for taurine in the retina have not been fully clarified, in recent years a definite connection has been demonstrated between the availability of taurine to the retina and the integrity of that organ (cf. 43), for depletion of taurine under various circumstances has been found to lead to abnormal electroretinograms or to degeneration of photoreceptor cells. This has been demonstrated in a number of species: initially in cats fed a taurine-free casein diet (44, 45), later in infant rhesus monkeys fed commercial human infant formula almost devoid of taurine (46), in rats receiving the taurine transport inhibitor guanidinoethanesulfonic acid (47) and in human children receiving taurine-free total parenteral nutrition (48). Retinal abnormalities also have been found in patients with intestinal disease and bacterial overgrowth and in rats with bacterial overgrowth in surgically-prepared blind loops; in these conditions, bacterial catabolism of taurine depletes the host (49). In most cases, supplementation with taurine reverses or prevents the retinal disturbance (44, 46, 48).

Several possible functions for taurine in the retina have been suggested: 1) Taurine may act in some species as an inhibitory neural transmitter (4). This is supported by the observations that photoreceptors release taurine after photoexcitation (50, 51) and intravitreal administration of taurine reversibly inhibits the b-wave of the electroretinogram (52). 2) Taurine may regulate osmotic pressure in the eye (10), since taurine exposure inhibits the swelling and disorganization of isolated rod outer segments induced by prolonged illumination (53). 3) Taurine may regulate retinal calcium homeostasis. At low calcium concentrations, taurine stimulates and ATP-dependent uptake of calcium (54, 55); however, it decreases uptake at high calcium concentrations (12, 54). 4) Taurine inhibits the phosphorylation of specific retinal membrane proteins (55). Although the mechanism(s) is not completely understood, taurine may be influencing a kinase or altering the ability of the retinal membrane to be phosphorylated via direct membrane interaction. 5) Taurine reverses melatonin- and serotonin-induced inhibition of phagocytosis in the retinal pigmented epithelium (56). Thus, taurine may be involved in the normal phagocytic function of the retina. 6) Taurine may stabilize retinal membranes either directly in maintaining structure after retinoid exposure (57) or by preventing increased lipid peroxidation after exposure to oxidant-generating systems (50). Finally, 7) Taurine may scavenge hypochlorous acid which is generated by a peroxidase present in retinal pigmented epithelial cells (58). This antioxidant property of taurine may attenuate the biocidal capability of retinal pigmented epithelium during the phagocytosis of shed rod outer segments.

Genetic diseases are excellent models for determining the function of biochemical products. A group of inherited diseases in which taurine may play a major role in pathogenesis are the various forms of retinitis pigmentosa. The degenerative changes that occur in the retina of taurine-deficient cats resemble those found in the retina of patients with retinitis pigmentosa, suggesting that this genetically and clinically heterogeneous group of human diseases might be related to a defect in taurine availability or utilization (38, 48, 49). Recently, it has been shown that cultured human cells possess a high-affinity, active uptake system for taurine (59). Taurine uptake in cells from patients with retinitis pigmentosa differed significantly from that of control cell lines (60). The mean V_{max} for affected cell lines is lower than the mean or control cell lines, whereas the mean appartent K_m for the affected cell lines is greater than that for control cell lines. Individual affected cell lines differ from the controls in having a low V_{max} or a high apparent K_m, or in some cases both. These findings indicate that a defective taurine uptake could be a contributory factor in the pathogenesis of the retinitis pigmentosa group of disorders. The defect may be in taurine uptake per se, or may be a deficiency in some component of the cell membrane.

This study suggests that a defective taurine uptake system is one component of multifactorial disturbances that lead to pigmentary degeneration of the retina. In the retina itself, it may be that certain cell types are especially vulnerable to a relatively mild taurine deficiency or that particular cells have a more severely affected uptake system. How taurine could act in maintaining the viability and functioning of such cells remains to be ascertained. It is clear, however, that defective uptake of taurine under conditions in which this amino acid is limiting is a biochemical defect manifest in non-retinal cells that can be propagated in culture. More extensive biochemical studies of taurine uptake are needed in families in which the mode of inheritance and the functional attributes are carefully defined. Clinical disease in the different genetic forms of retinitis pigmentosa might be the result of any of a number of possible defects in taurine transport, all leading to a relative unavailability of taurine at a site where its presence is essential for function. Nutritional deficiency severe enough to result in reduced plasma concentrations might then be an exacerbating factor.

SUMMARY

Taurine has been implicated in numerous physiological functions, pathological conditions and pharmacological actions (cf. 13). The biochemical mechanism(s) that underline these actions remain obscure. Its ubiquity among a variety of animals (1); abundance as a free amino acid in most body fluids and tissues; efficient mechanism of cellular transport and tissue conservation; inability to be translationally incorporated into protein or used for energy production, suggest that taurine has fundamental biochemical functions. Whether these functions are the result of a single mode of action of the result of multiple modes of action remain to be determined.

REFERENCES

1) JACOBSEN, J. G., SMITH, L. H. (1968). *Physiol. Rev.* **48:** 424—511.

2) STURMAN, J. A., HAYES, K. C. (1980). *Adv. Nutr. Res.* **3:** 231—299.

3) STURMAN, J. A. (1983). *Prog. Clin. Biol. Res.* **125:** 281—295.

4) PASANTES-MORALES, H., KLETHI, J., LEDIG, M., MANDEL, P. (1972). *Brain Res.* **41:** 494—497.

5) MACAIONE, S., RUGGIERI, P., DeLUCA, F., TUCCI, G. (1974). *J. Neurochem.* **22:** 881—891.

6) VOADEN, M. J., LAKE. N., MARSHALL, J., MORJAVIA, B. (1977). *Exp. Eye Res.* **25:** 249—257.

7) MANDEL, P., PASANTES-MORALES, H., URBAN, P. F. (1976). In: *Transmitters in the Visual Process* (Onting, S. L., ed.). Pergamon Press, pp. 89—105.

8) PASANTES-MORALES, H., ARZATE, N. E., CRUZ, C. (1982). In: *Taurine in Nutrition and Neurology* (Huxtable, R. J., Pasantes-Morales, H., eds.). Plenum Press, pp. 273—292.

9) OJA, S. S., KONTRO, P. (1978). In: *Taurine and Neurological Disorders* (Barbeau, A., Huxtable, R. J., eds.). Raven Press, pp. 181—200.

10) THURSTON, J. H., HAUHART, R. E., DIRGO, J. A. (1980). *Life Sci.* **26:** 1561—1568.

11) VAN GELDER, N. M., BARBEAU, A. (1985). In: *Taurine: Biological Actions and Clinical Perspectives* (Oja, S. S., Ahtee, L., Kontro, P., Paasonen, M. K., eds.). Alan R. Liss, pp. 149—163.

12) PASANTES-MORALES, H., CRUZ, C. (1985). In: *Taurine: Biological Actions and Clinical Perspectives* (Oja, S. S., Ahtee, L., Kontro, P., Paasonen, M. K., eds.). Alan R. Liss, pp. 371—381.

13) WRIGHT, C. E., TALLAN, H. H., LIN, Y. Y., GAULL, G. E. (1986). *Ann. Rev. Biochem.* **55:** 427—453.

14) SAKURAI, H., TAKESHIMA, S. (1983). In: *Sulfur Amino Acids: Biochemical and Clinical Aspects* (Kuriyama, K., Huxtable, Rl J., Iwata H., eds.). Alan R. Liss, pp. 398—399.

15) HEPNER, G. W., STURMAN, J. A., HOFMANN, A. F., THOMAS, R. J. (1973). *J. Clin. Invest.* **52:** 433—440.

16) EMUDIANUGHE, T. S., CALDWELL, J., SMITH, R. L. (1983). *Xenobiotica* **13:** 133—138.

17) SKARE, K. L., SIETSEMA, W. K., DeLUCA, H. F. (1982). *J. Nutr.* **112:** 1626—1629.

18) LAHDESMAKI, P., MARNELA, K. M. (1985). In: *Taurine: Biological Actions and Clinical Perspectives* (Oja, S. S., Ahtee, L., Kontro, P., Paasonen, M. K., eds.). Alan R. Liss, pp. 105—114.

19) MARNELA, K.-M., TIMONEN, M., LAHDESMAKI, P. (1984). *J. Neurochem.* **43:** 1650—1653.

20) LIAU, D.-F., MELLY, M. A., HASH, J. H. (1974). *J. Bacteriol.* **119:** 913—922.

21) ABBANAT, D. R., LEADBETTER, E. R., GODCHAUX, W., ESCHER, A. (1986). *Nature* **324:** 367—369.

22) BABANO, Y., KAYA, K., SASAKI, N., NOZAWA, Y. (1986). *Biochim. Biophys. Acta* **884:** 599—601.

23) WRIGHT, C. E., LIN, T., LIN, Y. Y., STURMAN, J. A., GAULL, G. E. (1985). In: *Taurine: Bioligical Actions and Clinical Perspectives* (Oja, S. S., Ahtee, L., Kontro, P., Paasonen, M. K., eds.). Alan R. Liss, pp. 137—147.

24) OJA, S. S., KONTRO, P. (1983). In: *Handbook of Neurochemistry,* **Vol. 3,** 2nd edition (Lajtha, A., ed.). Plenum Press, pp. 501—533.

25) STURMAN, J. A., GAULL, G. E. (1975). *J. Neurochem.* **25:** 831—835.

26) STURMAN, J. A., RASSIN, D. K., GAULL, G. E. (1978). In: *Taurine and Neurological Disorders* (Barbeau, A., Huxtable, R. J., eds.). Raven Press, pp. 49—71.

27) STURMAN, J. A., MORETZ, R. C., FRENCH, J. H., WISNIEWSKI, H. M. (1985). *J. Neurosci. Res.* **13:** 405—416.

28) STURMAN, J. A., MORETZ, R. C., FRENCH, J. H., WISNIEWSKI, H. M. (1985). *J. Neurosci. Res.* **13:** 521—528.

29) STURMAN, J. A., WEN, G. Y., WISNIEWSKI, H. M., HAYES, K. C. (1981). *Histochemistry* **72:** 34—350.

30) PALAKAL, T., MORETZ, R., WISNIEWSKI, H., STURMAN, J. (1986). *J. Neurosci. Res.* **15:** 223—239.

31) ORR, H. T., COHEN, A. I., LOWRY, O. H. (1976). *J. Neurochem.* **26:** 609—611.

32) PASANTES-MORALES, H., BONAVENTURE, N., WIOLAND, N., MANDEL, P. (1973). *Int. J. Neurosci.* **5:** 235—241.

33) POURCHO, R. G. (1977). *Exp. Eye Res.* **25:** 119—127.

34) STEINBERG, R. H., MILLER, S. S. (1979). In: *The Retinal Pigment Epithelium* (Zinn, K. M., Marmor, M. F., eds.). Harvard University Press, pp. 205—225.

35) MACAIONE, S., TUCCI, G., DeLUCA, G., DI GIORGIO, R. M. (1976). *J. Neurochem.* **27:** 1411—1415.

36) PASANTES-MORALES, H., LOPEZ-COLOME, A. M., SALCEDA, R., MANDEL, P. (1976). *J. Neurochem.* **27:** 1103—1106.

37) DI GIORGIO, R. M., MACAIONE, S., DeLUCA, G. (1977). *Life Sci.* **20:** 1657—1662.

38) SCHMIDT, S. Y., BERSON, E. L. (1978). *Exp. Eye Res.* **27:** 191—198.

39) SCHMIDT, S. Y. (1980). Exp. Eye Res. **31:** 373—379.

40) MILLER, S. S., STEINBERG, R. H. (1979). *J. Gen. Physiol.* **14:** 237—259.

41) LOPEZ-COLOME, A. M., PASANTES-MORALES, H. (1980). *J. Neurochem.* **34:** 1047—1052.

42) LOMBARDINI, J. B., PRIEM, S. D. (1983). *Exp. Eye Res.* **37:** 239—250.

43) STURMAN, J. A., HAYES, K. C. (1980). In: *Advances in Nutritional Research,* **Vol. 3** (Draper, H. H., ed.). Plenum Press, pp. 231—299.

44) HAYES, K. C., CAREY, R. E., SCHMIDT, S. Y. (1975). *Science* **188:** 949—951.

45) SCHMIDT, S. Y., BERSON, E. L., HAYES, K. C. (1976). *Invest. Ophthalmol.* **15:** 47—52.

46) NEURINGER, M., STURMAN, J. A., WEN, G. Y., WISNIEWSKI, H. M. (1985). In: *Taurine: Biological Actions and Clinical Perspectives* (Oja, S. S., Ahtee, L., Kontro, P., Paasonen, M. K., eds.). Alan R. Liss, pp. 53—62.

47) LAKE, N. (1981). *Life Sci.* **29:** 445—448.

48) GEGGEL, H. S., AMENT, M. E., HECKENLIVELY, J. R., MARTIN, D. A., KOPPLE, J. D. (1985). *N. Engl. J. Med.* **312:** 142—146.

49) SHEIKH, K. (1981). Gastroenterology **80:** 1363.

50) PASANTES-MORALES, H., ADEME, R. M., QUESADA, O. (1981). *J. Neurosci. Res.* **6:** 337—348.

51) PASANTES-MORALES, H., QUESADA, O., CARABEZ, A. (1981). *J. Neurochem.* **36:** 1583—1586.

52) GUIDOTTI, A., BADIAMI, G., PEPEAU, G. (1972). *J. Neurochem.* **19:** 431—435.

53) PAASONEN, M. K., HIMBERG, J.-J., PENTTILA, OL, SOLATUNTURI, E., YLITALO, P. (1980). In: *Pharmacological Control of Heart and Circulation* (Tardos, L., Szekeres, L., Papp, J. G., eds.). Pergamon Press, pp. 221—225.

54) LOPEZ-COLOME, A. M., PASANTES-MORALES, H. (1981). *Exp. Eye Res.* **32.** 771—780.

55) LOMBARDINI, J. B. (1985). *J. Neurochem.* **45:** 268—275.

56) OGINO, N., MATSUMURA, M., SHIRAKAWA, H., TSUKAHARA, I. (1983). *Ophtalmic Res.* **15:** 72—89.

57) CRUZ, C., PASANTES-MORALES, H. (1983). *J. Neurochem.* **41:** Suppl. S 134.

58) ARMSTRONG, D., SANTANGELO, G., CONNOLE, E. (1981). *Cur. Eye Res.* **1:** 225—242.

59) TALLAN, H. H., JACOBSON, E., WRIGHT, C. E., SCHNEIDMAN, K., GAULL, G. E. (1983). *Life Sci.* **33:** 1853—1860.

60) WRIGHT, C. E., TALLAN, H. H., GAULL, G. E. (1985). *Invest. Ophthalmol. Vis. Sci.* **26:** 132.

SUMMARY AND DISCUSSION-REPORT OF CHAPTER VII

Abel Lajtha and Robert Balàzs

SUMMARY

In spite of the diversity in the models used, the functions studied, the individual processes monitored and the effects described, all of these contributions have a common message. They all underline the important fact that the developing brain is very different from the adult brain, not only with respect to what happens in the brain, but also with respect to how the brain is affected from outside. Long lasting changes of individual factors in the external environment or in the supply of nutrients to the developing brain may have long lasting and often irreversible effects. Therefore, studies on the developing brain, much more than on the adult brain, should be made on a long-term basis to allow the detection of deficits or changes that become manifest only during later development. We have seen that multiple effects and multiple factors are important during brain development. In regard to protein synthesis in the developing brain, we need to learn much more about the factors that control not only the synthesis, but also the degradation, of proteins during brain development. As one might guess from the particular requirements of the growing brain, transport of substrates for protein synthesis must be very active. Not only is there higher transport capacity in the developing brain, but also there are several transport systems expressed that disappear in later life. Necessarily, with the increased transport capacity the blood-brain barrier in the developing brain is less selective. Certain substances, e. g., peptides and even proteins seem to enter the developing brain more readily than the adult brain. The developing brain is also more susceptible to changes in the balance of the individual nutrients supplied from the periphery. Small changes in this balance can have long-lasting and severe effects that are not seen if the adult brain is confronted with the same sort of imbalance. And finally, the particular role of certain compounds or metabolites as trophic factors or mediators of intercellular communication in the developing brain should be pointed out. There is an unexpected specificity of such factors for well-defined cell populations as well as defined states of maturity of these cells.

NATO ASI Series, Vol. H20
Amino Acid Availability and Brain Function in Health and Disease. Edited by G. Huether

DISCUSSION-REPORT

How can we explain the depletion of plasma amino acids in hyperphenylalaninaemia?

Rogers: The explanation given by Gerald Huether for the depletion of other amino acids in the blood of hyperphenylalaninaemic rats is a very stimulating idea. However, you would not expect this kind of amino acid pattern in the blood as a result of a stimulated protein synthesis. Actually hormonal changes produce the same kind of pattern, and we found, in models where we studied hepatic encephalopathy, that catecholamine infusion produces the same kind of pattern, as liver disease. The stimulation of insulin secretion will cause hypoglycemia and a stimulation of catecholamine secretion. Amino acids are then more rapidly taken up into tissues and trapped inside the cells. This effect is aggravated by phenylalanine, because phenylalanine inhibits the transport of amino acids from inside to outside the cell. Christensen recently published a review on this phenomenon (Christensen, H.W., 1987, Persp. Biol. Med. 30:186).

Huether: I think the concept of Christensen is valid as long as simple transport phenomena are concerned. The problem in these hyperphenylalaninaemic rats, however, is that the depletion of other amino acids in the blood and the accumulation of these amino acids in the peripheral tissues is seen only during the phase of most rapid growth. The same is true for untreated phenylketonuria. During early childhood the depletion of other amino acids is high. After the cessation of the most rapid growth period, however, this depletion disappears in spite of the fact that the elevation of phenylalanine persists. Hence, the depletion of other amino acids in the blood under the conditions of high levels of circulating phenylalanine is a developmental feature that happens to occur only during the phase of most rapid protein anabolism. Therefore it must be somehow related to a disturbed protein anabolism caused by the excess of phenylalanine.

Are long lasting irreversible changes of certain metabolic pathways brought about by nutritional shifts during early brain development?

Youdim: The effect of tryptophan administered during early embryonic development on the maturation of biochemical pathways for tryptophan utilization and serotonin synthesis described by Gerald Huether in chick embryos is rather exciting. I would like to add another example which is not directly related to amino acid availability and brain maturation, but which is certainly relevant, because it shows that brain function can be affected not only by changes in the brain's amino acid supply, but also by other nutritional factors governing the rate of amino acid utilization. The factor I am referring to is iron and its deficiency. Nutritional iron deficiency is the most prevalent nutritional disorder in the world and iron plays a key role for three of the enzymes of amino acid metabolism we have discussed in great detail, namely the aromatic amino acid hydroxylases. We have developed an animal model of iron deficiency to study the long-term consequences of early iron-deficiency on brain development. In the human brain, iron content increases during the first 3 years and reaches adult levels by the end of the first decade. In the rat, iron increases most rapidly during the first 4 weeks after birth. We therefore made rats iron deficient during the first 10 postnatal days. What we found was rather interesting. Early iron deficiency did not affect the activity of either tryptophan or tyrosine hydroxylase nor did it affect the brain levels of monoamines or their degradation products. However it severely affected the density of receptors. Dopamine receptor content as identified with radioligand binding was significantly diminished. And these animals loose the behavioural responses related to dopaminergic neurotransmission. Most interestingly, we found a depletion of dopamine receptors by iron deficiency, also in the brains of adult rats. This decreased receptor density of the adult is restored when the iron deficient rat is rehabilitated. This is not the case in the developing 10 day old rat brains. Here the loss of receptors persists even though the intake of iron is normalized in later life. The effect remains absolutely irreversible even after six months of supplementation with iron. So we concere that early deficiency can have irreversible and profound consequences on brain development and functional neurotransmission in later life. This is something we have to consider in the future, when we talk about dietary factors and their effects on brain development and brain function.

Are neurotransmitters informational substances in a global sense, and what are the consequences of such a perspective?

Gaull: The idea that neurotransmission is not the only function a neurotransmitter can have is rather fascinating. It raises a global question that involves much of what we have discussed with regard to the availability of aromatic amino acids for the formation of monoamines and for brain function. Is it possible that, in all those instances where we are not able to demonstrate a clear effect of an increased precursor availability on functional neurotransmission, we are only affecting the synthesis, the intracellular accumulation, and the nonquantal release of these monoamines? Is it possible that we do not affect neurotransmission, at least in some cases, but intra- or intercellular signalling by so called "transmitter substances". Somehow we have been pacing like racehorses over the last 15 years, always assuming that neurotransmitters in the nervous system are exclusively for the use and good of neurons. We are now about to learn a lot more about the role of other cells in the nervous system, and also about roles of neurotransmitters other than neurotransmission.

Sawatzki: Let me make one contribution on possible additional functions of hydroxylated aromatic amines. We know from very early phylogenetic development that these compounds are used by protozoa and by bacteria to collect and to store iron from the environment. It seems that, at least in bacteria this is the only reason that huge amounts of aromatic hydroxylated amino acid metabolites are formed. I wonder if, during early ontogenetic development in mammals, this very old phylogenetic function of monoamines may play a role. Maybe transferrin is not yet acting at very early stages of development and therefore monoamines are produced, released and taken up as carriers of iron or other divalent cations.

Gallo: It is not only the non-classic function of transmitters we ought to look for, but also the non-classic form of synaptic release of transmitters. There is good evidence that transmitters can be released in a non-classical manner. For example transmitters are known to be released at times when synapses are not yet formed. Acetylcholine, for instance, is released from growth cones. Nonsynaptic, non-classical release of dopamine seems to occur in the substantia nigra even in the adult brain.

Wolff: Some years ago we studied immunocytochemically the distribution of GABA in the developing cerebellum. In the adult cerebellum, antibodies to GABA very nicely stain the Purkinje cell axons, the axon terminals, and even the dendrites. But we never saw staining of the Purkinje cell perikaryon. During early cerebellar development, however, we found a rather intense staining of the perikaryon of the Purkinje cell and its outgrowing dendrites in those parts of the cerebellar "Anlage" where the first migrating granule cells passed the Purkinje cells. There is a very short period when you can stain the cell and the primordial dendrites these cells have at that time. A similar pattern is found in organ cultures of the cerebellum slices. Here, however, a very interesting phenomenon can be demonstrated: The transient staining of the Purkinje cell and its primodial dendrites can be made to vanish step by step by the stimulation of the cultured tissue using either electrical stimulation or high potassium in the medium. This suggests that GABA may be released from cells whose axons are not engaged in synaptic connections. At that time the axons do have a growth cone far down in the cerebellar "Anlage" and the dendrites have not yet really been developed. There are very few if any synapses on these young Purkinje cells. So GABA seems to be synthesized by these cells without having a chance of being used as a transmitter.

Balazs: The question we are addressing now is whether accredited transmitter substances may also serve as trophic agents during brain development. The answer from this meeting is affinative. The question refers to mechanisms. This is particularly important, since the mechanism of action of even the best known neural trophic factor, NGF, is not yet established. As to neurotransmitters a number of mechanisms using second messengers are plausible. What will happen after induction by the transmitter if the second messengers will depend on the state of maturation of the cells. In the adult cell the second messenger may cause alterations which will influence the firing of the cell. In developing nerve cells the same signals may affect nuclear functions determining ultimately the fate of the cell in terms of survival and differentiation. Recent observation on the possible trophic action of nervous transmitters in the developing nervous system do raise the attractive possibility that signalling systems which proved to be successful in evolution, are employed by cells, depending on their developmental state for purposes other than transmission.

Wolff: Is it not astonishing that we are now about to identify more and more trophic factors and that each of these factors seems to have its specific binding side or receptor acting through different messenger systems? We should ask ourselves how such a variety could have been stabilized throughout evolution. What we all would like to have is one signal specific for one differentiation step. What all of you actually show is a spectrum of effects caused by each factor. The specificity of the response would only depend on each cell's individual history, or in other words, on its actual state of maturation.

INDEX

NATO ASI Series H

Vol. 1: **Biology and Molecular Biology of Plant-Pathogen Interactions.** Edited by J. A. Bailey. 415 pages. 1986.

Vol. 2: **Glial-Neuronal Communication in Development and Regeneration.** Edited by H. H. Althaus and W. Seifert. 865 pages. 1987.

Vol. 3: **Nicotinic Acetylcholine Receptor: Structure and Function.** Edited by A. Maelicke. 489 pages. 1986.

Vol. 4: **Recognition in Microbe-Plant Symbiotic and Pathogenic Interactions.** Edited by B. Lugtenberg. 449 pages. 1986.

Vol. 5: **Mesenchymal-Epithelial Interactions in Neural Development.** Edited by J. R. Wolff, J. Sievers, and M. Berry. 428 pages. 1987.

Vol. 6: **Molecular Mechanisms of Desensitization to Signal Molecules.** Edited by T. M. Konijn, P. J. M. Van Haastert, H. Van der Starre, H. Van der Wel, and M. D. Houslay. 336 pages. 1987.

Vol. 7: **Gangliosides and Modulation of Neuronal Functions.** Edited by H. Rahmann. 647 pages. 1987.

Vol. 8: **Molecular and Cellular Aspects of Erythropoietin and Erythropoiesis.** Edited by I. N. Rich. 460 pages. 1987.

Vol. 9: **Modification of Cell to Cell Signals During Normal and Pathological Aging.** Edited by S. Govoni and F. Battaini. 297 pages. 1987.

Vol. 10: **Plant Hormone Receptors.** Edited by D. Klämbt. 319 pages. 1987.

Vol. 11: **Host-Parasite Cellular and Molecular Interactions in Protozoal Infections.** Edited by K.-P. Chang and D. Snary. 425 pages. 1987.

Vol. 12: **The Cell Surface in Signal Transduction.** Edited by E. Wagner, H. Greppin, and B. Millet. 243 pages. 1987.

Vol. 13: **Toxicology of Pesticides: Experimental, Clinical and Regulatory Perspectives.** Edited by L. G. Costa, C. L. Galli, and S. D. Murphy. 320 pages. 1987.

Vol. 14: **Genetics of Translation. New Approaches.** Edited by M. F. Tuite, M. Picard, and M. Bolotin-Fukuhara. 524 pages. 1988.

Vol. 15: **Photosensitisation. Molecular, Cellular and Medical Aspects.** Edited by G. Moreno, R. H. Pottier, and T. G. Truscott. 521 pages. 1988.

Vol. 16: **Membrane Biogenesis.** Edited by J. A. F. Op den Kamp. 477 pages. 1988.

Vol. 17: **Cell to Cell Signals in Plant, Animal and Microbial Symbiosis.** Edited by S. Scannerini, D. Smith, P. Bonfante-Fasolo, and V. Gianinazzi-Pearson. 414 pages. 1988.

Vol. 18: **Plant Cell Biotechnology.** Edited by M. S. S. Pais, F. Mavituna, and J. M. Novais. 540 pages. 1988.

Vol. 19: **Modulation of Synaptic Transmission and Plasticity in Nervous Systems.** Edited by G. Hertting and H.-C. Spatz. 457 pages. 1988.

Vol. 20: **Amino Acid Availability and Brain Function in Health and Disease.** Edited by G. Huether. 487 pages. 1988.

